AF478951

RODD'S CHEMISTRY OF CARBON COMPOUNDS

ELSEVIER SCIENCE B.V.
Sara Burgerhartstraat 25
P.O. Box 211, 1000 AE Amsterdam, The Netherlands

ISBN: 0-444-82758-7

Second Supplements to the 2nd Edition of

RODD'S CHEMISTRY OF CARBON COMPOUNDS

VOLUME I

ALIPHATIC COMPOUNDS
★

VOLUME II

ALICYCLIC COMPOUNDS
★

VOLUME III

AROMATIC COMPOUNDS
★

VOLUME IV

HETEROCYCLIC COMPOUNDS
★

VOLUME V

MISCELLANEOUS

GENERAL INDEX
★

Second Supplements to the 2nd Edition of

RODD'S CHEMISTRY OF CARBON COMPOUNDS

A modern comprehensive treatise

Edited by
MALCOLM SAINSBURY
School of Chemistry, The University of Bath,
Claverton Down, Bath BA2 7AY, England

Second Supplement to

VOLUME IV HETEROCYCLIC COMPOUNDS

Part B: Five-membered Monoheterocyclic Compounds:
Alkaloids, Dyes, Pigments

1997
ELSEVIER
Amsterdam – Lausanne – New York – Oxford – Shannon – Tokyo

Contributors to this Volume

G. BACH

Laboratorium für Zeitaufgelöste Spektroskopie (I.34),
Bundesanstalt für Materialforschung und Prüfung (BAM),
Außenstelle Adlershof, Rudower Chaussee 5, Haus 8.15,
D-12489 Berlin, Germany

S. DÄHNE

Laboratorium für Zeitaufgelöste Spektroskopie (I.34),
Bundesanstalt für Materialforschung und Prüfung (BAM),
Außenstelle Adlershof, Rudower Chaussee 5, Haus 8.15,
D-12489 Berlin, Germany

G.B. FODOR

Department of Chemistry, University of West Virginia, P.O. Box 6045,
Morgantown, WV 26506, U.S.A.

G.W. GRIBBLE

Department of Chemistry, Dartmouth College, 6128 Burke Laboratory,
Hanover, NH 03755-3564, U.S.A.

J.R. LEWIS

Department of Chemistry, University of Aberdeen, Meston Walk,
Old Aberdeen AB9 2UE, Scotland, U.K.

D.J. ROBINS

Department of Chemistry, University of Glasgow,
Glasgow G12 8QQ, Scotland, U.K.

M. SAINSBURY

School of Chemistry, University of Bath,
Claverton Down, Bath BA2 7AY, England

K. SMITH

Department of Chemistry, University of California,
Davis, CA 95616, U.S.A.

Preface to Volume IV B

It was a particular pleasure for me to edit this volume of the Second Supplement since research into colourful compounds, natural and synthetic, and alkaloids has been a major part of my own research interests. This volume has at its core aspects of the chemistry of pyrrole and pyridine, plus their benzo derivatives and reduced forms. For example, it begins with surveys of the natural occurrence, characterisation and synthesis of pyrrolidine and pyrrolizidine alkaloids and progresses on to consider the structures of the new indole alkaloids identified since 1985. A major undertaking! Alkaloids of the tropane type and those from plants of the Amaryllidacea family are surveyed next, and a skilfully crafted review of recent work in the porphyrin series is also featured.

In the last two chapters of this volume the emphasis changes slightly and a significant contribution is a discussion of chemistry of the cyanine dyestuffs and that of their allies. Chapter 13 has not been supplemented since much of the subjects that would have been dealt with here are largely covered in Chapters 12 and 15. In my opinion, the residual topics in Chapter 13 did not require attention at this time. Chapter 14 remains as before: a survey of the chemistry dyestuff indigo and related compounds.

Malcolm Sainsbury

Bath
October 1996

Contents
Volume IV B

Heterocyclic Compounds; Five-membered Monoheterocyclic Compounds: Alkaloids, Dyes, Pigments

Chapter 7. *Five-membered Monoheterocyclic Compounds Alkaloids (continued): Pyrrolidine Alkaloids*
by D.J. ROBINS

Chapter 8. *Five-membered Monoheterocyclic Compounds (continued): Pyrrolizidine Alkaloids*
by D.J. ROBINS

Chapter 9. *The Indole Alkaloids*
by G.W. GRIBBLE

Chapter 10. Five-membered Monoheterocyclic Compounds,
Amaryllidaceae Alkaloids
by J.R. LEWIS

Chapter 11. Tropane Alkaloids
by G. FODOR

Chapter 12. Pyrrole Pigments
by K.M. SMITH

Chapter 13. Five-membered Monoheterocyclic Compounds – The Pyrrole Pigments (continued)

This chapter has not been supplemented since the subjects that would have been dealt with here are largely covered in Chapters 12 and 15.

Chapter 14. Five-membered Monoheterocyclic Compounds: the Indigo Group
by M. SAINSBURY

Chapter 15. Cyanine Dyes and Related Compounds
by G. BACH and S. DÄHNE

List of Common Abbreviations and Symbols Used

A	acid
Å	Ångström units
Ac	acetyl
a	axial
as, $asymm$.	asymmetrical
at.	atmosphere
B	base
Bu	butyl
b.p.	boiling point
c, C	concentration
CD	circular dichroism
conc.	concentrated
crit.	critical
D	Debye unit, 1×10^{-18} e.s.u.
D	dissociation energy
D	dextro-rotatory; dextro configuration
d	density
dec., decomp	with decomposition
deriv.	derivative
E	energy; extinction; electromeric effect
E	entgegen (opposite) configuration
$E1$, $E2$	uni- and bi-molecular elimination mechanisms
E1cB	unimolecular elimination in conjugate base
ESR	electron spin resonance
Et	ethyl
e	nuclear charge; equatorial
f.p.	freezing point
G	free energy
GLC	gas liquid chromatography
g	spectroscopic splitting factor, 2.0023
H	applied magnetic field; heat content
h	Planck's constant
Hz	hertz
I	spin quantum number; intensity; inductive effect
IR	infrared
J	coupling constant in NMR spectra
J	Joule
K	dissociation constant
k	Boltzmann constant; velocity constant
kcal	kilocalories
L	laevorotatory, laevo configuration
M	molecular weight; molar; mesomeric effect
Me	methyl

m	mass; mole; molecule; *meta-*
m.p.	melting point
Ms	mesyl (methanesulphonyl)
[M]	molecular rotation
N	Avogadro number; normal
NMR	nuclear magnetic resonance
NOE	Nuclear Overhauser Effect
n	normal; refractive index; principal quantum number
o	*ortho-*
ORD	optical rotatory dispersion
P	polarisation; probability; orbital state
Pr	propyl
Ph	phenyl
p	*para-*; orbital
PMR	proton magnetic resonance
R	clockwise configuration
S	counterclockwise configuration; entropy; net spin of incompleted electronic shells; orbital state
S_N1, S_N2	uni- and bi-molecular nucleophilic substitution mechanism
S_Ni	internal nucleophilic substitution mechanism
s	symmetrical; orbital
sec	secondary
soln.	solution
symm.	symmetrical
T	absolute temperature
Tosyl	p-toluenesulphonyl
Trityl	triphenylmethyl
t	time
temp.	temperature (in degrees centrigrade)
tert	tertiary
UV	ultraviolet
v	velocity
Z	zusammen (together) configuration
α	optical rotation (in water unless otherwise stated)
$[\alpha]$	specific optical rotation
ϵ	dielectric constant; extinction coefficient
μ	dipole moment; magnetic moment
μ_B	Bohr magneton
μg	microgram
μm	micrometer
λ	wavelength
ν	frequency; wave number
χ, χ_d, χ_μ	magnetic; diamagnetic and paramagnetic susceptibilities
$\sim$	about
(+)	dextrorotatory
(−)	laevorotatory
−	negative charge
+	positive charge

Second Supplements to the 2nd Edition of Rodd's Chemistry of Carbon Compounds, Vol.IV B, edited by M. Sainsbury
© 1997 Elsevier Science B.V. All rights reserved.

Chapter 7

FIVE-MEMBERED MONOHETEROCYCLIC COMPOUNDS ALKALOIDS (CONTINUED) : PYRROLIDINE ALKALOIDS

DAVID J. ROBINS

Progress in this area has been reviewed periodically in "The Alkaloids", Vols. 11-13, Royal Society of Chemistry, London, 1981-3, and in "Natural Product Reports" (A.R. Pinder, 1984, **1**, 225; 1985, **2**, 181; 1986, **3**, 171; 1987, **4**, 527; 1989, **6**, 67; 1990, **7**, 447; 1992, **9**, 17; 1992, **9**, 491; A.O. Plunkett, 1994, **11**, 581). Another review is available (G. Massiot and C. Delaude in "The Alkaloids" ed. A. Brossi, Academic Press, New York, 1986, **27**, Ch.3).

1. Pyrrolidine Bases

Hygrine (1) is present in the plant *Crossostylis biflora* (D.H.G. Medina, M. Pusset, J. Pusset and H.-P. Husson, J. Nat. Prod., 1983, **46**, 398). A one pot synthesis of hygrine has been described involving the α-alkylation of a 2-hydroxypyrrolidine with a Wittig reagent (T. Nagasaka, H. Yamamoto, H. Hayashi, M. Watanabe and F. Hamaguchi, Heterocycles, 1989, **29**, 155).

CH_2COMe (1)   OH (2)

Synthesis of (+)-hygroline (2) was achieved using anodic oxidation of an α-amino acid (T. Shono, Y. Matsumura, K. Tsubata and K. Uchida, J. Org. Chem., 1986, **51**, 2590).

Norruspoline (3) and ruspolinone (4) were previously isolated from *Ruspolia hypercrateriformis*. Racemic forms of these two alkaloids have been synthesized from the same phenylsulphonyl derivative (5) (D.S. Brown, P. Charreau, T. Hansson and S.V. Ley, Tetrahedron, 1991, **47**, 1311). Alternative routes to ruspolinone (4) from (2S)-proline (K. Jones and K. Woo, Tetrahedron, 1991, **47**, 7179) and to (±)-norruspoline (3) are available (R.B. Herbert and P.C. Wormald, J. Chem. Res., 1982 (S), 299; (M), 3001).

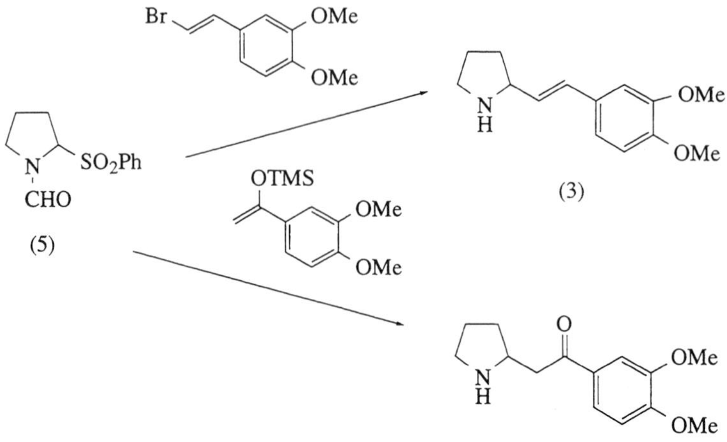

2

Darlinine (6), epidarlinine (7), dehydrodarlinine (8) and tetrahydrodarlingianine (9) have been isolated from the Queensland tree, *Darlingia darlingiana*. Synthesis of all four alkaloids has been carried out involving cycloaddition reactions with 1-pyrroline-1-oxide (J.J. Tufariello and J.M. Puglis, Tetrahedron Lett., 1986, **27**, 1265).

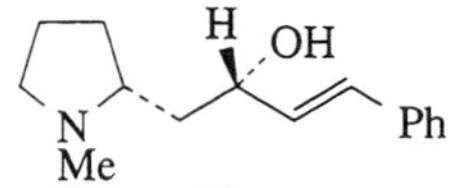

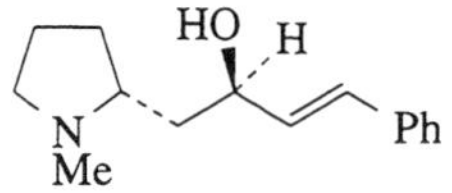

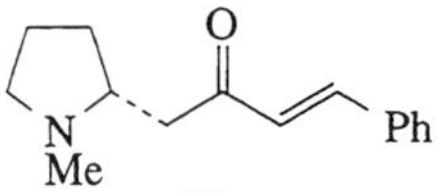

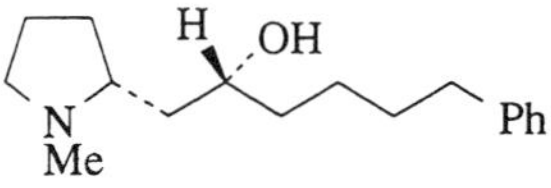

Synthesis of racemic hygrine (1), dehydrodarlinine (8), dehydrodarlingianine, and *N*-methylruspolinone have been achieved via an episulphide intermediate (R. Ghirlando, A.S. Howard, R.B. Katz and J.P. Michael, Tetrahedron, 1984, **40**, 2879).

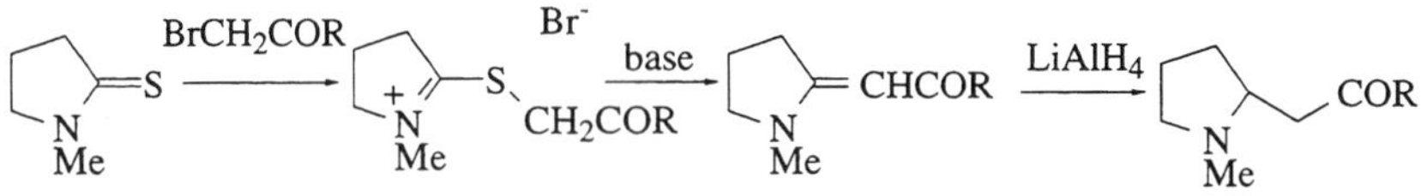

(-)-Dihydrocuscohygrine (10) is present in Peruvian coca leaves (C.E. Turner, M.A. Elsohly, L. Hanus and H.N. Elsohly, Phytochemistry, 1981, **20**, 1403).

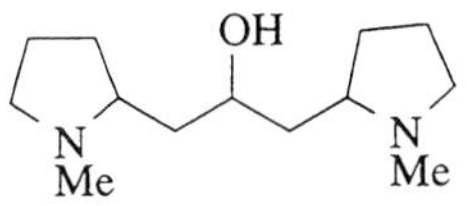

(10)

A review of the ant venom alkaloids present in *Solenopsis* and *Monomorium* is available (T.H. Jones, M.S. Blum and H.M. Fales, Tetrahedron, 1982, **38**, 1949). A number of 2,5-disubstituted pyrrolidines have been isolated from ant venoms (*Monomorium* spp.) (T.H. Jones, M.S. Blum, R.W. Howard, C.A. McDaniel, H.M. Fales, M.B. DuBois and J. Torres, J. Chem. Ecol., 1982, **8,** 285; T.H. Jones, M.S. Blum and H.M. Fales, Tetrahedron, 1982, **38,** 1949).

The ant venom alkaloid (12) has been synthesized enantioselectively from L-pyroglutamic acid methyl ester (11) (S. Rosset, J. Celerier and G. Lhommet, Tetrahedron Lett., 1991, **32**, 7521).

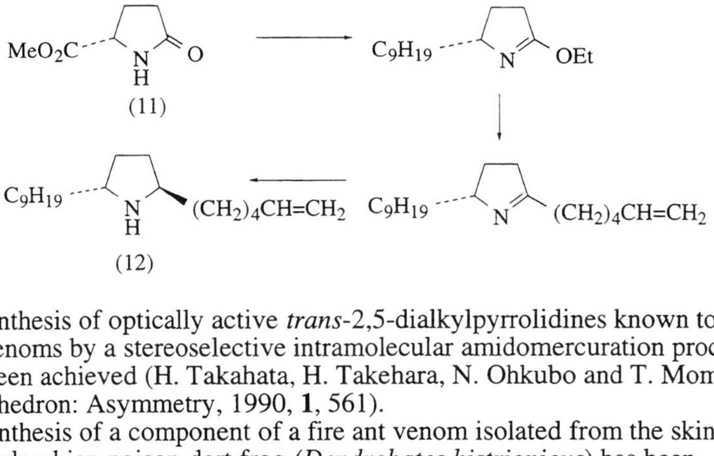

(11)

(12)

Synthesis of optically active *trans*-2,5-dialkylpyrrolidines known to be ant venoms by a stereoselective intramolecular amidomercuration process has been achieved (H. Takahata, H. Takehara, N. Ohkubo and T. Momose, Tetrahedron: Asymmetry, 1990, **1**, 561).

Synthesis of a component of a fire ant venom isolated from the skin of the Colombian poison-dart frog *(Dendrobates histrionicus)* has been reported (J.W. Daly, T.F. Spande, N. Whittaker, R.J. Highet, D. Feigl, N. Nishimori, T. Tokuyama and C.W. Myers, J. Nat. Prod., 1986, **49**, 265). Further ant venoms (13) and (14) were isolated from *Monomorium smithii* (T.H. Jones, A. Laddago, A.W. Don and M.S. Blum, J. Nat. Prod., 1990, **53**, 375). Related compounds such as (15) are found in the venom of *M. indicum* (T.H. Jones, M.S. Blum, P. Escoubas and T.M.M. Ali, J. Nat. Prod., 1989, **52**, 779).

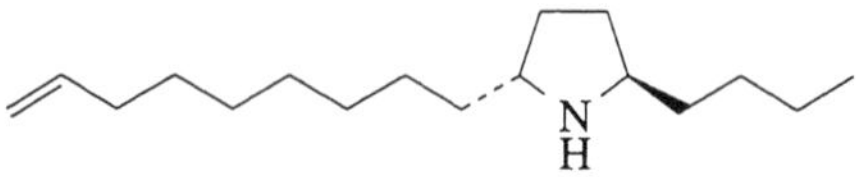

(13)

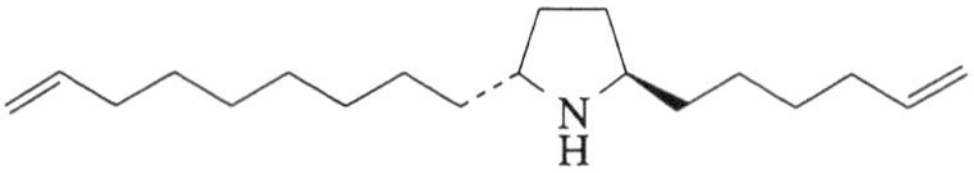

(14)

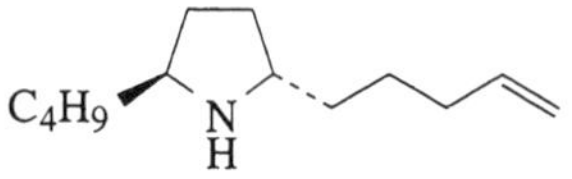

(15)

A stereocontrolled organopalladium route to 2,5-bisalkylpyrrolidines found in *M. latinode* venom is available (J.-E.Backvall, H.E. Schink and Z.F. Renko, J. Org. Chem., 1990, **55**, 826). An acyl nitronate has been used as a starting material for the preparation of several components of ant venoms (M. Miyasita, B.Z.E. Awen and A. Yoshikoshi, Chem. Lett., 1990, 239).

The 2,5-dialkylpyrrolidine (14) is a component of the venom of *Chelaner antarcticus* (T.H. Jones, R.J. Highet, A.W. Don and M.S. Blum, J. Org. Chem., 1986, **51**, 2712).

Venom components (16) of *S. fugax* and (17) from *S. molesta* and *S. texanas* have been prepared from 1-pyrroline-1-oxide using cycloaddition reactions (J.J. Tufariello and J.M. Puglis, Tetrahedron Lett., 1986, **27**, 1489).

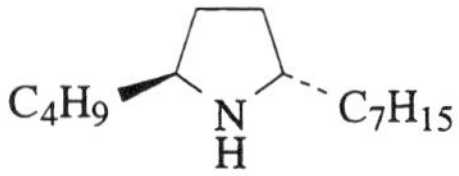

(16) 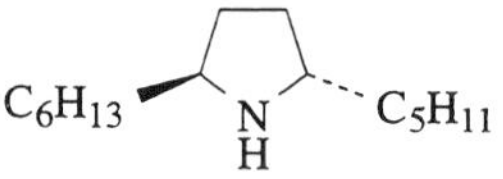 (17) 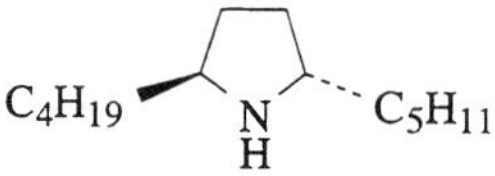 (18)

The synthesis of racemic *trans*-2,5-dialkylpyrrolidines including (16) present in venoms of various ants has been reported (W. Gessner, K. Takahashi, A. Brossi, M. Kowakski and M.A. Kaliner, Helv. Chim. Acta, 1987, **70**, 2003). A stereoselective route to the dialkylpyrrolidine (16) has been reported (P.Q. Huang, S. Arseniyadis and H.-P. Husson, Tetrahedron Lett., 1987, **28**, 547).

The skin of the poisonous frog *Dendrobates histrionicus* from Columbia contains the dialkylpyrrolidine (18) (J.W. Daly, T.F. Spande, N. Whittaker, R.J. Highet, D. Feigl, N. Nishimori, T. Tokuyama and C.W. Myers, J. Nat. Prod., 1986, **49**, 265). An enantioselective synthesis of (18) has been accomplished (N. Machinaga and C. Kibayashi, J. Org. Chem., 1991, **56**, 1386).

2-Ethyl-5-n-heptylpyrrolidine (19) is a component of the venom of the South African fire ant *Solenopsis punctaticeps*. It was synthesised as a mixture of *cis-* and *trans*-isomers (A.I. Meyers, P.D. Edwards, T.R. Bailey and G.F. Jagdmann Jr., J. Org. Chem., 1985, **50**, 1019).

(19)

(20)

The major alkaloid present in poison glands of the ants *Leptothoracini* (Myrmicinae) was the pyrrolidine (20) although only ng quantities were present in each gland (E. Reder, H.J. Veith and A. Buschinger, Helv. Chim. Acta, 1995, **78**, 73).

Extracts of adults of the Mexican bean beetle, *Epilachna varivestis*, (Coccinellidae) yielded two new pyrrolidine alkaloids (21) and (22) (A.B. Attygalle, S.C. Xu, K.D. McCormick, J. Meinwald, C.L. Blankespoor and T. Eisner, Tetrahedron, 1993, **49**, 9333).

(21)

(22)

Seeds of *Castanospermum australe* A. Cunn. yielded the pyrrolidine (23). The structure and relative configuration were established by spectroscopic studies and X-ray diffraction analysis (R.J. Nash, E.A. Bell, G.W.J. Fleet, R.H. Jones and J.M. Williams, J. Chem. Soc., Chem. Commun., 1985, 738).

(23)

(24)

The glycosidase inhibitor (24) present in fruits of *Angylocalyx boutiqueanus* Touss. (G.W.J. Fleet, S.J. Nicholas, P.W. Smith, S.V. Evans, L.E. Fellows and R.J. Nash, Tetrahedron Lett., 1985, **26**, 3127) and fronds of *Arachniodes standishii* (Moore) Ohwi (D.W.C. Jones, R.J. Nash, E.A. Bell and J.M. Williams, Tetrahedron Lett., 1985, **26**, 3125) has been

6

synthesized from D-xylose (G.W.J. Fleet and P.W. Smith, Tetrahedron, 1986, 42, 5685) .

Irniine (25) was isolated from the Moroccan tuber *Arisarum vulgare* and identified from 2D ^{1}H-^{13}C COSY studies (A. Melhaoui, A. Jossang and B. Bodo, J. Nat. Prod., 1992, **55**, 950). The tubers are eaten when food is scarce despite the fact that the alkaloid is toxic.

(25)

The bispyrrolidine, hypercratine (26), has been isolated from the roots of *Ruspolia hypercrateriformis* M.R. (G. Neukomm, F. Roessler, S. Johne and M. Hesse, Planta Med., 1983, **48**, 246).

(26)

(+)-Preussin (27) was isolated from fermentation cultures of *Aspergillus ochraceus* and *Preussia* spp. It has antifungal activity and was synthesized from D-glucose by a lengthy route (C.S. Pak and G.H. Lee, J. Org. Chem., 1991, **56**, 1128).

(27)

(28) R = OMe

(29) R = H

^{13}C NMR spectra of codonopsine (28) and codonopsinine (29) have been analysed (M.R. Yagudaev and S.F. Aripova, Chem. Nat. Compds. (Engl. transl.), 1989, **25**, 459). The absolute configuration of (-)-codonopsinine was assigned on the basis of a synthesis of the enantiomer from L-tartaric acid (H. Iida, N. Yamazaki and C. Kibayashi, Tetrahedron Lett., 1985, **26**, 3255) and was then revised to (2R,3R,4R,5R) after X-ray diffraction studies had been carried out (H. Iida, N. Yamazaki, C. Kibayashi and H. Nagase, Tetrahedron Lett., 1986, **27**, 5393).

Three syntheses of shihunine (30) have been described (V.G. Gore, M.D. Chordia and N.S. Narasimhan, Tetrahedron, 1990, **46**, 2483). Dihydroshihunine (31) is present in *Banisteriopsis caapi* Morton. and has been assigned the (S)-configuration from comparison of CD data with those of known related compounds (K. Kawanishi, Y. Uhara and Y. Hashimoto, J. Nat. Prod., 1982, **45**, 637).

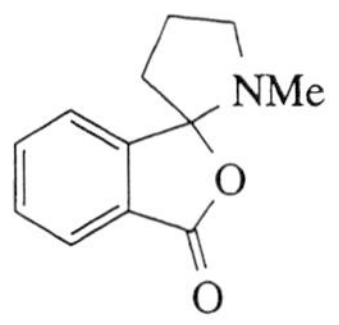

(30) 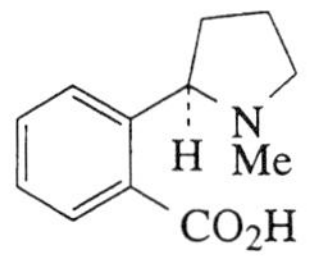 (31)

(+)-Polyzonimine (32) is an insect repellent present in the millipede *Polyzonium rosalbum* and has been synthesized (T. Sugahara, Y. Komatsu and S. Takano, Heterocycles, 1984, **21**, 551; J. Chem. Soc., Chem. Commun. 1984, 214).

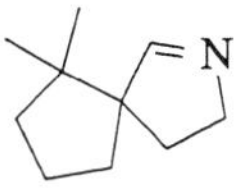

(32) 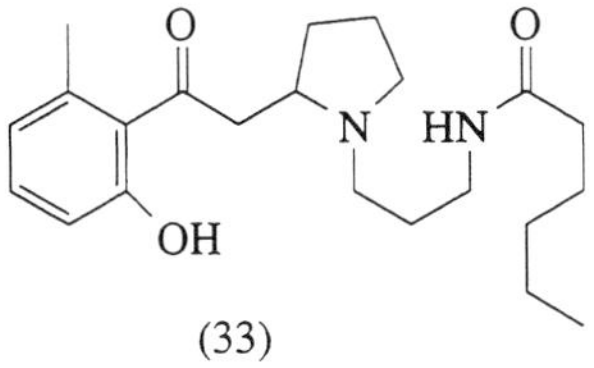 (33)

Peripentadenine (33) is present in the leaves and bark of the Queensland species *Peripentadenia mearsii* (Elaeocarpaceae) and its structure was established by spectroscopic data, degradative studies and synthesis of its O-methyl ether (J.A. Lamberton, Y.A. Geewananda, P. Guanawardana and I.R.C. Bick, J. Nat. Prod., 1983, **46**, 235).

Dinorperipentadenine (34) has also been isolated from this species (H. Ishibashi, T.S. So, T. Sato, K. Kuroda and M. Ikeda, J. Chem. Soc., Chem. Commun., 1989, 762).

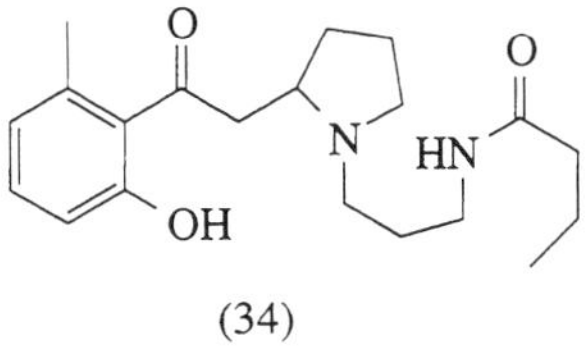

(34)

Vochysine (35) was isolated from the fruit of *Vochysia guainensis* (Aubl.) Poir. and its structure was established by spectroscopic data and by synthesis involving addition to 1-pyrroline (G. Baudouin, F. Tillequin, M.

Koch, M. Vuilhorgne, J.-Y. Lallemand and H. Jacquemin, J. Nat. Prod., 1983, **46**, 681).

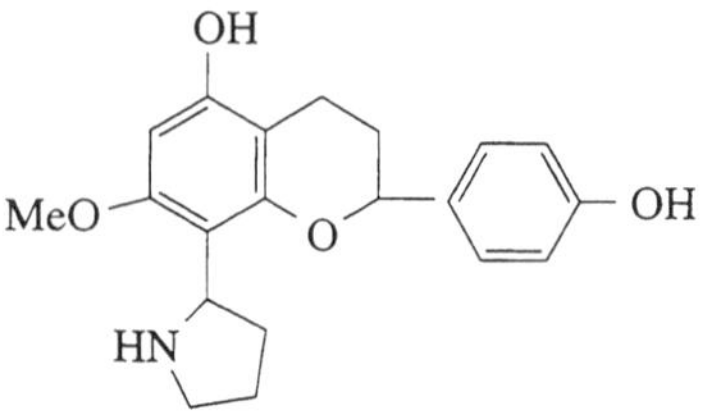

(35)

Synthesis of the alkaloid (36) present in *Derris elliptica* and *Lonchocarpus sericeus* has been achieved (G.W. Fleet and P.W. Smith, Tetrahedron Lett., 1985, **26**, 1469).

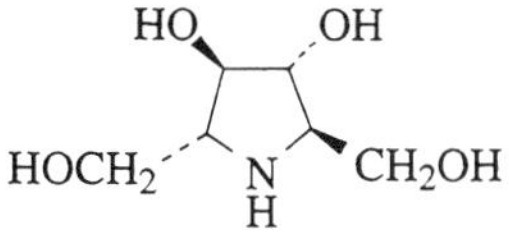

(36)

Guinesines A - C (37)- (39) isolated from the bark of the Brazilian shrub *Cassipourea guainensis* were identified by spectroscopic studies and relative configurations were established by X-ray diffraction studies. These compounds have insecticidal properties (A. Kato, M. Ichimaru, Y. Hashimoto and H. Mitsudera, Tetrahedron Lett., 1989, **30**, 3671). They were synthesised as a stereoisomeric mixture and separated chromatographically (H. Mitsudera, H. Uneme, Y. Okada, M. Numata and A. Kato, J. Heterocyclic Chem., 1990, **27**, 1361).

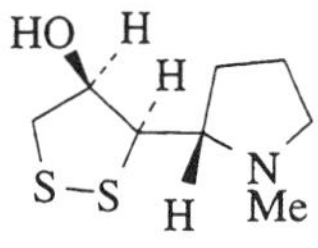

(37) 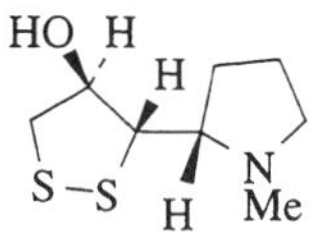   (38) 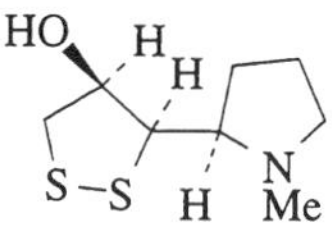   (39)

Amathamide A (40) and its geometrical isomer amathamide B were isolated from the bryozoa *Amathia wilsoni* Kirkpatrick. The absolute configurations of each compound were established by correlation with (S)-proline (A.J. Blackman and D.J. Matthews, Heterocycles, 1985, **23**, 2829). Further work has led to the identification of four more amathamides C - F (41)-(44) by spectroscopic studies (A.J. Blackman and R.D. Green, Aust. J. Chem., 1987, **40**, 1655).

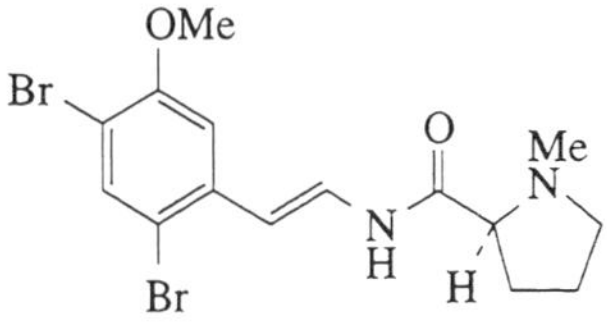

(40)

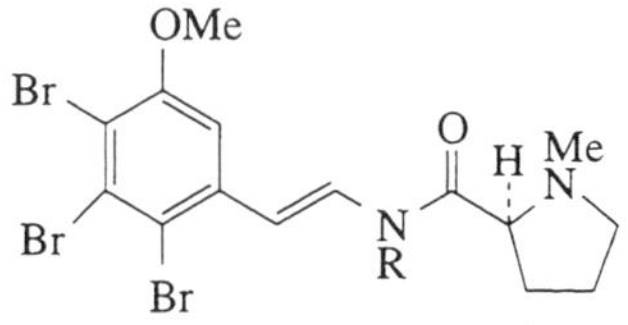

(41) R = Me

(43) R = H

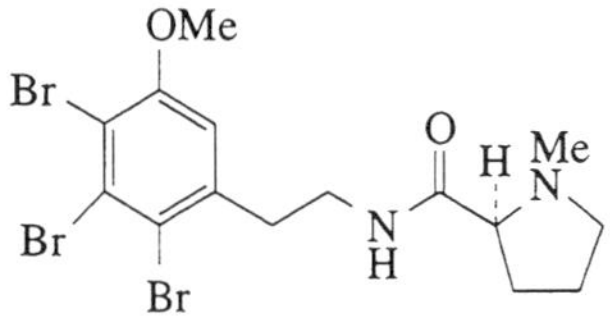

(42)

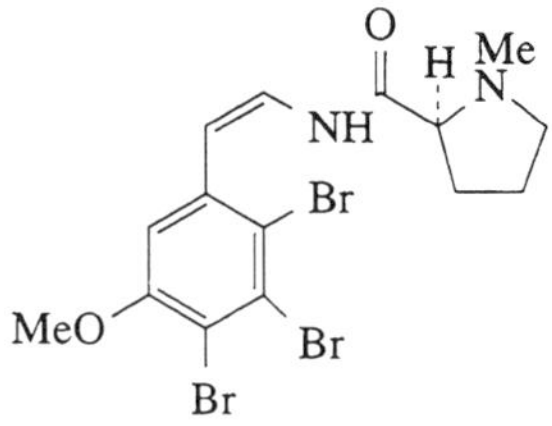

(44)

A new pyrrolidine alkaloid (45) was isolated from the leaves of *Schizanthus integrifolius* Phil. (O. Munoz, C. Schneider and E. Breitmaier, Leibigs Ann. Chem., 1994, 521).

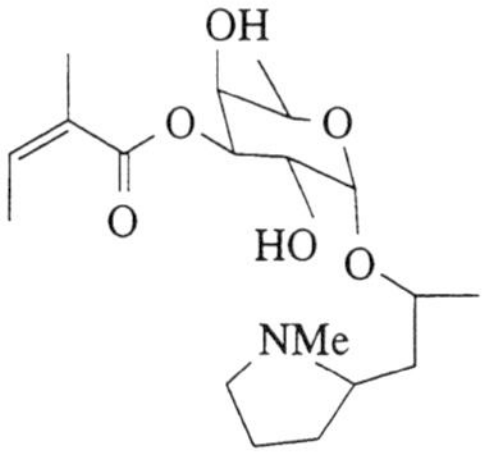

(45)

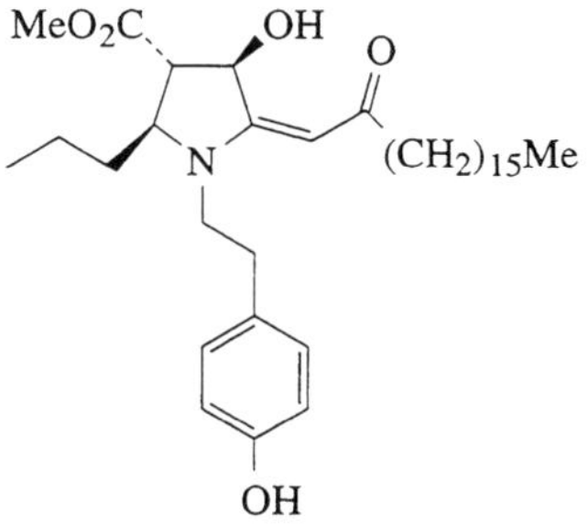

(46)

Plakoridine A (46) is a new alkaloid isolated from the Okinawan marine sponge *Plakortis* sp. Its structure including relative configuration was established by spectroscopic studies (S. Takeuchi, M. Ishibashi and J. Kobayashi, J. Org. Chem., 1994, **59**, 3712).

2 Pyrrolidones

Synthesis of (-)-4-hydroxypyrrolidin-2-one (47) which was previously isolated from the toadstool *Amanita muscaria* has shown that this compound has the (S)-configuration (E. Santaniello, R. Casati and F. Milani, J. Chem. Res. (S), 1984, 132).

(47)   (48)

The structure of lilidine (48) from *Lilium martagon* was established by NMR spectroscopy [N.D. Abdullaev, K. Samikov, T.P. Antsupova, M.R. Yagudaev and S.Y. Yunusov, Chem. Nat. Compds. (Engl. trans.) 1987, **23**, 576].

Two dimeric pyrrolinones (49) and (50) are present in the bulbs of *Lilium candicum* L. (M. Haladova, E. Eisenreichova, A. Buckova, J. Tomko, D. Uhrin and K. Ubik, Coll. Czech. Chem. Commun., 1991, **56**, 436).

(49)   (50)

Bulbs of a Chinese lily, *Lilium regale*, contain the succinimide derivative (51) (Y. Mimaki and Y. Sashida, Chem. Pharm. Bull., 1990, **38**, 541).

(51)

(+)-Lilaline (52) is present in flowers of *Lilium candicum* and the structure of this flavonoid pyrrolidone was established by spectroscopic methods (I. Masterova, D. Uhrin and J. Tomko, Phytochemistry, 1987, **26**, 1844).

Bulbs of *Lilium candidum* have yielded another bis-lactam (53) (E. Eisenreichova, M. Haladova, A. Buckova, J. Tomko, D. Uhrin and K. Ubik, Phytochemistry, 1992, **31**, 1084).

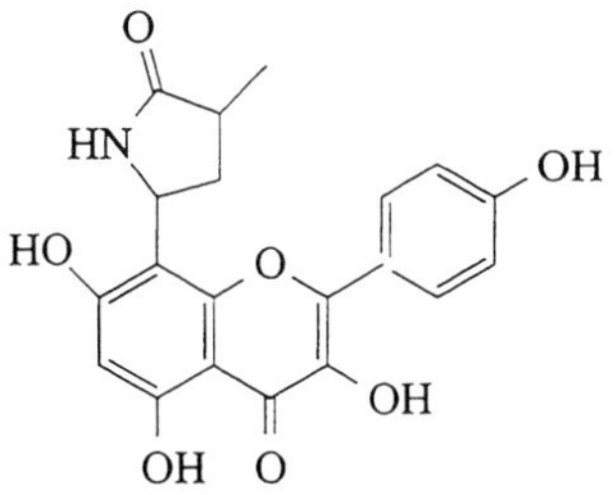

(52)

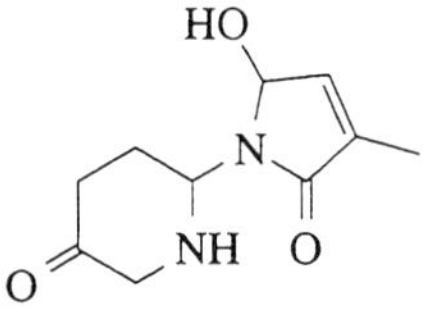

(53)

Aerial parts of *Hemerocallis fulva* var. *kwanzo* (Liliaceae) yielded three new pyrrolidones, fulvanines A - C (54)-(56) (T. Inoue, K. Iwagoe, T. Konishi, S. Kiyosawa and Y. Fujiwara, Chem. Pharm. Bull., 1990, **38**, 3187).

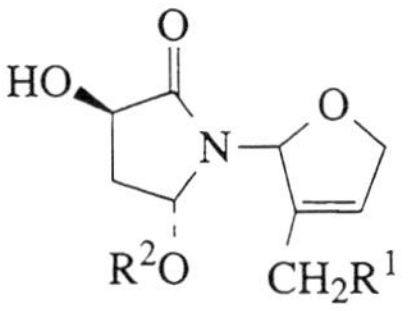

(54) $R^1 = OH, R^2 = H$

(55) $R^1 = R^2 = H$

(56) $R^1 = H, R^2 = Me$

Fusarin C (57) and its (8Z)-geometrical isomer (58) were isolated from *Gibberella fujikuroi* (A. Barrero, J.F. Sanchez, J.E. Oltra, N. Tamayo, E. Cerda-Olmedo, R. Candau, and J. Avalos, Phytochemistry, 1991, **30**, 2259).

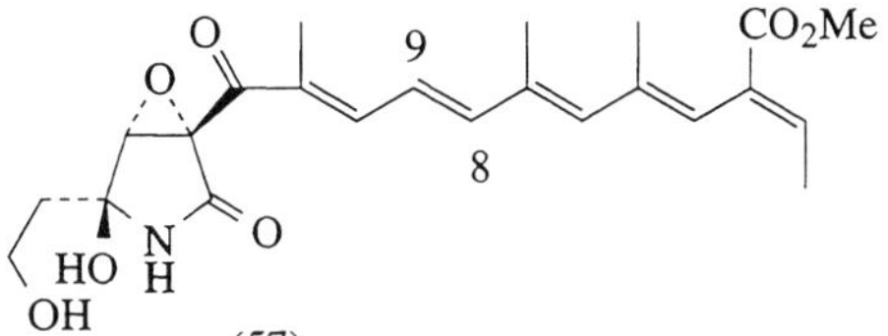

(57)

(58) 8,9 double bond (Z)

The structure of vasicol (59) from *Peganum harmala* was established by X-ray crystal structure analysis of a derivative (M.V. Telezhenetskaya, B. Tashkhodzhaev, M.R. Yagudaev, B. Ibragimov and S.Y. Yunusov, Chem. Nat. Compds. (Eng. Transl.), 1989, **25**, 14).

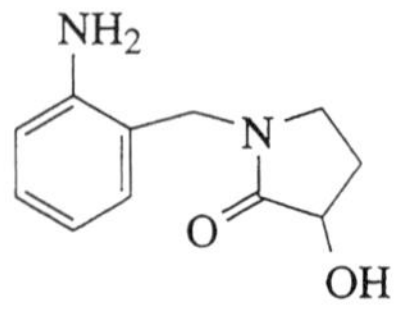

12

(59)

Jatropham (60) is present in *Jatropha macorhiza*. Synthesis of this antitumour compound has been achieved from succinimide (T. Nagasaka, S. Esumi, N. Ozawa, Y. Kosugi and F. Hamaguchi, Heterocycles, 1981, **16**, 1987).

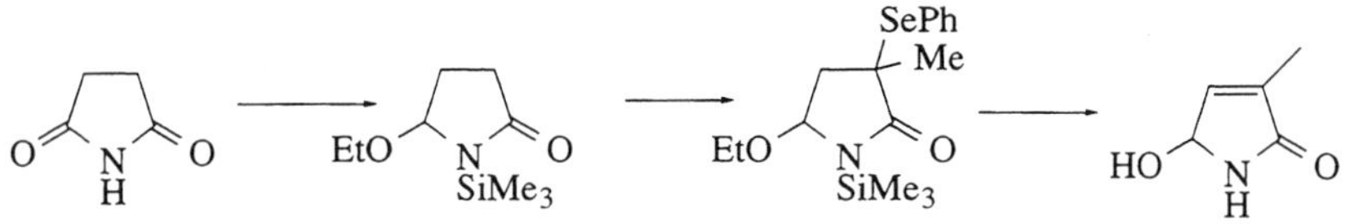

(60)

Clausenamide (61) was isolated as a racemate from the leaves of the Chinese plant, *Clausena lansium*. The structure and relative configuration were established by X-ray diffraction studies and by synthesis (W. Hartwig and L. Born, J. Org. Chem., 1987, **52**, 4352).

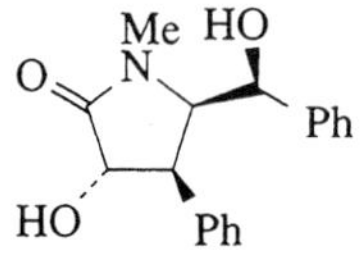

(61)

The dimeric pyrrolinone (63) was found in aerial parts of *Mercurialis leiocarpa* Sieb. et Zucc. (Y. Masui, C. Kawabe, K. Matsumoto, K. Abe and T. Miwa, Phytochemistry, 1986, **25**, 1470) and was synthesized from the dimer (62) utilising a benzil-benzilic acid rearrangement (K. Abe, T. Okada, Y. Masui and T. Miwa, Phytochemistry, 1989, **28**, 960).

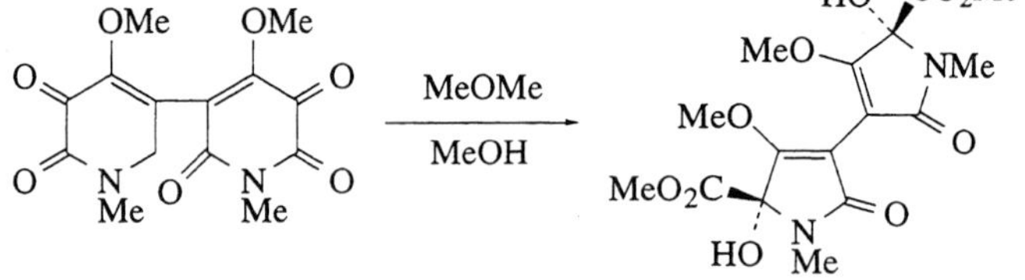

(62)   (63)

Radish seedlings grown under artificial light produce a growth inhibitor (64) which is a thiopyrrolidone (M. Sakoda, T. Hase and K. Hasegawa, Phytochemistry, 1990, **29**, 1031).

(64)

3 N-Acylpyrrolidines

Roxburghiline (65) (identical to odorine) is present in the leaves of *Aglaia roxburghiana* and *A. odorata*. The absolute configuration of this alkaloid was established by synthesis from (S)-proline (P.J. Babidge, R.A. Massy-Westropp, S.G. Pyne, D. Shiengthong, A. Ungphakoru and G. Veerachat, Aust. J. Chem., 1980, **33**, 1841). A synthesis of the racemate has also been reported (T. Nagasaka, H. Yamamoto, A. Sugiyama and F. Hamaguchi, Heterocycles, 1988, **27**, 2219).

(65)

The spicy components of peppers (Piperaceae) contain unsaturated N-acylpyrrolidines (66). A stereoselective synthesis of these compounds is available (A. Ohta, Y. Tonomura, J. Sawaki, N. Sato, H. Akiike, M. Ikuta and M. Shimazaki, Heterocycles, 1991, **32**, 965).

(66)

A new cinnamoylpyrrolidine with the unusual Z-geometry of the double bond has been isolated from leaves of *Piper peepuloides* and the structure

(67) was confirmed by synthesis (S. Shah, A.K. Kalla and K.L. Dhar, Phytochemistry, 1986, **25**, 1997). The corresponding E-isomer is present in roots of *Piper amalga* (X.A. Dominguez, J. Verde S, S. Sucar S. and R. Trevino, Phytochemistry, 1989, **25**, 239).

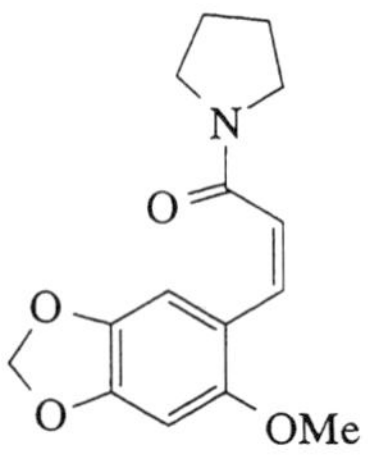

(67)

Several syntheses of trichonine (68) present in *Piper trichostachyon* have been reported. One involves use of two different arsonium ylides (L. Shi, J. Yang, M. Li and Y.-Z. Huang, Liebegs Ann. Chem., 1988, 377). Another gave predominantly (85%) the E,E-isomer (T. Moriyama, T. Mandai, M. Kawada, J. Otera and B.M. Trost, J. Org. Chem., 1986, **51**, 3896).

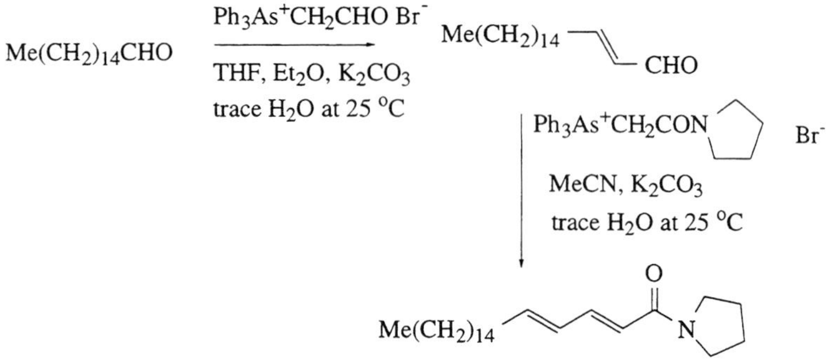

(68)

Me(CH$_2$)$_{14}$CH$_2$SO$_2$Ph

BuLi at 0°C

OHC

at -78 °C

SO$_2$Ph

Me(CH$_2$)$_{14}$

OAc

Bu^{t}OK, Bu^{t}OH
12 h RT

Me(CH$_2$)$_{14}$

(68)

Tricholeine (68a), a constituent of the same species has also been synthesized (M. Lounasmaa, J. Pusset and T. Sevenet, Phytochemistry, 1980, **19**, 953).

HO

OH

DHP, H$^+$

pyridinium chlorochromate, NaOAc

OHC

OTHP

+

CH=PPh$_3$

OTHP

p-TSA, MeOH
PBr$_3$, py
NaCH(CO$_2$Et)$_2$
NaOH, MeOH
POCl$_3$ then pyrrolidine, Et$_3$N

(68a)

Single crystal X-ray diffraction analysis has established the structure and relative stereochemistry of (-)-odorinol (69) (N. Hayashi, K.-H Lee, I.S. Hall, A.T. McPhail and H.-C.Huang, Phytochemistry, 1982, **21**, 2371).

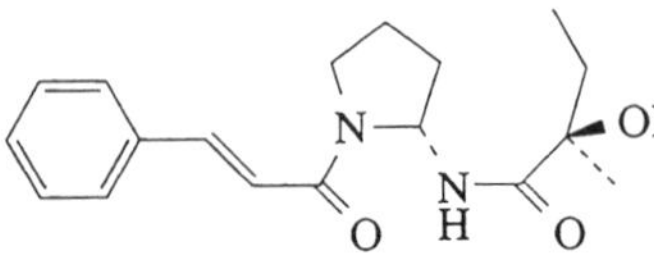

(69)

4,5-Dihydrookolasine (70) has been isolated from *Piper guineense* and synthesized by a route involving a Wittig-Horner reaction (S. Linke, J. Kurz and H.-J. Zeiler, Liebigs Ann. Chem., 1982, 1142).

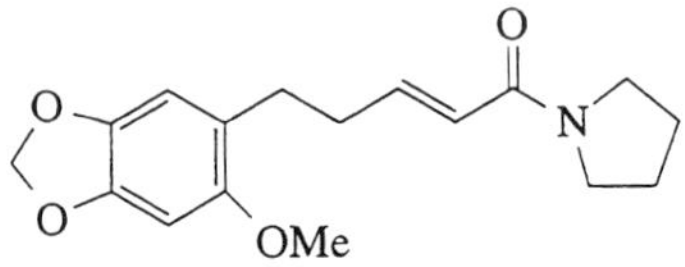

(70)

The ^{1}H and ^{13}C NMR spectra of a number of the N-acylpyrrolidinones from *Achillea* spp. have been reported (O. Hofer, H. Greger, W. Robien and A. Werner, Tetrahedron, 1986, **42**, 2707; H. Greger, M. Grenz and F. Bohlmann, Phytochemistry, 1982, **21**, 1071). Unsaturated N-acylpyrrolidines (71)-(76) were identified from *Achillea* spp. (H. Greger, C. Zdero and F. Bohlmann, Phytochemistry, 1984, **23**, 1503).

Roots of *Achillea ageratifolia* have yielded four new unsaturated amides (77)-(80) (H. Greger, C. Zdero and F. Bohlmann, Liebigs Ann. Chem., 1983, 1194).

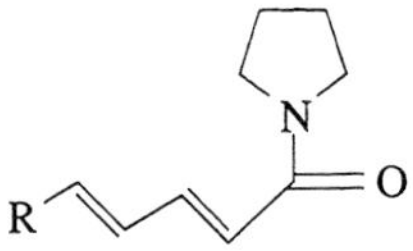

(71) R = Me(CH$_2$)$_8$

(72) R = Me(CH$_2$)$_2$CH=CHC≡C(CH$_2$)$_2$ (Z)

(73) R = Me(CH$_2$)$_2$(CH=CH)$_2$(CH$_2$)$_2$

(74) R = MeCH=CH(C≡C)$_2$(CH$_2$)$_2$ (E)

Me(CH$_2$)$_5$ (CH$_2$)$_8$

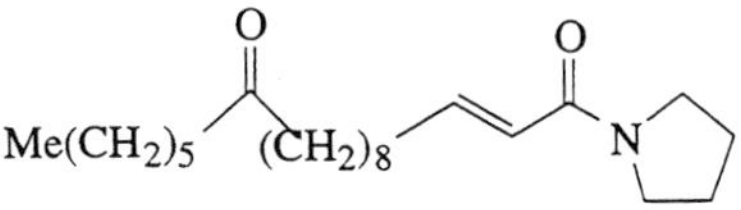

(75)

Me(CH$_2$)$_5$ (CH$_2$)$_{10}$

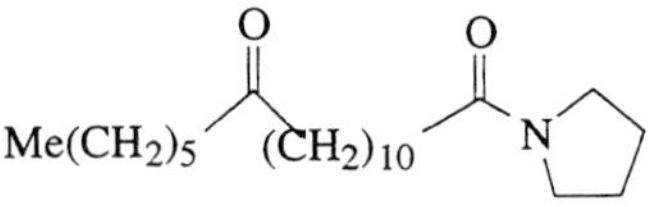

(76)

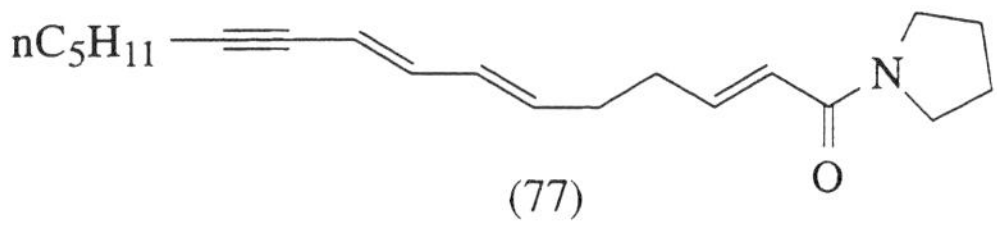

nC_5H_{11} (77)

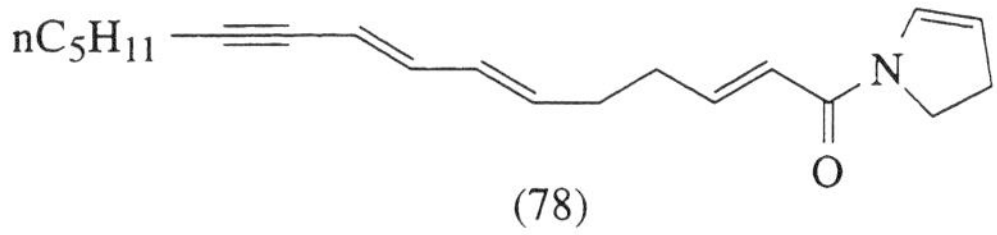

nC_5H_{11} (78)

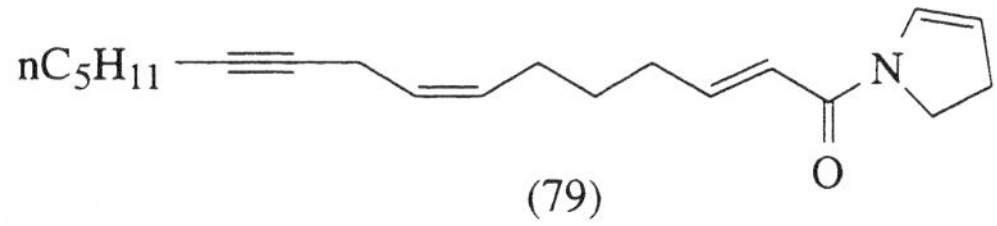

nC_5H_{11} (79)

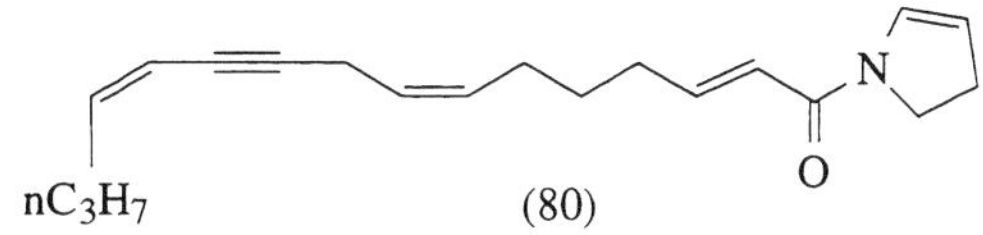

nC_3H_7 (80)

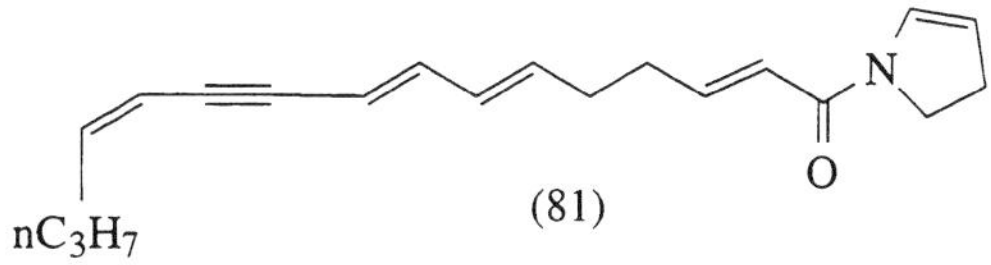

nC_3H_7 (81)

The N-acylpyrroline (81) is present in *A. tomentosa* L. (H. Greger, M. Grenz and F. Bohlmann, Phytochemistry, 1981, **20**, 2579).

Piriferine (82) is present in leaves of the Thailand plant *Aglaia pirifera* (E. Saifah, V. Jongbunprasert and C.J. Kelley, J. Nat. Prod., 1988, **51**, 80).

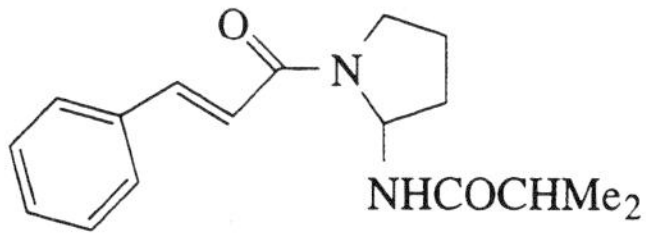

$NHCOCHMe_2$ (82)

(+)-Dysidin (83) was isolated from the sponge *Dysidea herbaceae*. It has been synthesized as a racemate (P.G. Willard and S.E. de Laszlo, J. Org. Chem., 1984, **49**, 3489), and as an enantiomer of the natural material (H. Kohler and H. Gerlach, Helv. Chim. Acta, 1984, **67**, 1783).

18

(83)

4 N-acylpyrrolidones

Squamolone (84) and the methoxy compound (85) are present in the
bark of the West African tree *Hexalobus crispiflorus* (H. Achenbach, C.
Renner and I. Addae-Mensah, Leibigs Ann. Chem., 1982, 1623).

(84)   (85)

The N-acylpyrrolinone (86) was isolated from aerial parts of *Piper
demeraranum* (A. Maxwell and D. Rampersad, J. Nat. Prod., 1989, **52**,
891).

(86)

Four new N-acylpyrrolinones named mirabimides A - D (87)-(90) were
extracted from the blue green alga *Scytonema mirabile* and their structures
were established by spectroscopic methods (S. Carmeli, R.E. Moore and
G.M.L. Patterson, Tetrahedron, 1991, **47**, 2087).

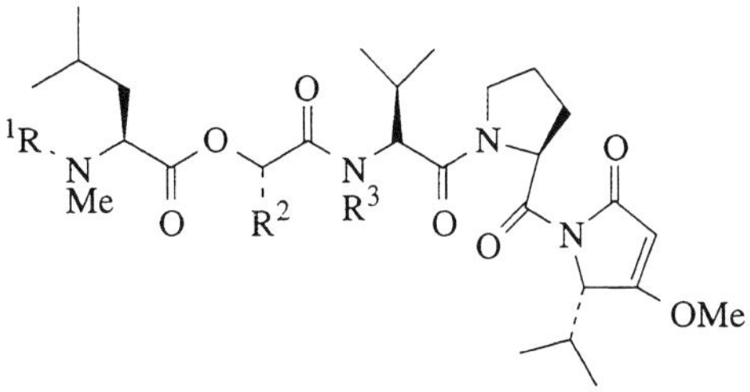

(87) $R^1 = R^3 = Me$, $R^2 = CHMeEt$

(88) $R^1 = Me$, $R^2 = CHMeEt$, $R^3 = H$

(89) $R^1 = Me$, $R^2 = Pr^i$, $R^3 = Me$

(90) $R^1 = Ac$, $R^2 = CHMeEt$, $R^3 = Me$

Mirabimide E is a solid tumour selective cytotoxin also isolated from *Scytonema mirabile* . The unusual structure (91) for mirabimide E containing a tetrachlorinated ethylene group was established by spectroscopic studies, stereoselective synthesis of three degradation products and finally by total synthesis (S. Paik, S. Carmeli, J. Cullingham, R.E. Moore, G.M.L. Patterson and M.A. Tins, J. Amer. Chem. Soc., 1994, **116**, 8116).

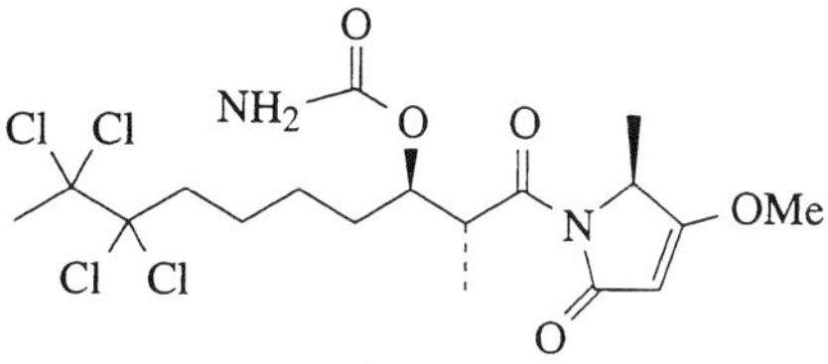

(91)

Dysidamide (92) from the Red Sea sponge *Dysidea herbacea* was identified by spectroscopic data and identification of its hydrolysis products (S. Carmely, T. Gebreyesus, Y. Kashman, B.W. Skelton, A.H.White and T. Yosief, Aust. J. Chem., 1990, **43**, 1881).

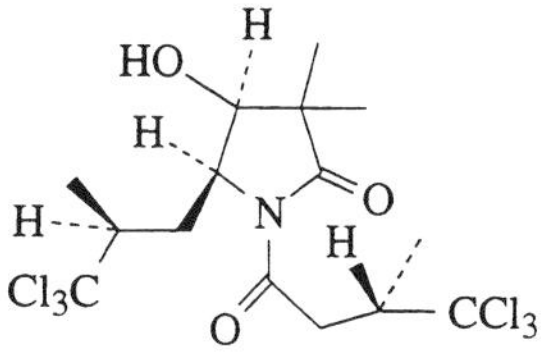

(92)

Chapter 8

FIVE-MEMBERED MONOHETEROCYCLIC COMPOUNDS (CONTINUED) : PYRROLIZIDINE ALKALOIDS

DAVID J. ROBINS

Progress in this area has been reviewed annually in "The Alkaloids", (D.J. Robins, Vols. 11-13, Royal Society of Chemistry, London, 1981-3) and in "Natural Product Reports" (D.J. Robins, 1984, **1**, 235; 1985, **2**, 213; 1986, **3**, 297; 1987, **4**, 577; 1989, **6**, 221; 1990, **7**, 377; 1991, **8**, 213; 1992, **9**, 313; 1993, **10**, 487; 1994, **11**, 613; 1995, **12**, 413). A book on pyrrolizidine alkaloids covering isolation and identification of the alkaloids, ^{13}C NMR spectroscopy, mass spectrometry and biological properties is available ("Naturally Occurring Pyrrolizidine Alkaloids", ed. A.F.M. Rizk, CRC Press, Boca Raton, Fla., USA, 1990). An authoritative book by A.R. Mattocks dealing mainly with the biological activities of pyrrolizidine alkaloids has been published ("Chemistry and Toxicology of Pyrrolizidine Alkaloids", Academic Press, London, 1986). Other useful reviews are available (D.J. Robins, Fortschr. Chem. Org. Naturst., 1982, **41**, 115; L. W. Smith and C.C.J. Culvenor, J. Nat. Prod., 1981, **44,** 129; J.T. Wrobel in "The Alkaloids", ed. A. Brossi, Academic Press, New York, 1985, **26**, Ch. 7; E. Roeder, Pharm. Unserer Zeit., 1984, **13**, 33; T. Hartmann and L. Witte in "Alkaloids, Chemical and Biological Perspectives", ed. S.W. Pelletier, 1995, **9**, Ch. 4; E. Roeder, Pharmazie, 1995, **50**, 83).

Extensive NMR and mass spectral data are now available for pyrrolizidine alkaloids. The ^{1}H NMR spectra of more than 350 pyrrolizidine alkaloids have been recorded (C.G. Logie, M.R. Grue and J.R. Liddell, Phytochemistry, 1994, **37**, 43) and two sets of ^{13}C NMR spectroscopic data are listed for more than 130 pyrrolizidine alkaloids (E. Roeder, Phytochemistry, 1990, **29**, 11) and for 30 alkaloids (A.J. Jones, C.C.J. Culvenor and L.W. Smith, Aust. J. Chem., 1982, **35**, 1173). The mode of attachment of the necic diacid to the necine diol in macrocyclic pyrrolizidine alkaloids has often been deduced by mass spectrometry. This method can be unreliable, and the NMR selective population inversion technique depending upon long range couplings to establish bond connectivities is now advocated (M.W. Bredenkamp and A. Wiechers, Tetrahedron Lett., 1987, **28**, 3725). Macrocyclic pyrrolizidine alkaloids exhibit strong [M-H]⁻ and [M+OH]⁻ ions in negative ion chemical ionisation mass spectrometry together with much fragmentation but ions corresponding to the intact necine and necic acid components could usually be distinguished (H.J. Huizing, T.W.J. Gadella and E. Kliphuis, Plant Syst. Evol., 1982, **140**, 279). Fast atom bombardment mass spectra of macrocyclic pyrrolizidine alkaloids containing retronecine N-oxide gave a characteristic [M+H]$^+$ ion, together with the series of fragments m/z 136, 120, 118, 106, 94, and 80 (J.J. Karchesy, M.L. Deinzer and D.A. Griffin, J. Agric. Food Chem., 1984, **32**, 1056). Positive ion chemical ionisation mass spectra of pyrrolizidine alkaloids using methane produced mainly [M+H]$^+$ ions (J.J. Karchesy, M.L. Deinzer and D.A. Griffin, Biomed. Mass Spectrom., 1984, **11**, 455). Tandem mass spectrometry has been used to provide information about pyrrolizidine alkaloids present in crude methanolic extracts of dried plants (W.F. Haddon and R.J. Molyneux in "Tandem Mass Spectrometry", ed. F.W. McLafferty, Wiley, New York, 1983, Ch. 24.

The range of analytical methods available for the determination of pyrrolizidine alkaloids continues to widen. A useful field test for the detection of toxic pyrrolizidine alkaloids involves the conversion of the 1,2-unsaturated necine component into the corresponding pyrrole which gives a magenta coloured derivative with Ehrlich's reagent (A.R. Mattocks and R. Jukes, J. Nat. Prod., 1987, **50**, 161). Microwave irradiation has been used to extract pyrrolizidine alkaloids from *Senecio* species (C. Bicchi, F. Beliardo and P. Rubiolo, Lab. 2000, 1992, **6**, 36). Supercritical fluid extraction with methanol or carbon dioxide is recommended for rapid extraction, simple work up and improved recovery of pyrrolizidine alkaloids [C. Bicchi, P. Rubiolo, C. Frattini, P. Sandra and F. David, J. Nat. Prod., 1991, **54**, 941; S.T. Schaeffer, L.H. Zalkow and A.S. Teja, Biotechnol. Bioeng., 1989, **34**, 1357; S.T. Schaeffer, L.H. Zalkow and A.S. Teja, ACS Symp. Ser., 1989, **406** (Supercrit. Fluid Sci. Technol.), 416; G. Holzer, L.H. Zalkow and C.F. Asibal, J. Chromatogr., 1987, 400, 317]. A most sensitive method for detection of pyrrolizidine alkaloids in the ppb range utilizes a two step enzyme-linked immunosorbent assay (ELISA) (D.M. Roseman, X. Wu, L.A. Milco, M. Bober, R.B. Miller and M.J. Kurth, J. Agric. Food Chem., 1992, **40**, 1008). Another ELISA which is specific for macrocyclic 12-membered unsaturated pyrrolizidine alkaloids was recently reported (E. Roeder and T. Pflueger, Nat. Toxins, 1995, **3**, 305). Separation of mono- and di-ester pyrrolizidine alkaloids can be carried out by ion-pair adsorption TLC using chloride as the counter ion (H.J. Huizing and T.M. Malingre, J. Chromatogr., 1981, **205**, 218). Many HPLC systems for separation of pyrrolizidine alkaloids have been reported (B. Sener, H. Temizer, A. Temizer and A.E. Karakaya, J. Pharm. Belg., 1986, **41**, 115; H.J. Segall, J. Liq. Chromatogr., 1979, **2**, 1319; G. Tittel, H. Hinz and H. Wagner, Planta Med., 1979, **37**, 1). Many underivatized pyrrolizidine alkaloids from plants and insects have been analysed by capillary gas chromatography-mass spectrometry (L. Witte, P.

Rubiolo, C. Bicchi and T. Hartmann, Phytochemistry, 1992, **32**, 187). Columns coated with polydimethylsiloxane are recommended and about 100 retention indices were quoted.

$$\text{(1)}$$

A substantial number of X-ray crystal structures of pyrrolizidine alkaloids have been determined and some generalities have emerged. The ester carbonyl groups are antiparallel in all 12-membered macrocyclic pyrrolizidine alkaloids which contain retronecine (1) such as jacoline (2) and jaconine (R.W. Gable, M.F. Mackay and C.C.J. Culvenor, Acta Crystallogr., Sect. C, 1988, **44**, 1942); or rosmarinecine present in rosmarinine (3) (A.A. Freer, H.A. Kelly and D.J. Robins, Acta Crystallogr., Sect. C., 1986, **42**, 1348); or platynecine as part of hygrophylline (4) (M.F. Mackay, P. Mitrprachachon and C.C.J. Culvenor, Acta Crystallogr., Sect. C, 1985, **41**, 395); or otonecine which appears in otosenine (5) (H. Wiedenfeld, E. Roeder and F. Knoch, Acta Crystallogr., Sect. C, 1990, **46**, 1345). Transannular distances between the nitrogen and carbon of the carbonyl group in other macrocyclic diesters containing otonecine have been compared. Most eleven-membered macrocyclic diesters of retronecine such as incanine (6) [B. Tashkhodzhaev, M.V. Telezhenetskaya and S. Yu. Yunusov, Khim. Prir. Soedin., 1979, 363 (Chem. Abstr., 1980, **92**, 111199)] have ester carbonyl groups that are *syn* -parallel and pointing in the same direction as H-8, except for trichodesmine (7) [B. Tashkhodzhaev, M.R. Yagudaev and S. Yu. Yunusov, Khim. Prir. Soedin., 1979, 368 (Chem. Abstr., 1980, **92**, 111194)] and grantianine (H. Stoeckli-Evans and D.J. Robins, Acta Crystallogr., Sect. C, 1984, **40**, 1445) which have the different antiparallel conformation of the macrocycle.

The biosynthesis of pyrrolizidine alkaloids has been studied extensively by Robins and co-workers and is the subject of a recent review (D.J. Robins, in "The Alkaloids", ed. G.A. Cordell, Academic Press, New York, 1995, **46**, 1). The importance of pyrrolizidine alkaloids in plant-insect interaction has been reviewed (M. Wink, Bioforum, 1993, **73**, 871).

1. Synthesis of necines

The numbering scheme used for the pyrrolizidine nucleus is shown for retronecine (1). The past 15 years have seen rapid growth in the number of syntheses of pyrrolizidine alkaloids. The first syntheses of optically active pyrrolizidine bases (necines) have appeared and macrocyclic pyrrolizidine alkaloids have been prepared for the first time. An extraordinary number of different synthetic routes to necines have been published and this area has been reviewed [M. Ikeda, T. Sato and H. Ishibashi, Heterocycles, 1988, **27**, 1465 and Y. Nishimura, Stud. Nat. Prod. Chem., 1988, **1** (Stereosel. Synth., Part A), 227]. Efforts are now being concentrated on producing the necines in optically active form (see the review by W.-M. Dai, Y. Nagao and E. Fujita, Heterocycles, 1990, **30**, 1231) and this will be the emphasis for this review.

(a) Unhydroxylated compounds

Cassipourine (10) is present in *Cassipourea* species from the family Rhizophoraceae. Racemic material has been prepared from the β-epoxide (8) via the α-epithio-compound (9) (J.T. Wrobel and J.A. Glinski, Can. J. Chem., 1981, **59**, 1101).

(8)   (9)   (10)

Loline (12) has been isolated from *Lolium cuneatum* and *Festuca arundinacea* and it was synthesised for the first time using a cycloaddition process involving a nitrone to generate the pyrrolizidine nucleus (11) (J.J. Tufariello, H. Meckler and K. Winzenberg, J. Org. Chem., 1986, **51**, 3556).

(11)   (12)

(b) Monohydroxylated necines

Tussilagine (13) was isolated from *Tussilago farfara* L. (coltsfoot) (E. Roeder, H. Wiedenfeld and E.-J. Jost, Planta Med., 1981, **43**, 99) and the structure was revised after an X-ray crystal structure analysis was carried out [H. Wiedenfeld, E. Roeder, A. Kirfel and G. Will, Arch. Pharm. (Weinheim, Ger.), 1983, **316**, 367]. (±)-Tussilagine (13) together with the C-2 epimer has been synthesised by 1,3-dipolar cycloaddition of a nitrone [E. Roeder, H. Wiedenfeld and E.-J. Jost, Arch. Pharm. (Weinheim, Ger.), 1984, **317**, 403]. Tussilagine is in fact an artefact formed from the corresponding acid during the Soxhlet extraction of the plant with methanol (C.M. Passreiter, Phytochemistry, 1992, **31**, 4135).

(13)

The first synthesis of all the optically active forms of isoretronecanol, trachelanthamidine and supinidine was achieved by Robins and Sakdarat from 2*S*,4*R*-4-hydroxyproline (D.J. Robins and S. Sakdarat, J. Chem. Soc., Perkin Trans. 1, 1981, 909). (-)-Isoretronecanol (15), (-)-trachelanthamidine (16) and (-)-supinidine (17) have been prepared from L-proline by homologation of the acid, N-alkylation and Dieckmann cyclisation to give the pyrrolizidine nucleus (14) (H. Rueger and M. Benn, Heterocycles, 1982, **19**, 1677).

Proline has also been used in a route to (-)-isoretronecanol (15) and (-)-trachelanthamidine (16) (T. Moriwake, S. Hamano and S. Saito, Heterocycles, 1988, **27**, 1135).

Alternative routes to (-)-trachelanthamidine (16) from L-prolinol involve intramolecular cyclization processes with an alkene (H. Ishibashi, H. Ozeki and M. Ikeda, J. Chem. Soc., Chem. Commun., 1986, 654; T. Sato, K. Matsubayashi, K. Tsujimoto and M. Ikeda, Heterocycles, 1993, **36**, 1205). A ruthenium catalysed chlorine atom transfer cyclization process has also been used to prepare (-)-trachelanthamidine from L-prolinol (H. Ishibashi, N. Uemura, H. Nakatani M. Okazaki, T. Sato, N. Nakamura and M. Ikeda, J. Org. Chem., 1993, **58**, 2360). A different approach to the synthesis of (-)-trachelanthamidine (16) made use of a chiral auxiliary (18) to control the formation of the adjacent chiral centres in (19) (Y. Nagao, W.-M. Dai, M. Ochiai, S. Tsukagoshi and E. Fujita, J. Org. Chem., 1990, **55**, 1148; Y. Nagao, W.-M. Dai, M. Ochiai, S. Tsukagoshi and E. Fujita, J. Am. Chem. Soc., 1988, **10**, 289). This route was extended to the synthesis of (-)-supinidine (17) (Y. Nagao, W.-M. Dai and M. Ochiai, Tetrahedron Lett., 1988, **29**, 6133; Y. Nagao, W.-M. Dai, M. Ochiai and M. Shiro, Tetrahedron, 1990, **46**, 6361).

The iodoacetamide (20), prepared from L-proline, was subjected to radical annulation to generate the pyrrolizidinone (21) and eventually (-)-trachelanthamidine (16) (R.S. Jolly and T. Livinghouse, J. Am. Chem. Soc., 1988, **110**, 7536).

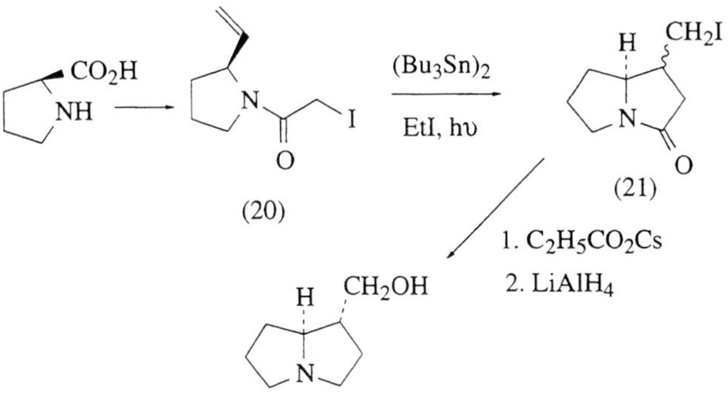

A 5-exo-trig radical cyclization to form the C-1 to C-2 bond in (-)-trachelanthamidine (16) also used L-proline as starting material (J.A. Seijas, M.P. Vazquez-Tato, L. Castedo, R.J. Estevez, M.G. Onega and M. Ruiz, Tetrahedron, 1992, **48**, 1637).

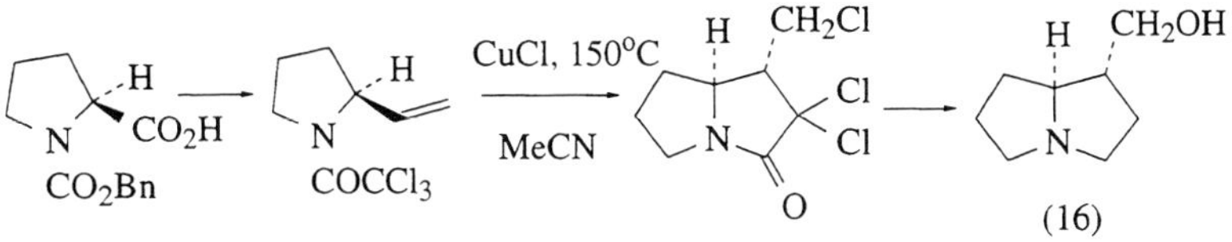

The enantiomeric (+)-trachelanthamidine was prepared from maleimide using the chiral auxiliary 10-mercaptoisoborneol (Y. Arai, T. Kontani and T. Koizumi, Chem. Lett., 1991, 2135).

A route to (-)-isoretronecanol (15) starting from L-proline makes use of a π-allyltricarbonyliron carbon complex (22) to generate the C-2 to C-3 bond (J.G. Knight and S.V. Ley, Tetrahedron Lett., 1991, **32**, 7119).

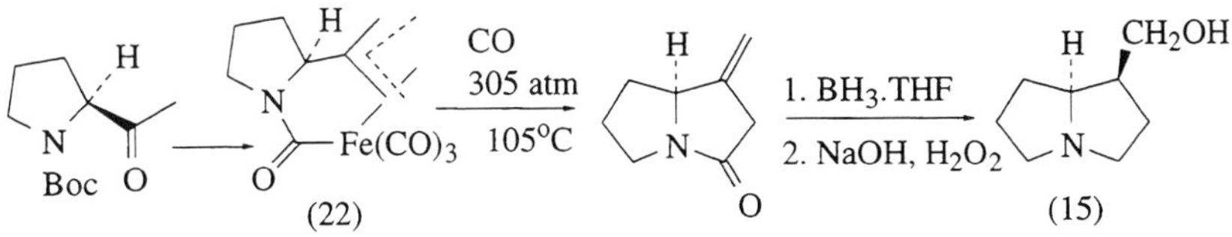

28

(-)-Isoretronecanol (15) can also be made from the cyclopropane diester (23) incorporating *(S)*-α-methylbenzylamine as chiral auxiliary in (24). Catalytic reduction gave the *cis*-diastereoisomer and hydrogenolysis led to the pyrrolizidinone (25) (G. Haviari, J.P. Celerier, H. Petit and G. Lhommet, Tetrahedron Lett., 1993, **34**, 1599).

A recent route to (-)-supinidine (17) started from L-proline and used a thermal oxime alkene cycloaddition process in the key step (A. Hassner, S. Singh, R. Sharma and R. Maurya, Tetrahedron, 1993, **49**, 2317).

Sharpless asymmetric oxidation has been used to generate the optically active alkene (26), epoxide (27) and diol (28), all of which were converted into (-)-supinidine (17) (H. Takahata, Y. Banba and T. Momose, Tetrahedron, 1991, 47, 7635).

Loroquine (31) was synthesised in eight steps including a Dieckmann cyclization to produce the dihydropyrrolizine (30) from pyrrole ester (29) (E. Roeder, J.-P.B. Bartkowski and T. Bourauel, Liebigs Ann. Chem., 1993, 711).

(c) Dihydroxylated necines

(-)-Petasinecine (34) was first synthesised from L-proline via the enolester (32). Catalytic hydrogenation of this ester gave a mixture of C-1 epimers which was separated chromatographically and the minor product (33) was reduced to yield (-)-petasinecine (H. Rueger and M. Benn, Heterocycles, 1983, **20**, 235).

An alternative route to (-)-petasinecine (34) also starts from L-proline but utilises an Ireland-Claisen type of rearrangement to give the pyrrolizidinone (35) (J. Mulzer and M. Shanyoor, Tetrahedron Lett., 1993, **34**, 6545).

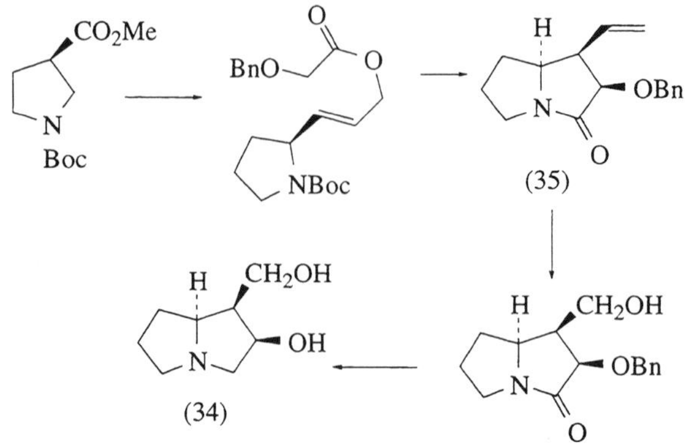

L-proline was also used to make (-)-macronecine (36) which is the enantiomer of the naturally occurring material (H. Ito, Y. Ikeuchi, T. Taguchi and Y. Hanzawa, J. Am. Chem. Soc., 1994, **116**, 5469).

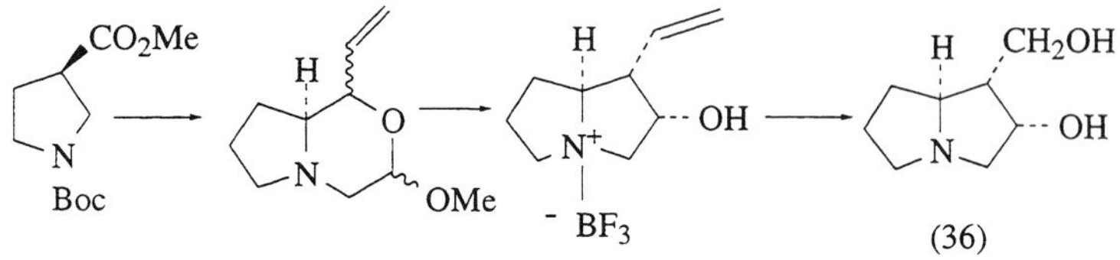

Curassanecine (37) is the first necine to contain a 1-hydroxy group. Confirmation of this unusual structure was provided by its synthesis in racemic form from N-acetylpyrrolidine (S. Mohanraj, P. Subramanian and W. Herz, Phytochemistry, 1982, **21**, 1775).

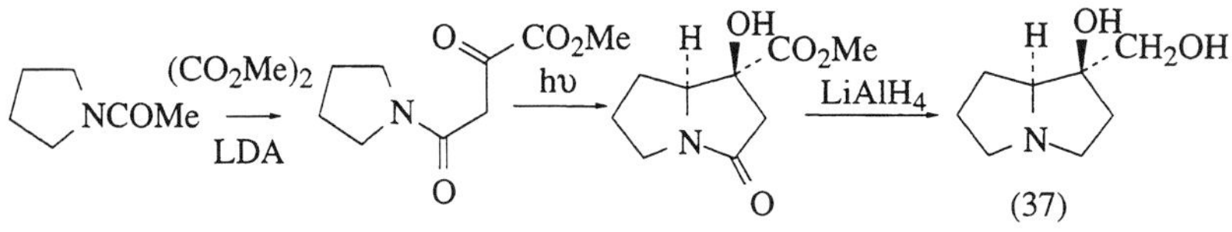

The protected amine (38) was prepared from D-glucose and converted into platynecine (39) (G.W.J. Fleet, J.A. Seijas and M.P. Vazques-Tato, Tetrahedron, 1991, **47**, 525).

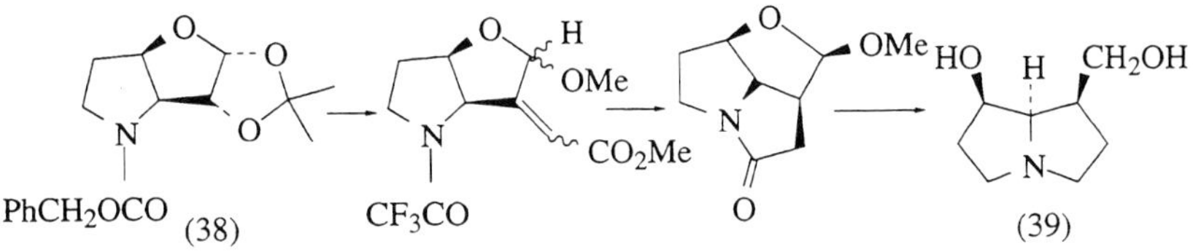

R-Malic acid was used in the synthesis of (-)-hastanecine (42). The optically active imide (41) was reduced and the product rearranged to give an acylium ion which underwent an aza-Cope rearrangement to afford the pyrrolizidinone (41) (D.J. Hart and T.-K. Yang, J. Chem. Soc., Chem. Commun., 1983, 135). Tandem cycloaddition reactions of nitroalkenes have been employed in the synthesis of (-)-hastanecine (42) (S.E. Denmark and A. Thorarensen, J. Org. Chem., 1994, **59**, 5672).

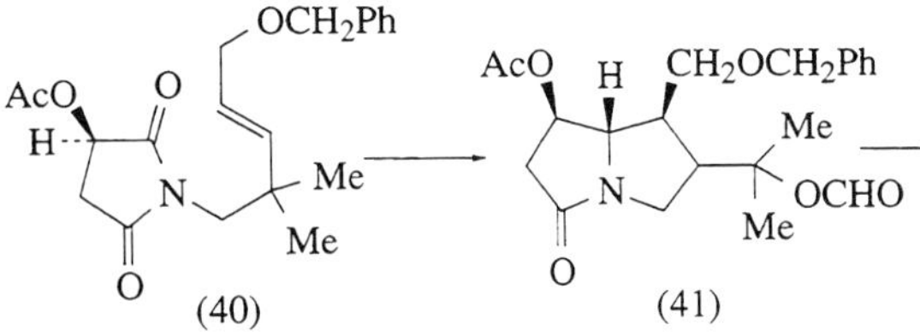

Ketenedithioacetals were used as terminators for cyclization of α-acylium ions derived from S-malic acid to give the key pyrrolizidinone intermediate (43). This was used to produce seven different pyrrolizidine diols including (+)-hastanecine (44) (A.R. Chamberlin and J.Y.L. Chung, J. Org. Chem., 1985, **50**, 4425).

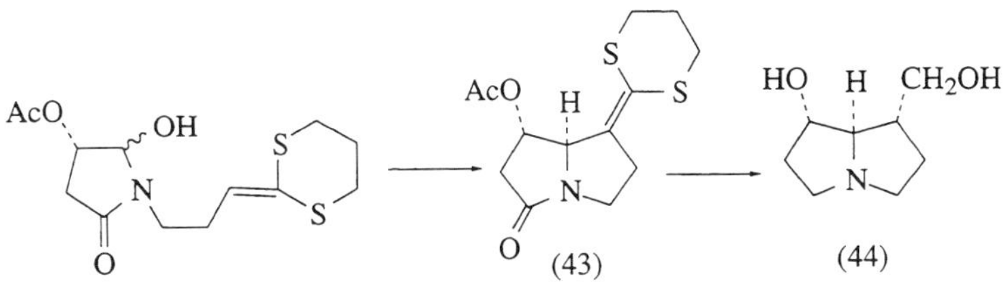

Heliotridine and retronecine have had most synthetic attention directed towards them. A recent synthesis of (+)-retronecine (1) involved the preparation of (45) from 2S,4R-4-hydroxyproline and generation of an azomethine ylid from (45) which underwent 1,3-dipolar cycloaddition with methyl propiolate to give (46) as major isomer (G. Pandey and G. Lakshmaiah, Synlett, 1994, 277).

32

Other routes to (+)-retronecine (1) start from *S*-malic acid (T. Kametani, H. Yukawa and T. Honda, J. Chem. Soc., Chem. Commun., 1988, 685) and D-glucose (Y. Nishimura, S. Konde and H. Umezawa, J. Org. Chem., 1985, 50, 5210).

(+)-Heliotridine (50) has been prepared from *S*-malic acid via the imide (47) involving an intramolecular carbenoid displacement process to produce the pyrrolidone (48). Intramolecular allylation led to the pyrrolizidinone (49) (T. Kametani, H. Yukawa and T. Honda, J. Chem. Soc., Perkin Trans. 1, 1990, 571).

The (+)-Geissman-Waiss lactone (53) is a useful intermediate which has been successfully used in routes to retronecine, platynecine and croalbinecine. A short route to this lactone involves Sharpless asymmetric oxidation of the allylic alcohol (51) followed by Pd (II) catalysed intramolecular amidocarbonylation to afford the lactone (52) (H. Takahata, Y. Banba and T. Momose, Tetrahedron: Asymmetry, 1991, 2, 445).

Other routes are available utilizing optically active tin enolates (Y. Nagao, W.-M. Dai, M. Ochiai and M. Shiro, J. Org. Chem., 1989, 54, 5211); starting from *S*-pyroglutamic acid (N. Ikota and A. Hanaki, Heterocycles, 1988, 27, 2535); or D-glucose (M.K. Gurjar, V.J. Patil and S.M. Pawar, Indian J. Chem., Sect. B, 1987, 26, 1115); using Bakers' yeast reduction (J. Cooper, P.T. Gallagher and D.W. Knight, J. Chem. Soc., Chem.

Commun., 1988, 509); 2S,4R-4-hydroxyproline (H. Rueger and M.H. Benn, Heterocycles, 1982, **19**, 23); carbohydrate precursors (J.G. Buchanan, G. Singh and R.H. Wightman, J. Chem. Soc., Chem. Commun., 1984, 1299; Y. Nishimura, S. Kondo and H. Umezawa, J. Org. Chem., 1985, **50**, 5210); diethyl (+)-tartrate (K. Shishido, Y. Sukegawa, K. Fukumoto and T. Kametani, Heterocycles, 1985, **23**, 1629); and S-malic acid (A.R. Chamberlin and J.Y.L. Chung, J. Org. Chem, 1985, 50, 4425).

The first synthesis of (±)-otonecine (55) was achieved from a hydroxy ester (54) which had previously been used in the synthesis of (±)-retronecine (1) (H. Niwa, Y. Uosaki and K. Yamada, Tetrahedron Lett., 1983, **24**, 5731).

(d) Trihydroxylated necines

The absolute configuration at C-1 of hadinecine (56) was established by preparing it from (+)-retronecine (1) (R. Hanselmann and M. Benn, Tetrahedron Lett., 1993, **34**, 3511).

2S,4R-4-Hydroxyproline was the starting point for the first synthesis of (+)-croalbinecine (57) (H. Rueger and M. Benn, Heterocycles, 1983, **20**, 1331) via the lactone (53).

The first synthesis of (+)-crotanecine (59) was also achieved from 2S,4R-4-hydroxyproline via the bicyclic lactone (58) (V.K. Yadav, H. Rueger and M. Benn, Heterocycles, 1984, **22**, 2735). An alternative route from 2,3-O-isopropylidene-D-erythrose to (+)-crotanecine (59) is also available (R.B. Bennett and J.K. Cha, Tetrahedron Lett., 1990, **31**, 5437).

D-Glucosamine was the starting material for the synthesis of (-)-rosmarinecine (60) (K. Tatsuta, H. Takahashi, Y. Amemiya and M. Kinoshita, J. Am. Chem. Soc., 1983, **105**, 4096). Tandem cycloaddition reactions were used in another route to (-)-rosmarinecine (60) (S.E. Denmark, A. Thorarensen and D.S. Middleton, J. Org. Chem., 1995, **60**, 3574).

(60)   (61)   (62)

(e) Tetrahydroxylated necines

Alexines are polyhydroxylated pyrrolizidine alkaloids with a hydroxymethyl group at C-3 rather than C-1. Some of them are glycosidase inhibitors. Alexine (61) was prepared from D-glucose (G.W.J. Fleet, M. Haraldsson, R.J. Nash and L.E. Fellows, Tetrahedron Lett., 1988, **29**, 5441). Routes to 1,7a-diepialexine (62) have been devised from S-pyroglutamic acid (N. Ikota, Tetrahedron Lett., 1992, **33**, 2553), and a readily available heptonolactone (S. Choi, I. Bruce, A.J. Fairbanks, G.W.J. Fleet, A.H. Jones, R.J. Nash and L.E. Fellows, Tetrahedron Lett., 1991, **32**, 5517). 7,7a-Dipialexine was made from L-xylose (W.H. Pearson and J.V. Hines, Tetrahedron Lett., 1991, **32**, 5513).

2. Synthesis of macrocyclic pyrrolizidine alkaloids

Dicrotaline (63) was prepared from (+)-retronecine (1) and 3-hydroxy-3-methylglutaric anhydride and cyclization was achieved via the pyridine-2-thiolesters. One of the diastereoisomeric products was identical to dicrotaline and the absolute configuration at C-13 was established by chemoselective degradation to R-mevalonolactone (K. Brown, J.A. Devlin and D.J. Robins, J. Chem. Soc., Perkin Trans. 1, 1983, 1819; J.A. Devlin and D.J. Robins, J. Chem. Soc., Chem. Commun., 1981, 1272).

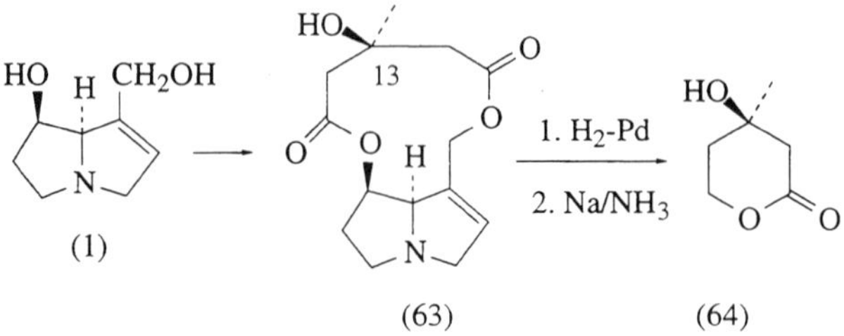

The cyclic stannoxane (65) derived from (+)-retronecine (1) has been used
to prepare dicrotaline (63) (H. Niwa, O. Okamoto, H. Ishiwata, A. Kuroda,
Y. Uosaki and K. Yamada, Bull. Chem. Soc. Jpn., 1988, 61, 3017),
integerrimine (66) (H. Niwa, Y. Miyachi, Y. Uosaki, A. Kuroda, H.
Ishiwata and K. Yamada, Tetrahedron Lett., 1986, **27**, 4609; H. Niwa, Y.
Miyachi, O. Okamoto, Y. Uosaki, A. Kuroda, H. Ishiwata and K. Yamada,
Tetrahedron, 1992, **48**, 393), monocrotaline (67) (H. Niwa, T. Ogawa, O.
Okamoto and K. Yamada, Tetrahedron, 1992, **48**, 10531; H. Niwa, O.
Okamoto and K. Yamada, Tetrahedron Lett., 1988, **29**, 5139), and
senecionine (H. Niwa, T. Sakata and K. Yamada, Bull. Chem. Soc. Jpn.,
1994, **67**, 1990). The same group has also carried out the synthesis of the
12-membered pyrrolizidine alkaloid (+)-yamataimine (H. Niwa, K.
Kunitani, T. Nagoya and K. Yamada, Bull. Chem. Soc. Jpn., 1994, **67**,
3094).

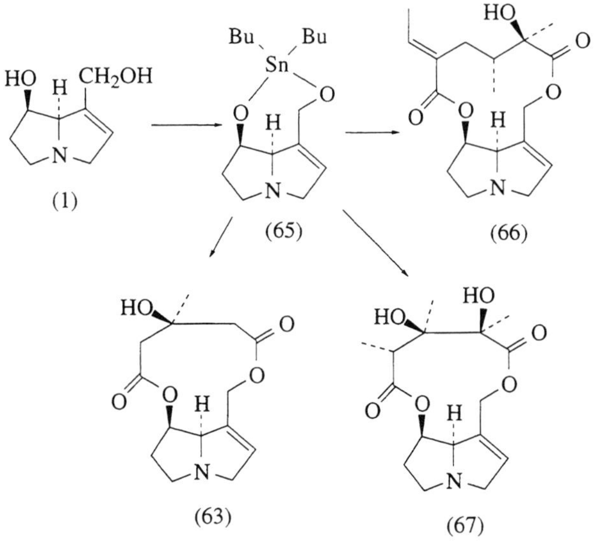

Other routes leading to integerrimine (66) include the synthesis of (+)-integerrinecic acid from *R*-pulegone (J.D. White, J.C. Amedio, S. Gut and L.R. Jayasinghe, J. Org. Chem., 1992, **57**, 2270; J.D. White and S. Ohira, J. Org. Chem., 1986, **51**, 5492) or using Sharpless asymmetric epoxidation (H. Niwa, Y. Miyachi, Y. Uosaki and K. Yamada, Tetrahedron Lett., 1986, **27**, 4601). The first synthesis of (±)-integerrimine (66) was achieved after separate synthesis of the necine and necic acid components (K. Narasaka, T. Sakakura, T. Uchimaru, and D. Guedin-Vuong, J. Am. Chem. Soc., 1984, **106**, 2954; K. Narasaka, T. Sakakura, T. Uchimaru, K. Morimoto and T. Mukaiyama, Chem. Lett., 1982, 445).

Synthesis of the 11-membered pyrrolizidine alkaloids crispatine (68) and fulvine (69) in racemic forms has also been reported (E. Vedejs and S.D. Larsen, J. Am. Chem. Soc., 1984, **106**, 3030).

(68)   (69)

3. Structures of pyrrolizidine alkaloids

New alkaloids, revised structures and additional stereochemical details are subdivided into the following groups: (a) unesterified necines; (b) monoesters; (c) acyclic diesters; and (d) cyclic diesters.

(a) Unesterified necines

helibractinecine

Heliotropium bracteatum R.Br.

A.J. Lakshmanan and S. Shanmugasundaram, Phytochemistry, 1994, **36**, 245.

clazamycin

Streptomyces puniceus

D.D. Buechter and D.E. Thurston, J. Nat. Prod., 1987, **50**, 360

curassanecine

Heliotropium curassavicum

S. Mohanraj, P.S. Subramanian and W. Herz, Phytochemistry, 1982, **21**, 1775

Streptomyces olivaceus

S.J. Box and D.F. Corbett, Tetrahedron Lett., 1981, **22**, 3293 pyrrolam A

R = OMe pyrrolam B

R = OH pyrrolam C

R = OCHMe(OEt)
pyrrolam D

Streptomyces olivaceus

R. Grote, A. Zeek, J. Stumpfel
and H. Zahner, Liebig's Ann.
Chem., 1990, 525

asparagamine A

Asparagus racemosus Willd.

T. Sekine, F. Ikegami, N. Fukasawa,
Y. Kashiwagi, T. Aizawa, Y. Fujii, N.
Ruangrungsi and I. Murakoshi, J.
Chem. Soc., Perkin Trans. 1, 1995,
391; T. Sekine, N. Fukasawa, Y.
Kashiwagi, N. Ruangrungsi and I.
Murakoshi, Chem. Pharm. Bull., 1994,
42, 1360

Alexines

alexine
Alexa leiopetala
R.J. Nash, L.E. Fellows, J.V. Dring,
G.W.J. Fleet, A.E. Derome, T.A.
Hamor, A.M. Scofield and D.J.
Watkin, Tetrahedron Lett., 1988, **29**,
2487

3,7a-diepialexine
Castanospermum australe A. Cunn
R.J. Nash, L.E. Fellows, J.V.
Dring, G.W.J. Fleet, A.E. Derome,
M.P. Baird, M.P. Hegarty and
A.M. Scofield, Tetrahedron, 1988,
44, 5959

7a-epialexine (australine)
Castanospermum australe A. Cunn

R.J. Molyneux, M. Benson, R.
Wong, J.E. Tropea and A.D. Elbein,
J. Nat. Prod., 1988, **51**, 1198

1,7a-diepialexine (epiaustraline)
Castanospermum australe A. Cunn

R.J. Nash, L.E. Fellows, J.V. Dring,
G.W.J. Fleet, A.Girdhar, N.G.
Ramsden, J.M. Peach, M.P. Hegarty and
A.M. Scofield, Phytochemistry, 1990,
29, 111; C.M. Harris, T.M. Harris, R.J.
Molyneux, J.E. Tropea and A.D. Elbein,
Tetrahedron Lett., 1989, **30**, 5685

7,7a-diepialexine

Castanospermum australe A. Cunn
Alexa leiopetala

R.J. Nash, L.E. Fellows, J.V. Dring, G.W.J. Fleet, A.Girdhar, N.G. Ramsden, J.M. Peach, M.P. Hegarty and A.M. Scofield, Phytochemistry, 1990, **29**, 111

or enantiomer

7a-epialexaflorine

Alexa grandiflora

A.C. de S. Pereira, M.A.C.Kaplan, J.G.S. Maia, O.R. Gottlieb, R.J. Nash, G.Fleet, L. Pearce, D.J. Watkin and A.M. Scofield, Tetrahedron, 1991, **47**, 5637

casuarine

Casuarina equisetifolia L.

R.J. Nash, P.I. Thomas, R.D. Waigh, G.W.J. Fleet, M.R. Wormald, P.M. de Q. Lilley and D.J. Watkin, Tetrahedron Lett., 1994, **35**, 7849

(b) Monoesters

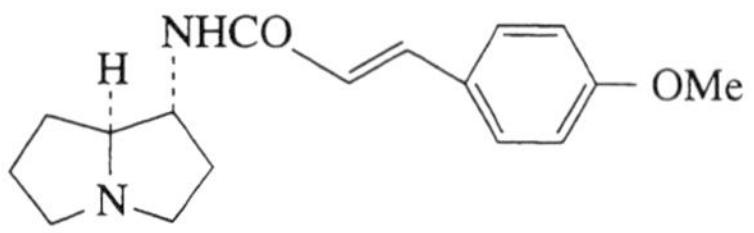

absouline

Hugonia oreogena
Hugonia penicillanthemum

K. Ikhiri, A. Ahmond, C. Poupat, P. Potier, J. Pusset and T. Sevenet, J. Nat. Prod., 1987, **50**, 626

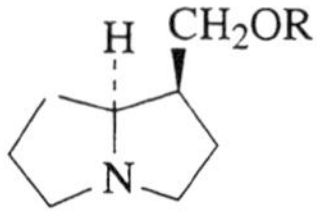

+ N-oxide

S. Mohanraj, P.S. Subramanian and W. Herz, Phytochemistry, 1982, **21**, 1775

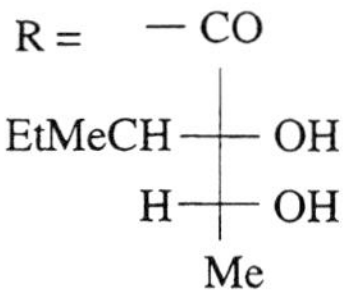

heliocurassavine
Heliotropium curassavicum

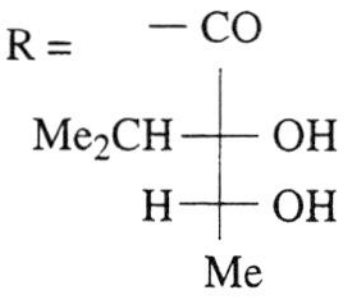

heliocoromandaline
H.curassavicum

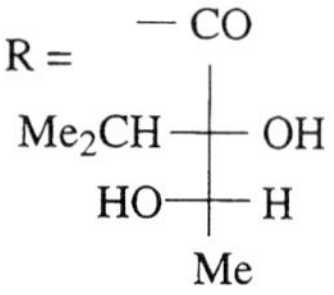

heliocurassavicine
H. curassavicum

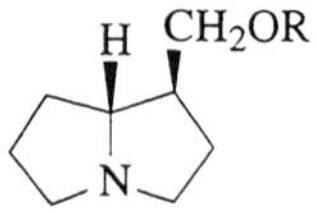

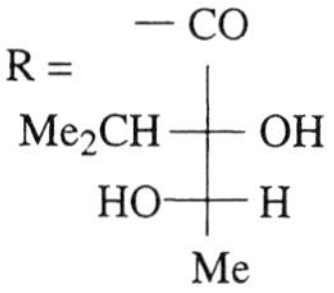

heliocurassavinine
H. curassavicum

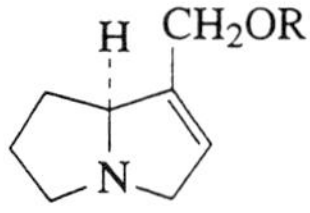

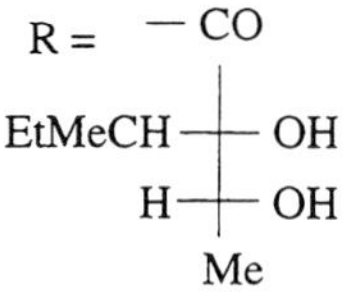

curassavinine
H. curassavicum

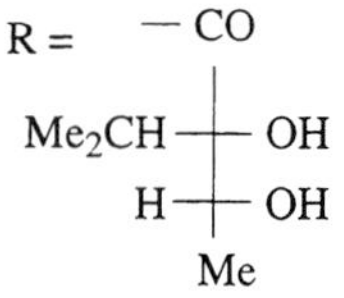

coromandalinine
H. curassavicum

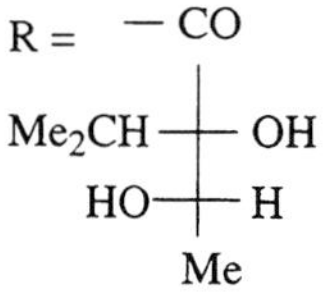

heliovinine
H. curassavicum

ehretinine (artefact?)

Ehretia aspera Willd.

O.P. Suri, R.S. Jamwal, K.A. Suri and C.K. Atal, Phytochemistry, 1980, **19**, 1273

parsonine

Parsonsia laevigata Alston

F. Abe and T. Yamauchi, Chem. Pharm. Bull., 1987, **35**, 4661

farfugine

Farfugium japonicum

H. Niwa, H. Ishiwata, A. Kuroda and K. Yamada, Chem. Lett., 1983, 789

racemonine

Senecio racemosus

W. Ahmed, Z. Ahmed, A. Malik, F. Ergun and B. Sener, Heterocycles, 1991, **32**, 1729

R^1=angelyl, R^2=H
7-angelylplatynecine

R^1 =H, R^2 =angelyl
9-angelylplatynecine

Castilleja rhexifolia aff. *miniata*

M.R. Roby and F.R. Stermitz, J. Nat. Prod., 1984, **47**, 846 racemozine

Senecio racemosus

W. Ahmed, A.Q. Khan, A. Malik, F. Ergun, and B. Sener, Fitoterapia, 1993, **64**, 361

7-*O*-angelylturneforcidine

Senecio integrifolius var. *fauriri*

E. Roeder and K. Liu, Phytochemistry, 1991, **30**, 1734

heliocurassavicine N-oxide

Heliotropium curassavicum L.

G. Marquina, A. Laguna, H. Velez and L.E. Fernandez, Rev. Cubana Farm., 1992, **26**, 52

crotalarine lactone

Crotalaria aegyptiaca Benth.

E. Roeder, T. Sarg, S. El-Dahmy, and A.A. Ghani, Phytochemistry, 1993, **34**, 1421

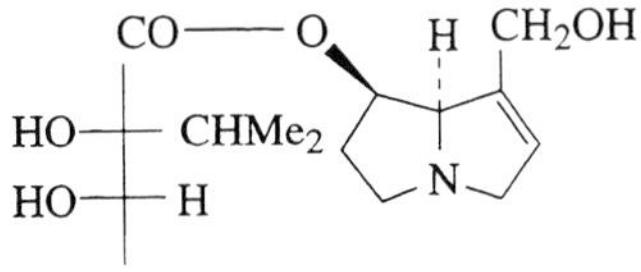

tessellatine

Amsinckia tessellata var. *gloriosa*

R.B. Kelley and J.N. Seiber,
Phytochemistry, 1992, **31**, 2513

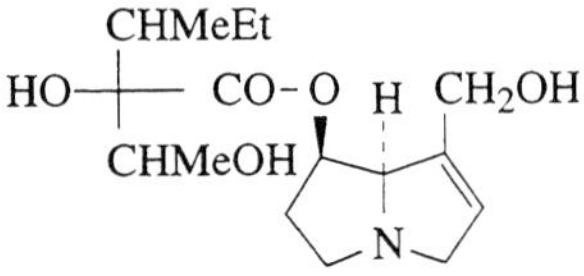

heliospathine

Heliotropium spathulatum

E. Roeder, E. Breitmaier, H.
Birecka, M.W. Frohlich and A.
Badzies-Crombach, Phytochemistry,
1991, **30**, 1703

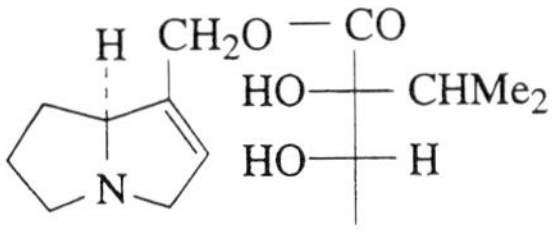

heliospathuline

Heliotropium spathulatum

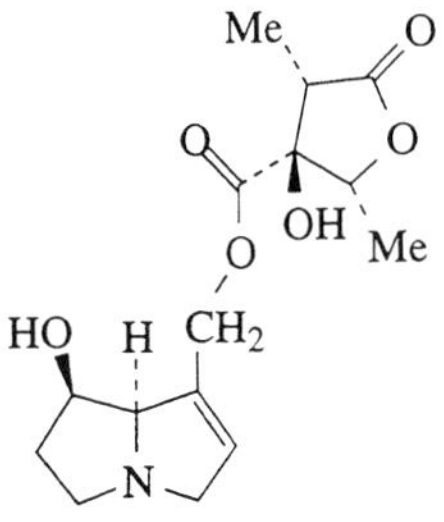

hackelidine

Hackelia californica
(Gray) Johnston

Y. Li, Huaxue Xuebao, 1990,
48, 415 (Chem. Abstr., 1990,
113, 168 971)

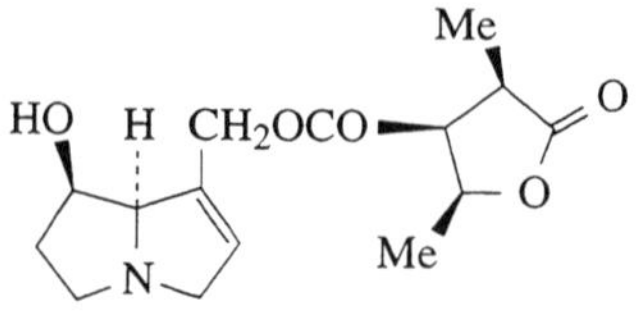

46

9-*O*-latifolylretronecine
Hackelia californica (Gray) Johnston

K.M. L'Empereur, Y. Li, and F.R. Stermitz, J. Nat. Prod., 1989, **52**, 360

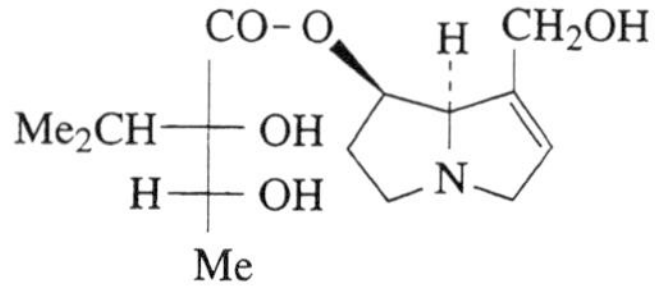

iso-lycopsamine

Heliotropium keralense

S. Ravi, A.J. Lakshmanan and W. Herz, Phytochemistry, 1990, 29, 361

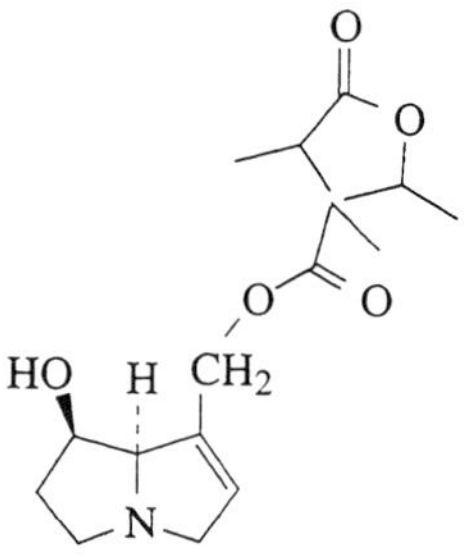

assamicadine

Crotalaria assamica Benth.

D. Cheng, Y. Lui, T.T. Chu, Y. Cui, J. Cheng and E. Roeder, J. Nat. Prod., 1989, **52**, 1153

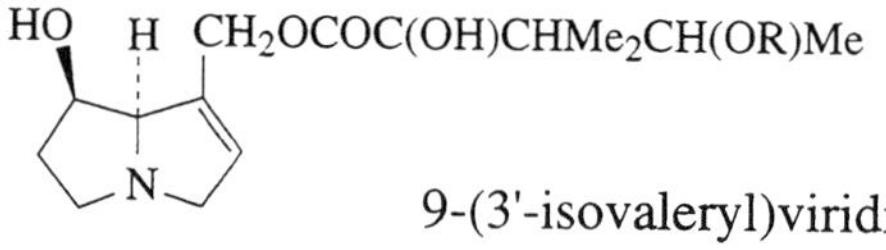

9-(3'-isovaleryl)viridiflorylretronecine

$R = COCH_2CHMe_2$ *Heliotropium argentinum* and *H. curassavicum*

J.G. Davicinio, M.J. Pestchanker and O.S. Giordano, Phytochemistry, 1988, **27**, 960

$R = COMe$ acetate of 9-viridiflorylretronecine

Heliotropium argentinum

3'-acetyl-lycopsamine

Amsinckia menziesii (Lehm.) Nels. et Macbr.

J.N. Roitman, Aust. J. Chem., 1983, **36**, 769.

scorpioidine

Myosotis scorpioides L.

J.F. Resch, D.F. Rosberger, J. Meinwald and J.W. Appling, J. Nat. Prod., 1982, **45**, 358.

R = O-tiglyl

5'-acetyleuropine

Heliotropium rotundifolium Sieber ex Lehm.

C.F. Asibal, L.T. Gelbaum and L.H. Zalkow, J. Nat. Prod., 1989, **52**, 726

cynoglossamine

Cynoglossum creticum

C.F. Asibal, J.A. Glinski, L.T. Gelbaum and L.H. Zalkow, J. Nat. Prod., 1989, **52**, 109

R = (E)-COCH=CH—⟨ ⟩—OH

helifoline

Heliotropium ovalifolium Forsk.

S. Mohanraj, P. Kulanthaivel, P.S. Subramanian and W. Herz, Phytochemistry, 1981, **20**, 1991

48

(c) Acyclic diesters

neosarracine

Senecio kaschkarovii C. Winkl.

D.-L. Cheng, J.-K Niu and E. Roeder, Phytochemistry, 1992, **31**, 3671

Senecio chrysocoma

M.R. Grue and J.R. Liddell, Phytochemistry, 1993, **34**, 1517

racemodine

Senecio racemosus DC

W. Ahmad, A.Q. Khan, A,. Malik, F. Ergun and B. Sener, Phytochemistry, 1993, 32, 224.

R^1=Me, R^2=OH   ipanguline A
R^1=OH, R^2=Me isoipanguline A

Ipomoea hederifolia L.

K. Jennett-Siems, M. Kaloga and E. Eich, Phytochemistry, 1993, **34**, 437

R^1=Me, R^2=OH   ipanguline B
R^1=OH, R^2=Me   isoipanguline B

Ipomoea hederifolia L.

K. Jennett-Siems, M. Kaloga and E. Eich, Phytochemistry, 1993, **34**, 437 punctanecine

Liatris punctata Hook

E.W. Mead, M. Looker, D.R. Gardner and F.R. Stermitz, Phytochemistry, 1992, **31**, 3255

threo-2",3"-dihydroxyechiumine

erythro-3"-chloro-2"-hydroxyechiumine

2",3"-epoxyechiumine

Cryptantha clevelandii Greene

F.R. Stermitz, M.A. Pass, R.B. Kelley and J.R. Liddell, Phytochemistry, 1993, **33**, 383

symviridine

Symphytum asperum Lepechin
S. officinale L.
S. x *uplandicum* Nyman

E. Roeder, T. Bourauel and V. Neuberger, Phytochemistry, 1992, **31**, 4041

3',7-diacetylintermedine

Amsinckia menziesii
var. *intermedia*

R.B. Kelley and J.N. Seiber, Phytochemistry, 1992, **31**, 2513 echiupinine

Echium pininana Webb et Berth.

E. Roeder, K. Liu and T. Bourauel, Phytochemistry, 1992, **30**, 3107

+ N-oxide

myoscorpine N-oxide

Echium pininana Webb et Berth.

E. Roeder, K. Liu and T. Bourauel, Phytochemistry, 1991, **30,** 3107

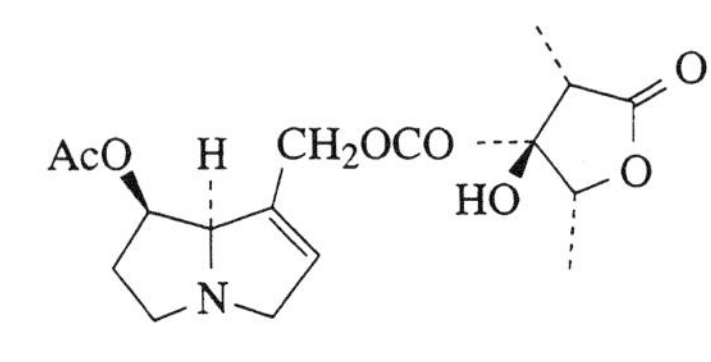

latifoline

Cynoglossum latifolium

C.C.J. Culvenor and M.F. Mackay, Aust. J. Chem., 1992, **45**, 451

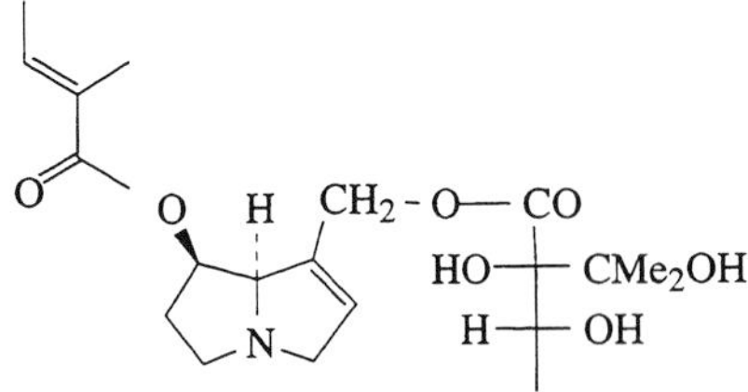

7-acetylhackelidine

Hackelia californica (Gray) Johnston

Y. Li, Huaxue Xuebao, 1990, **48**, 415 (Chem. Abstr., 1990, **113**, 168971)

hydroxymyoscorpine

Lithospermum erythrorhizon Siebold et Zuccharini

J.F. Resch, D.F. Rosberger and J. Meinwald, J. Nat. Prod., 1982, **45**, 360

Me₂COH

R=H lithosenine
R=Ac acetyllithosenine

Lithospermum officinale

E. Roeder, L. Krenn and H. Wiedenfeld, Phytochemistry, 1994, **37**, 275

longitubine
Hackelia longituba I.M. Johnston

J.N. Roitman, Aust. J. Chem., 1988, **41**, 1827

doriasenine
Senecio doria L.

E. Roeder, H. Wiedenfeld and A. Pfitzer, Phytochemistry, 1989, **27**, 4000

Senecio triangularis Hook.
J.N. Roitman, Aust. J. Chem., 1983, **36**, 1203

Alkanna tinctoria Tausch
E. Roeder, H. Wiedenfeld and R. Schraut, Phytochemistry, 1984, **23**, 2125

R=COC(CH₂OH)ᶻCHMe
triangularine

R=COC(CH₂OH)(OH)CH(OH)Me
dihydroxytriangularine

52

R=COC(CH$_2$OH)$\overset{E}{=}$CHMe

sencalenine

Senecio cacaliaster Lam

E. Roeder, H. Wiedenfeld and R. Britz-Kirstgen, Phytochemistry, 1984, **23**, 1761

neotriangularine

Senecio triangularis Hook.

J.N. Roitman, Aust. J. Chem., 1983, **36**, 1203

3',7-diacetyllycopsamine

Amsinckia menziesii (Lehm.) Nels. et Macbr.

J.N. Roitman, Aust. J. Chem., 1983, **36**, 769

R^1=Ac, R^2=O-tiglyl, R^3=H
7-acetylscorpioidine

R^1=tiglyl, R^2=H, R^3=OH
myoscorpine

Myosotis scorpioides L.

J.F. Resch, D.F. Rosberger, J. Meinwald and J.W. Appling, J. Nat. Prod., 1982, **45**, 358.

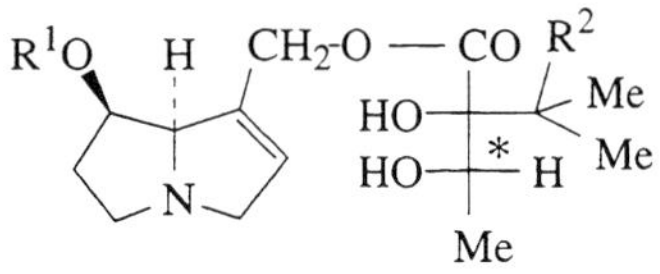

R^1=Ac, R^2=H 7-acetyllycopsamine

R^1=Ac, R^2=H 7-acetylintermedine

opposite stereochemistry at C*

R^1=angelyl, R^2=H   symlandine

R^1=Ac, R^2=OH   uplandicine

  stereochemistry of acid unknown

Symphytum x *uplandicum* Nyman

C.C.J. Culvenor, M. Clarke, J.A. Edgar, J.L. Frahn, M.V. Jago, J.E. Peterson and L.W. Smith, Experientia, 1980, **36**, 377; C.C.J. Culvenor, J.A. Edgar, J.L. Frahn and L.W. Smith, Aust. J. Chem., 1980, **33**, 1105

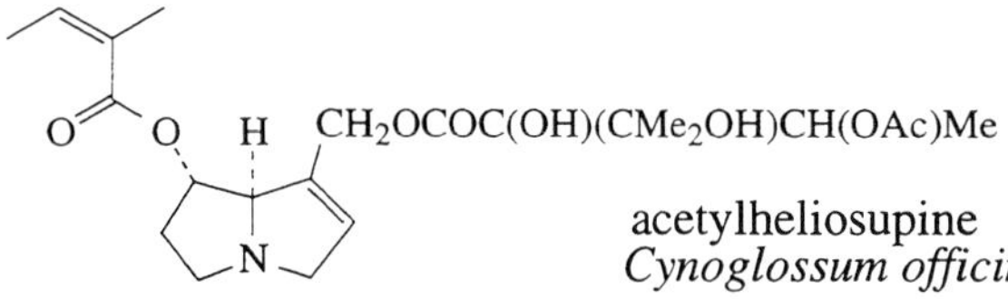

acetylheliosupine
Cynoglossum officinale L.
Myosotis sylvatica Hoffm.

J.F. Resch and J. Meinwald, Phytochemistry, 1982, **21**, 2430

7-acetyleuropine
Heliotropium bevei

M. Riena, A.H. Merioli, R. Cabrera and A. Gonzalez-Coloma, Phytochemistry, 1995, 38, 355

N-methyl-7,9-*O*,*O*-diangeloyl-1-hydroxyplatynecinium chloride
Senecio integrifolius var. *fauriri*

E. Roeder and K. Liu, Phytochemistry, 1991, **30**, 1734

Jacmaia incana (Sw.) B. Nord.

F. Bohlmann, R.K. Gupta and J. Jakupovic, Phytochemistry, 1991, **20**, 831

(d) Cyclic diesters

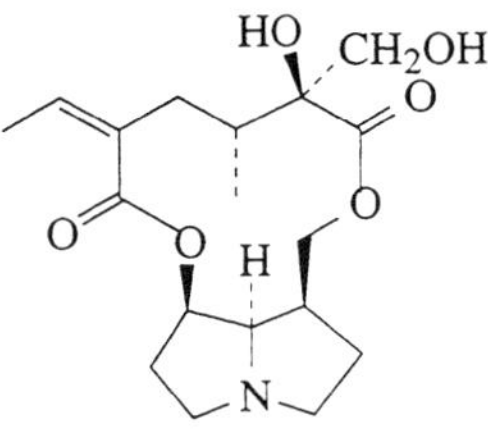

crotaleschenine
Crotalaria leschenaulti
L.W. Smith, J.A. Edgar, R.I. Willing, R.W. Gable, M.F. Mackay, O.P. Suri, C.K. Atal and C.C.J. Culvenor, Aust. J. Chem., 1988, **41**, 429

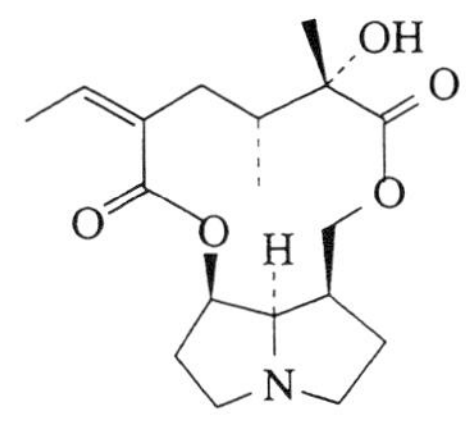

neoplatyphylline N-oxide

Cacalia hupehensis

Z. Miao, W.Wu and R. Feng, Bopuxue Zazhi, 1993, **10**, 55 (Chem Abstr., 1994, **120**, 164605)

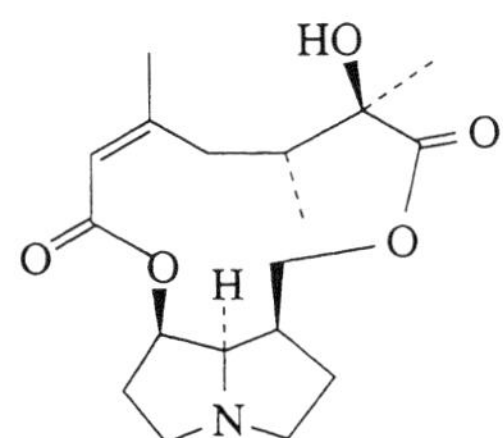

dihydroretrorsine

Senecio subulatus Don ex Hook. et Arn. var. *erectus*

M.J. Pestchanker, M.S. Ascheria and O.S. Giordano, Planta Med., 1985, 165

ligularinine
Ligularia dentata
Y. Asada and T. Furuya, Chem. Pharm. Bull., 1984, **32**, 475

bulgarsenine

H. Stoeckli-Evans, Acta Crystallogr., Sect. B, 1980, **36**, 3150

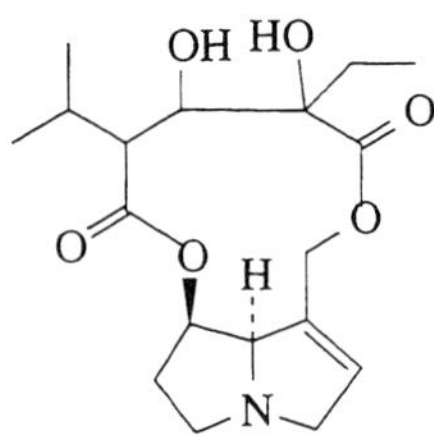

nemorensine

Senecio nemorensis

M.P. Dillon, N. Clee, F. Stappenbeck and J.D. White, J. Chem. Soc., Chem. Commun., 1995, 1645

aucherine

Senecio integrifolius (L.) Clairv. ssp. *aucheri* (DC.) Matthews

B. Sener, F. Ergun, S. Kusmenoglu and A.E. Karakaya, Gazi Univ. Eczacilik Fak. Derg., 1988, **5**, 157 (Chem. Abstr., 1989, **111**, 211907)

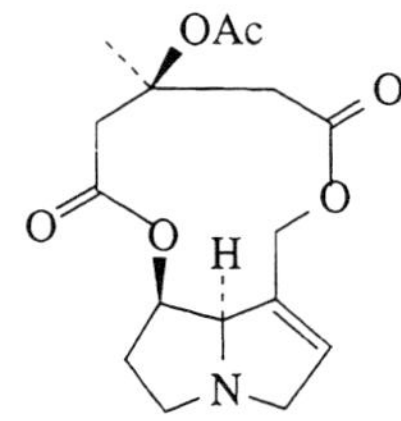

othonnine

Senecio othonnae Bieb.

B. Sener, F. Ergun, S. Kusmenoglu and A.E. Karakaya, Gazi Univ. Eczacilik Fak. Derg., 1988, **5**, 101 (Chem. Abstr., 1989, **110**, 132140)

acetyldicrotaline

Crotalaria lachnosema Stapf.

A.R. Mattocks and N. Nwude, Phytochemistry, 1988, **27**, 3289

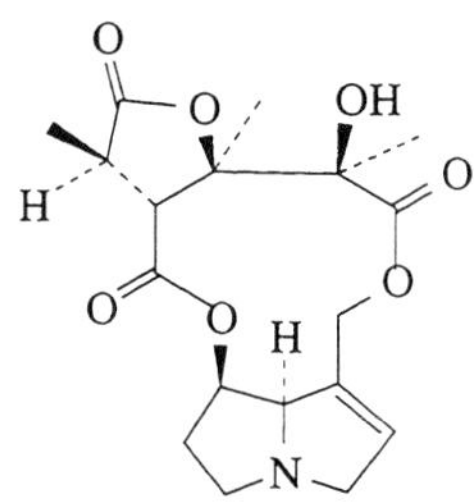

retusine

Crotalaria retusa

L.W. Smith, J.A. Edgar, R.I. Willing, R.W. Gable, M.F. Mackay, O.P. Suri, C.K. Atal and C.C.J. Culvenor, Aust. J. Chem., 1988, **41**, 429

desoxyaxillaridine

Crotalaria scassellatii Chiov.

H. Wiedenfeld, E. Roeder and E. Anders, Phytochemistry, 1985, **24**, 376

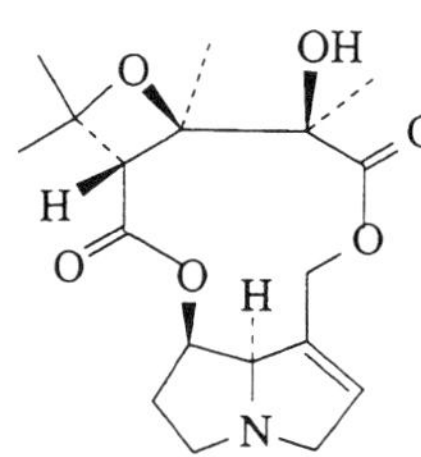

grantianine

Crotalaria globifera

K. Brown, J.A. Devlin and D.J. Robins, Phytochemistry, 1984, **23**, 457

H. Stoeckli-Evans and D.J. Robins, Acta Crystallogr., Sect. C, 1984, **40**, 1445

grantaline

Crotalaria virgulata subsp. *grantiana*

L.W. Smith and C.C.J. Culvenor, Phytochemistry, 1984, **23**, 473

M.F. Mackay and C.C.J. Culvenor, Acta Crystallogr., Sect. C, 1983, **39**, 1227

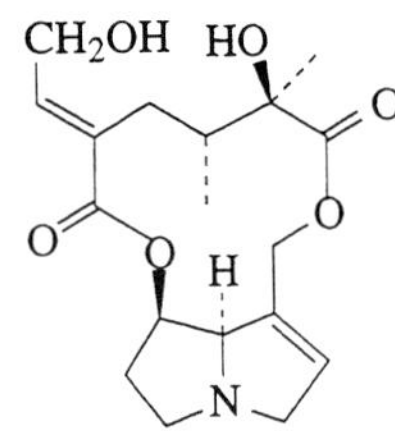

adonifoline

Senecio adonidifolius

L. Witte, L. Ernst, V. Wray
and T. Hartmann,
Phytochemistry, 1992, **31**,
1027

seneciphyllinine

Gynura segetum (Lour.) Merr.

S.Q. Yuan, G.M. Gu and
T.T. Wei, Yaoxue Xuebao,
1990, **25**, 191 (Chem.
Abstr., 1990, **113**, 112479)

21-hydroxyintegerrimine

Senecio argunensis

K. Liu and E. Roeder,
Phytochemistry, 1991,
30, 1303

eruciflorine

Senecio erucifolius

L. Witte, L. Ernst, H. Adam and
T. Hartmann, Phytochemistry,
1992, **31**, 559

O-acetyljacoline
Cirsium wallichii DC.
R.K.S. Negi, T.M. Fakhir and
T.R. Rajagopalan, Indian J.
Chem., Sect. B, 1989, **28**, 524

acetylintegerrimine
Crotalaria naragutensis Hutch.
A.R. Mattocks and N. Nwude,
Phytochemistry, 1988, **27**, 3289

Senecio adonidifolius
J.G. Urones, P.B. Barcala, I.S.
Marcos, R.R. Moro, M.L. Esteban
and A.F. Rodriguez,
Phytochemistry, 1988, **27**, 1507

sceleratine
Senecio latifolius DC.
M.W. Bredenkamp, A. Wiechers
and P.H. van Rooyen,
Tetrahedron Lett., 1985, **26**
5721

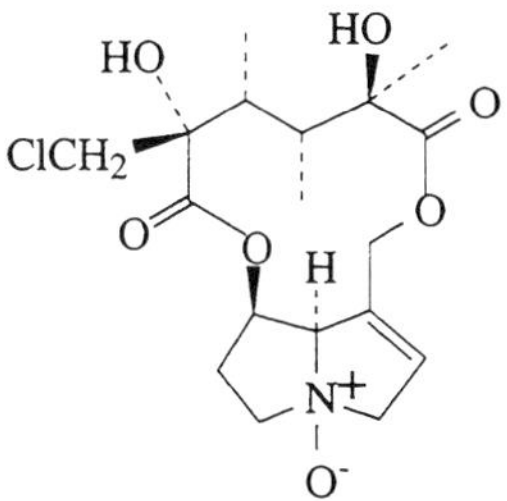

merenskine N-oxide
Senecio latifolius DC.
M.W. Bredenkamp, A.
Wiechers and P.H. van
Rooyen, Tetrahedron Lett.,
1985, **26** 929

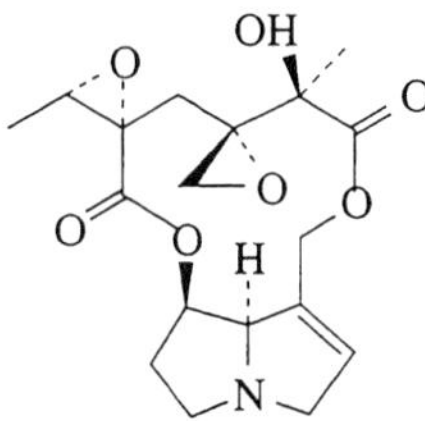

R=H gynuramine
R=Ac acetylgynuramine

Gynura scandens
O. Hoffm.

H. Weidenfeld,
Phytochemistry, 1982,
21, 2767

senecivernine
Senecio vernalis Wald. et Kit.

E. Roeder, H. Wiedenfeld and
U. Pastewka, Planta Med.,
1979, **37**, 131

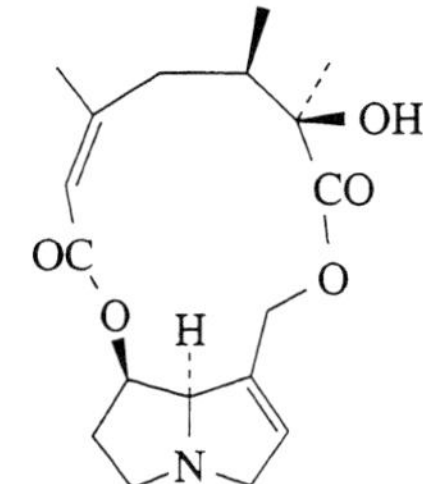

senecicannabine
Senecio cannabifolius

Y. Asada, T. Furuya, M.
Shiro and H. Nakai,
Tetrahedron Lett., 1982, **23**,
189

doronenine
Senecio doronicum L.

E. Roeder, H. Wiedenfeld
and M. Frisse,
Phytochemistry, 1980, **19**,
1275.

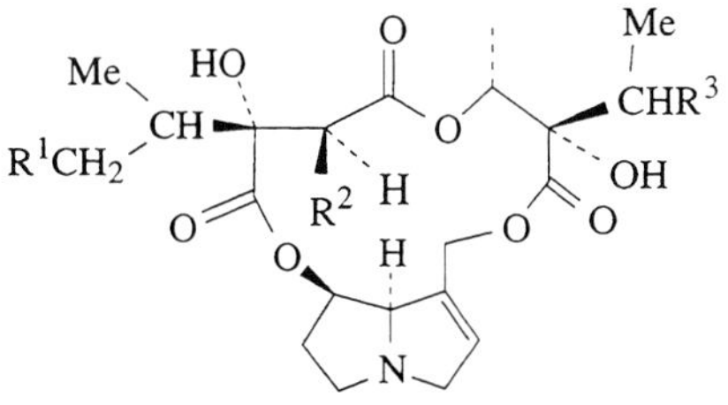

Parsonsia laevigata Alston

$R^1=R^3=H, R^2=OH$
parsonsianine

$R^1=Me, R^2=OH, R^3=H$
parsonsianidine
$R^1=R^3=Me, R^2=OH$
17-methylparsonsianidine

$R^1=R^2=R^3=H$
14-deoxyparsonsianine
$R^1=Me, R^2=R^3=H$
14-deoxyparsonsianidine

F. Abe, T. Nagao, H. Okabe, T. Yamauchi, M. Marubayashi and I. Ueda, Chem. Pharm. Bull., 1990, **38**, 2127

F. Abe, T. Nagao, H. Okabe and T. Yamauchi, Phytochemistry, 1991, **30**, 1737

F. Abe, T. Nagao, S. Yaga, and K. Minato, Chem. Pharm. Bull., 1991, **39**, 1576

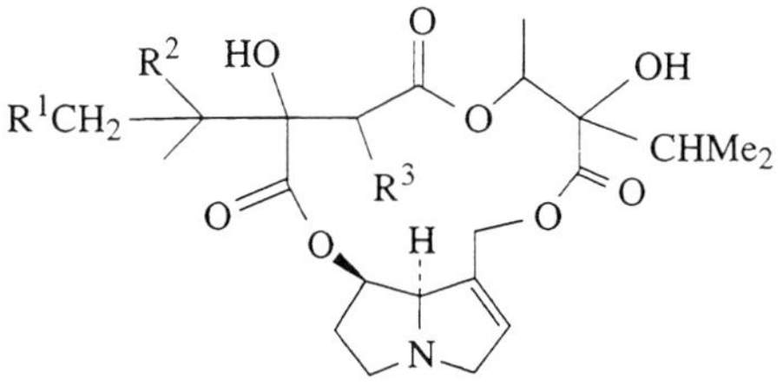

$R=R^2=R^3=H$
parsonsine
$R^1=Me, R^2=R^3=H$
heterophylline

$R^1=R^2=H, R^3=OH$
spiraline
$R^1=Me, R^2=H, R^3=OH$
spiranine
$R^1=Me, R^2=R^3=OH$
spiracine

Parsonsia heterophylla A. Cunn.
Parsonsia spiralis Wall.

Parsonsia spiralis Wall.

J.A. Edgar, N.J. Eggers, A.J. Jones and G.B. Russell, Tetrahedron Lett., 1980, **21**, 2657

62

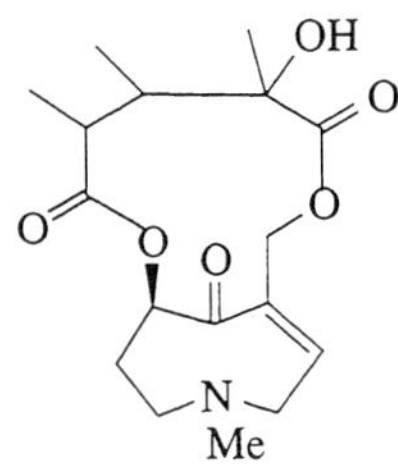

cropodine
Crotalaria candicans W. & A.
C.N. Haskar, O.P. Suri, R. S. Jamwal and C.K. Atal, Indian J. Chem., Sect. B, 1982, **21**, 492

crocandine and isocrocandine
Crotalaria candicans W. & A.
M.A. Siddiqi, K.A. Suri, O.P. Suri, and C.K. Atal, Phytochemistry, 1979, **18**, 1413

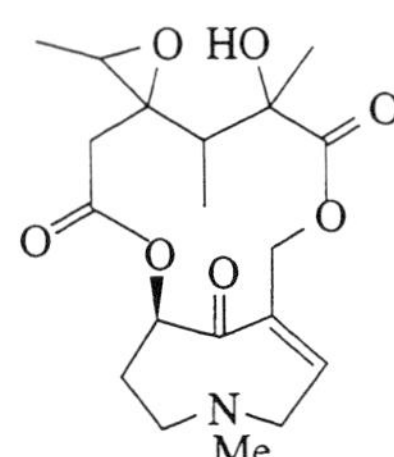

croaegyptine

Crotalaria aegyptiaca Benth.
E. Roeder, T. Sarg, S. El-Dahmy and A.A. Ghani, Phytochemistry, 1993, **34**, 1421

senecioracenine

Senecio racemosus DC.
W. Ahmad, A.Q. Khan, A. Malik, F. Ergun and B. Sener, J. Nat. Prod., 1992, **55**, 1764

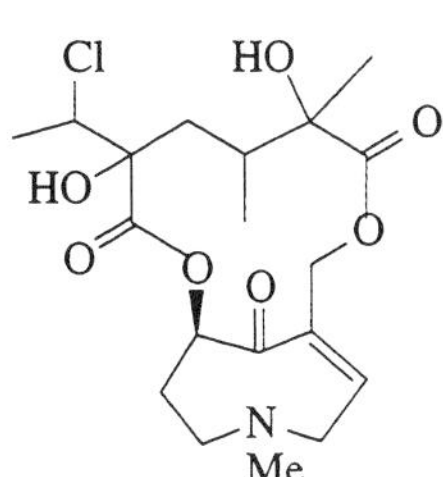

acetylanonamine
Senecio anonymus Wood
C.F. Asibal, L.H. Zalkow and
L.T. Gelbaum, J. Nat. Prod.,
1991, **54**, 1425

anonamine
Senecio anonymus Wood
L.H. Zalkow, C.F. Asibal,
J.A. Glinski, S.J Bonetti, L.T.
Gelbaum, D. VanDerveer and
G. Powis, J. Nat. Prod.,
1988, **51**, 690

hydroxyneosenkirkine
Senecio anonymus Wood
L.H. Zalkow, C.F. Asibal, J.A.
Glinski, S.J Bonetti, L.T.
Gelbaum, D. VanDerveer and G.
Powis, J. Nat. Prod., 1988, **51**,
690

desacetyldoronine
Senecio inaequidens DC.
C. Bicchi, R. Caniato, R. Tabacchi
and G. Tsoupras, J. Nat. Prod.,
1989, **52**, 32

64

emiline

Emilia flammea Cass.

R.H. Barbour and D.J. Robins,
Phytochemistry, 1987, **26**, 2430

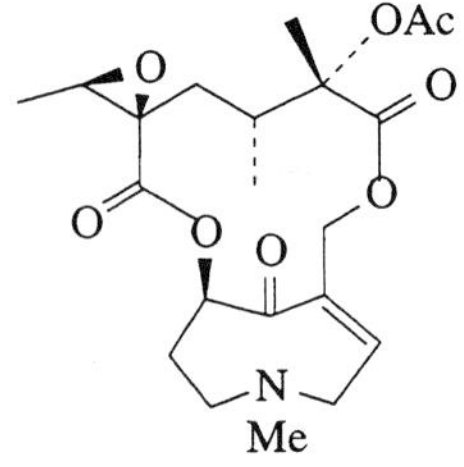

neoligularidine

Ligularia dentata

Y. Asada and T. Furuya, Chem.
Pharm. Bull., 1984, **32**, 475

ligularizine

Ligularia dentata

Y. Asada and T. Furuya,
Chem. Pharm. Bull., 1984, **32**,
475

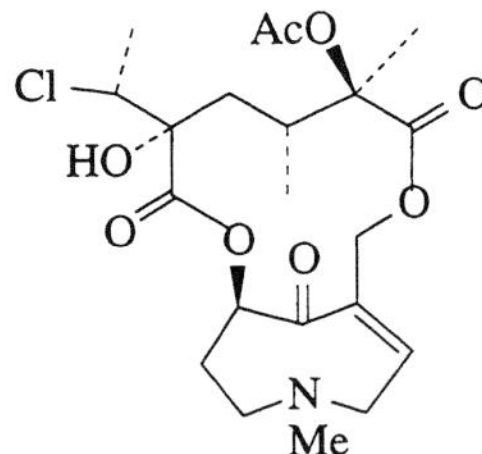

doronine

Senecio clevelandii E.L. Greene

R.Y. Wong and J.N. Roitman,
Acta Crystallogr., Sect. C, 1984,
40, 163

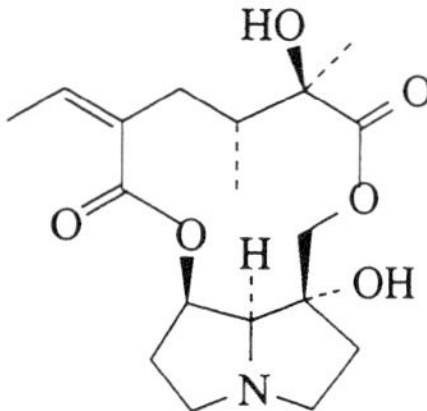

R=H
7-epidesacetylsenaetnine

R=Ac
7-episenaetnine

Senecio portalesianus

J. Jakupovic, M. Grenz, F. Bohlmann and H.M. Niemeyer, Phytochemistry, 1991, **30**, 2691

desacetylsenaetnine

Senecio magnificus F. Muell.

C. Zdero, F. Bohlmann, R.M. King and L. Haegi, Phytochemistry, 1990, **29**, 509

hadiensine

Senecio hadiensis

O. Were, M. Benn and R.M. Munavu, J. Nat. Prod., 1991, **54**, 491

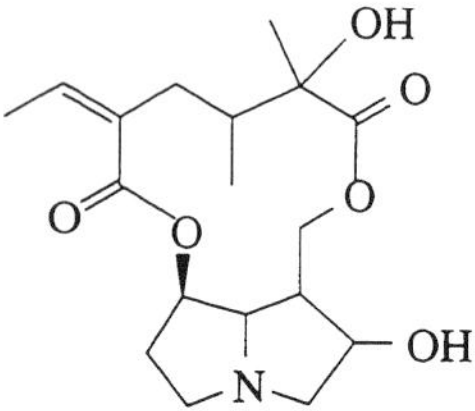

$R^1=R^2=H$
rosmarinine
$R^1=H, R^2=Ac$
12-*O*-acetylrosmarinine
$R^1=OH, R^2=H$
petitianine

Senecio hadiensis

O. Were, M. Benn and R.M. Munavu, J. Nat. Prod., 1991, **54**, 491

neorosmarinine

Senecio hadiensis

O. Were, M. Benn and R.M. Munavu, J. Nat. Prod., 1991, **54**, 491

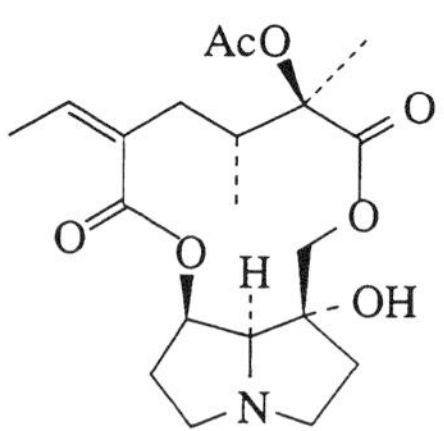

isorosmarinine

Senecio pterophorus

J.R. Liddell and C.G. Logie, Phytochemistry, 1993, **34**, 1629

12-*O*-acetylhadiensine

Senecio hadiensis

O. Were, M. Benn and R.M. Munavu, J. Nat. Prod., 1991, **54**, 491

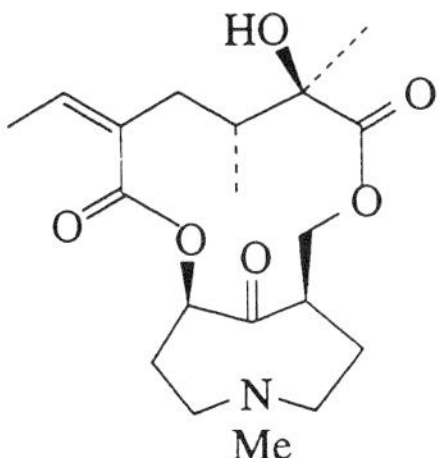

12-*O*-acetylneohadiensine

Senecio hadiensis
O. Were, M. Benn and R.M. Munavu, J. Nat. Prod., 1991, **54**, 491

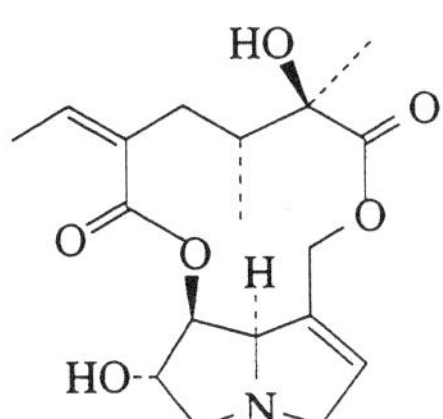

trans-anacrotine

Crotalaria capensis

G.H. Verdoorn and B.-E. van Wyk, Phytochemistry, 1992, **31**, 369

1,2-dihydrosenkirkine

Senecio integrifolius var. *fauriri*

E. Roeder and K. Liu, Phytochemistry, 1991, **30**, 1734

uspallatine

Senecio uspallatensis

M.J. Pestchanker, M.S. Ascheri and O.S. Giordano, Phytochemistry, 1985, **24**, 1622

*Second Supplements to the 2nd Edition of Rodd's Chemistry
of Carbon Compounds, Vol.IV B*, edited by M. Sainsbury
© 1997 Elsevier Science B.V. All rights reserved.

Chapter 9

THE INDOLE ALKALOIDS[‡]

G. W. GRIBBLE

Introduction

Since the publication in 1985 of Vol. IVB, 1st Suppl., there has been an explosion of activity in all areas of natural products discovery, isolation, and characterization, including the enormous field of indole alkaloids. This has been due largely to improved identification techniques (e.g., 2D NMR), powerful bioassay methods, and a revitalization in the interest in natural products as a new source of pharmaceuticals. The present chapter covers the literature from 1980 to early 1996. Moreover, a fascination with marine natural products has led to the discovery of numerous indole-containing marine natural products without counterparts in the terrestrial kingdom.

Despite the fact that several novel structural types have been characterized since the last volume, the author has retained the previous format, with the addition of only a few new subcategories. However, the huge quantity of material has necessitated the omission of synthesis, biosynthesis, and structure determination. Rather, the focus is on new indole-containing natural products and interesting biological activity. This tremendous increase in the number of indole natural products has occurred in the areas of plant alkaloids, fungal metabolites, and marine natural products.

The reader is advised to consult several excellent sources for further information on indole alkaloids and related natural products. These include "Monoterpenoid Indole Alkaloids," Supp. to Vol. 25 Part IV, ed. J.E. Saxton, John Wiley, New York, 1994; the "New Natural Products" section in *Heterocycles*; *Natural Products Updates* and *Natural Product Reports*, both published by the Royal Society of Chemistry; "Dictionary of Alkaloids," I.W. Southon and J. Buckingham, Chapman and Hall, London, 1989; "The Alkaloids," Academic Press, San Diego, now edited by G.A. Cordell; "Progress in the Chemistry of Organic Natural Products," Springer-Verlag, New York; "Studies in Natural Products Chemistry,"

[‡] This chapter is dedicated to Dr. Shin-ichiro Sakai for his 40 years of pioneering and brilliant indole alkaloid chemistry.

Elsevier, Amsterdam, edited by A. Rahman; "Alkaloids: Chemical and Biological Perspectives," Wiley-Interscience, New York, edited by S.W. Pelletier. These latter four series of monographs afford excellent, current reviews by experts in the field. Of relevance to the present review are discussions in the series "The Alkaloids" on Marine Alkaloids (C. Christophersen, Vol. 24, p. 25, 1985; J. Kobayashi and M. Ishibashi, Vol. 41, p. 41, 1992), Sulfur-Containing Alkaloids (J.T. Wróbel, Vol. 26, p. 53, 1985; J.T. Wróbel and K. Wojtasiewicz, Vol. 42, p. 249, 1992), Simple Indole, β-Carboline, and Carbazole Alkaloids (H.-P. Husson, Vol. 26, p. 1, 1985), Cinchona Alkaloids (R. Verpoorte, *et al.*, Vol. 34, p. 331, 1988), *Tabernaemontana* Alkaloids (B. Danieli and G. Palmisano, Vol. 27, p. 1, 1986), African *Strychnos* Alkaloids (G. Massiot and C. Delaude, Vol. 34, p. 211, 1988), *Strychnos* and *Gardneria* Alkaloids (N. Aimi, *et al.*, Vol. 36, p. 1, 1989), Ellipticine Alkaloids (G.W. Gribble, Vol. 39, p. 239, 1990), Alkaloids from Mushrooms (R. Antkowiak and W.Z. Antkowiak, Vol. 40, p. 189, 1991), Microbial and *In Vitro* Enzymic Transformations of Alkaloids (H.L. Holland, Vol. 18, p. 323, 1981), Metabolic Transformations of Alkaloids (J.P.N. Rosazza and M.W. Duffel, Vol. 27, p. 323, 1986), Plant Biotechnology for the Production of Alkaloids (R. Verpoorte, *et al.*, Vol. 40, p. 1, 1991), Biosynthesis in *Rauwolfia serpentina* (J. Stöckigt, Vol. 47, p. 115, 1995), and Allelochemical Properties of Alkaloids (M. Wink, Vol. 43, p. 1, 1993). This last review is an excellent discourse on why there are the more than 10,000 known alkaloids, and the role they play as defensive compounds. The series, "Alkaloids: Chemical and Biological Perspectives," has recent coverage of *Tabernaemontana* Alkaloids (T.A. van Beek and M.A.J.T. van Gessel, Vol. 6, Chapter 2, 1988), Alkaloids from Cell Cultures of *Aspidosperma Quebracho-Blanco* (P. Obitz, *et al.*, Vol. 9, Chapter 5, 1995), and Indole Alkaloids Chemotaxonomy of Apocynaceae, Loganiaceae, and Rubiaceae Families (M.V. Kisakürek, *et al.*, Vol. 1, Chapter 5, 1983). Recent volumes of "Studies in Natural Products Chemistry" include chapters on *Tabernaemontana* Alkaloid Identification by Mass Spectrometry (R. van der Heijden and R. Verpoorte, Vol. 5, p. 69, 1989), *Strychnos dinklagei* Alkaloids (S. Michel, *et al.*, Vol. 6, p. 503, 1990), and NMR in Trace Indole Alkaloid Identification (J. Schripsema and R. Verpoorte, Vol. 9, p. 163, 1991). The venerable series, "Progress in the Chemistry of Organic Natural Products," covers the Production of Indole Alkaloids in *Catharanthus roseus* Cell Suspension Cultures (M. Lounasmaa and J. Galambos, Vol. 55, p. 89, 1989), Halogenated Indoles, Carbazoles, Carbolines, and Indolo[2,3-*a*]carbazoles (G.W. Gribble, Vol. 68, p. 1, 1996). Other recent notable reviews of indole natural products are 2-Acylindole Alkaloids (D.G.I. Kingston and O. Ekundayo, *J. Nat. Prod.*, 1981, **44**, 509), Marine Indoles (C. Christophersen, in "Marine Natural Products," Vol. V, ed. P.J. Scheuer, Chapter 5, Academic Press, New York, 1983), Amino Acid Metabolites of Ascidians (B.S. Davidson, *Chem. Rev.*, 1993, **93**, 1771), Marine Pyridoacridine

Alkaloids (T.F. Molinski, *Chem. Rev.*, 1993, **93**, 1825), Isocyanide and Cyanide Natural Products (P.J. Scheuer, *Acct. Chem. Res.*, 1992, **25**, 433). Other reviews of a more specific nature will be inserted in the text as appropriate. Where exhaustive reviews of a particular subject are available, coverage herein will be limited to the period of time not covered by these other reviews.

1. Alkaloids Lacking a Tryptamine Unit

(a) Simple indoles

A very large number of new simple indoles have been discovered since 1980, especially from fungal and marine sources. The simple indoles, 2,4-dimethylindole, 4-hydroxymethyl-2-methylindole, and 4-methoxymethyl-2-methylindole, were isolated from the fruiting bodies of the mushrooms *Tricholoma sciodes* and *T. virgatum* (L. Garlaschelli, *et al.*, *Tetrahedron*, 1994, **50**, 3571). The peptide lascivol may be a precursor to these novel indole natural products. The blue-green alga *Lyngbya majuscula*, which is a rich source of novel metabolites, produces 1,7-dimethylindole-3-carbalde-hyde (J.S. Todd and W.H. Gerwick, *J. Nat. Prod.*, 1995, **58**, 586). Salvadoricine (2-acetyl-3-methylindole) is found in *Salvadora persica* (S. Malik, *et al.*, *Tetrahedron Lett.*, 1987, **28**, 163), and the fungus *Lactarius deliciosus* (W.A. Ayer and L.S. Trifonov, *J. Nat. Prod.*, 1994, **57**, 839), the sponge *Tedania ignis* (R.L. Dillman and J.H. Cardellina II, *J. Nat. Prod.*, 1991, **54**, 1056), and the red alga *Prionitis lanceolata* (M. Bernart and W.H. Gerwick, *Phytochemistry*, 1990, **29**, 3697) have yielded the plant growth regulator (1). The 5-hydroxyl derivative hyrtiosin A (2) has been isolated from the marine sponge *Hyrtios erecta* (J. Kobayashi, *et al.*, *Tetrahedron*, 1990, **46**, 7699). A careful study of cruciferous vegetables such as broccoli, kale, collards, and brussels sprouts has determined that 3-hydroxymethylindole (3) is present in only trace amounts. The other indoles (4) and (5) are the major indoles present (M.E. Wall, *et al.*, *J. Nat. Prod.*, 1988, **51**, 129). In addition to the well-known plant growth inhibitor indole-3-acetic acid, the almond tree pathogen *Pseudomonas amygdali* produces methyl indole-3-acetate (6), the first report of its occurrence as a microbial metabolite (A. Evidente, *et al.*, *Experientia*, 1993, **49**, 182). The sponge *Tedania ignis* has also yielded the indoles (7)-(9), the first two of which may be acetone aldol-condensation artefacts (R.L. Dillman and J.H. Cardellina, II, *J. Nat. Prod.*, 1991, **54**, 1056). The interesting glucoside (10) has been found in red wine (V.A. Marinos, *et al.*, *Phytochemistry*, 1992, **31**, 2755).

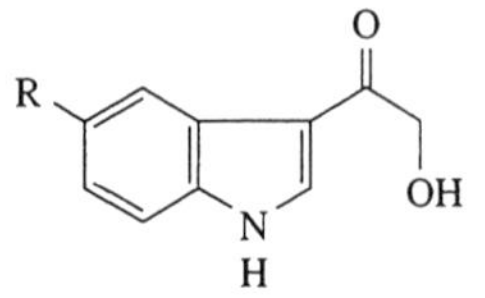

(1) R = H
(2) R = OH

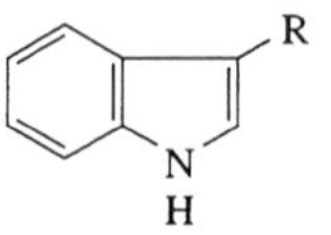

(3) R = CH$_2$OH
(4) R = CH$_2$CN
(5) R = CHO
(6) R = CH$_2$CO$_2$Me
(7) R = (*E*)CH=CHAc
(8) R = (*E*)CH$_2$CH = CHAc

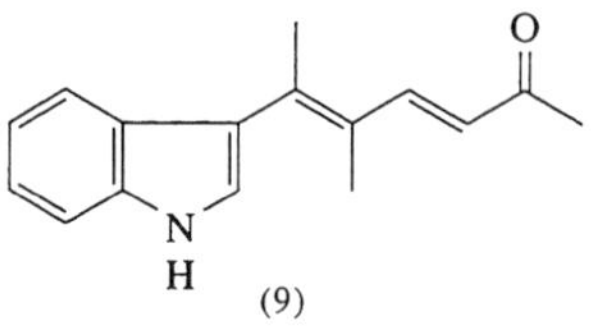

(9)

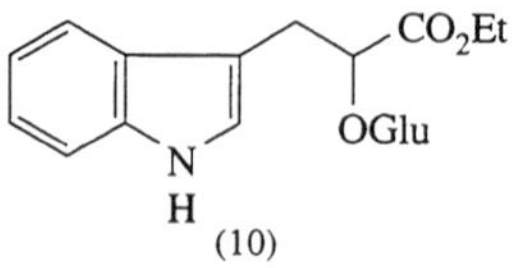

(10)

The plant *Dithyrea wislizenii* produces the insect antifeedant dithyreanitrile (11) (R.G. Powell, *et al.*, *Experientia*, 1991, **47**, 304), and the most unusual sulfur-containing indoles (12) and (13) have been identified in the hyperthermophilic archaea *Thermococcus tadjuricus* and *T. acidamino-vorans* (M. Ritzau, *et al.*, *Ann.*, 1993, 871). Epoxyindole (14) as well as the novel oxime sulfates (15) and (16) are found in *Moricandia arvensis* (A. Belkhiri and G.B. Lockwood, *Phytochemistry*, 1990, **29**, 1315).

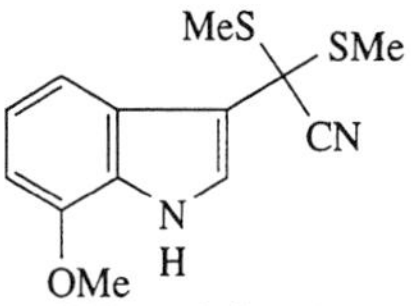

Dithyreanitrile (11)

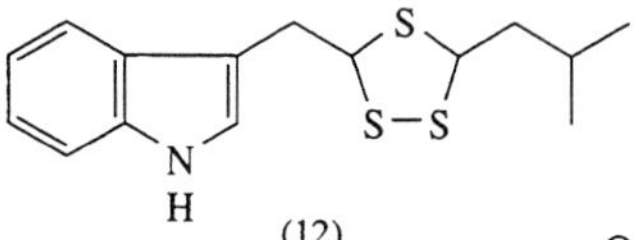

(12)

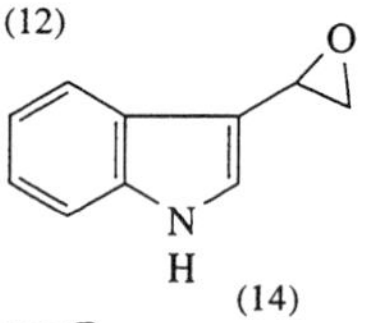

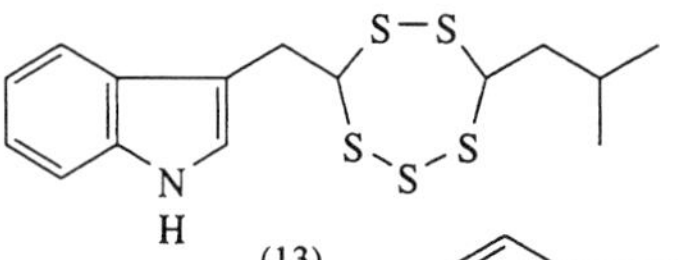

(13)

(14)

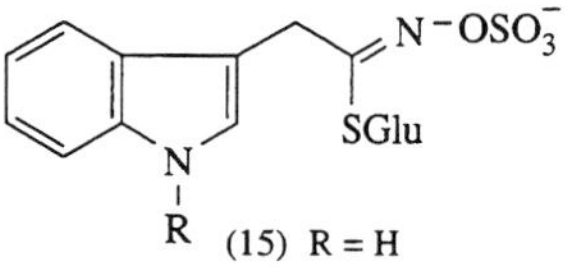

(15) R = H
(16) R = OMe

A *Bacillus* sp. has afforded sattazolin (17) and the corresponding methyl ether (G. Lampis, *et al.*, *J. Antibiot.*, 1995, **48**, 967). This metabolite exhibits selective inhibition of protein synthesis in Herpes virus infected cells. The related bruceolline F (18) is produced by *Brucea mollis* var. *tonkinensis* (Y. Ouyang, *et al.*, *Phytochemistry*, 1994, **37**, 575). Diol (18) as the free indole had been previously isolated as tanakine (19) along with tanakamine (20) from the small tree *Limonia acidissima* (M.H.A. Zarga, *J. Nat. Prod.*, 1986, **49**, 901). The related paniculidines A (21), B (22), and

C (23) are produced by the plant *Murraya paniculata* (T. Kinoshita, *et al.*, *Phytochemistry*, 1989, **28**, 147; C. Ito, *et al.*, *J. Chem. Soc., Perkin Trans. 1*, 1990, 2047). Several potent antibacterials (24)-(27) related to sattazolin have been isolated from *Xenorhabdus bovienii* (J. Li, *et al.*, *J. Nat. Prod.*, 1995, **58**, 1081).

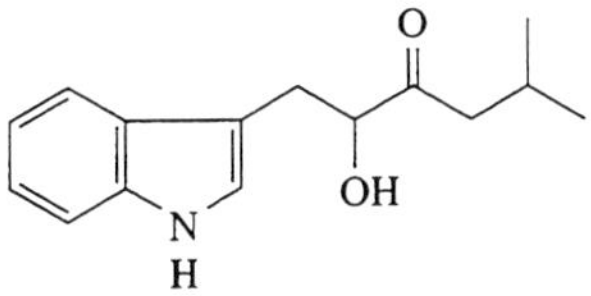

Sattazolin (17)

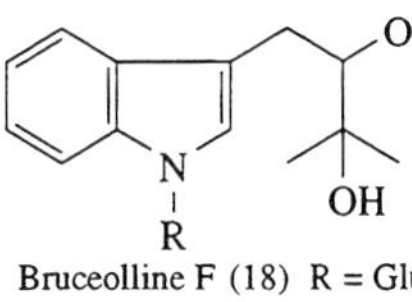

Bruceolline F (18) R = Glu
Tanakine   (19) R = H

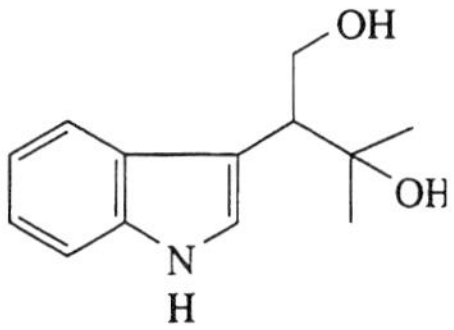

Tanakamine (20)

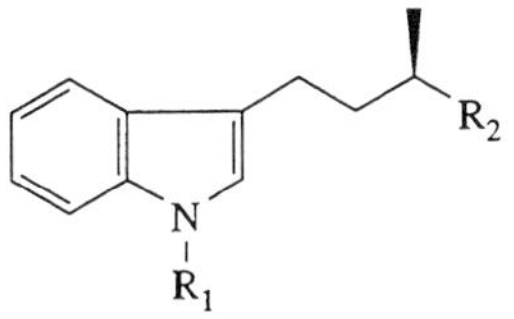

Paniculidine A (21) R_1 = H, R_2 = CO$_2$Me
Paniculidine B (22) R_1 = OMe, R_2 = CH$_2$OH
Paniculidine C (23) R_1 = H, R_2 = CH$_2$OH

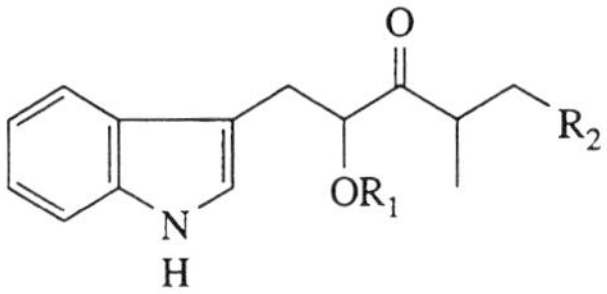

(24) R_1 = R_2 = H
(25) R_1 = Ac, R_2 = H
(26) R_1 = H, R_2 = Me
(27) R_1 = Ac, R_2 = Me

A *Streptomyces* sp. has afforded diolmycins A_1 (28) and the epimeric A_2, both of which possess anticoccidial activity (N. Tabata, *et al.*, *J. Antibiot.*, 1993, **46**, 762). Several novel cyclopent[*g*]indoles have been found in marine organisms. The Western Australian sponge *Axinella* sp. has afforded the three herbindoles A-C (29)-(31), which are cytotoxic and fish antifeedants (R. Herb, *et al.*, *Tetrahedron*, 1990, **46**, 3089), and the sponge *Trikentrion flabelliforme* produces a series of five related trikentrins, such as (32) (R.J Capon, *et al.*, *Tetrahedron*, 1986, **42**, 6545). Nostodione A (33) is a mitotic arrester found in a blue-green alga (A. Kobayashi, *et al.*,

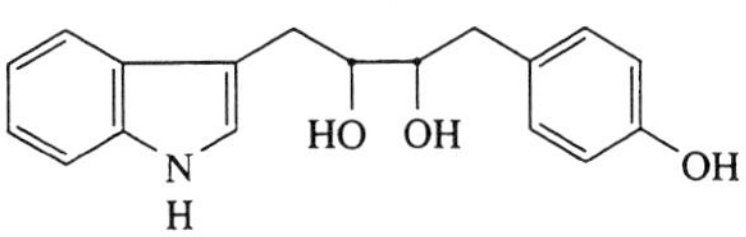

Diolmycin A_1 (28)

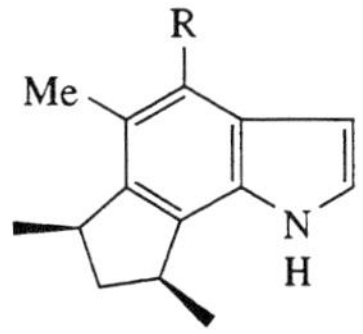

Herbindole A (29) R = Me
Herbindole B (30) R = Et
Herbindole C (31) R = (*E*) CH=CHEt

74

trans-Trikentrin B (32)   Nostodione A (33)   (34) R = H₂
(35) R = O

International Research Congress of Natural Products, Nova Scotia, July 31-
August 4, 1994, Abstract P:183), and bruceollines D (34) and E (35) are
produced by *Brucea mollis* var. *tonkinensis* (Y. Ouyang, *et al.*, *Phytochemistry*, 1994, **37**, 575).

Although most halogenated indoles are found in the oceans, the plant
growth hormone 4-chloroindole-3-acetic acid (36) is ubiquitous in *Vicieae*.
This compound, which is 10-100 times better as an auxin than indole-3-
acetic acid, and the methyl ester (37) are found in the seeds of peas, vetch,
lentil, and other plants (A. Ernstsen and G. Sandberg, *Physiol. Plant*,
1986, **68**, 511; M. Katayama, *et al.*, *Plant Cell Physiol.*, 1987, **28**, 383).
Indeed, the concentration of (36) and (37) in *Vicia amurensis* is higher than
that of the corresponding nonchlorinated compounds. Interestingly, the
fava bean (*Vicia faba*) is a source of 4-chloro-6-methoxyindole, which
seems to be a precursor to a potent gastric carcinogen (G. Büchi, *et al.*, *J.
Am. Chem. Soc.*, 1986, **108**, 4115; N.K. Brown, *et al.*, *Chem. Res.
Toxicol.*, 1992, **5**, 797). Marine sources have yielded close to 100 indole
natural products, most of which contain bromine. New examples reported
since 1980 include 3,6-dibromoindole and 6-methoxy-3,5,7-tribromoindole
from the acorn worm *Ptychodera flava* and 3,4,6-tribromoindole from the
acorn worm *Balanoglossus carnosus* (T. Higa, *et al.*, *Comp. Biochem.
Physiol.* , 1980, **65B**, 525). The related *Glossobalanus* sp. has also
yielded 4,6-dibromoindole and 4,6-dibromo-2-methylindole (T. Higa, *et
al.*, *Experientia*, 1985, **41**, 1487).

The sponge *Oceanapia bartschi* has furnished 6-bromo-5-hydroxy-3-
indolecarbaldehyde (F. Cafieri, *et al.*, *Z. Naturforsch.*, 1993, **48B**, 1408),
and the deep water sponge *Pseudosuberites hyalinus* produces the new 6-
bromoindoles (38)-(41) (T. Rasmussen, *et al.*, *J. Nat. Prod.*, 1993, **56**,
1553). Similarly, the Coral Sea sponge *Pleroma menoui* contains (42) and
(43) (G. Guella, *et al.*, *Z. Naturforsch.*, 1989, **44C**, 914). Indole acrylate
(44) is found in the sponge *Iotrochota* sp. (G. Dellar, *et al.*, *J. Chem. Soc.,
Perkin Trans. 1*, 1981, 1679), and the corresponding acid penaresin (45),
which is a potent sarcoplasmic reticulum calcium inducer, is found in the
Okinawan sponge *Penares* sp. (J. Kobayashi, *et al.*, *Heterocycles*, 1990,

31, 2205). The tribromoindole (46) from the bryozoan *Zoobotryon verticillatum* delays metamorphosis in fertilized sea urchin eggs (M.J. Ortega, *et al.*, *J. Nat. Prod.*, 1993, **56**, 633). This organism was previously found to contain gramine derivatives (47) and (48). The latter appears to be the first alkaloid *N*-oxide to be isolated from marine sources (A. Sato and W. Fenical, *Tetrahedron Lett.*, 1983, **24**, 481). The red alga *Laurencia brongniartii* is a rich source of polybrominated indoles such as the sulfur-containing itomanindoles A (49) and B (50), in addition to the new 2,4,6-tribromoindole and 2,3,4,6-tetrabromoindole (J. Tanaka, *et al.*, *Tetrahedron*, 1989, **45**, 7301). The bryozoan *Amathia convoluta* contains convolutamydines A-D (51)-(54) (H. Zhang, *et al.*, *Tetrahedron*, 1995, **51**, 5523), and the tunicate *Polycitorella mariae* from Fiji produces citorellamine (55) (D.M. Roll and C.M. Ireland, *Tetrahedron Lett.*, 1985, **26**, 4303), which structure was subsequently revised (R.M. Moriarty, *et al.*, *Tetrahedron Lett.*, 1987, **28**, 749).

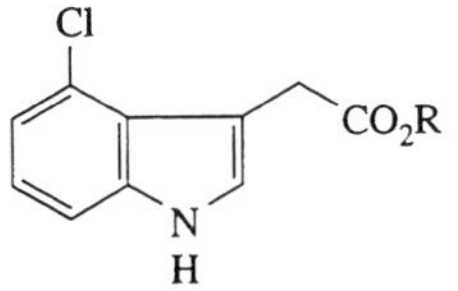

(36) R = H
(37) R = Me

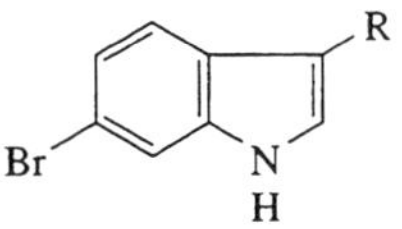

(38) R = CH_2CN
(39) R = CH_2CONH_2
(40) R = CH_2CO_2Me
(41) R = CO_2H
(42) R = CO_2Et
(43) R = $COCH_2OH$
(44) R = (*E*)CH=CHCO$_2$Me
(45) R = (*E*)CH=CHCO$_2$H

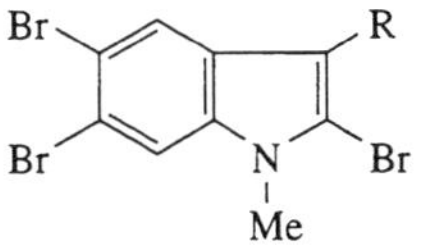

(46) R = CHO
(47) R = CH_2NMe_2
(48) R = $CH_2N(O)Me_2$

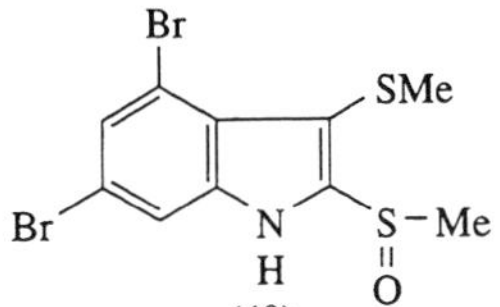

(49)

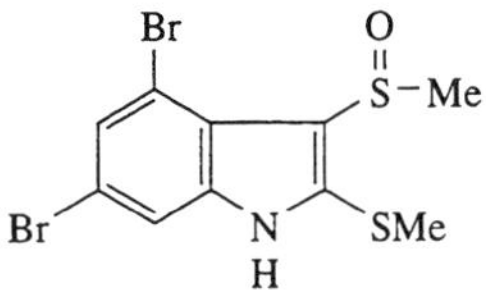

(50)

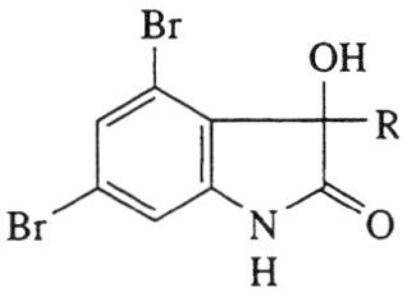

(51) R = CH_2Ac
(52) R = CH_2CH_2Cl
(53) R = CH_3
(54) R = CH=CH$_2$

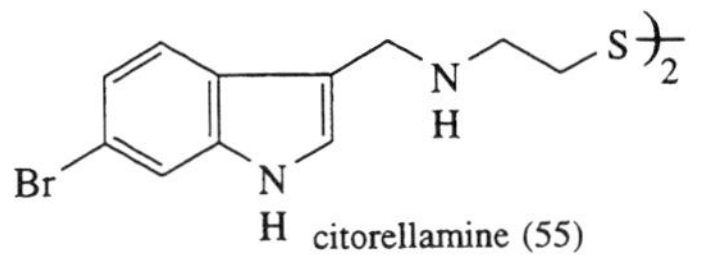

citorellamine (55)

In recent years, several new indoloterpenes, both simple and complex, have been discovered from natural sources, in addition to the few that were mentioned earlier.

The plant *Pamburus missionis* contains the simple ketone (56) (V. Kumar, *et al.*, *Phytochemistry*, 1994, **36**, 879) and the isomeric leiocarpone (57) is found in *Esenbeckia leiocarpa* along with a pair of diastereomeric leiocarpatriols A and B (58) (F.D. Monache, *et al.*, *Gazz. Chim. Ital.*, 1990, **120**, 387). The simple 3-γ,γ-dimethylallylindole is found in *Monodora tenuifolia* (A.O. Adeoye, *et al.*, *J. Nat. Prod.*, 1986, **49**, 534), and cultures of *Penicillium daleae* have afforded *N*-isoprenylgramine (59) (Y.K.T. Lam, *et al.*, *J. Antibiot.*, 1994, **47**, 724). The organism *Uvaria pandensis* contains 3-farnesylindole (M.H.H. Nkunya, *et al.*, *Phytochemistry*, 1987, **26**, 2402), and two derived diols (M.H.H. Nkunya and H. Weenan, *Phytochemistry*, 1989, **28**, 2217). A plethora of diisoprenylated indoles and related compounds (diols, fatty acids, esters) are produced by *Hexalobus crispiflorus*, *H. monopetalus*, and *Isolona maitlandii* (H. Achenbach, *et al.*, *Heterocycles*, 1984, **22**, 2501; *Ann.*, 1995, 1327; *Phytochemistry*, 1995, **40**, 967). An example is 3,6-hexalobine B (60). The red alga *Chondria* sp. has been found to contain the simple imide (61) as well as indole-3-acrylic acid and the 7-hydroxy derivative (J.A. Palermo, *et al.*, *Tetrahedron Lett.*, 1992, **33**, 3097). Murrayacarine is a simple keto ester from *Murraya paniculata* var. *omphalocarpa* (T.-S. Wu, *et al.*, *Phytochemistry*, 1989, **28**, 2873).

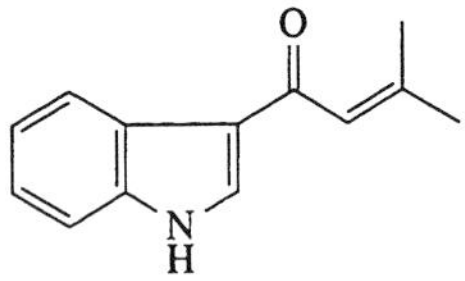

(56)

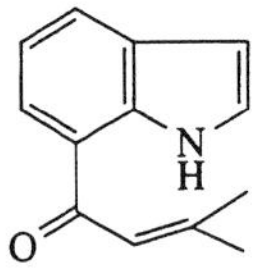

leiocarpone (57)

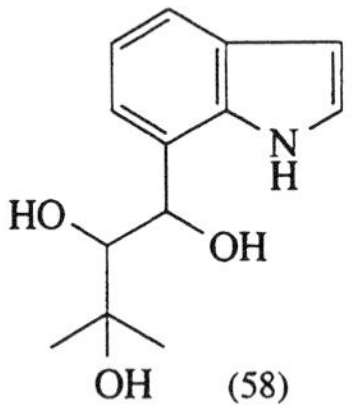

(58)

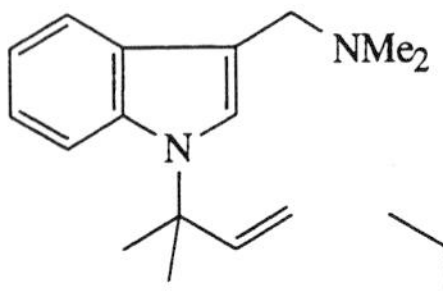

(59)

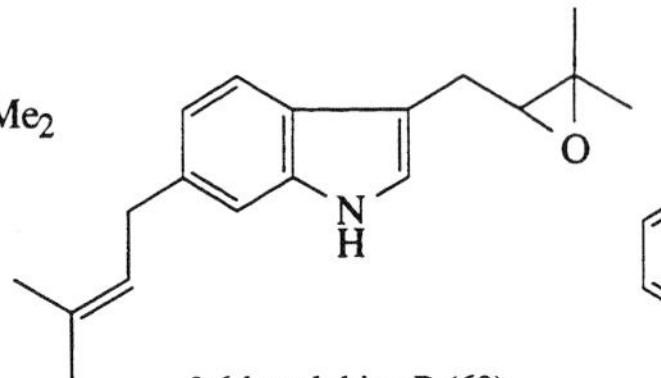

3,6-hexalobine B (60)

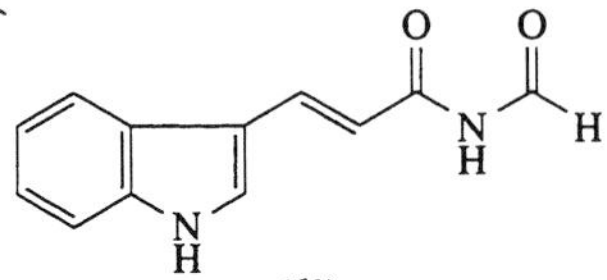

(61)

The number of complex indoloterpenes continues to grow and several reviews have appeared: Indolosesquiterpene Alkaloids of the Annonaceae (P.G. Waterman in "Alkaloids: Chemical and Biological Perspectives, "Vol. 3, Chapter 2, 1985), Tremorgenic Mycotoxins (P.S. Steyn and R. Vleggaar, *Prog. Chem. Org. Nat. Prod.*, 1985, **48**, 1) and *Aspergillus* Alkaloids (Y. Yamamoto and K. Arai in "The Alkaloids," Vol. 29, Chapter 4, 1986).

The novel indolosesquiterpenes neopolyalthenol (62) and isopolyalthenol (attached at C-3 of indole) are found in *Polyalthia suaveolens* (N. Kunesch, *et al.*, *Tetrahedron Lett.*, 1985, **26**, 4937). The mould *Aspergillus flavus* produces several compounds related to aflavinine (J.B. Gloer, *et al.*, *J. Org. Chem.*, 1988, **53**, 5457), and *A. nomius* has yielded nominine (63) (J.B. Gloer, *et al.*, *J. Org. Chem.*, 1989, **54**, 2530). Four related cytotoxic radarins, e.g., (64) and (65), are produced by *A. sulphureus* (J.A. Laakso, *et al.*, *J. Org. Chem.*, 1992, **57**, 138). A compound that is structurally similar to nominine (63) is emeniveol, which is produced by *Emericella nivea* and is a pollen growth inhibitor (Y. Kimura, *et al.*, *Tetrahedron Lett.*, 1992, **33**, 6987). The fungi *Emericella desertorum* and *E. striata* produce the emindoles, such as emindole DA (66), metabolites related to the earlier described paxilline (K. Nozawa, *et al.*, *J. Chem. Soc., Perkin Trans. 1*, 1988, 1689). The first example of an indoloditerpenoid with a 1,1-dimethylallyl moiety attached to the C-2 position of indole is emindole PA (67), which was characterized from cultures of *E. purpurea* (K. Kawai, *et al.*, *J. Chem. Soc., Perkin Trans. 1*, 1994, 1673). Polyavolensin (68), polyavolensinol (69), and polyavolensinone (70), which were first isolated in 1980 from *Polyathia suaveolens* (D.A. Okorie, *Tetrahedron*, 1980, **36**, 2005), have had their structures modified more recently to those shown following an X-ray study (C.P. Falshaw, *et al.*, *Tetrahedron*, 1982, **38**, 2311). The latter compound has also been isolated as greenwayodendrin-3-one from *Greenwayodendron suaveolens* along with three other related metabolites (C.M. Hasan, *et al.*, *J. Chem. Soc., Perkin Trans. 1*, 1982, 2807). It appears that both sets of compounds are identical.

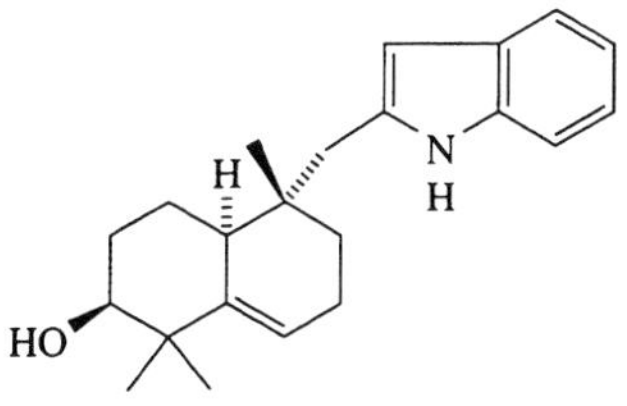

Neopolyalthenol (62)

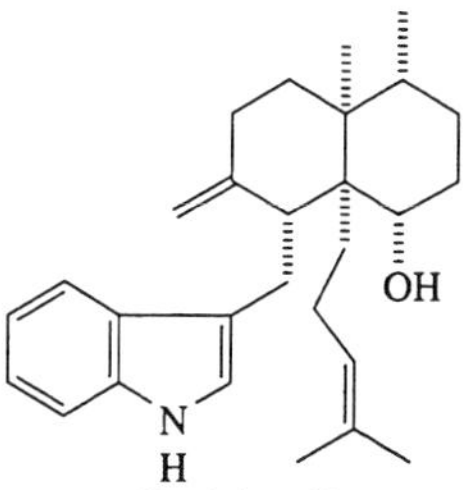

Nominine (63)

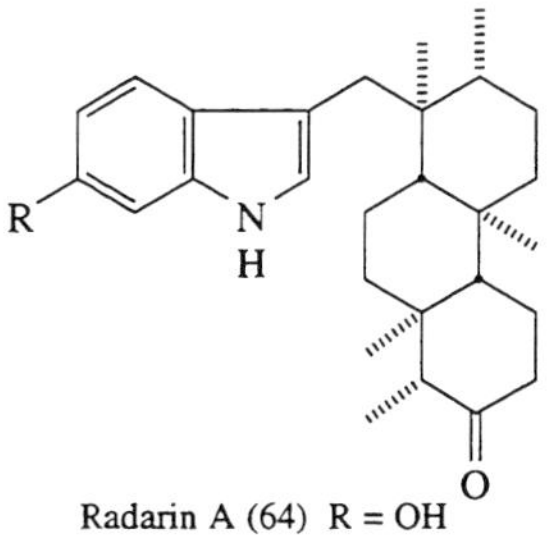

Radarin A (64) R = OH
Radarin C (65) R = H

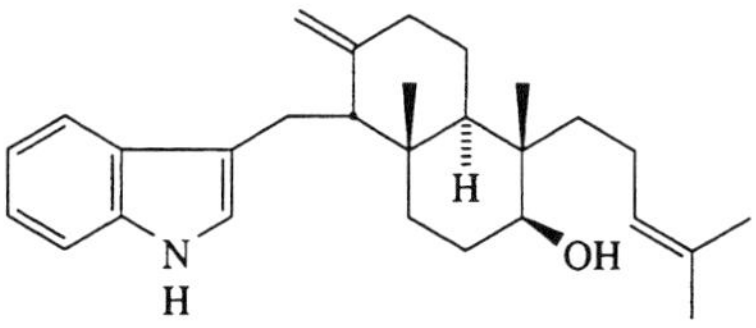

Emindole DA (66)

Emindole PA (67)

Polyavolensin (68) R = Ac
Polyavolensinol (69) R = H

Polyavolen-
sinone (70)

Both the number of new tremorgenic mycotoxins and their astonishing structural complexity are staggering.

The isolation of 14α-hydroxypaspalinine and a *N,N*-dimethylvaline ester from *Aspergillus nomius* (G.M. Staub, *et al.*, *Tetrahedron Lett.*, 1993, **34**, 2569) and 7α-hydroxy-13-desoxypaxilline and 10β-hydroxy-13-desoxy-paxilline from *Penicillium paxilli* (P.G. Mantle and C.M. Weedon, *Phytochemistry*, 1994, **36**, 1209) have been reported. The interesting ring-contracted paxinorol (71), which may be derived from paxilline, has also been isolated from *P. paxilli* (C.O. Miles, *et al.*, *J. Org. Chem.*, 1995, **60**, 6067). The mould *P. crustosum*, which was found on contaminated bread intended for Tokyo school lunches, produces two indoloterpenes related to paxilline-paspaline that may be biogenetic precursors of the penitrems (*vide infra*) (T. Hosoe, *et al.*, *Chem. Pharm. Bull.*, 1990, **38**, 3473), which are also found in this common mould (mouldy buns, cheese, bread, nuts, etc.). The fungus *Aspergillus flavus* produces β-aflatrem, an isomer of aflatrem in which the 1,1-dimethyl group is at the indole C-6 position (M.R. TePaske, *et al.*, *J. Nat. Prod.*, 1992, **55**, 1080). A series of a dozen novel acyl-CoA: cholesterol acyltransferase (ACAT) inhibitors, the terpendoles A-L, are produced by *Albophoma yamanashiensis* and are related to paspaline (X.-H. Huang, *et al.*, *J. Antibiot.*, 1995, **48**, 5; H. Tonroda, *et al.*, *ibid.*, 1995, **48**, 793). One example is terpendole L (72).

Paxinorol (71)

Terpendole L (72)

The prototypical tremorgenic mycotoxins are the penitrems A-F, which are produced by *Penicillium crustosum* (A.E. de Jesus, *et al.*, *J. Chem. Soc., Perkin Trans. 1*, 1983, 1847, 1857), and have resulted in numerous incidents of farm animal poisoning. An example is penitrem A (73). A close analogue of the penitrems is pennigritrem, which contains an oxetane ring and is produced by *P. nigricans* (J. Penn, *et al.*, *ibid.*, 1992, 23), and *P. janthinellum*, which is associated with the livestock neuromuscular disease "ryegrass staggers," produces several penitrem-related janthitrems, such as janthitrem E (74) (A.E. de Jesus, *et al.*, *ibid.*, 1984, 697; J. Penn, *et al.*, *Phytochemistry*, 1993, **32**, 1431). New compounds of the janthitrem class, the shearinines A-C, have been identified from cultures of *Eupenicillium shearii*, along with a novel paxilline analogue (G.N. Belofsky, *et al.*, *Tetrahedron*, 1995, **51**, 3959). Two possible biogenetic precursors to the penitrems have been reported in *P. crustosum* (T. Yamaguchi, *et al.*, *Phytochemistry*, 1993, **32**, 1177), and, similarly, three penitrem analogues, the sulpinines A-C and secopenitrem B, are formed by *Aspergillus sulphureus* (J.A. Laakso, *et al.*, *J. Org. Chem.*, 1992, **57**, 2066).

Penitrem A (73)

Janthitrem E (74)

Several new isatin derivatives, in addition to the convolutamydines (*vide supra*), have been discovered since the last volume was published. Thus, isatin itself is an antifungal agent produced by *Alteromonas* sp. (M.S. Gil-Turnes, *et al.*, *Science*, 1989, **246**, 116), a compound which protects shrimp embryos from a pathogenic fungus. It has also been reported that isatin is a natural MAO inhibitor in humans and rats (V. Glover, *et al.*, *J. Neurochem.*, 1988, **51**, 656). The microbe *Streptomyces albus* produces 6-isopentenylisatin (U. Gräfe and L. Radics, *J. Antibiot.*, 1986, **39**, 162). The tumorigenic plant *Melochia tomentosa* produces melosatins C (75) and

D (76), in addition to the previously isolated A and B (G.J. Kapadia, *et al.*, *Tetrahedron*, 1980, **36**, 2441; *Planta Med.*, 1993, **59**, 568). Melochicorine (77) is found in *M. corchorifolia* (R.S. Bhakuni, *et al.*, *Phytochemistry*, 1991, **30**, 3159). The novel indisocin (78) and the *N*-methyl analogue are produced by *Nocardia blackwellii* (K. Isshiki, *et al.*, *J. Antibiot.*, 1987, **40**, 1195), and the quite remarkable murrayaculatine (79) has been reported to occur in *Murraya paniculata* (T.-S. Wu, *et al.*, *Phytochemistry*, 1994, **37**, 287). Indolone (80), an unstable compound isolated from the mushroom *Pleurotus salmoneostramineus*, plays an important role in the photochemical generation of oxygen from water and, thus, it may be involved in photosynthesis (S. Takekuma, *et al.*, *J. Am. Chem. Soc.*, 1994, **116**, 8849). A series of uvarindoles, e.g., (81) and (82), are produced by *Uvaria angolensis* (P.G. Waterman and I. Mohammad, *Chem. Commun.*, 1984, 1280).

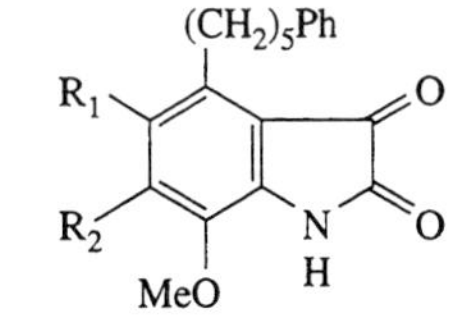

Melosatin C (75) $R_1 = R_2 = H$
Melosatin D (76) $R_1 = OH$, $R_2 = OMe$

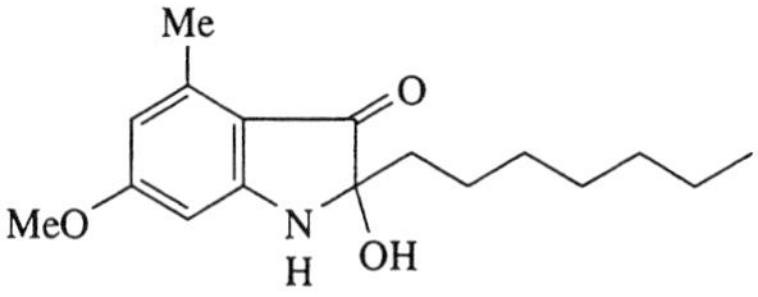

Melochicorine (77)

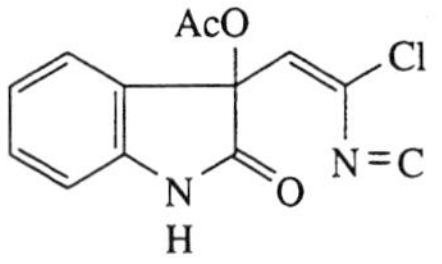

Indisocin (78)

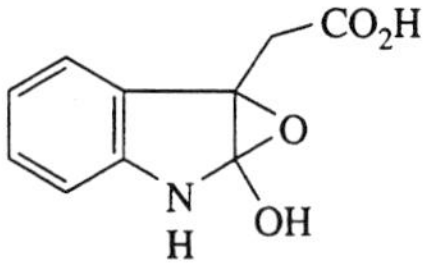

Murrayaculatine (79)

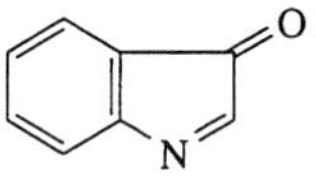

(80)

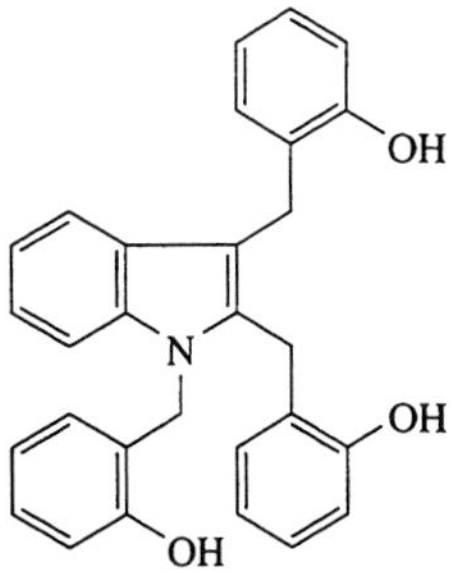

Uvarindole A (81)

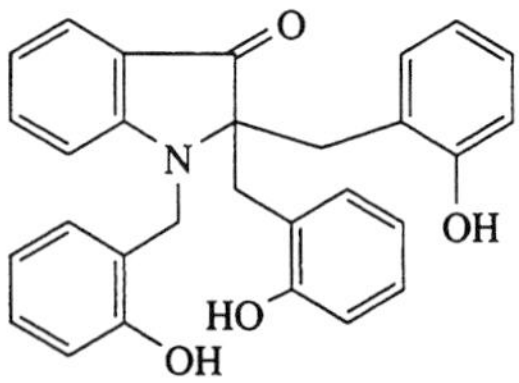

Uvarindole D (82)

The plant *Schlerochiton ilicifolius* has yielded the high-intensity sweetener monatin (83) which is 1400 times sweeter than sucrose (R. Vleggaar, *et al.*, *J. Chem. Soc., Perkin Trans. 1*, 1992, 3095). Pallidin (84), which is found in the sponge *Rhaphisia pallida*, is the first *N*-carboxyindole alkaloid to be isolated (J. Su, *et al.*, *J. Nat. Prod.*, 1996, **59**, 504). Several simple gramine derivatives have been discovered in recent years. The major pasture grass of the temperate regions of the world *Phalaris aquatica* contains the new 5-methoxy-, 7-methoxy-, and 5,7-dimethoxygramine (D.P. Mulveno, *et al.*, *Phytochemistry*, 1983, **22**, 2885). The deoxysugar (85) is found in *Actinomadura* sp. (T. Ito, *et al.*, *J. Antibiot.*, 1984, **37**, 931), and *Lupinus arbustus* subsp. *calcaratus* has yielded gramodendrine (86), a structure confirmed by synthesis (W.J. Keller and G.M. Hatfield, *Phytochemistry*, 1982, **21**, 1415). Red beets (*Beta vulgaris*) contain the novel neobetanin (87) (D. Alard, *et al.*, *Phytochemistry*, 1985, **24**, 2383).

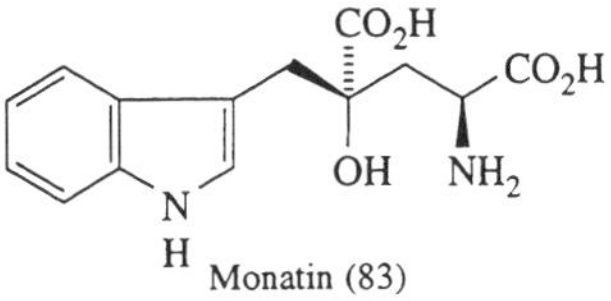

Monatin (83)

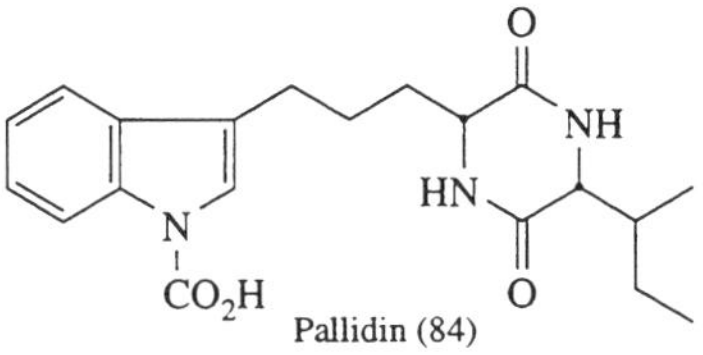

Pallidin (84)

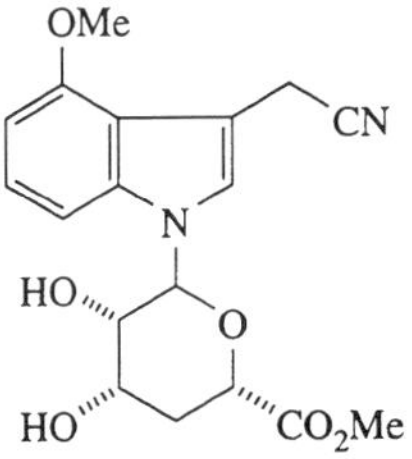

(85)

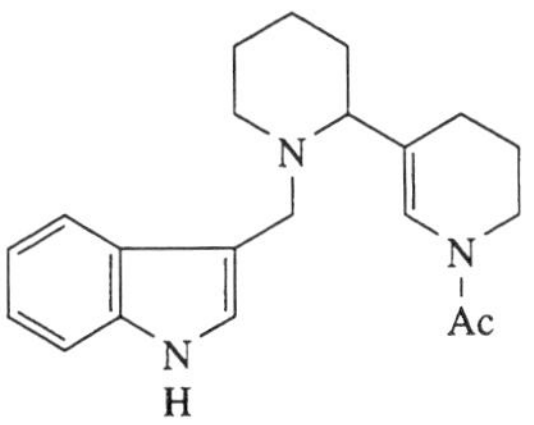

Gramodendrine (86)

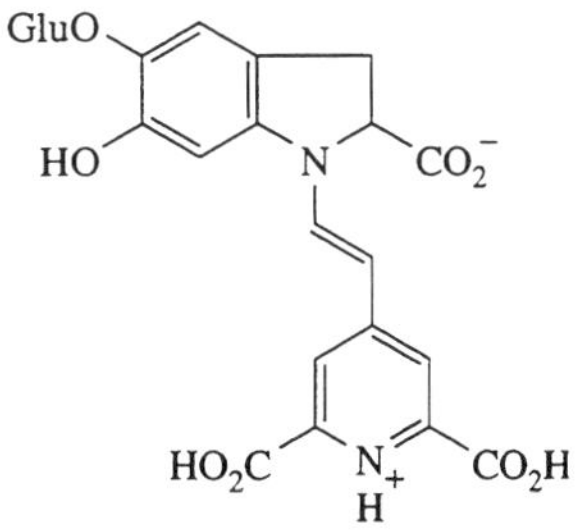

Neobetanin (87)

Ailanindole (88) is a novel metabolite from *Ailanthus malabarica* (H. Aono, *et al.*, *Phytochemistry*, 1994, **37**, 579), and the Korean mushroom *Agrocybe cylindracea* contains agrocybenine (89) (H. Koshino, *et al.*, *Tetrahedron Lett.*, 1996, **37**, 4549). Monomargine (90) is a novel cytotoxic alkaloid from *Monocarpia marginalis* (K. Mahmood, *et al.*, *Tetrahedron Lett.*, 1993, **34**, 1795). The simple oxazole alkaloid pimprinethine (91) is formed in cultures of *Streptomyces cinnamomeus* (M. Noltemeyer, *et al.*, *J. Antibiot.*, 1982, **35**, 549). This alkaloid and the related pimprinaphine are produced by *Streptoverticillium olivoreticuli* (Y. Koyama, *et al.*, *Agric. Biol. Chem.*, 1981, **45**, 1285), and the related WS-30581A and B are found in *S. waksmanii* (K. Umehara, *J. Antibiot.*, 1984, **37**, 1153). BE 10988 (92) is a topoisomerase II inhibitor isolated from a culture broth (H. Suda, *et al.*, *Tetrahedron Lett.*, 1991, **32**, 2791). The simple thiazole camalexin (93) and a methoxyl analogue are produced by leaves of *Camelina sativa* following elicitation by the fungus *Alternaria brassicae* (L.M. Browne, *et al.*, *Tetrahedron*, 1991, **47**, 3909). The plant growth regulator thienodolin (94), which is produced by *Streptomyces albogriseolus*, contains the rare thieno[2,3-*b*]indole ring system (K. Kanbe, *et al.*, *Biosci. Biotechnol. Biochem.*, 1993, **57**, 632, 636).

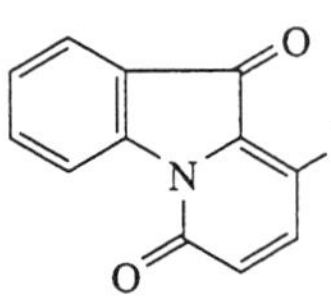

Ailanindole (88)

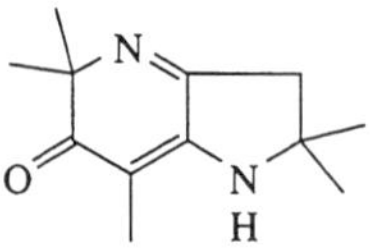

Agrocybenine (89)

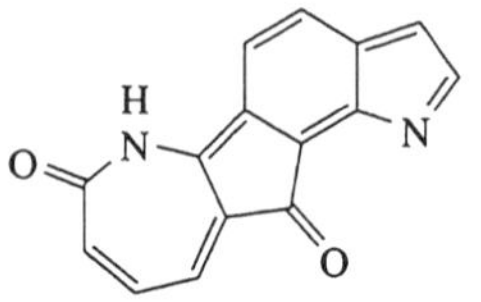

Monomargine (90)

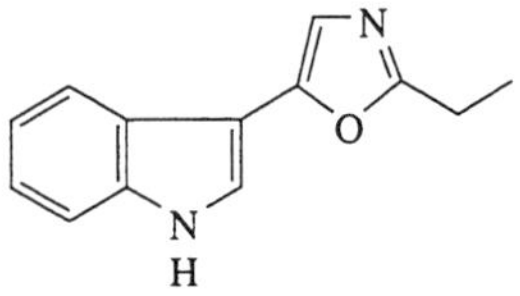

Pimprinethine (91)

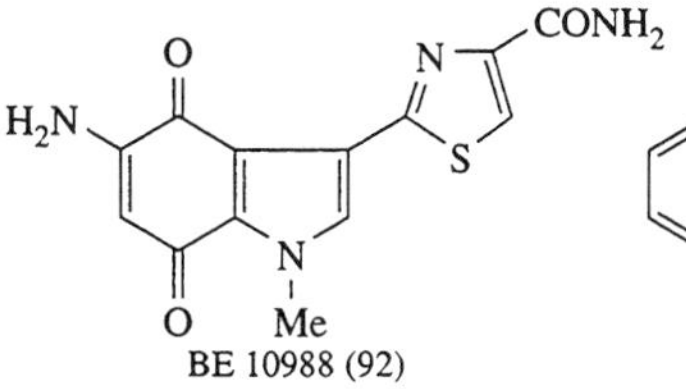

BE 10988 (92)

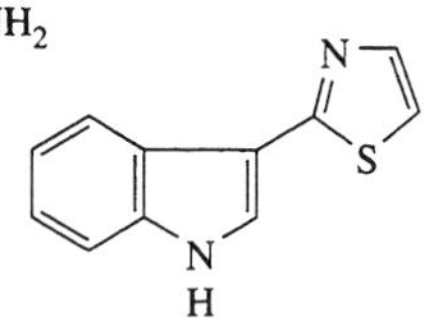

Camalexin (93)

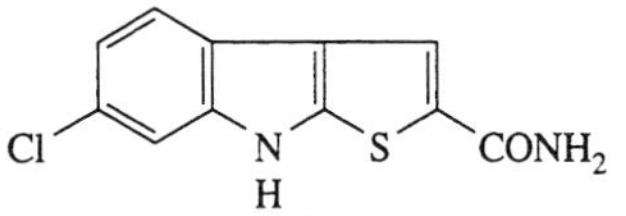

Thienodolin (94)

When Chinese cabbage (*Brassica campestris*) is inoculated with *Pseudomonas cichorii*, the phytoalexins brassinin (95), (96), and cyclobrassinin (97) are produced (M. Takasugi, *et al.*, *Chem. Commun.*, 1986, 1077). Similarly, white cabbage (*B. oleracea*) has furnished 4-methoxybrassinin (K. Monde, *et al.*, *Phytochemistry*, 1990, **29**, 1499). Subsequent studies have identified brassicanals A (98), B (99), and C (100), and dioxibrassinin (101) (K. Monde, *et al.*, *Chem. Lett.*, 1990, 209; *Phytochemistry*, 1991, **30**, 2915). The carbonyl analogue of (96) has also been identified (M. Takasugi, *et al.*, *Bull. Chem. Soc. Japan*, 1988, **61**, 285), as have been two derivatives of cyclobrassinin from *B. juncea* and *B. campestris* (M. Devys, *et al.*, *Phytochemistry*, 1990, **29**, 1087; K. Monde, *et al.*, *Heterocycles*, 1994, **38**, 263). The novel spirobrassinin (102), which is the first oxindole phytoalexin, is found in the daikon *Rhaphanus sativus* (M. Takasugi, *et al.*, *Chem. Lett.*, 1987, 1631).

Brassinin (95) R = H
(96) R = OMe

Cyclobrassinin (97)

Brassicanal A   (98) R = SMe
Brassicanal B   (99) R = SCH$_2$Ac
Brassicanal C (100) R = S(O)OMe

Dioxibrassinin (101)

Spirobrassinin (102)

A few relatively simple gliotoxin analogues have been described since the last volume appeared. Gliotoxin G (103), the tetrasulfide analogue of gliotoxin, is produced by *Aspergillus fumigatus* (P. Waring, *et al.*, *Tetrahedron Lett.*, 1986, **27**, 735), and gliotoxin E (104) is found in cultures of *A. fumigatus*, the thermophilic fungus *Thermoascus crustaceus*, and *Penicillium terlikowski* (P. Waring, *et al.*, *Aust. J. Chem.*, 1987, **40**, 991). The organism *Gliocladium deliquescens* produces (105) (J.R. Hanson and M.A. O'Leary, *J. Chem. Soc., Perkins Trans. 1*, 1981, 218).

Gliotoxin G (103) x - 4
Gliotoxin E (104) x = 3

(105)

(b) Carbazole derivatives

There has been a wealth of new carbazole derivatives discovered in natural sources over the past 15 years. Since several excellent reviews of this area are available (P. Bhattacharyya and D.P. Chakraborty, *Prog. Chem. Org. Nat. Prod.*, 1987, **52**, 159; D.P. Chakraborty and S. Roy, *ibid.*, 1991, **57**, 71; D.P. Chakraborty in "The Alkaloids," Vol. 44, p. 257, (1993), coverage here will be generally limited to the literature from 1993 to the present. Indeed, some 100 new natural carbazoles were discovered during the period 1980-1992.

Carbazole itself has been isolated from a plant for the first time (*Glycosmis pentaphylla*) (B.K. Chowdhury, *et al.*, *Phytochemistry*, 1987, **26**, 2138), and 1-methylcarbazole is found in the sponge *Tedania ignis* (R.L.

84

Dillman and J.H. Cardellina II, *J. Nat. Prod.*, 1991, **54**, 1056). The first carbazole to be isolated from mammals is 3-chlorocarbazole from bovine urine (K.-C. Luk, *et al.*, *ibid.*, 1983, **46**, 852). This compound has diazepam-like activity. The isolation of 2-hydroxy-3-methylcarbazole from *Murraya koenigii* lends support to the proposal that the numerous pyrano-carbazole alkaloids are probably formed from the former carbazole (P. Bhattacharyya, *et al.*, *Chem. Ind.*, 1986, 246). The isomeric 2-hydroxy-7-methylcarbazole is found in *Cimicifuga simplex* (G. Kusano, *et al.*, *Heterocycles*, 1993, **36**, 2367). Clausenalene (106) was isolated from *Clausena heptaphylla*, a plant with a rich history of carbazole production (P. Bhattacharyya, *et al.*, *Phytochemistry*, 1993, **33**, 248). This species also has yielded clausenal (107), which has antifungal and antibacterial activity (A. Chakraborty, *et al.*, *ibid.*, 1995, **38**, 787). Clausenol (108) and clausenine (109) are found in *C. anisata* (A. Chakraborty, *et al.*, *ibid.*, 1995, **40**, 295), and clausine L (110) was characterized from *C. excavata* (T.-S. Wu, *et al.*, *ibid.*, 1993, **32**, 449). This latter plant also produces the cytotoxic clausenaquinone A (111), which is a potent inhibitor of platelet aggregation (T.-S. Wu, *et al.*, *Bioorg. Med. Chem. Lett.*, 1994, **4**, 2395). The plant *C. lansium* has furnished (112) (V. Kumar, *et al.*, *Phytochemistry*, 1995, **40**, 1563).

Clausenalene (106)

Clausenal (107)

Clausenol (108)   R = OH
Clausenine (109)   R = OMe

Clausine L (110)

Clausenaquinone A (111)

(112)

The plant *Murraya euchrestifolia* is also an amazing source of carbazole alkaloids. New examples include murrayamines D and E (113) (T.-S. Wu, *et al.*, *Phytochemistry*, 1995, **40**, 1817), and O and P (114) (T.-S. Wu, *et al.*, *Tetrahedron Lett.*, 1995, **36**, 5385). The latter two compounds are novel cannabinol-skeletal carbazoles. Several new carbazoles have also been reported from *M. koenigii*, such as girinimbilol (115) and mahanim-binol (116) (J. Reisch, *et al.*, *Phytochemistry*, 1994, **36**, 1073; P. Bhattacharyya, *et al.*, *ibid.*, 1994, **36**, 1085). The latter two carbazoles may be non-cyclized biogenetic precursors of girinimbine and mahanimbine, respectively. Furthermore, mukoenines A-C are new carbazole alkaloids from this plant (C. Ito, *et al.*, *Chem. Pharm. Bull.*, 1993, **41**, 2096). The number of novel antioxidant carbazole quinones and related compounds continues to grow. The potent 5-lipoxygenase inhibitors epocarbazolins A (117) and B (118) are produced by *Streptomyces anulatus* (Y. Nihei, *et al.*, *J. Antibiot.*, 1993, **46**, 25). The potent neuronal cell protecting substance and radical scavenger carquinostatin A (119) is produced by *S. exfoliatus* (K. Shinya, *et al.*, *Tetrahedron Lett.*, 1993, **34**, 4943). The neuronal protector lavanduquinocin from *S. viridochromogenes* is similar to carquinostatin A except for a different terpene side chain (K. Shin-ya, *et al.*, *J. Antibiot.*, 1995, **48**, 574). Six closely related carbazoquinocins A-F (e.g., (120)), which are comparable to Vitamin E as antioxidants, have been isolated from cultures of *S. violaceus* (M. Tanaka, *et al.*, *ibid.*, 1995, **48**, 326), and differ only in the alkyl side chain at C-1.

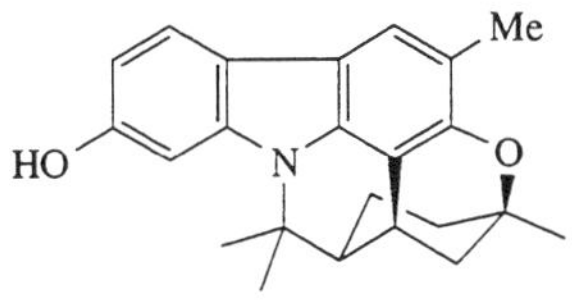

Murrayamine E (113)

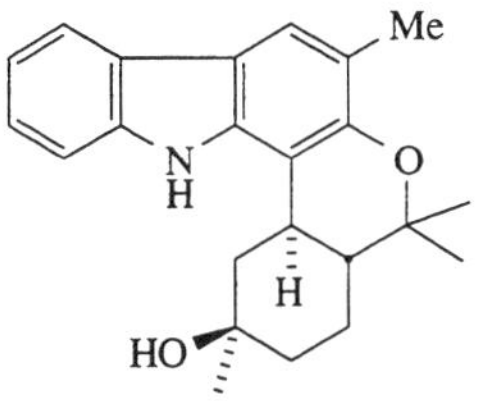

Murrayamine P (114)

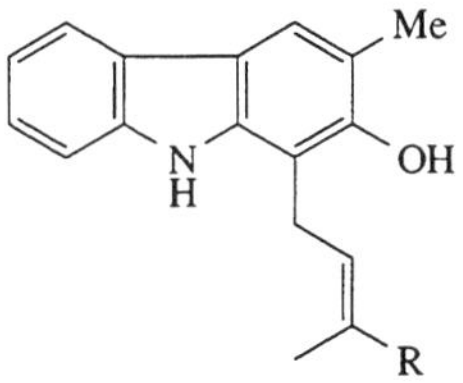

Girinimbilol (115)
R = Me
Mahanimbinol (116)
R = $(CH)_2C = CMe_2$

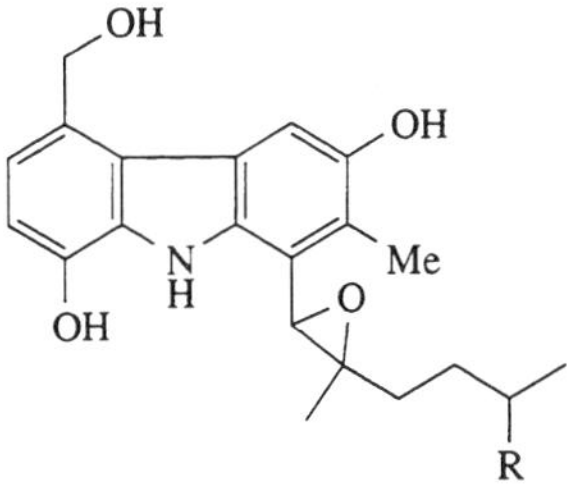

Epocarbazolin A (117) R = Me
Epocarbazolin B (118) R = Et

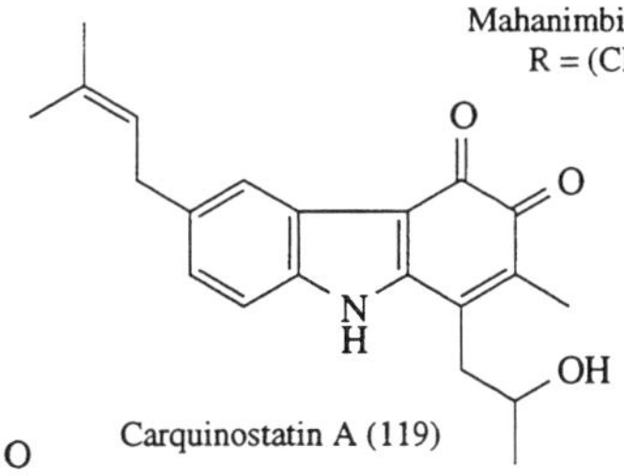

Carquinostatin A (119)

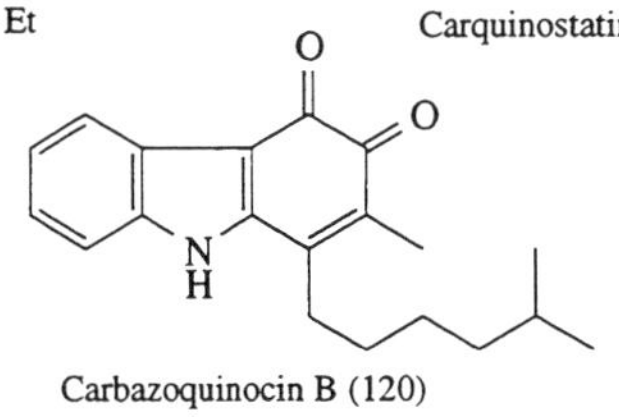

Carbazoquinocin B (120)

Purpurone (121) is a novel ATP-citrate lyase inhibitor from the sponge *Iotrochota* sp. (G.W. Chan, *et al.*, *J. Org. Chem.*, 1993, **58**, 2544). The unique pyrrolo[2,3-*c*]carbazole spiro lactone skeleton is embodied in the potent reductase inhibitor (122), which, along with two other similar compounds, has been discovered in the sponge *Dictyodendrilla* sp. (A. Sato, *et al.*, *J. Org. Chem.*, 1993, **58**, 7632).

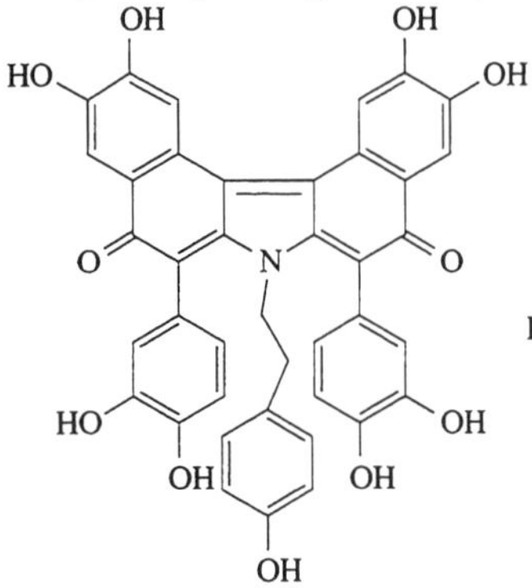

Purpurone (121)

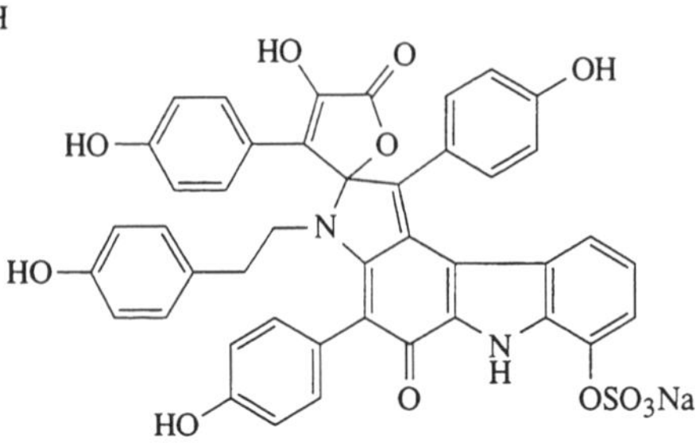

(122)

2. Alkaloids Containing a Tryptamine Unit

(a) Compounds without an isoprene moiety

(i) Simple tryptamine and tryptophan derivatives

A large number of new compounds in this category have been discovered since 1980. The only class of compounds purposely not considered here are the new tryptophan-containing peptides, of which a large number have been discovered. A number of other tryptophan-derived metabolites are discussed in the section on Mould Metabolites. A review has appeared on the Chemistry and Reactions of Cyclic Tautomers of Tryptamines and Tryptophans (T. Hino and M. Nakagawa, in "The Alkaloids," Vol. 34, Chapter 1, 1988).

Konbamidin (123) is a cytotoxic (*R*) tryptophan derivative found in an Okinawan sponge *Ircinia* sp. (H. Shinonaga, *et al.*, *J. Nat. Prod.*, 1994, **57**, 1603. Green robusta coffee beans (*Coffea canephora*) contain caffeoyltryptophan (124) (H. Morishita, *et al.*, *Phytochemistry*, 1987, **26**, 1195). The small tree *Limonia acidissima* produces N_b-acetyl-N_b-methyltryptamine (M.H.A. Zarga, *J. Nat. Prod.*, 1986, **49**, 901), and *Melicope leptococca* has furnished (125) (A.L. Skaltsounis, *et al.*, *ibid.*, 1983, **46**, 732). Two possible precursors to pimprinine, (126) and (127), have been isolated for the first time from *Streptomyces ramulosus* (Y. Chen, *et al.*, *J. Antibiot.*, 1983, **36**, 916). Similarly, both almazole C (128) and its possible precursor peptides, e.g., prealmazole C (129), have been discovered in a Senegalese seaweed (G. Guella, *et al.*, *Helv. Chim. Acta*, 1994, **77**, 1999). The tunicate *Dendrodoa grossularia* contains the unusual (130), which appears to be the first example of a 4*H*-imidazol-4-one (M. Guyot and M. Meyer,

Tetrahedron Lett., 1986, **27**, 2621). Dehydroborrecapine (131) is a new alkaloid from *Borreria capitata* (A. Jossang, *et al.*, *Planta Med.*, 1981, **43**, 301).

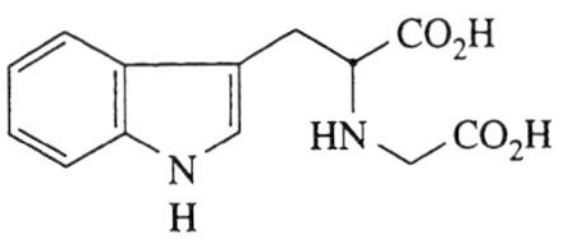

Konbamidin (123)

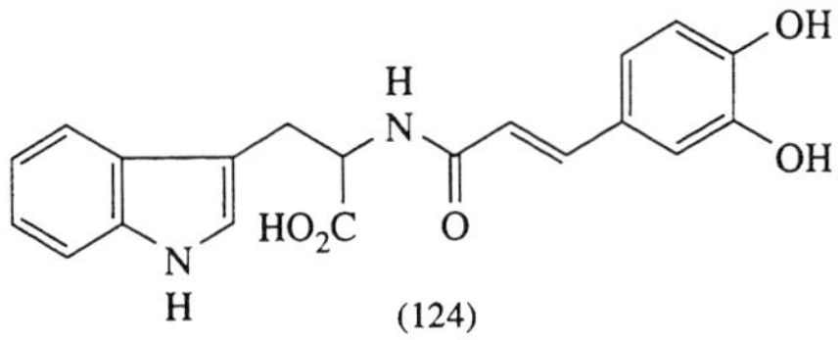

(124)

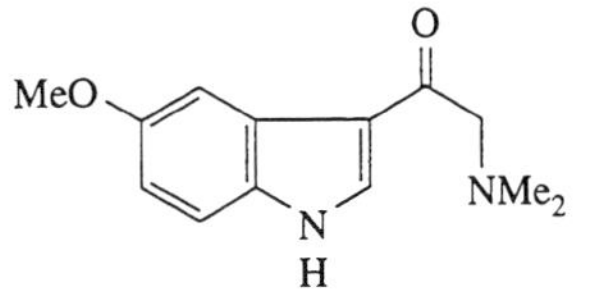

(125)

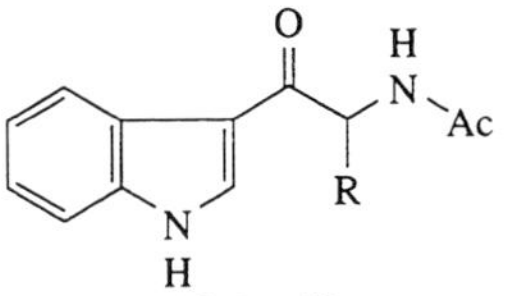

(126) R = H
(127) R = OH

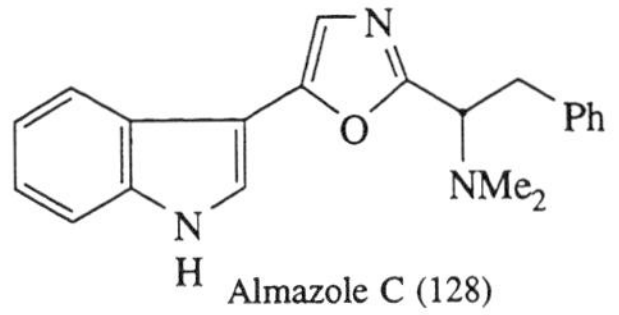

Almazole C (128)

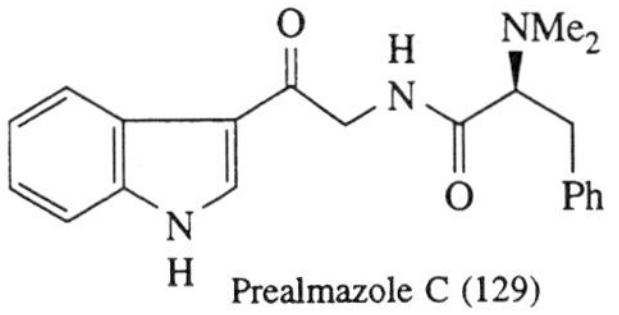

Prealmazole C (129)

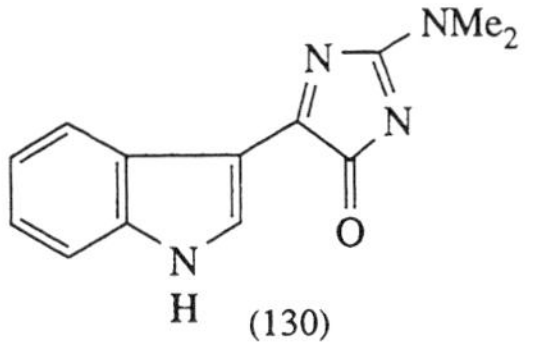

(130)

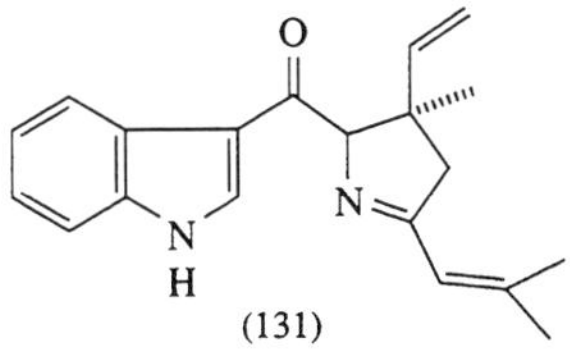

(131)

A new inhibitor of superoxide anion generation is OPC-15160, which degrades to the more active OPC-15161 (132), and is produced by the fungus *Thielavia minor* (Y. Nakano, *et al.*, *J. Antibiot.*, 1991, **44**, 52). The plant *Evodia rutaecarpa* produces the seco alkaloids (133) and goshuyu-amide II (134) (N. Shoji, *et al.*, *J. Nat. Prod.*, 1988, **51**, 161; 1989, **52**, 1160). The novel alkaloid margaritarine (135) has been isolated from *Margaritaria indica* (D. Arbain, *et al.*, *J. Chem. Soc., Perkin Trans. 1*, 1991, 1863). Oxindole derivative (136) is found in *Cinnamomum tripliner-vis* (H. Ripperger, *et al.*, *Phytochemistry*, 1981, **20**, 1453), and horsfiline (137) has been isolated from *Horsfieldia superba* (A. Jossang, *et al.*, *J. Org. Chem.*, 1991, **56**, 6527).

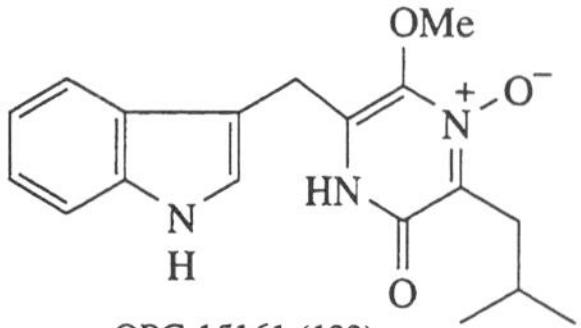

OPC-15161 (132)

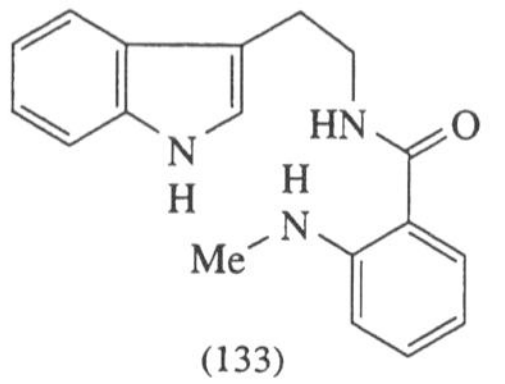

(133)

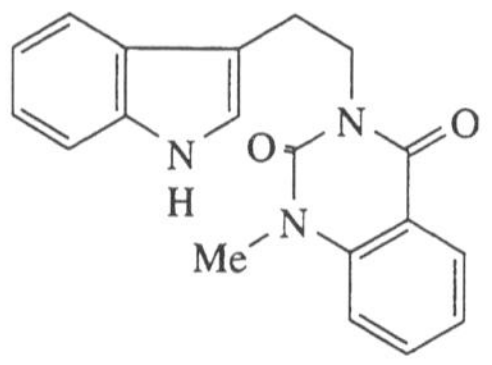

Goshuyuamide II (134)

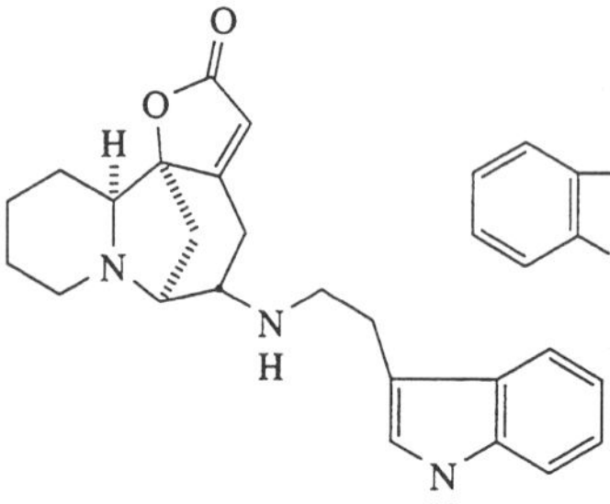

(136)

Margaritarine (135)

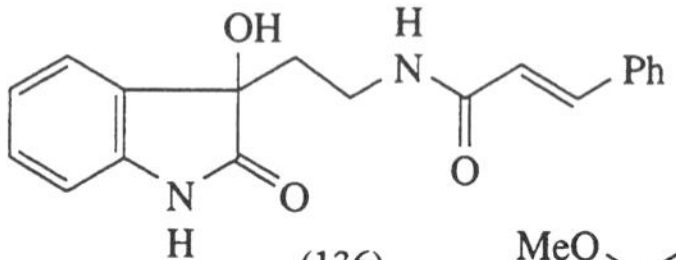

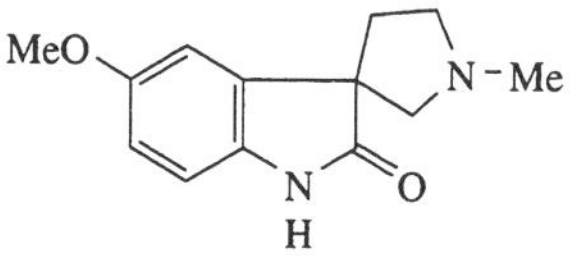

Horsfiline (137)

Funnel-web spiders such as *Agelenopsis aperta* and *Argiope* sp. produce in their venom several novel polyamines such as agel 489a (138). These agelenotoxins and argiotoxins are potent excitatory amino acid antagonists (V.J. Jasys, *et al.*, *J. Org. Chem.*, 1992, **57**, 1814; N.A. Saccomano, *et al.*, *Ann. Rep. Med. Chem.*, 1989, **24**, 287). Some related compounds, monodontamide D (139) and E (140) are found in the mollusc *Monodonta labio* and have some serine protease inhibitory activity (H. Niwa, *et al.*, *Tetrahedron*, 1994, **50**, 6805). Three other unique marine peptides are hemiasterlin (141) from the sponge *Hemiasterella minor* (R. Talpir, *et al.*, *Tetrahedron Lett.*, 1994, **35**, 4453), and hemiasterlins A (142) and B (143) from the sponge *Cymbastela* sp. Some related larger peptides are also

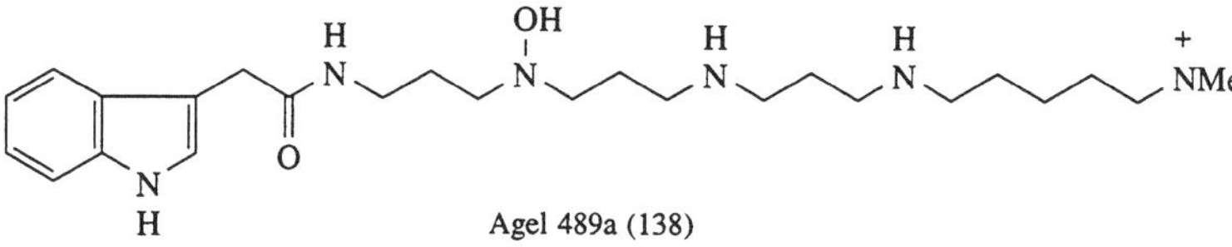

Agel 489a (138)

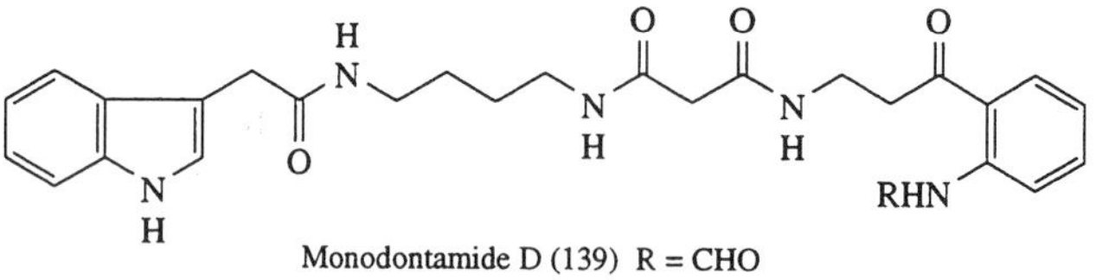

Monodontamide D (139) R = CHO
Monodontamide E (140) R = H

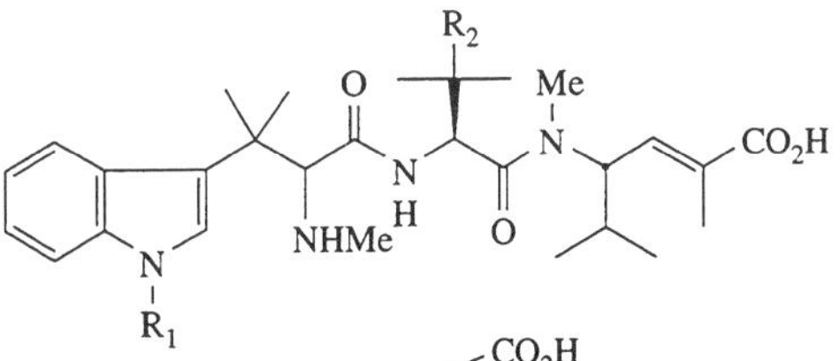

Hemiasterlin (141) $R_1 = R_2 = Me$
Hemiasterlin A (142) $R_1 = H$, $R_2 = Me$
Hemiasterlin B (143) $R_1 = R_2 = H$

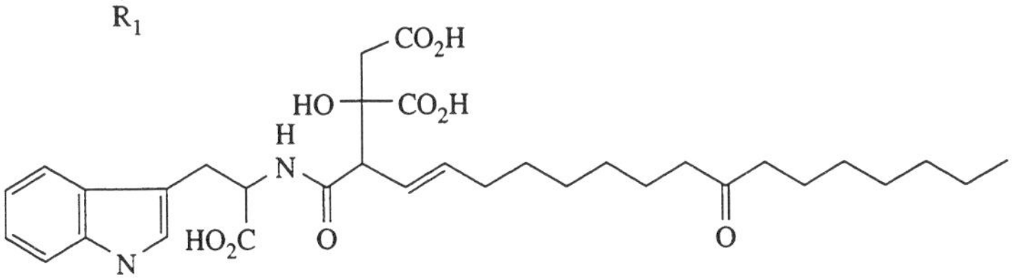

Viridiofungin C (144)

found in the latter organism (J.E. Coleman, *et al.*, *Tetrahedron*, 1995, **51**, 10653). Viridiofungin C (144) from *Trichoderma viride* is a potent antifungal compound and inhibitor of squalene synthase (G.H. Harris, *et al.*, *Tetrahedron Lett.*, 1993, **34**, 5235).

The red alga *Martensia fragilis* produces several alkaloids which represent the first examples of basic indole alkaloids to be isolated from a marine eukaryotic plant. These compounds are fragilamide (145) and martensine A, B (146) and 10-epimartensine A (M.P. Kirkup and R.E. Moore, *Tetrahedron Lett.*, 1983, **24**, 2087). The related igzamide (147), which is one of a large number of known marine bromine-containing tryptamine derivatives, was isolated from the sponge *Plocamissma igzo* (E. Dumdei and R.J. Andersen, *J. Nat. Prod.*, 1993, **56**, 792). For example, 6-bromohypaphorine is found in the sponge *Aplysina* sp. (K. Kondo, *et al.*, *ibid.*, 1994, **57**, 1008), 6-bromo-N_b-methyl-N_b-formyltryptamine and the 2-dimethylallyl derivative, flustrabromine, are produced by the bryozoan *Flustra foliacea* (P. Wulff, *et al.*, *Comp. Biochem. Physiol.*, 1982, **71B**, 523; *J. Chem. Soc., Perkin Trans. 1*, 1981, 2895), 6-bromotryptamine is found in the tunicate *Didemnum candidum* (E. Fahy, *et al.*, *J. Nat. Prod.*, 1991, **54**, 564), and the deep sea sponge *Discodermia polydiscus* has afforded the antitumor dibromo (148) (H.H. Sun and S. Sakemi, *J. Org. Chem.*, 1991, **56**, 4307). The sponge *Chelonaplysilla* sp. from Palau produces the novel chelonins A (149), B, and bromochelonin B (150). These metabolites have antimicrobial activity and chelonin A is the first natural product to contain a 2,6-disubstituted morpholine (S.C. Bobzin and D.J. Faulkner, *ibid.*, 1991, **56**, 4403). The red alga *Martensia denticulata* has yielded the anti-photooxidative denticins A-C, which are novel indole peptides containing trimethylammonium and benzenesulfonate groups (H. Murakami, *et al.*, *Biosci. Biotechnol. Biochem.*, 1994, **58**, 535).

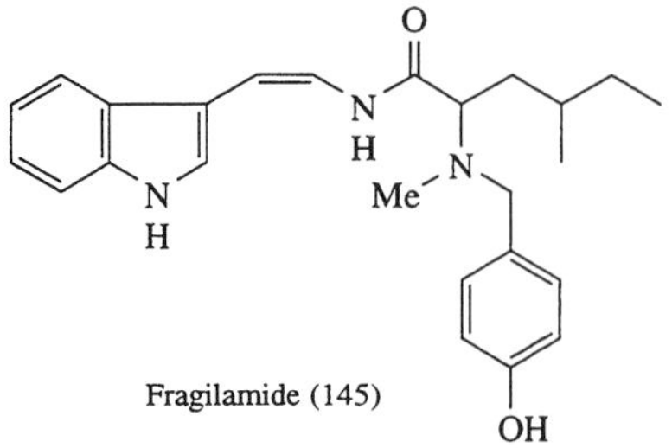

Fragilamide (145)

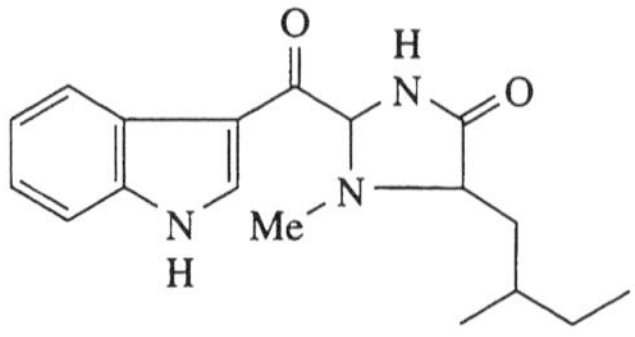

Martensine B (146)

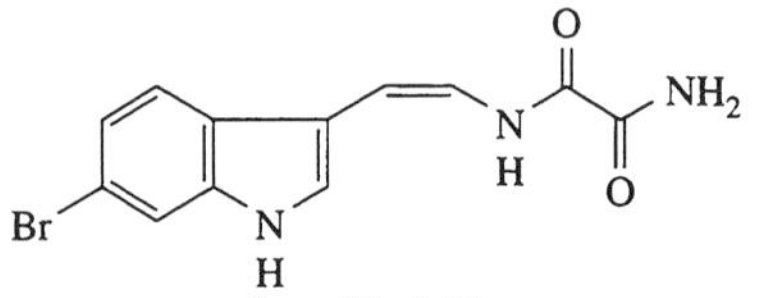

Igzamide (147)

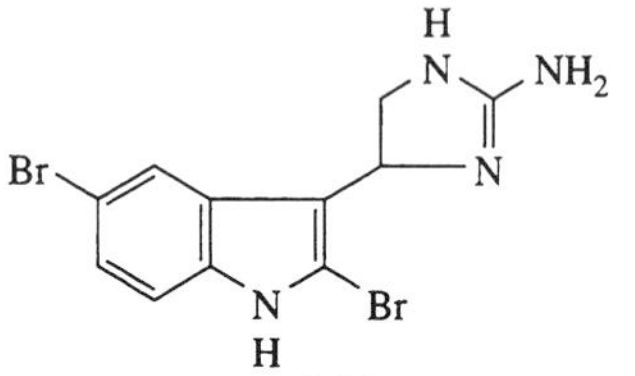

(148)

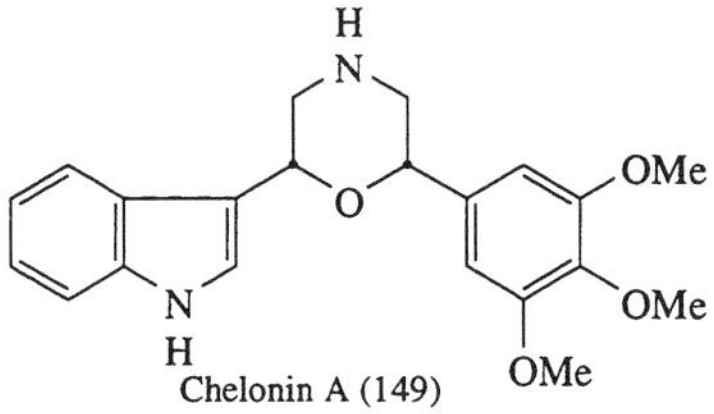

Chelonin A (149)

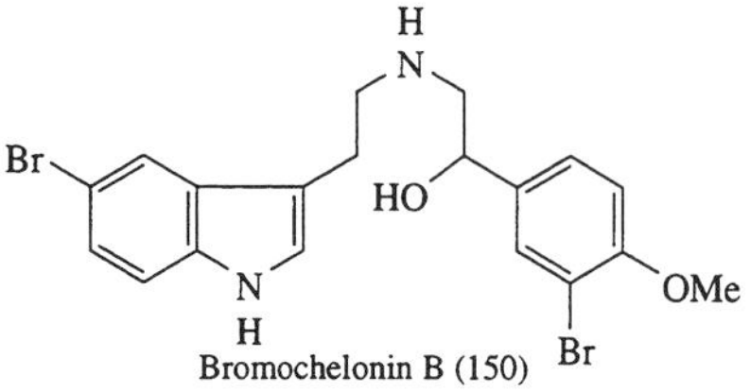

Bromochelonin B (150)

The marine ascidian *Polyandrocarpa* sp. produces polyandrocarpamides A-C (151)-(153) (N. Lindquist and W. Fenical, *Tetrahedron Lett.*, 1990, **31**, 2521), and the Japanese Ivory shell (*Babylonia japonica*) has afforded neosurugatoxin and prosurugatoxin, which are related structurally to surugatoxin as reported in the last volume (T. Kosuge, *et al.*, *ibid.*, 1981, **22**, 3147; *Chem. Pharm. Bull.*, 1985, **33**, 2890). Neosurugatoxin has 100 times more antinicotinic activity than surugatoxin. Several new aplysinopsin-type metabolites have been isolated since the last review. The coral *Dendrophyllia* sp. has yielded (154) (G. Guella, *et al.*, *Helv. Chim. Acta*, 1989, **72**, 1444) and barettin (155) is produced by *Geodia baretti* (G. Lidgren, *et al.*, *Tetrahedron Lett.*, 1986, **27**, 3283). A significant number of marine metabolites are known in which the tryptamine nitrogen is cyclized to the indole C-4 position. These include the complex pyrrolo[1,7]-phenanthroline discorhabdins from sponges of the *Latrunculia* genus (N.B. Perry, *et al.*, *Tetrahedron*, 1988, **44**, 1727; A. Yang, *et al.*, *J. Nat. Prod.*, 1995, **58**, 1596), the structurally similar prianosins (J. Kobayashi, *et al.*, *Tetrahedron Lett.*, 1987, **28**, 4939; *ibid.*, 1991, **32**, 1227), the pyrrolo-quinoline batzellines and isobatzellines from *Batzella* sp. sponges (S. Sakemi, *et al.*, *Tetrahedron Lett.*, 1989, **30**, 2517; H.H. Sun, *et al.*, *J.*

Org. Chem., 1990, **55**, 4964), the related damirones from *Damiria* sp. sponge (D.B. Stierle and D.J. Faulkner, *J. Nat. Prod.*, 1991, **54**, 1131), the makaluvamines from *Zyzzya* and *Histodermella* sponges (D.C. Radisky, *et al.*, *J. Am. Chem. Soc.*, 1993, **115**, 1632; J.R. Carney, *et al.*, *Tetrahedron*, 1993, **49**, 8483), and the epinardins from deep-water green demosponges from the Indian Ocean (M. D'Ambrosio, *et al.*, *Tetrahedron*, 1996, **52**, 8899). These novel metabolites have a range of interesting biological activity (cytotoxicity, topoisomerase II inhibition, antifungal and antimicrobial activity) and two examples, (156) and (157), are shown.

Polyandrocarpamide A (151) R = Br
Polyandrocarpamide B (152) R = I
Polyandrocarpamide C (153) R = H

(154)

Barettin (155)

Damirone A (156)

Makaluvamine A (157)

Marine bryozoans or "moss animals," produce an array of stunningly complex tryptamine-derived β-lactam metabolites. Chartellines A (158), B, and C, and chartellamides A and B are manufactured by *Chartella papyracea* (U. Anthoni, *et al.*, *J. Org. Chem.*, 1987, **52**, 4709, 5638). Hinckdentine A (159) has been isolated from *Hincksinoflustra denticulata* (A.J. Blackman, *et al.*, *Tetrahedron Lett.*, 1987, **28**, 5561). This remarkable compound contains the novel pyrimidino[3',4':1,2]pyrrolo[2,3-*d*]azepine ring system.

Chartelline A (158)

Hinckdentine A (159)

92

Seeds of *Pisum sativum* and *Vicia faba* contain 4-chlorotryptophan and two N_b-malonyl derivatives (A. Fock, *et al.*, *Phytochemistry*, 1992, **31**, 2327; Y. Sakagami, *et al.*, *Tetrahedron Lett.*, 1993, **34**, 1057).

(ii) Eserine types

A review on Alkaloids of the Calabar Bean has appeared (S. Takano and K. Ogasawara in "The Alkaloids," Vol. 36, Chapter 5, 1989). Excepting mould metabolites, which are covered separately, only a few new examples of this ring system have been reported since the preceding volume.

Leuconoxine (160) is found in *Leuconotis eugenifolius* (F. Abe and T. Yamauchi, *Phytochemistry*, 1994, **35**, 169), and a series of alkaloids, e.g., (161), have been isolated from *Hunteria umbellata* (E.A. Adegoke and B. Alo, *ibid.*, 1986, **25**, 1461). The novel alkaloid saifine (162) is found in the roots of *Rhazya stricta* (A. Rahman, *et al.*, *Nat. Prod. Lett.*, 1995, **5**, 245). Raucubaine is an unusual polycyclic alkaloid with an eserine core found in *Rauwolfia salicifolia* (J.P. Kutney, *et al.*, *Heterocycles*, 1980, **14**, 1309). Australian frogs of *Pseudophryne* genus produce an array of pseudophrynamines, such as (163) (J.W. Daly, *et al.*, *J. Nat. Prod.*, 1990, **53**, 407). The sponge *Phakellia carteri* contains two cyclic peptides, e.g., phakellistatin 3, which contain the eserine ring system shown in (164) (G.R. Pettit, *et al.*, *J. Org. Chem.*, 1994, **59**, 1593), and *Streptomyces rugosporus* has afforded pyrroindomycins A and B, which contain the novel pyrrolo[2,3-*b*]indole ring system (165) linked to an unbranched deoxytrisaccharide and a polyketide tetramic acid unit (W. Ding, *et al.*, *J. Antibiot.*, 1994, **47**, 1250). These important antibiotics have potent activity against vancomycin-resistant bacteria.

Leuconoxine (160)

(161)

Saifine (162)

(163)

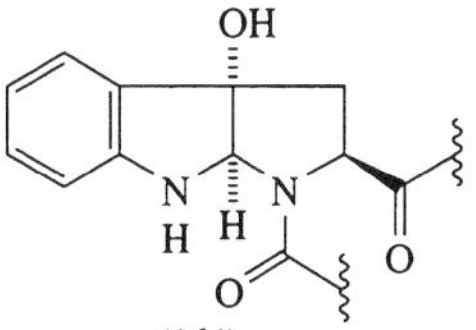

(164)

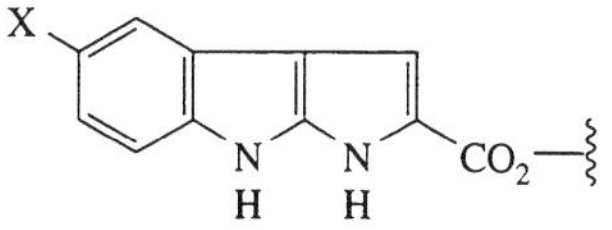

(165) X = H, Cl

The bryozoan *Flustra foliacea* continues to be a fantastic source of novel flustramine metabolites, and a dozen new examples have been reported since 1980 (J.S. Carlé and C. Christophersen, *J. Org. Chem.*, 1981, **46**, 3440; P. Wulff, *et al.*, *Comp. Biochem. Physiol.*, 1982, **71B**, 523; J.L.C. Wright, *J. Nat. Prod.*, 1984, **47**, 893; P. Keil, *et al.*, *Acta Chem. Scand.*, 1986, **B40**, 555; M.V. Laycock, *et al.*, *Can. J. Chem.*, 1986, **64**, 1312; P.B. Holst, *et al.*, *J. Nat. Prod.*, 1994, **57**, 997, 1310). A recent example is flustramine E (166), which has activity against the fungi *Botrytis cinera* and *Rhizotonia solani*. The tunicate *Ciona savignyi* produces urochorda-mines A (167) and B (epimer), which are novel pteridine alkaloids. Interestingly, only A is active (at 2 ng/mL) in promoting larval settlement and metamorphosis of its own larvae (S. Tsukamoto, *et al.*, *Tetrahedron Lett.*, 1993, **34**, 4819). The bryozoan *Securiflustra securifrons* has yielded the unusual securamines A-D, e.g., (168) and (169) (L. Rahbaek, *et al.*, *J. Org. Chem.*, 1996, **61**, 887).

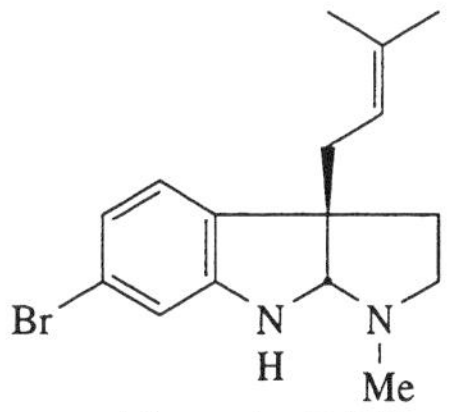

Flustramine E (166)

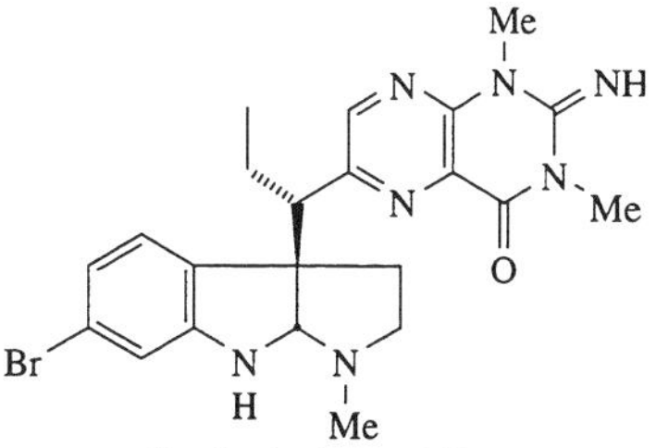

Urochordamine A (167)

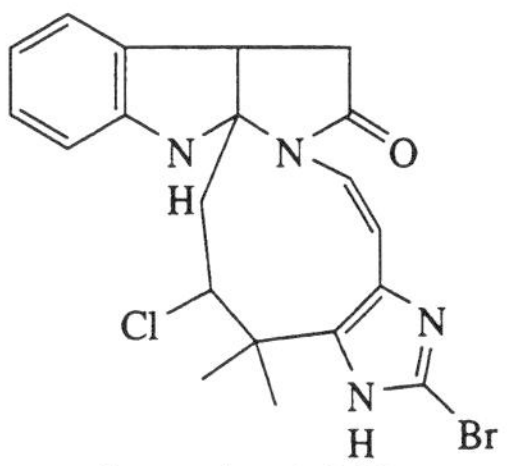

Securamine A (168)

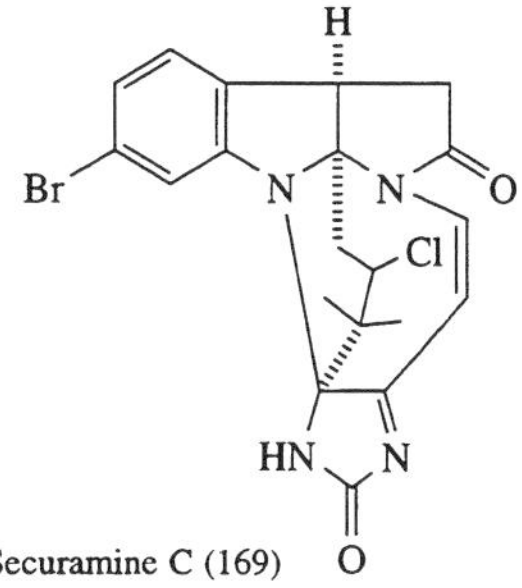

Securamine C (169)

94

iii. β-Carboline and reduced β-carbolines

Although this topic was not covered in the last volume, a very large number of simple β-carbolines have been discovered since 1980.

Xestoamine (170) is a sponge metabolite from *Xestospongia* sp. (J.-C. Quirion, *et al.*, *J. Nat. Prod.*, 1992, **55**, 1505), and the structure of nazlinin (171), an alkaloid isolated from *Nitraria schoberi* (L. Üstünes, *et al.*, *J. Nat. Prod.*, 1991, **54**, 959), has been revised following its synthesis (M.J. Wanner, *et al.*, *Chem. Commun.*, 1993, 174). Isoborrerine (172) is found in *Flindersia fournieri* (F. Tillequin and M. Koch, *Phytochemistry*, 1980, **19**, 1282), and the synthetic product (173) has now been isolated from *Adhatoda vasica* (M.P. Jain, *et al.*, *ibid.*, 1980, **19**, 1980). The simple (174) is found in *Alstonia venenata* (J. Banerji, *et al.*, *ibid.*, 1982, **21**, 2765). Harmine carboxamide (175) is produced by *Strychnos potatorum* (G. Massiot, *et al.*, *ibid.*, 1992, **31**, 2873), and lycii alkaloid, which is found in *Lycium chinense* and several other plants, has been revised to structure (176) (F. Bracher and D. Hildebrand, *Ann.*, 1993, 1335). This same 1-acetyl-β-carboline (176) has been isolated from the sponge *Tedania ignis* (R.L. Dillman and J.H. Cardellina II, *J. Nat. Prod.*, 1991, **54**, 1056). Carbolines (177), (178), and the N_b-oxides of both are found in *Hannoa klaineana* (L. Lumonadio and M. Vanhaelen, *Phytochemistry*, 1984, **23**, 453). Hydroxyethyl analogue (179) has been isolated from *Soulamea fraxinifolia* (B. Charles, *et al.*, *J. Nat. Prod.*, 1986, **49**, 303). The Chinese medicinal plant *Arenaria kansuensis* produces arenarines A (180), B (181), C, and D. This is the first report of β-carbolines from genus *Arenaria* (F. Wu, *et al.*, *Chem. Pharm. Bull.*, 1989, **37**, 1808). The plant *Ailanthus altissima* contains methyl 6-methoxy-β-carboline-1-carboxylate, which does not appear to be an artefact formed from 1-methoxycathin-6-one (E. Varga, *et al.*, *Planta Med.*, 1980, **40**, 337).

Xestoamine (170) R = NMe₂
Nazlinin (171)   R = (CH₂)₄NH₂

Isoborrerine (172)

(175) R = CONH₂
(176) R = COCH₃
(177) R = CH₂CH₃
(178) R = CH₂CH₂CO₂Et
(179) R = CH₂CH₂OH
(180) R = COCH₂OMe
(181) R = CHOHCH₂OMe

(173)

(174)

Isoharmine (182) is found in *Peganum harmala* (M.T. Ayoub and L.J. Rashan, *Phytochemistry*, 1991, **30**, 1046), and the dimethoxylated carbolines (183)-(186) are produced by *Roemeria hybrida* (B. Gözler and M. Shamma, *J. Nat. Prod.*, 1990, **53**, 740). In addition to yielding several known β-carbolines, the bryozoan *Cribricellina cribraria* contains (187) and (188) (M.R. Prinsep, *et al.*, *ibid.*, 1991, **54**, 1068). Three novel chlorinated bauerines A-C (189)-(191) have been characterized from the blue-green alga *Dichothrix baueriana* (L.K. Larsen, *et al.*, *ibid.*, 1994, **57**, 419). These compounds display activity against Herpes simplex virus. The plant *Evodia rutaecarpa*, the fruit of which is used as a Chinese drug, has furnished goshuyuamide I (192) (N. Shoji, *et al.*, *ibid.*, 1989, **52**, 1160). Cecilin (193), which is structurally related to roecarboline, is found in *Aniba santalodora* (L.M.G. Aguiar, *et al.*, *Phytochemistry*, 1980, **19**, 1859), and the novel cyclized harmalidine (194) is produced by *Peganum harmala* (S. Siddiqui, *et al.*, *ibid.*, 1987, **26**, 1548). A *Streptomyces* sp. produces several oxopropalines, such as (195), some of which are attached to α-rhamnose (N. Abe, *et al.*, *J. Antibiot.*, 1993, **46**, 1678).

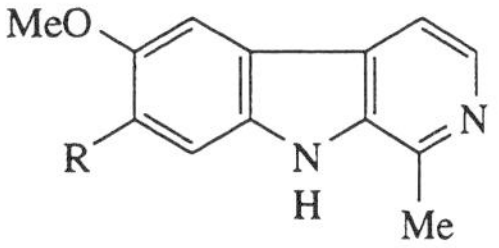

Isoharmine (182)   R = H
Roeharmine (183)   R = OMe

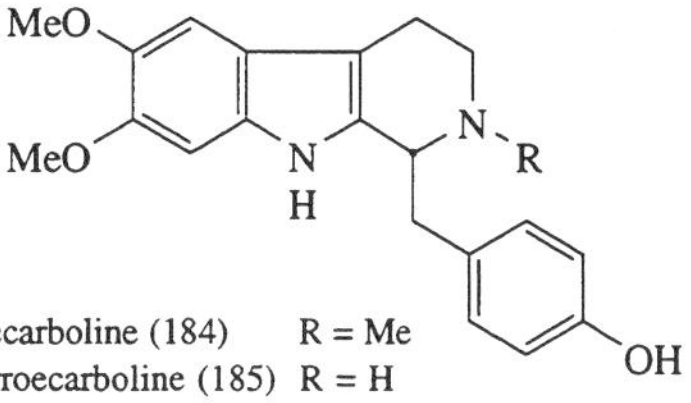

Roecarboline (184)   R = Me
Norroecarboline (185)   R = H

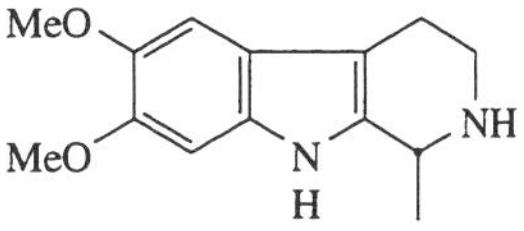

(186)

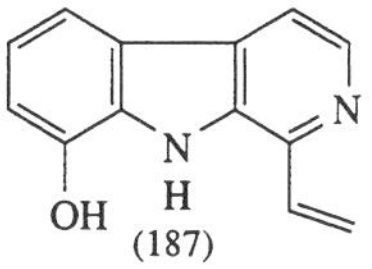

(187)

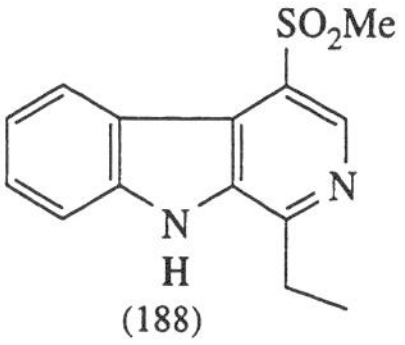

(188)

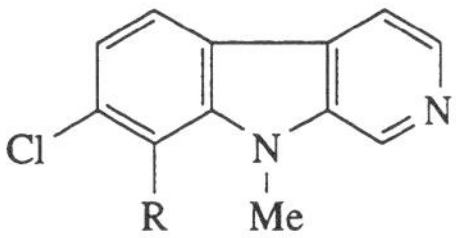

Bauerine A (189)   R = H
Bauerine B (190)   R = Cl

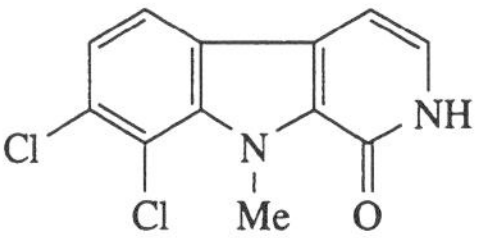

Bauerine C (191)

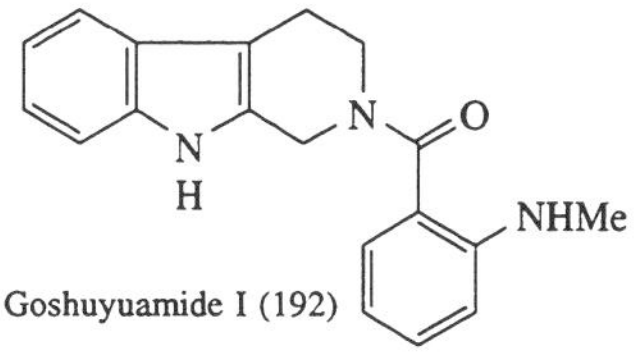

Goshuyuamide I (192)

96

Cecilin (193)

Harmalidine (194)

Oxopropaline D (195)

The plant *Annona montana* has furnished the pyrimidine-containing annomontine (196) and 6-methoxyannomontine (M. Leboeuf, *et al.*, *J. Chem. Soc., Perkin Trans. 1*, 1982, 1205), and a series of quinoline-substituted carbolines, e.g., dihydroisokomarovine (197), have been identified in *Nitraria komarovii* (T.S. Tulyaganov and S. Yu. Yunusov, *Chem. Nat. Cpds.*, 1990, **26**, 49). The most famous example of this type of compound is lavendamycin (198), isolated from *Streptomyces lavendulae* (T.W. Doyle, *et al.*, *Tetrahedron Lett.*, 1981, **22**, 4595; D.M. Balitz, *et al.*, *J. Antibiot.*, 1982, **35**, 259). The furocarbolines (199) and (200) are found in soy sauce and other sources (S. Nakatsuka, *et al.*, *Tetrahedron Lett.*, 1986, **27**, 3399). The first γ-carboline to be isolated from plant sources is isoperlolyrine (201) (*Gloriosa superba*) (S. Dvorackova, *et al.*, *Coll. Czech. Chem. Commun.*, 1984, **49**, 1536). Salted radish roots (*Raphanus sativus*) have yielded (202) (Y. Ozawa, *et al.*, *Agric. Biol. Chem.*, 1990, **54**, 1241), and another novel β-carboline is hyrtiomanzamine (203) from the Red Sea sponge *Hyrtios erecta*. This compound displays immunosuppressive activity (M.L. Bourguet-Kondracki, *et al.*, *Tetrahedron Lett.*, 1996, **37**, 3457). The new quinazolinocarboline alkaloid 7,8-dehydro-1-hydroxyrutaecarpine is found in *Vepris louisii* (J.F. Ayafor, *et al.*, *Phytochemistry*, 1982, **21**, 2733), and 7-hydroxyrutaecarpine is found in *Tetradium glabrifolium* and *T. ruticarpum* (T.-S. Wu, *et al.*, *Heterocycles*, 1995, **41**, 1071). Cuscutamine is a novel lactam from *Cuscuta chinensis* (S. Yahara, *et al.*, *Phytochemistry*, 1994, **37**, 1755).

Annomontine (196)

Dihydroisokomarovine (197)

Lavendamycin (198)

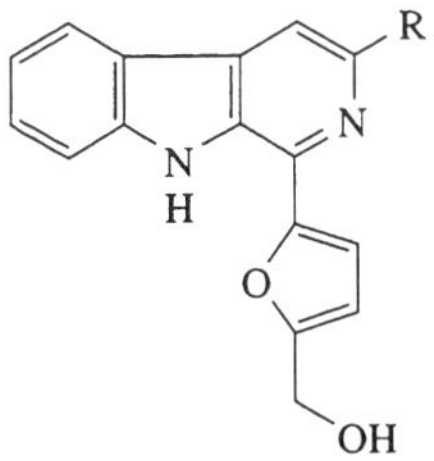

Flazin (199) R = CO₂H
Perlolidin (200) R = H

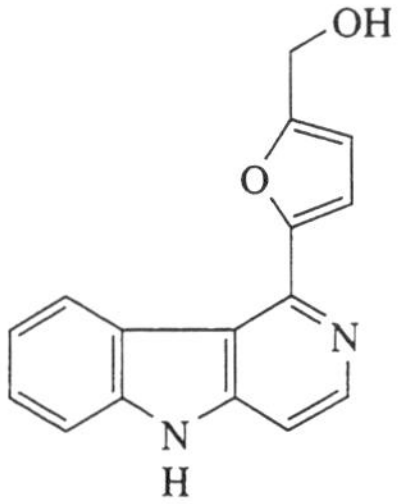

(201)

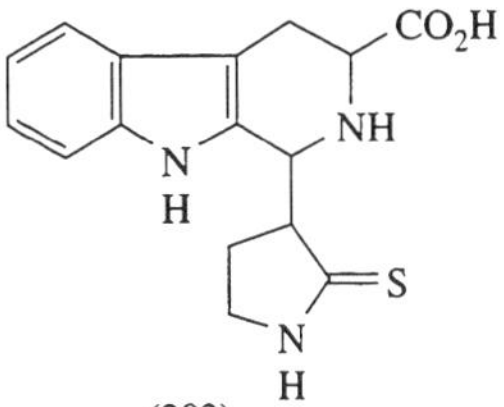

(202)

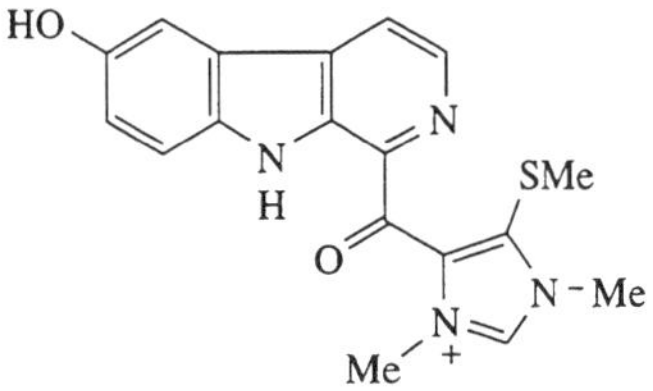

Hyrtiomanzamine (203)

Fascaplysin (204) is one of several related pentacyclic carbolines found in the sponge *Fascaplysinopis reticulata* (D.M. Roll, *et al.*, *J. Org. Chem.*, 1988, **53**, 3276; C. Jiménez, *et al.*, *ibid.*, 1991, **56**, 3403). Novel terpenoid carboxylates are found with fascaplysin. The West African medicinal plant *Cryptolepis sanguinolenta* has provided several unique tetracyclic carboline alkaloids. These are the δ-carboline cryptolepine (205) (S.Y. Ablordeppey, *et al.*, *Planta Med.*, 1990, **56**, 416), the demethyl analogue quindoline (T.D. Spitzer, *et al.*, *J. Heterocycl. Chem.*, 1991, **28**, 2065), and isocryptolepine (206) (J.-L. Pousset, *et al.*, *Phytochemistry*, 1995, **39**, 735). The New Caledonian gorgonian *Villagorgia rubra* has afforded the acetylcholine antagonists villagorgin A (207) and B (A. Espada, *et al.*, *Tetrahedron Lett.*, 1993, **34**, 7773). Carbolines (208) and (209) (and their epimers) are produced by *Clerodendron trichotomum* (Y. Toyoda, *et al.*, *Chem. Lett.*, 1982, 903). The Turkish plant *Roemeria hybrida* contains ten proaporphine-tryptamine dimers, one of which is roehybridine (210) (B. Gözler, *et al.*, *J. Nat. Prod.*, 1990, **53**, 675).

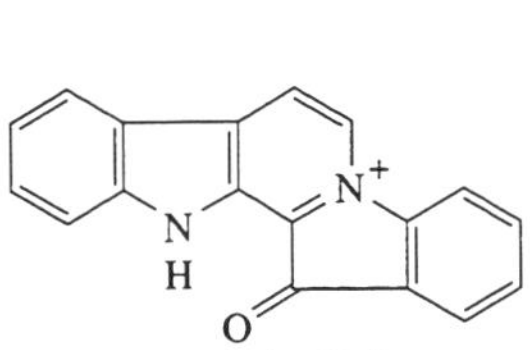

Fascaplysin (204)

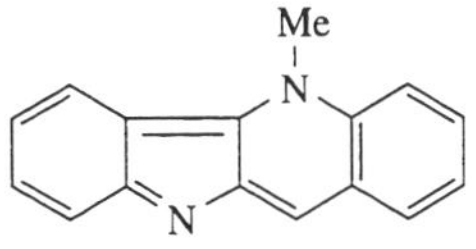

Cryptolepine (205)

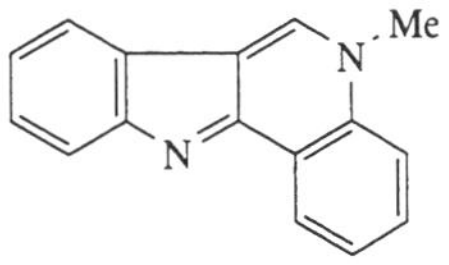

Isocryptolepine (206)

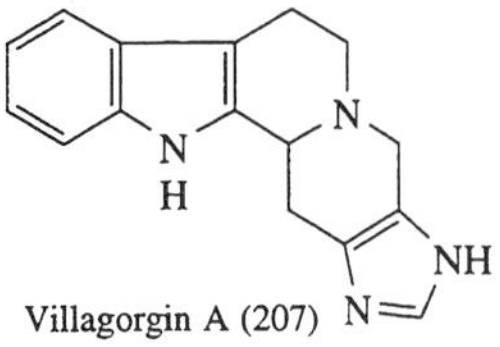

Villagorgin A (207)

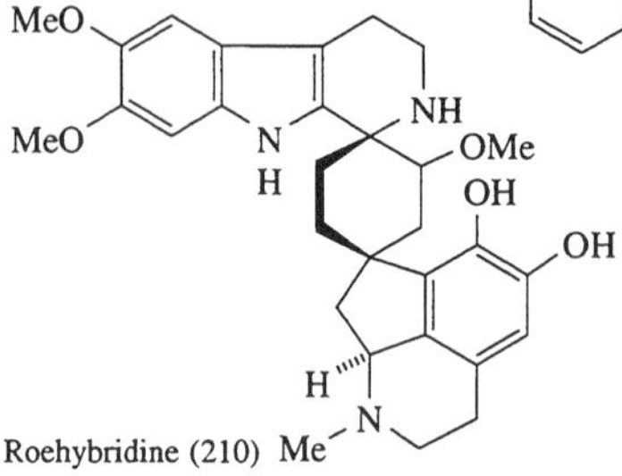

Roehybridine (210)

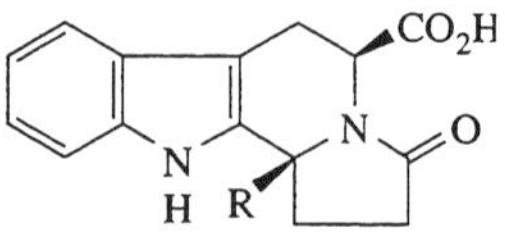

(208) R = H
(209) R = CO₂H

Since an excellent review on the canthin-6-one alkaloids is available (T. Ohmoto and K. Koike in "The Alkaloids," Vol. 36, Chapter 3, 1989), coverage of these alkaloids will start with the 1989 literature.

The plant *Picrasma quassioides* is an amazing source for the production of carboline alkaloids. Recent examples include picrasidine W (211) and Q (212) (H.-Y. Li. *et al., Chem. Pharm. Bull.*, 1993, **41**, 1807). Several novel simple carbolines (213)-(215) are also found in this plant (K. Koike, *et al., Phytochemistry*, 1990, **29**, 3060). *Eurycoma longifolia* has yielded 9-hydroxycanthin-6-one and the corresponding N(3)-oxide (L.B.S. Kardone, *et al., J. Nat. Prod.*, 1991, **54**, 1360), *Picrolemma granatensis* contains 8-hydroxy-9-methoxycanthin-6-one and 9-methoxycanthin-6-one N(3) oxide (E. Rodrigues Fo. *et al., Phytochemistry*, 1992, **31**, 2499), and *Aerva lanata* has produced 10-methoxy- and 10-hydroxycanthin-6-one (aervin) (G. Zapesochnaya, *et al., Planta Med.*, 1992, **58**, 192). The plant *Quassia amara* contains 2-methoxycanthin-6-one (V.C.O. Njar, *et al., ibid.*, 1993, **59**, 259), and *Eurycoma longifolia* contains four new dioxygenated canthin-6-ones in addition to (216) (K. Mitsunaga, *et al., Phytochemistry*, 1994, **35**, 799). In addition to producing the novel 11-hydroxycanthin-6-one N(3)-oxide, *Brucea mollis* var. *tonkinensis* has afforded two novel canthin-6-one diglucosides at C-11 and C-5, bruceollines A and B, respectively (Y. Ouyang, *et al., ibid.*, 1994, **36**, 1543).

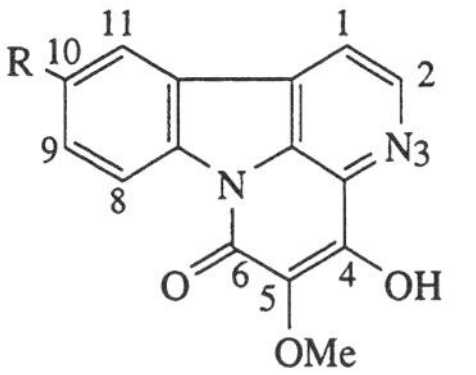

Picrasidine W (211) R = OH
Picrasidine Q (212) R = H

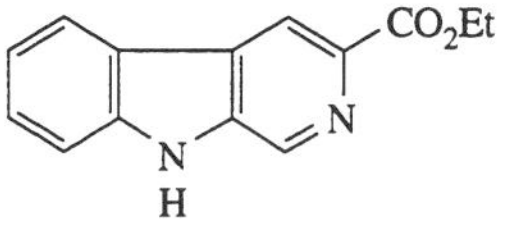

(213)

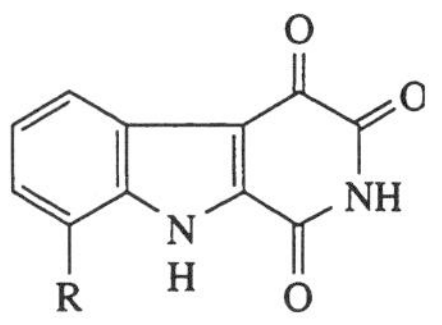

(214) R = H
(215) R = OMe

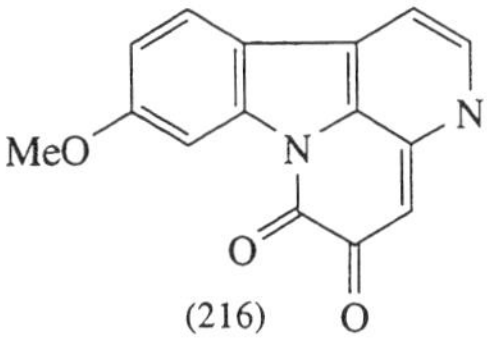

(216)

Not surprisingly, a large of brominated carbolines are found in marine organisms, and *Eudistoma* tunicates are extraordinary biofactories for the production of these unique compounds. For example, the Okinawan *Eudistoma glaucus* has yielded six such metabolites, e.g., eudistomidin A (217) (J. Kobayashi, *et al., Tetrahedron Lett.*, 1986, **27**, 1191; *J. Org. Chem.*, 1990, **55**, 3666; O. Murata, *et al., Tetrahedron Lett.*, 1991, **32**, 3539), *E. olivaceum* produces at least 20 eudistomins, e.g., (218) (K.L. Rinehart, *et al., J. Am. Chem. Soc.*, 1987, **109**, 3378; K.F. Kinzer and J.H. Cardellina, *Tetrahedron Lett.*, 1987, **28**, 925), and both *E. fragum* (C. Debitus, *et al., J. Nat. Prod.*, 1988, **51**, 799) and *E. album* produce unique metabolites, such as eudistalbin A (219), which has potent cytotoxicity, isolated from the latter tunicate (S.A. Adesanya, *et al., ibid.*, 1992, **55**, 525). Biosynthetic studies demonstrate that both 5-bromotryptophan and 5-bromotryptamine are incorporated into eudistomin H (G.Q. Shen and B.J. Baker, *Tetrahedron Lett.*, 1994, **35**, 4923). Eudistomins have also been discovered in the New Zealand ascidian *Ritterella sigillinoides*, and eudistomin K sulfoxide is active against both Polio and Herpes simplex viruses (R.J. Lake, *et al., Tetrahedron Lett.*, 1988, **29**, 2255; *Aust. J. Chem.*, 1989, **42**, 1201). Lissoclin C (220) is found in the ascidian *Lissoclinum* sp. (P.A. Searle and T.F. Molinski, *J. Org. Chem.*, 1994, **59**, 6600) and the tunicate *Pseudodistoma arborescens* has yielded arborescidines A-D, e.g., (221) and (222) (M. Chbani, *et al., J. Nat. Prod.*, 1993, **56**, 99). The hydroid *Aglaophenia pluma* contains a mixture of several novel brominated β-carbolines, the major component of which is (223) (A. Aiello, *et al., Tetrahedron*, 1987, **43**, 5929). The tunicate *Didemnum* sp. produces the novel imidazole-carbolines didemnolines A-D, e.g., (224) (R.W. Schumacher and B.S. Davidson, *Tetrahedron*, 1995, **51**, 10125).

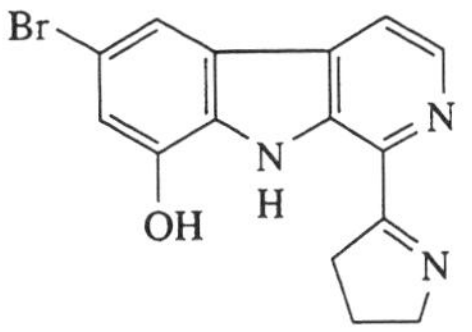

Eudistomidin A (217)

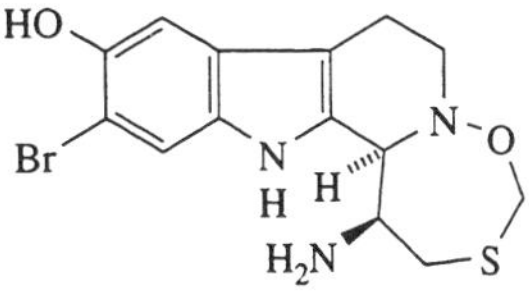

Eudistomin C (218)

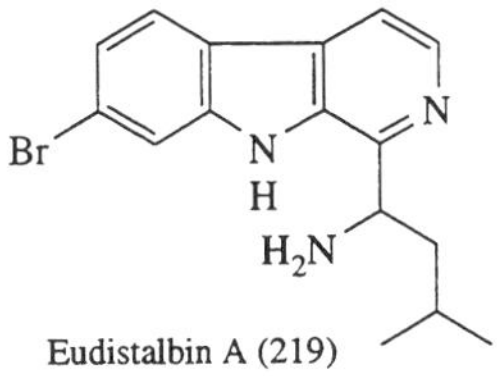

Eudistalbin A (219)

100

Lissoclin C (220)

Arborescidine A (221)

Arborescidine B (222)

(223)

Didemnoline A (224)

Perhaps the most exciting development in β-carboline natural products research in recent years was the discovery of the manzamines in various sponges. The number of these extraordinarily complex metabolites continues to expand.

Manzamine A (225), the prototype of this class, was isolated from a *Haliclona* sp. sponge (R. Sakai, *et al.*, *J. Am. Chem. Soc.*, 1986, **108**, 6404), and, at about the same time, keromamine A was found in a *Pellina* sp. sponge (H. Nakamura, *et al.*, *Tetrahedron Lett.*, 1987, **28**, 621). There followed manzamines B and C (226) from the same sponge (R. Sakai, *et al.*, *ibid.*, 1987, **28**, 5493), and several other oxygenated and other manzamine A derivatives from sponges *Amphimedon* sp. (J. Kobayashi, *J. Nat. Prod.*, 1994, **57**, 1737), *Pachypellina* sp. (T. Ichiba, *et al.*, *ibid.*, 1994, **57**, 168), *Petrosia* sp. and *Cribochalina* sp. (P. Crews, *et al.*, *Tetrahedron*, 1994, **50**, 13567), *Xestospongia* sp. (M. Kobayashi, *et al.*, *ibid.*, 1995, **51**, 3727; T. Ichiba, *et al.*, *Tetrahedron Lett.*, 1988, **29**, 3083), and manzamines H and J from *Ircinia* sp. (K. Kondo, *et al.*, *J. Org. Chem.*, 1992, **57**, 2480).

Manzamine A (225)

Manzamine C (226)

(b) Compounds containing an isoprene- (but not terpene-) derived moiety

(i) The ergot alkaloids

An excellent review covers the ergot alkaloid literature through 1988 (I. Ninomiya and T. Kiguchi, in "The Alkaloids," Vol. 38, Chapter 1, 1990), so coverage herein will commence after 1988.

Two new fructosides of chanoclavine have been isolated from *Claviceps fusiformis* (M. Flieger, *et al.*, *J. Nat. Prod.*, 1990, **53**, 171) and *C. paspali* has afforded 8-hydroxyergine (227) as a pair of epimers (M. Flieger, *et al.*, *ibid.*, 1989, **52**, 1003). When sandbur grass weed (*Cenchrus echinatus*) is infected with the fungus *Balansia obtecta*, it poses a severe threat to livestock. Ergobalansine (228) has been identified as a metabolite of this fungus (R.G. Powell, *et al.*, *ibid.*, 1990, **53**, 1272). The C-8 epimer ergobalansinine is produced by seeds of *Ipomoea piurensis*, a morning glory species from Ecuador (K. Jenett-Siems, *et al.*, *ibid.*, 1994, **57**, 1304). The related ergobine (229) is found in *Claviceps purpurea* (N.C. Perellino, *et al.*, *ibid.*, 1993, **56**, 489) and *C. paspali* has furnished the novel epimeric *cis-* and *trans-*10-hydroxypaspalic acid amides (M. Flieger, *et al.*, *ibid.*, 1993, **56**, 810).

8-Hydroxyergine (227)

Ergobalansine (228)

Ergobine (229)

(ii) Mould metabolites

An extraordinary large number of new mould metabolites have been described since the last coverage. These tryptophan-diketopiperazine compounds usually contain isopentenyl and amino acid units, although exceptions are known.

The simple terezine D (230) is produced by *Sporormiella teretispora* (Y.

Wang, *et al.*, *J. Nat. Prod.*, 1995, **58**, 93), and the first sulfur-containing diketopiperazines from marine sources are maremycins A (231) and B (232) from a marine *Streptomyces* sp. (W. Balk-Bindseil, *et al.*, *Ann.*, 1995, 1291). Another simple example is FR 900452 (233) from *S. phaeofaciens* (S. Takase, *et al.*, *J. Org. Chem.*, 1987, **52**, 3485). Tryprostatins A (234) and B (235) from *Aspergillus fumigatus*, which inhibit the mammalian cell cycle, may be shunt metabolites of verruculogen and fumitremorgin (C.-B. Cui, *et al.*, *J. Antibiot.*, 1995, **48**, 1382). This same mould produces 12,13-dihydroxyfumitremorgin C (236) (W.-R. Abraham and H.-A. Arfmann, *Phytochemistry*, 1990, **29**, 1025). Seven new fumiquinazolines A-G, e.g., (237), were isolated from *A. fumigatus* found growing in the saltwater fish *Pseudolabrus japonicus* (C. Takahashi, *et al.*, *J. Chem. Soc., Perkin Trans. 1*, 1995, 2345). Structurally similar metabolites to the fumi-quinazolines, but which cannot be shown for reasons of space, are fiscalins A-C (*Neosartorya fischeri*) (S.-M. Wong, *et al.*, *J. Antibiot.*, 1993, **46**, 545), asperlicins A-E (*Aspergillus alliaceus*) (J.M. Liesch, *et al.*, *ibid.*, 1985, **38**, 1638; 1988, **41**, 878), glyantrypine (*A. clavatus*) (J. Penn, *et al.*, *J. Chem. Soc., Perkin Trans. 1*, 1992, 1495), spiroquinazoline (*A. flavipes*), tilivalline (*Klebsiella pneumoniae* var. *oxytoca*) (N. Mohr and H. Budzikiewicz, *Tetrahedron*, 1982, **38**, 147), and citreoindole (*Penicillium citreo-viride*) (K. Matsunaga, *et al.*, *Tetrahedron Lett.*, 1991, **32**, 6883). These metabolites typically are either Substance P inhibitors or antagonists of cholecystokinin.

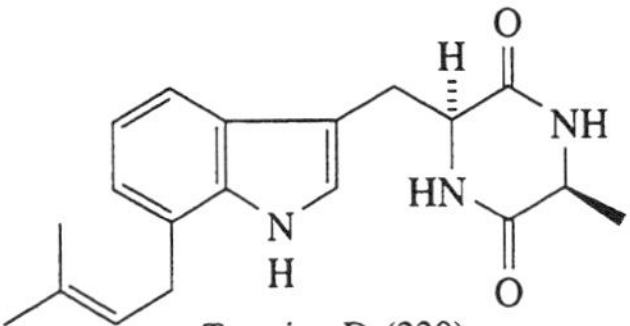

Terezine D (230)

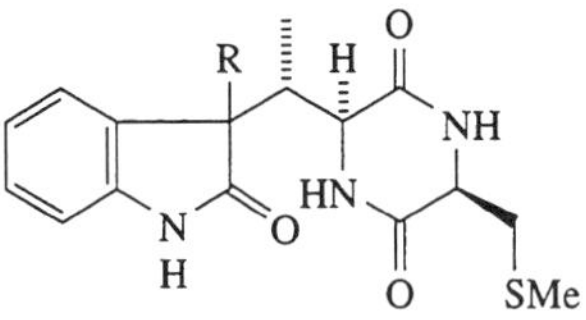

Maremycin A (231) R = α-OH

Maremycin B (232) R = β-OH

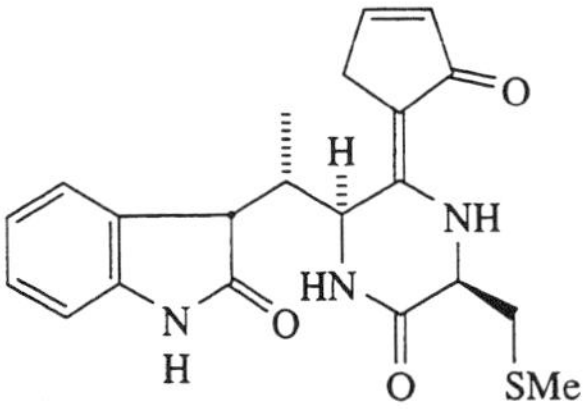

FR 900452 (233)

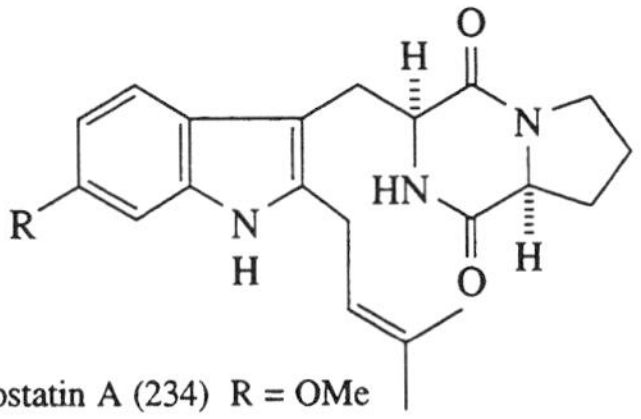

Tryprostatin A (234) R = OMe

Tryprostatin B (235) R = H

(236)

Fumiquinazoline A (237)

The novel (238) has been isolated from *Aspergillus ochraceus* (F.S. de Guzman, *et al.*, *J. Nat. Prod.*, 1992, **55**, 931). The N_a-formylroquefortine has been found in cultures of *Penicillium verrucosum* var. *cyclopium* (A. Musuku, *et al.*, *ibid.*, 1994, **57**, 983), and both meleagrin (*P. meleagrinum*) (K. Kawai, *et al.*, *Chem. Pharm. Bull.* 1984, **34**, 94), and neoxaline (*Aspergillus japonicus*) (Y. Konda, *et al.*, *ibid.*, 1980, **28**, 2987) are close derivatives of oxaline. Several new eserine-type mould metabolites are known, including verrucofortine (239) from the mouldy hay fungus *Penicillium verrucosum* var. *cyclopium* (R.P. Hodge, *et al.*, *J. Nat. Prod.*, 1988, **51**, 66), fructigenines A (240) and B(239) (= verrucofortine) from *P. fructigenum* (K. Arai, *et al.*, *Chem. Pharm. Bull.*, 1989, **37**, 2937), three new ardeemins from *Aspergillus fischeri* var. *brasiliensis* (J.E. Hochlowsky, *et al.*, *J. Antibiot.*, 1993, **46**, 380), and aszonalenin from *A. zonatus* (Y. Kimura, *et al.*, *Tetrahedron Lett.*, 1982, **23**, 225). A new oxygenated derivative of verruculogen, which causes severe tremorgenic reactions in mice, is produced by *Penicillium verruculosum* (M. Uramoto, *et al.*, *Heterocycles*, 1982, **17**, 349). Okaramines A-F, e.g., (241), have been isolated from *Penicillium simplicissimum* (S. Murao, *et al.*, *Agric. Biol. Chem.*, 1988, **52**, 885; H. Hayashi, *et al.*, *ibid.*, 1988, **52**, 2131; 1989, **53**, 461; 1992, **55**, 3143; International Research Congress of Natural Products, Nova Scotia, July 31-August 4, 1994, Abs. P. 82). Okaramine B (241) is insecticidal against silkworms at 0.1 ppm.

(238)

Verrucofortine (239) R = *i*-Pr
Fructigenine A (240) R = Pr

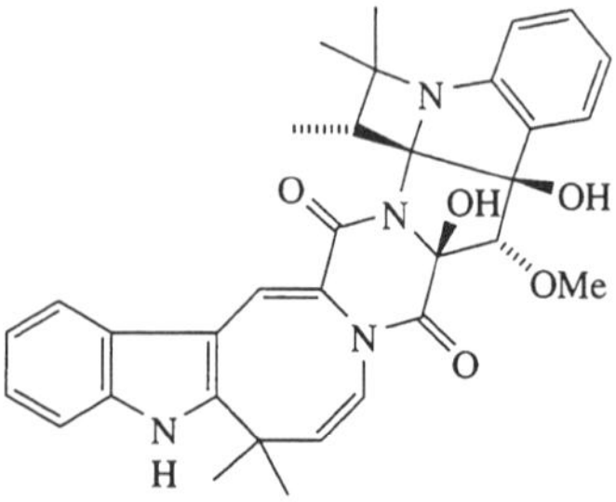

Okaramine B (241)

The absolute configuration of paraherquamide (*Penicillium paraherquei*), first isolated in 1981, has been determined (242) (T.A. Blizzard, *et al.*, *J. Org. Chem.*, 1989, **54**, 2657). A number of closely related analogues have recently been isolated from *P. charlesii* (J.G. Ondeyka, *et al.*, *J. Antibiot.*, 1990, **43**, 1375), and *Penicillium* sp. (S.E. Blanchflower, *et al.*, *ibid.*, 1991, **44**, 492; 1993, **46**, 1355). The latter study reports the isolation of VM 55599 (243), which may be a biogenetic precursor to the brevianamide, paraherquamide, and marcfortine families of mould metabolites.

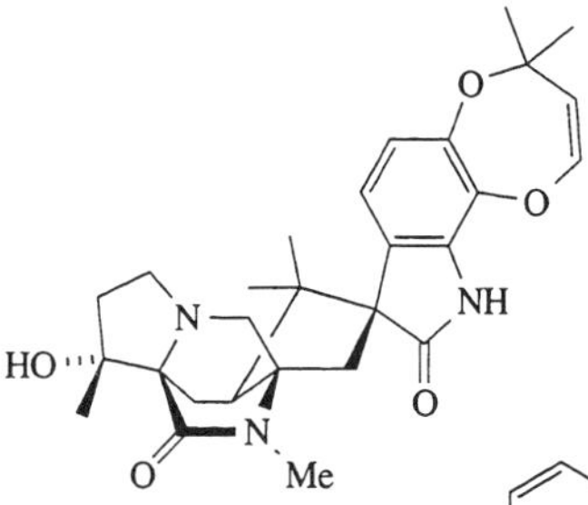

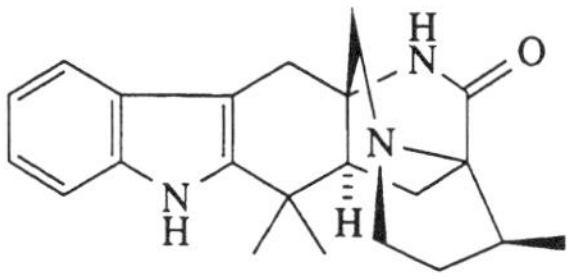

VM 55599 (243)

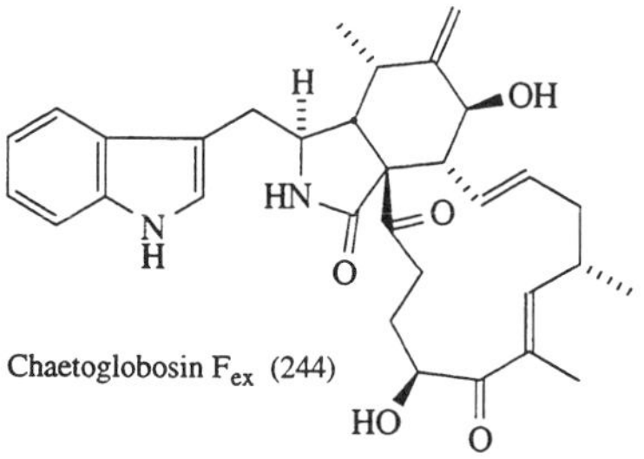

Paraherquamide (242)

Chaetoglobosin F$_{ex}$ (244)

Some tryptamine-derived fungal metabolites do not fit the structural criteria of the compounds discussed above. These include the cytochalasans and several new examples of this group have been discovered from *Chaeto-mium globosum* (A. Probst and C. Tamm, *Helv. Chim. Acta*, 1981, **64**, 2056), *Diplodia macrospora* (A. Probst and C. Tamm, *ibid.*, 1982, **65**, 1543), and *Chaetomium subaffine*, which has yielded chaetoglobosin F$_{ex}$ (244) (H. Oikawa, *et al.*, *Biosci. Biotechnol. Biochem.*, 1993, **57**, 628).

Another biologically important group of tryptophan metabolites is the tryptophan-valine derived indolactam (245) class of natural products. Teleocidin B (246) was first described in 1962 from several strains of *Streptomyces* (M. Takashima, *et al.*, *Agric. Biol. Chem.*, 1962, **26**, 660), and this was followed by the isolation of lyngbyatoxin A (247) from the blue-green alga *Lyngbya majuscula* (J.H. Cardellina II, *et al.*, *Science*, 1979, **204**, 193). This latter organism is responsible for the severe dermatitis, "swimmer's itch," in Hawaii. Both teleocidin B and lyngbyatoxin are highly inflammatory, vesicatory, and acutely toxic to mice and fish. The absolute configurations of these compounds (shown) were established later (S. Sakai, *et al.*, *Tetrahedron Lett.*, 1986, **27**, 5219; Y. Endo, *et al.*, *Tetrahedron*, 1986, **42**, 5905; *Chem. Pharm. Bull.*, 1984, **32**, 358).

Indolactam V (245)

Teleocidin B (246)
(= Olivoretin D)

Lyngbyatoxin A (247) (19*R*)
(= Teleocidin A-1)

A number of related compounds have been described subsequently. Indolactam V (245) was isolated from *Streptoverticillium blastmyceticum* (K. Irie, *et al.*, *Agric. Biol. Chem.*, 1984, **48**, 1269). Olivoretin A is the methyl ether of olivoretin D which is identical to teleocidin B (*S. olivoreticuli*) (S. Sakai, *et al.*, *Chem. Pharm. Bull.*, 1984, **32**, 354), and the same microorganism produces olivoretin B, which is the C-25 epimer of olivoretin A, and olivoretin C (248) (Y. Hitotsuyanagi, *et al.*, *ibid.*, 1984, **32**, 3774). The other possible teleocidins (B-1, B-2, B-3), involving epimers at C-19 and C-25, are produced by *Streptomyces mediocidicus* (Y. Hitotsuyanagi, *et al.*, *ibid.*, 1984, **32**, 4233). This organism also produces teleocidin B-4, which is the same as teleocidin B (= olivoretin D) (246), and des-*N*-methylteleocidin B-4 (S. Sakai, *et al.*, *ibid.*, 1986, **34**, 4883). This latter study also discovered olivoretin E (249) (*Streptoverticillium olivoreticuli*) containing a *tert*-butyl group. Lyngbyatoxins B and C are hydroxylated at C-25 and C-26 in (247) (N. Aimi, *J. Nat. Prod.*, 1990, **53**, 1593). Blastmycetins B (250), C (C-3 epimer of B), and F (251) are found in *S. blastmyceticum* (K. Irie, *et al.*, *Agric. Biol. Chem.*, 1987, **51**, 285; *J. Nat. Prod.*, 1994, **57**, 363). This organism also produces the novel 7-geranyl-indolactam V, which is a strong tumor promoter, and indolactam I, which

106

has a *sec*-butyl group in place of isopropyl in indolactam V (245) (K. Irie, *et al.*, *Tetrahedron Lett.*, 1990, **31**, 7337). An actinomycetes has yielded 14-*O*-acetylindolactam V (K. Irie, *et al.*, *Agric. Biol. Chem.*, 1984, **48**, 1269). Pendolmycin (252) and methylpendolmycin (253) are found in cultures of *Nocardiopsis* sp. (Y. Yamashita, *et al.*, *J. Nat. Prod.*, 1988, **51**, 1184; H.H. Sun, *et al.*, *ibid.*, 1991, **54**, 1440). The latter compound inhibits phorbol ester binding to protein kinase C.

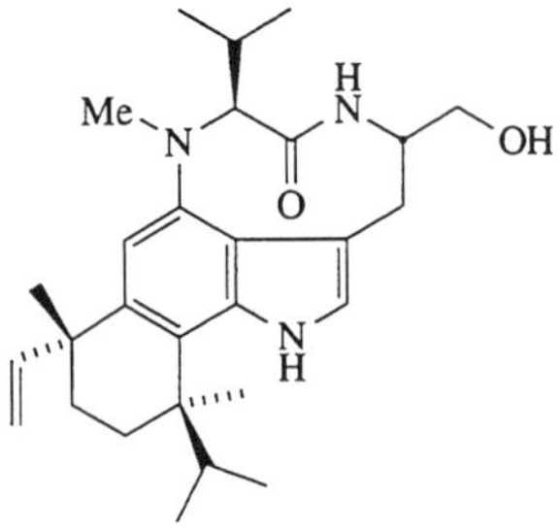

Olivoretin C (248)

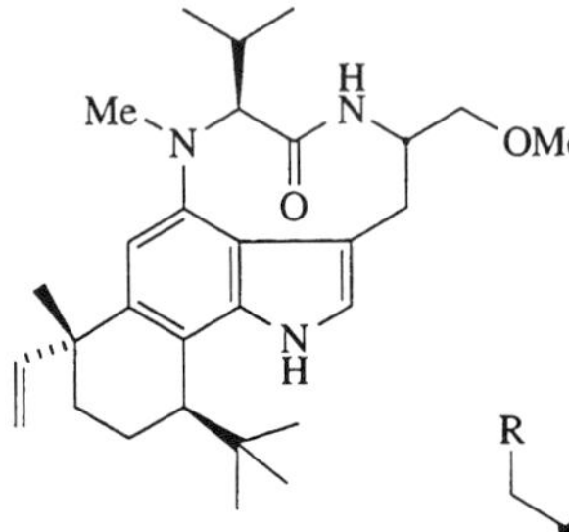

Olivoretin E (249)

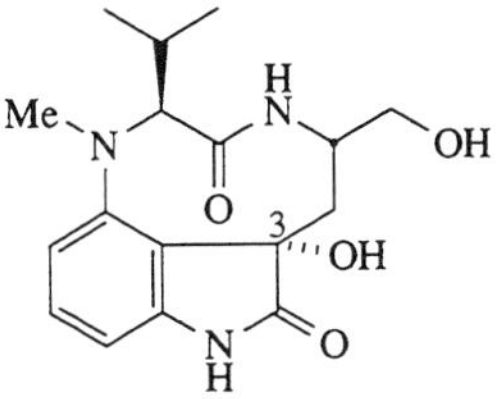

Blastmycetin B (250)

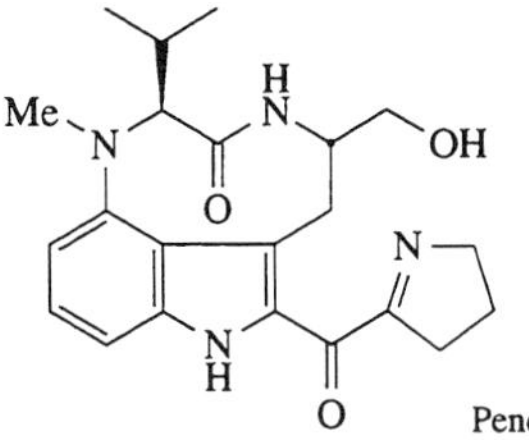

Blastmycetin F (251)

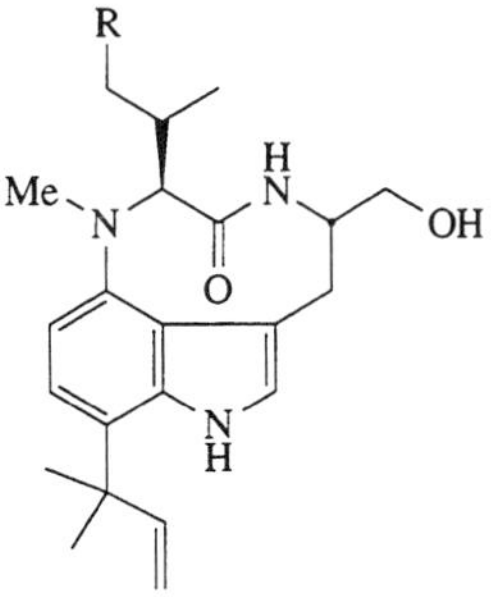

Pendolmycin (252)   R = H
Methylpendolmycin (253) R = Me

(c) Compounds containing a terpene-derived moiety

(i) Alkaloids of the Corynanthé-Strychnos unit
  (1) Glycoalkaloids
Several new and biogenetically significant glycoalkaloids have been reported since the last volume. Palicoside (254) has been isolated from the Brazilian plant *Palicourea marcgravii* and the structure established by chemical conversion (120 °C, DMSO) to strictosamide. Thus, palicoside is the N_b-methyl carboxylic acid of strictosidine (H. Morita, *et al.*, *Planta Med.*, 1989, **55**, 288). The first example of a bioside congener of a monoterpene indole alkaloid glucoside is the 6'-α-D-glucoside of strictosidinic acid, called hunterioside, which was found in *Hunteria zeylanica* (S. Subhadhira-sakul, *et al.*, *Chem. Pharm. Bull.*, 1994, **42**, 991). Several new carboline derivatives have also been discovered. The plant *Nauclea diderrichii* produces desoxycordifolinic acid (255) (A.O. Adeoye and R.D. Waigh,

Phytochemistry, 1983, **22**, 2097), and the novel isopauridianthoside (256), isolated from *Pauridiantha lyallii*, is the first example of a naturally occurring epimer of secologanoside (J. Levesque, *et al.*, *J. Nat. Prod.*, 1983, **46**, 619). This same plant has yielded several lyaloside derivatives in which a feruloyl or sinapoyl group is attached to the 6'-position of glucose (J. Levesque, *et al.*, *Tetrahedron*, 1982, **38**, 1417). Similarly, *Uncaria rhynchophylla* produces rhynchophine (6'-feruloylvincoside lactam) (257) (N. Aimi, *et al.*, *Chem. Pharm. Bull.*, 1982, **30**, 4046). The plant *Nauclea orientalis* has furnished the new alkaloids 10-hydroxystrictosamide and 6'-*O*-acetylstrictosamide (C.A.J. Erdelmeier, *et al.*, *Planta Med.*, 1991, **57**, 149), and 10-hydroxyvincoside lactam is found in *Alangium lamarckii* (A. Itoh, *et al.*, *J. Nat. Prod.*, 1995, **58**, 1228). Finally, the unusual glucoalkaloid glabratine (258), which is related to vallesiachotamine, is isolated from *Uncaria glabrata* (D. Arbain, *Aust. J. Chem.*, 1993, **46**, 863), and the ophiorines (259) are found in *Ophiorrhiza japonica* and *O. kuroiwai* (N. Aimi, *et al.*, *Tetrahedron Lett.*, 1985, **26**, 5299). On treatment with CH_2N_2, they undergo a *retro*-Michael reaction to afford lyaloside. The novel glucoalkaloid sickingine (260), which is proposed to be a biosynthetic derivative of 5α-carboxystrictosidine, is found in *Sickinga tinctoria* and *S. williamsii* (R. Aquino, *et al.*, *Phytochemistry*, 1994, **37**, 1471).

Palicoside (254)

Desoxycordifolinic acid (255)

Isopauridianthoside (256)

Rhynchopine (257)

Glabratine (258)

Ophiorine A (259a) R_1 = H, R_2 = CO_2^-
Ophiorine B (259b) R_1 = CO_2^-, R_2 = H

Sickingine (260)

(2) Yohimbine, heteroyohimbine and secoyohimbine types and related oxindoles

This large and important group of alkaloids has been discussed in several reviews (C. Szántay, *et al.*, in "The Alkaloids," Vol. 27, Chapter 2, 1986; E.W. Baxter and P.S. Mariano, in "Alkaloids: Chemical and Biological Perspectives," Vol. 8, Chapter 3, 1992; I. Ninomiya and O. Miyata, in "Studies in Natural Products Chemistry," Vol. 3, p. 399, 1989; M. Lounasmaa in *ibid.*, Vol. 14, p. 703, 1994).

Several new yohimbine alkaloids have been isolated in recent years. The plant *Aspidosperma pruinosum* has afforded 10-methoxyyohimbine (261) (D.S. Nunes, *et al.*, *Phytochemistry*, 1992, **31**, 2507), 3-*epi*-β-yohimbine is found in *Rauwolfia linearifolia* (J.A. Pèrez, *et al.*, *ibid.*, 1991, **30**, 1352), and *Amsonia elliptica* has yielded 17-α-*O*-methylyohimbine (M. Sauerwein and K. Shimomura, *ibid.*, 1990, **29**, 3377). Somewhat earlier, it was found that *Aspidosperma oblongum* produced 13 new yohimbine and related alkaloids, including (261), 10-methoxy-β-yohimbine, 10-methoxy-17-*epi*-alloyohimbine, 19,20-dehydro-β-yohimbine, 19,20-dehydro-α-yohimbine, 3,4-dehydro-β-yohimbine, β-yohimbine oxindole, β-yohimbine pseudoindoxyl, and β-yohimbine *N*-oxide (G.M.T. Robert, *et al.*, *J. Nat. Prod.*, 1983, **46**, 708). The plant *Alstonia venenata* contains both the novel 16-*epi*-venenatine (262) and 16-*epi*-alstovenine (263) (A. Chatterjee, *et al.*, *Phytochemistry*, 1981, **20**, 1981). Most interesting is the isolation of yohambinine (264), the first example of a C-5 methyl yohimbanoid, from *Rauwolfia serpentina* (S. Siddiqui, *et al.*, *Tetrahedron Lett.*, 1987, **28**,

1311). The plant *Hunteria zeylanica* has furnished *epi*-yohimbol, one of the reduction products of yohimbone (L.S.R. Arambewela and F. Khuong-Huu, *Phytochemistry*, 1981, **20**, 349).

(261)

16-*epi*-Venenatine (262) C-3β

16-*epi*-Alstovenine (263) C-3α

Yohambinine (264)

A number of new heteroyohimbinoids have been isolated since the last review. The previously synthesized 17-hydroxy-16-decarbomethoxy-dihydro-*epi*-ajmalicine (265) has been found in *Hunteria zeylanica* (L.S.R. Arambewela and F. Khuong-Huu, *Phytochemistry*, 1981, **20**, 349). Amsosinine (266) is a novel hydroxylated dihydroajmalicine alkaloid from *Amsonia sinensis* (H. Liu, *et al.*, *Planta Med.*, 1991, **57**, 566; *Chin. Chem. Lett.*, 1991, **2**, 297; *Chem. Abstr.*, 1991, **115**, 131980). The structure of the alkaloid isolated from *Uncaria attenuata* (D. Ponglux, *et al.*, *Phytochemistry*, 1980, **19**, 2013) has been revised from 14β- to 14α-hydroxy-rauniticine based on synthesis (E. Yamanaka, *et al.*, *Chem. Pharm. Bull.*, 1986, **34**, 3713). The previously known synthetic product 11-methoxy-tetrahydroalstonine has been discovered in *Vinca major* (G. Mukhopadhyay, *et al.*, *Phytochemistry*, 1991, **30**, 2447). Ajmalicidine (267) is a novel N_a-carbomethoxy alkaloid found in *Rauwolfia serpentina* (S. Siddiqui, *et al.*, *ibid.*, 1987, **26**, 875), and magniflorine (268) (*Hamelia magniflora*) is a novel lactone (A. Rumbero and P. Vásquez, *Tetrahedron Lett.*, 1991, **32**, 5153). Serpenticine (269) is a new zwitterionic alkaloid from *Rauwolfia vomitoria* (A. Malik, *et al.*, *Heterocycles*, 1981, **16**, 1727).

(265)

Amsosinine (266)

Ajmalicidine (267)

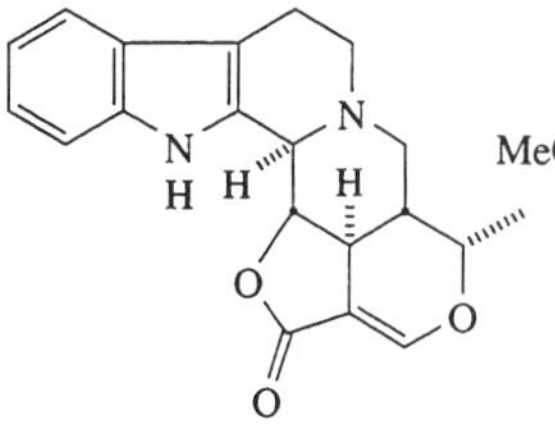

110

Magniflorine (268) 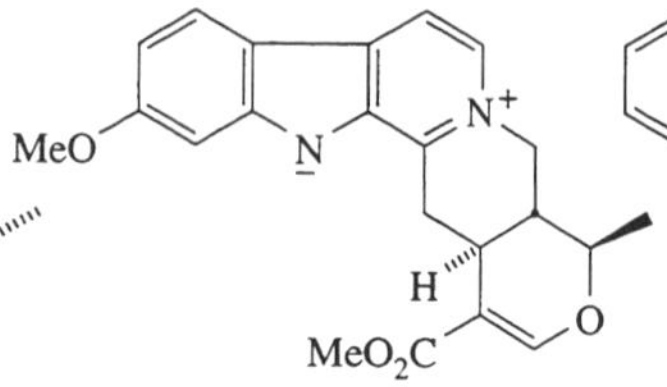   Serpenticine (269) 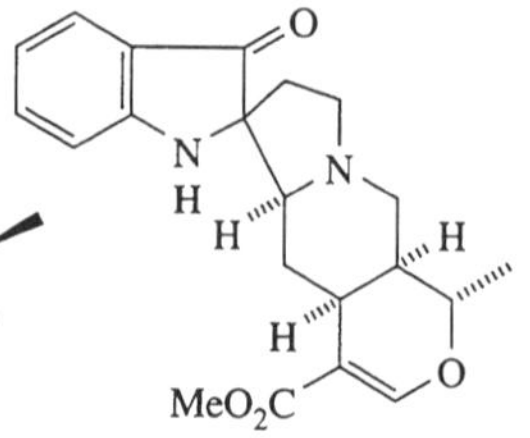   Aricine Pseudoindoxyl (270)

Several new oxindoles and pseudoindoxyls have been discovered of both
the heteroyohimbine and secoyohimbine types. For example, *Uncaria
elliptica* contains rauniticine oxindole A, rauniticine pseudoindoxyl, 3-
isorauniticine pseudoindoxyl, and akuammigine pseudoindoxyl (J.D.
Phillipson and N. Supavita, *Phytochemistry*, 1983, **22**, 1809). Five new
oxindoles have been isolated from *U. sinensis*, specifically those of ptero-
podic acid, isopteropodic acid, mitraphyllic acid, rhynchophyllic acid, and
isorhynchophyllic acid (H.-M. Liu and X.-Z. Feng, *ibid.*, 1993, **33**, 707),
and the oxindole catharinensine is found in *Peschiera catharinensis* (A.R.
Araujo, *et al.*, *ibid.*, 1984, **23**, 2359). Aricine pseudoindoxyl (270) is
found in *Aspidosperma oblongum* (G.M.T. Robert, *et al.*, *J. Nat. Prod.*,
1983, **46**, 708) and tetraphylline pseudoindoxyl has been isolated from
Neisosperma glomerata (E. Seguin, *et al.*, *ibid.*, 1984, **47**, 687).

A study using *Catharanthus roseus* cell suspensions has demonstrated that
deuterium-labeled iridodial is a precursor to ajmalicine (J. Balsevich and G.
Bishop, *Heterocycles*, 1989, **29**, 921).

A number of "missing" alkaloids of the secoyohimbine (corynane) class
have now been found in nature. In addition, several new oxygenated
members of this family have been discovered.

The productive *Aspidosperma oblongum* contains 10-methoxysitsirikine
(271) and a benzene-ring methoxyantirhine (G.M.T. Robert, *et al.*, *J. Nat.
Prod.*, 1983, **46**, 708), while *A. marcgravianum* contains 12 new alkaloids, most of which are in the secoyohimbine family (G.M.T. Robert, *et
al.*, *ibid.*, 1983, **46**, 694). These include 18,19-dihydroantirhine (272),
isogeissoschizol, 10-methoxyisogeissoschizol, 16(*S*)-isositsirikine *N*-
oxide, and isoantirhine. The plants *Rhazya stricta* and *Catharanthus roseus*
both contain 16-*epi*-(*Z*)-isositsirikine (273), which has antitumor activity
(S. Mukhopadhyay, *et al.*, *ibid.*, 1983, **46**, 409), and the plant *Alstonia
sphaerocapitata* has yielded (*Z*)-isositsirikine, epimeric at C-16 (C. Caron,
et al., *Phytochemistry*, 1984, **23**, 2355). Bhimberine is the C-3β-(*E*)-
isomer of (273) and is found in *Rhazya stricta* (A. Rahman, *et al.*, *Heterocycles*, 1986, **24**, 703), and the same plant has yielded rhazimanine, which
is the C-16 epimer of bhimberine (A. Rahman, *et al.*, *Phytochemistry*,
1986, **25**, 1731), and 16(*R*)-19,20-(*E*)-isositsirikine acetate (A. Rahman, *et
al.*, *ibid.*, 1991, **30**, 1285). The plant *Ochrosia moorei* has yielded eight

new alkaloids, including 10-hydroxydihydrocorynantheol (274), 10,11-dimethoxy-19,20-dihydro-16(*S*),20(*R*)-sitsirikine, and some *N*-oxides in the corynantheol family (A. Ahond, *et al.*, *Lloydia*, 1981, **44**, 193). The related *O. alyxioides* contains the new alkaloids 10-methoxycorynantheol, and 10-hydroxy- and 10-methoxyantirhine (N. Boughandjioua, *et al.*, *J. Nat. Prod.*, 1989, **52**, 1107). The related alkaloids 20-*epi*-antirhine and 19(*S*)-hydroxydihydrocorynantheol (275) are found in *Antirhea portoricensis* (B. Weniger, *et al.*, *ibid.*, 1994, **57**, 287).

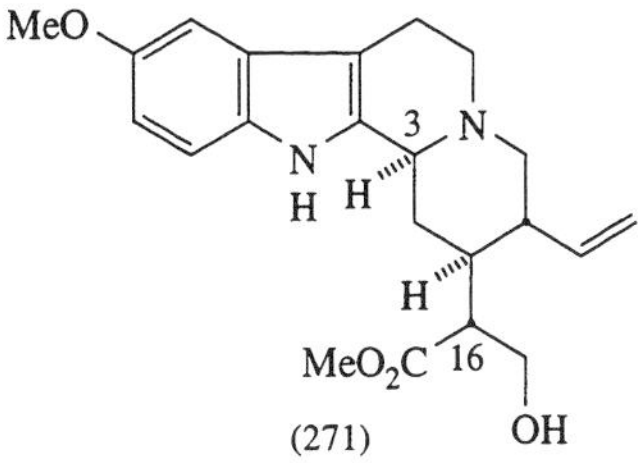

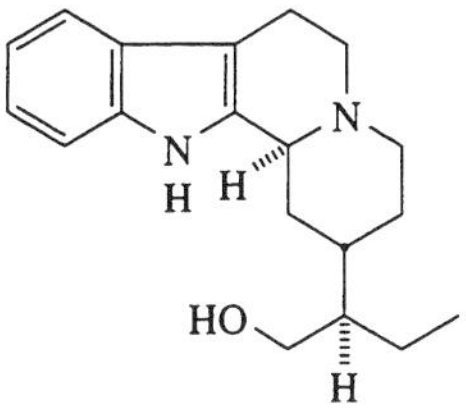

The Malaysian plant *Uncaria callophylla* has afforded the new gambireine (276) and isogambirine (277) (T.-S. Kam, *Phytochemistry*, 1992, **31**, 2031). The lactone of vallesiachotamine (278) is found in *Cephaelis dichroa* (P.N. Solis, *et al.*, *ibid.*, 1993, **33**, 1117). The unusual 2,7-dihydroapogeissoschizine (279) was characterized from *Strychnos gossweileri* (J. Quetin-Leclercq, *et al.*, *ibid.*, 1994, **35**, 533), and the similarly oxidized 7α-hydroxy-7*H*-mitragynine (280) has been isolated from *Mitragyna speciosa* (D. Ponglux, *et al.*, *Planta Med.*, 1994, **60**, 580). This alkaloid is not an artefact. This same plant has also been the source of the new alkaloids mitragynaline (281), corynantheidinaline (282), mitragynalinic acid, and corynantheidinalinic acid (P.J. Houghton, *et al.*, *Phytochemistry*, 1991, **30**, 347).

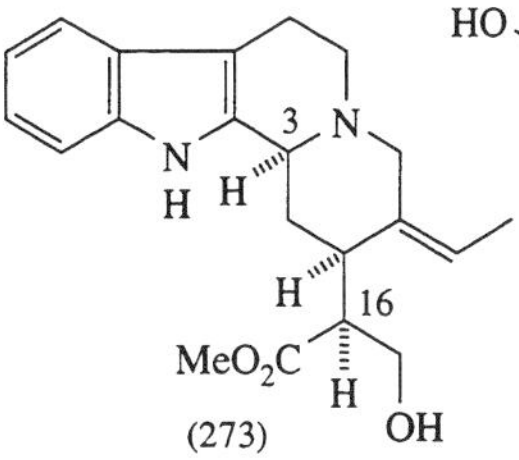

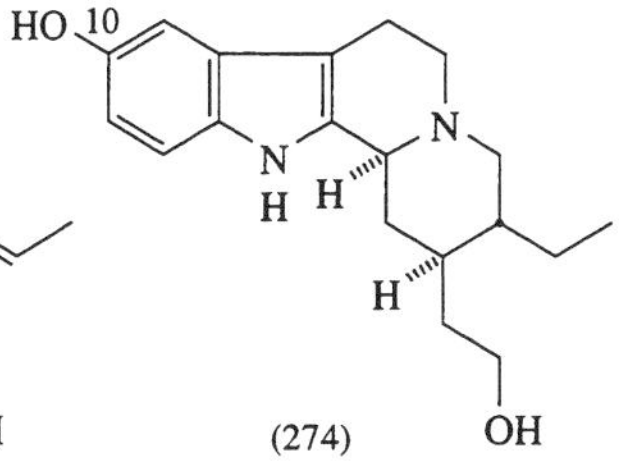

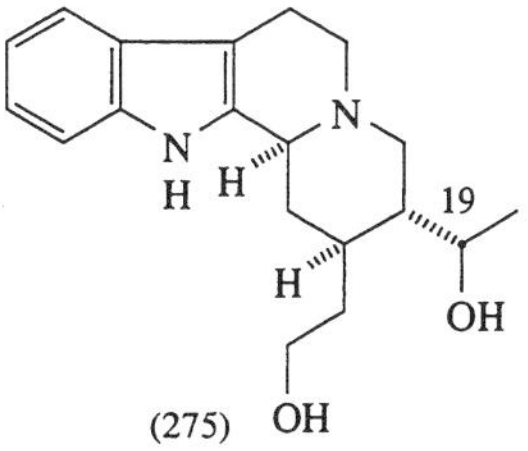

Several new quaternary *Corynanthé* alkaloids have been reported, including lercheine (283) (dihydro-3-*epi*-corynantheol methochloride) from *Lerchea bracteata* (D. Arbain, *et al.*, *J. Chem. Soc., Perkin Trans. 1*, 1992, 3039), 10-methoxycorynantheol α-methochloride from *Neisosperma glomerata* (E. Seguin, *et al.*, *J. Nat. Prod.*, 1984, **47**, 687), and 10-methoxygeissoschizol α-methochloride from *Aspidosperma pruinosum*

112

(D.S. Nunes, *et al.*, *Phytochemistry*, 1992, **31**, 2507).

Gambireine (276)
R$_1$ = OH, R$_2$ = H, R$_3$ = CH=CH$_2$
Isogambirine (277)
R$_1$ = H, R$_2$ = OH, R$_3$ = Et

(278)

(279)

(280)

Mitragynaline (281)   R = OMe
Corynantheidinaline (282) R = H

(283)

The indolo[2,3-*a*]quinolizine alkaloids have been reviewed (G.W.
Gribble, in "Studies in Natural Products Chemistry," Vol. 1, p. 123,
1988). Several new examples of these zwitterionic alkaloids have been
described, such as 3,4,5,6-tetradehydro-18,19-dihydrocorynantheol from
Aspidosperma marcgravianum (G.M.T. Robert, *et al.*, *J. Nat. Prod.*, 1983,
46, 694), 3,4,5,6-tetradehydrositsirikine from *A. oblongum* (G.M.T.
Robert, *et al.*, *ibid.*, 1983, **46**, 708), matadine (284) from *Strychnos
gossweileri* (J. Quertin-Leclercq, *et al.*, *Phytochemistry*, 1991, **30**, 1697),
and (285)-(288) from *Neisosperma glomerata* (E. Seguin, *et al.*, *J. Nat.
Prod.*, 1984, **47**, 687.) The novel ring system present in melinonine E
(289) from *Strychnos melinoniana* (R.P. Borris, *et al.*, *Helv. Chim. Acta*,
1984, **67**, 455) is also found in the ketone strychnoxanthine (290) from *S.
gossweileri* (C. Coune, *et al.*, *Planta Med.*, 1984, 93), both isolated as HCl
salts. *Mitragyna speciosa* contains 3-dehydromitragynine (P.J. Houghton
and I.M. Said, *Phytochemistry*, 1986, **25**, 2910), and the novel javacarboline (291) is found in *Picrasma javanica* (K. Koike, *et al.*, *Heterocycles*,
1994, **38**, 1413).

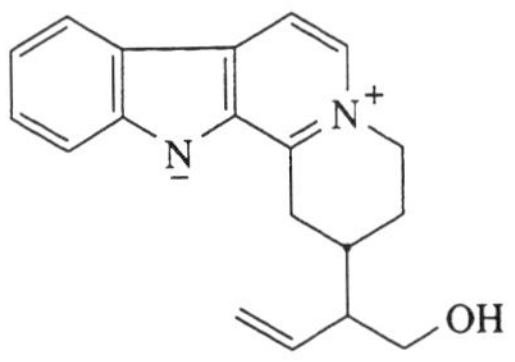

Matadine (284)

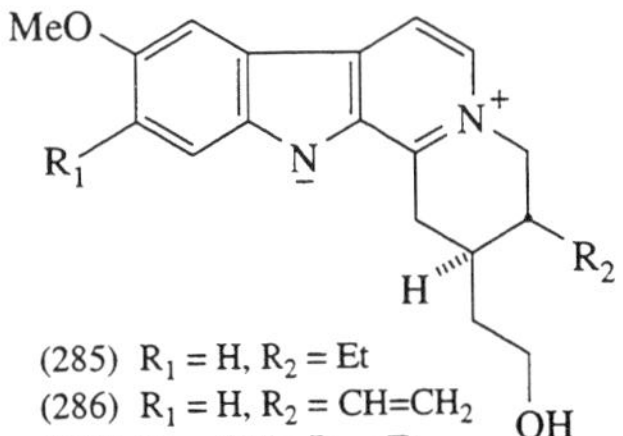

(285) R_1 = H, R_2 = Et
(286) R_1 = H, R_2 = CH=CH$_2$
(287) R_1 = OMe, R_2 = Et
(288) R_1 = OMe, R_2 = CH=CH$_2$

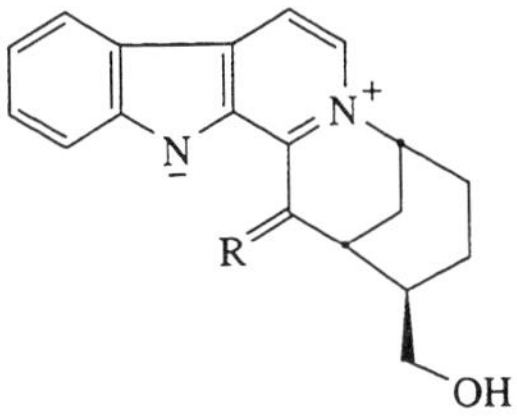

Melinonine E (289)   R = H$_2$
Strychnoxanthine (290) R = O

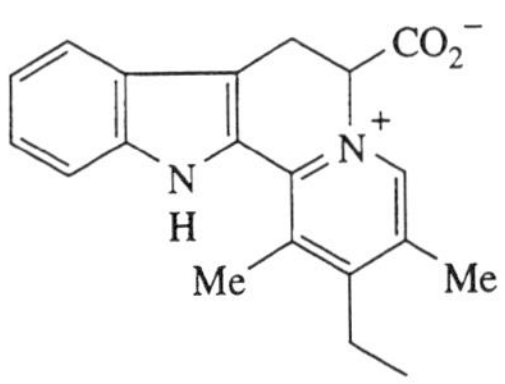

Javacarboline (291)

Several new alkaloids with a pyridine E-ring have been described since 1981. For example, 19-*O*-acetylangusatoline and 3,14-dihydroangustine (292) are found in *Nauclea pobeguinii* (M. Zeches, *J. Nat. Prod.*, 1985, **48**, 42), and 19-*O*-methylangustoline, which is cytotoxic, has been isolated from *Camptotheca acuminata* (L.-Z. Lin and G.A. Cordell, *Phytochemistry*, 1990, **29**, 2744). Normalindine (293) and the epimeric norepimalindine are found in *Strychnos johnsonii* (G. Massiot, *et al.*, *ibid.*, 1987, **26**, 2839). The plant *Nauclea orientalis* produces 10-hydroxyangustine (C.A.J. Erdelmeier, *et al.*, *Planta Med.*, 1992, **58**, 43). Nauclefidine (294) and nauclefoline (295), which are deaza analogues of this group of alkaloids, are found in *N. offinalis* (L. Mao, *et al.*, *ibid.*, 1984, **50**, 459). Malindine (296) is a novel muscle-relaxant alkaloid from *Strychnos decussata* (A.A. Olaniyi, *et al.*, *Planta Med.*, 1981, **43**, 353), which is a quaternary derivative of normalindine (293). The C-21 epimer isomalindine is found in *S. usambarensis* (M. Caprasse, *et al.*, *ibid.*, 1984, **50**, 27).

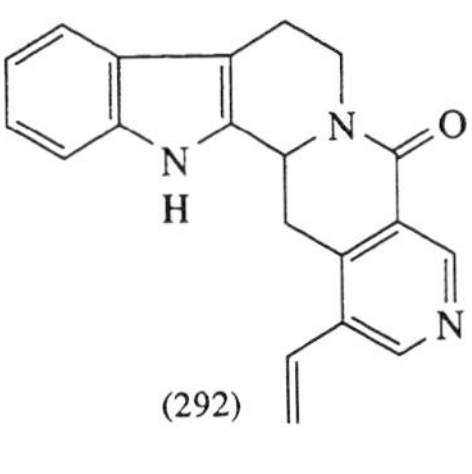

(292)

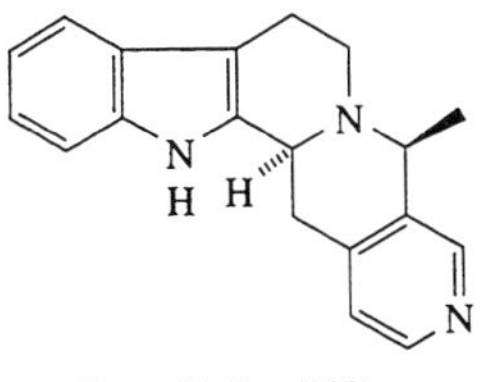

Normalindine (293)

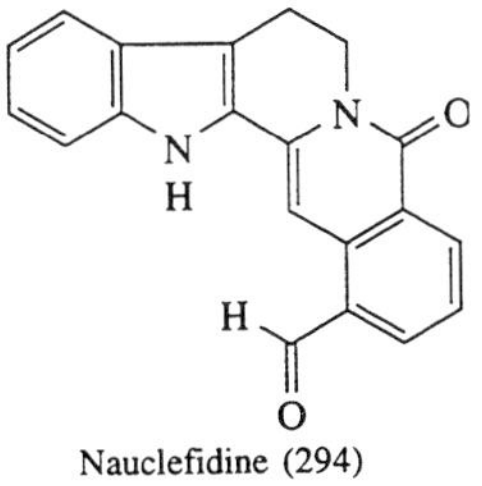

Nauclefidine (294)

114

Nauclefoline (295)   Malindine (296)

A proposed intermediate guettardine (297) which links *Cinchona* alkaloids
with *Corynanthé* alkaloids has been found in *Guettarda heterosepala* (M.H.
Brillanceau, *et al.*, *Tetrahedron Lett.*, 1984, **25**, 2767). Conversion of
(297) to dihydrocorynantheol was readily accomplished (1. TsCl; 2.
LiAlH$_4$). The plant *Cinchona ledgeriana* has yielded 10-methoxycinchona-
mine (Th. Mulder-Krieger, *et al.*, *Planta Med.*, 1982, **46**, 19), and the
novel apodihydrocinchonamine (298) was isolated from *Isertia haenkeana*
(M. Bruix, *et al.*, *Phytochemistry*, 1993, **33**, 1257). The structure of the
previously isolated C-alkaloid-O from Calabash curare has been determined
as (299) (R.P. Borris, *et al.*, *Helv. Chim. Acta*, 1983, **66**, 405), and
ophiorrhizine (300) is found in *Ophiorrhiza major* (D. Arbain, *et al.*, *J.
Chem. Soc., Perkin Trans. 1*, 1992, 663); the absolute configuration is as
shown by chemical correlation with cinchonine (M. Ohba, *et al.*, *Heterocycles*, 1994, **38**, 1741).

Guettardine (297)   (298)   (299)

Ophiorrhizine (300)

The novel cinchonaminone (301), which is a strong monoamine oxidase
inhibitor, has been isolated from *Cinchona succirubra* (N. Mitsui, *et al.*,
Chem. Pharm. Bull., 1989, **37**, 363). The plant *Neolaugeria resinosa*
produces neolaugerine (302) and the related isoneolaugerine and 15-
hydroxyneolaugerine, compounds which may be biogenetically related to

the *Cinchona* alkaloids (B. Weniger, *et al.*, *Phytochemistry*, 1993, **32**, 1587). The plant *Alstonia lanceolifera* produces the 10-methoxydeplancheine (303) (N. Petitfrere-Auvray, *et al.*, *ibid.*, 1981, **20**, 1987), an alkaloid difficult to categorize.

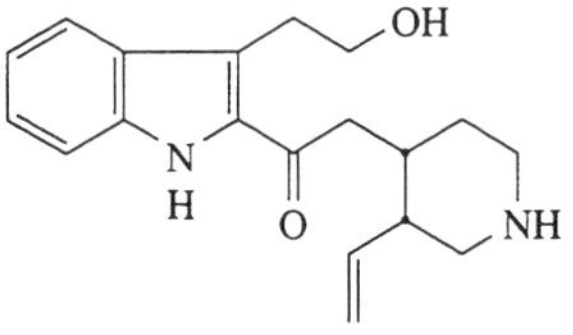

Cinchonaminone (301)

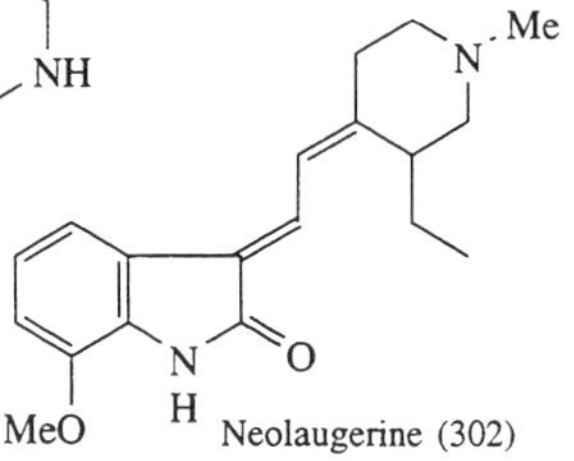

Neolaugerine (302)

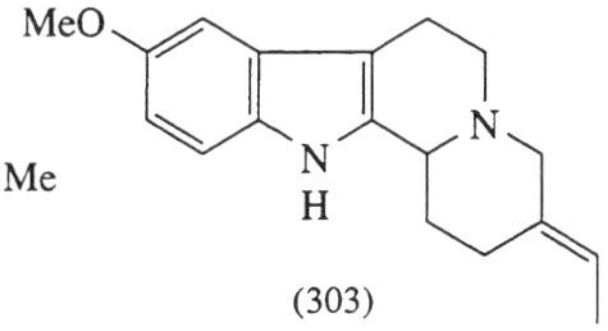

(303)

A few alkaloids of the pleiocarpamine type have been identified. For example, the most unusual 2,7-dihydroxypleiocarpamine has been isolated from *Alstonia plumosa* (M.J. Jacquier, *et al.*, *Phytochemistry*, 1982, **21**, 2973), and strictine from *Rhazya stricta* is a 19-oxopleiocarpamine derivative (A. Rahman and S. Khanum, *Heterocycles*, 1987, **26**, 2125). Normavacurine has been revised to a pleiocarpamine structure from an earlier proposed strychnan type (*Strychnos minfiensis*, *S. potatorum*, and *S. longicaudata*) (G. Massiot, *et al.*, *Heterocycles*, 1989, **29**, 1435). The novel minfiensine with N-1 bonded to C-3 was also uncovered in this study. Flurocarpamine N_b-oxide from *Catharanthus roseus* is a pleiocarpamine pseudoindoxyl (A. Rahman and M. Bashir, *Planta Med.*, 1983, **49**, 124).

(3) Picraline type alkaloids

The picraline group of alkaloids is distinguished by a bond between C-7 and C-16. Several new derivatives of the various picraline subtypes have been discovered. For example, several novel strictamines are newly identified, including 5α,10,11-trimethoxystrictamine (304) from *Alstonia macrophylla* (F. Abe, *et al.*, *Phytochemistry*, 1994, **35**, 253), gomaline (18-hydroxystrictamine) from *Catharanthus roseus* (A. Rahman, *et al.*, *Heterocycles*, 1984, **22**, 85), 10-hydroxystrictamine from *Alstonia macrophylla* (A. Rahman, *et al.*, *Nat. Prod. Lett.*, 1994, **5**, 201), 11-methoxystrictamine from *Rauwolfia sumatrana* (S. Subhadhirasakul, *et al.*, *Chem. Pharm. Bull.*, 1994, **42**, 1427), and strictamine N_b-oxide from *Rhazya stricta* (A. Rahman and S. Khanum, *Phytochemistry*, 1984, **23**, 709). Alstolenine (305) is a akuammiline tri-*O*-methyl gallate derivative found in *Alstonia venenata* (P. Majumder and A. Basu, *ibid.*, 1982, **21**, 2389). Methyl 12-hydroxyakuammilan-17-oate has been isolated from *Rauwolfia sumatrana*

116

(D. Arbain, *et al.*, *Aust. J. Chem.*, 1991, **44**, 1007). Several new akuammiline derivatives (5-hydroxymethyl-, deacetyl-, 1,2β-dihydro-, and deacetyl-1,2β-dihydroakuammiline) are found in *R. oreogiton* (B.A. Akinloye and W.E. Court, *Phytochemistry*, 1980, **19**, 2741). Alkaloids of the picraline type are also known with a saturated N-1, C-2 bond, and newly discovered examples include 10-hydroxycathofoline (306) from *Vinca major* (J. Balsevich, *et al.*, *Planta Med.*, 1982, **44**, 91), cathofoline N_b-oxide from *Alstonia macrophylla* (F. Abe, *et al.*, *Phytochemistry*, 1994, **35**, 249), 18-hydroxycabucraline from *Tonduzia pittieri* (A.-M. Morfaux, *et al.*, *ibid.*, 1990, **29**, 3345), and the novel 10-substituted monoterpene cabucraline derivative gentiacraline (307) from *Alstonia undulata* (D. Guillaume, *et al.*, *ibid.*, 1984, **23**, 2407). The N_b-oxide of cabucraline is found in *A. plumosa* along with 10-formylcabucraline and the novel 3,4-*seco*-3,14-dehydrocabucraline (M.J. Jacquier, *ibid.*, 1982, **21**, 2973).

Picraline itself has an ether bridge between C-2 and C-5, and several new alkaloids of this type have been characterized. The plant *Alstonia lanceolata* has yielded 10,11-dimethoxy-1-methylpicraline (J. Vercauteren, *et al.*, *Phytochemistry*, 1981, **20**, 1411), and *A. lanceolifera* has yielded several new picraline alkaloids, including 10,11-dimethoxy-1-methyldeacetylpicraline, 10,11-dimethoxy-1-methyldeacetylpicraline benzoate, and 10,11-dimethoxy-1-methyldeacetylpicraline 3',4',5'-trimethoxybenzoate (N. Petitfrere-Auvray, *et al.*, *ibid.*, 1981, **20**, 1987). *A. macrophylla* contains *N*-methylpicralines (308) and (309) (F. Abe, *et al.*, *ibid.*, 1994, **35**, 253). Picratidine (N_a-methylpicraline) itself is found in *Picralima nitida* from Ghana (R. Ansa-Asamoah, *et al.*, *J. Nat. Prod.*, 1990, **53**, 975). Tabernulosine (310) and the 12-demethoxy analogue (311) are picrinine types present in *Tabernaemontana glanulosa* (H. Achenbach, *et al.*, *Ann.*, 1982, 830). The former alkaloid displays significant antihypertonic activity in rats. The N_b-oxide of pseudoakuammigine, which has an ether bridge between C-2 and C-17, is a new alkaloid found in *Alstonia angustifolia* (W. Hu, *et al.*, *Planta Med.*, 1989, **55**, 463).

The echitamine-corymine sub-family of alkaloids is characterized by a bond between C-2, rather than C-3, and N-4, and with an oxygen often at C-3. Several new such alkaloids have been characterized. For example, 3-*epi*-dihydrocorymine 17-acetate is found in *Hunteria zeylanica* (C. Lavaud, *et al.*, *Phytochemistry*, 1982, **21**, 445), and this plant has yielded also two N_a-demethylcorymines, e.g., (312) (S. Subhadhirasakul, *et al.*, *Chem. Pharm. Bull.*, 1994, **42**, 2645). Echitaminic acid (313) and 17-*O*-acetyl-N_b-demethylechitamine are found in *Alstonia glaucescens* (N. Keawpradub, *et al.*, *Phytochemistry*, 1994, **37**, 1745). The widespread plant *A. macrophylla* has been the source of the novel alstonamide (314) and

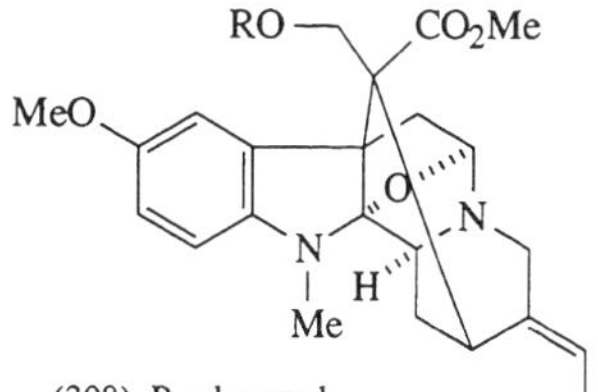

(308) R = benzoyl
(309) R = 3,4-dimethoxybenzoyl

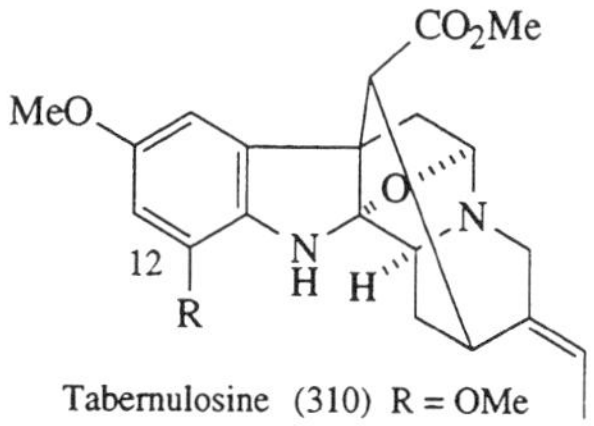

Tabernulosine (310) R = OMe
(311) R = H

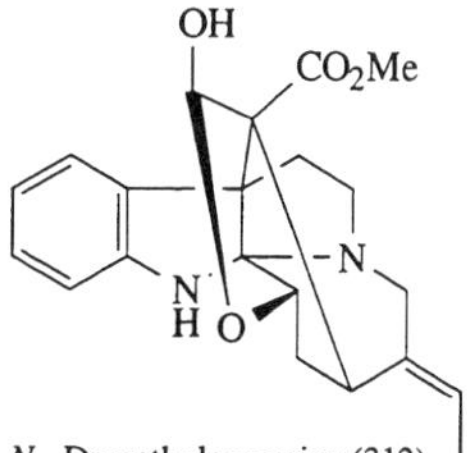

N_a-Demethylcorymine (312)

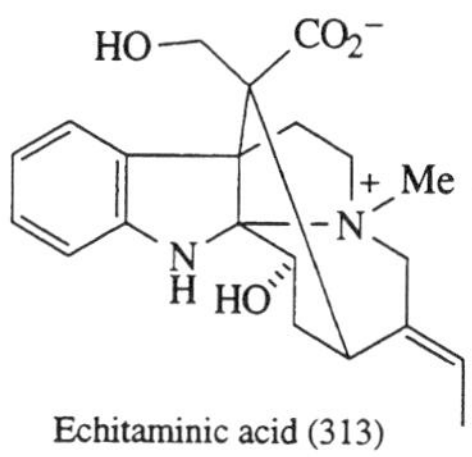

Echitaminic acid (313)

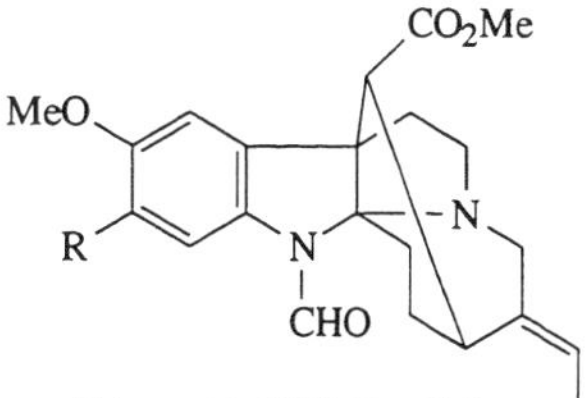

Alstonamide (314) R = OMe
(315) R = H

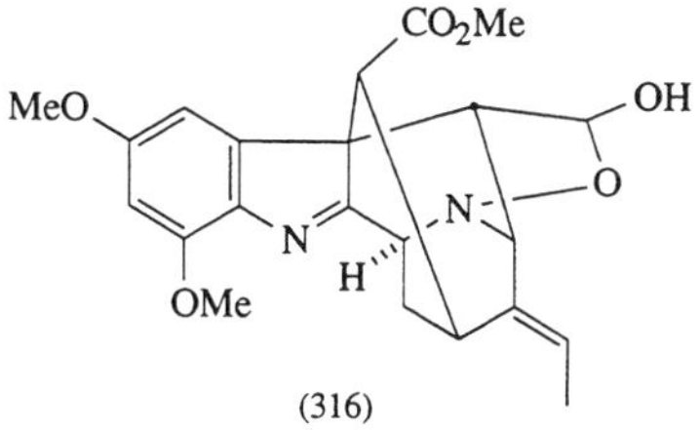

(316)

118

demethoxyalstonamide (315) (A. Rahman, *et al.*, *J. Nat. Prod.*, 1991, **54**, 750), and *Tabernaemontana glandulosa* has yielded the complex 10,12-dimethoxynareline (316) (H. Achenbach, *et al.*, *Phytochemistry*, 1994, **37**, 1737).

(4) Ajmaline-sarpagine type alkaloids

New ajmaline-sarpagine alkaloids, which numbered 102 as of 1981, continue to be discovered. An excellent review of these fascinating natural products has appeared (A. Koskinen and M. Lounasmaa, *Prog. Chem. Org. Nat. Prod.*, 1983, **43**, 267), and reviews covering the synthesis of the macroline family of sarpagine alkaloids are available (Y. Bi, *et al.*, "Studies in Natural Products Chemistry," Vol. 13, p. 383, 1993; L.K. Hamaker and J.M. Cook, "Alkaloids: Chemical and Biological Perspectives," Vol. 9, Chapter 2, 1995).

New examples of the sarpagine class include several pericyclivine derivatives. The New Caledonian *Alstonia undulata* has yielded 10-hydroxypericyclivine (317), N(1)-methyl-10-hydroxypericyclivine, 10-methoxypericyclivine, N(1)-methyl-10-methoxypericyclivine, N(1)-methyl-16-*epi*-pericyclivine, and voachalotinal (318) (T.-M. Pinchon, *et al.*, *Phytochemistry*, 1990, **29**, 3341). *Tabernaemontana divaricata* contains 11-methoxy-N(1)-methyl-19,20-dihydropericyclivine (L.S.R. Arambewela and T. Ranatunge, *ibid.*, 1991, **30**, 1740), and *Peschiera buchtieni* has yielded N(1)-methylpericyclivine and 18-hydroxyaffinisine (319) (M. Azoug, *et al.*, *ibid.*, 1995, **39**, 1223). The related 18-hydroxylochnerine (320) is found in *Rauwolfia biauriculata* (J. Abaul, *et al.*, *J. Nat. Prod.*, 1986, **49**, 829). Difforine, which was isolated from *Vinca difformis*, is an alkaloid formally formed from vellosimine and acetone, but it does not appear to be an isolation artefact (J. Garnier and J. Mahuteau, *Planta Med.*, 1986, 66). The plant *Alstonia venenata* has afforded 19,20-dihydropolyneuridine, although the stereochemistry at C-20 is uncertain (P. Majumder and A. Basu, *Phytochemistry*, 1982, **21**, 2389). Alstoumerine (321) is an unusual unsaturated alkaloid from *A. macrophylla* (A. Rahman, *et al.*, *J. Nat. Prod.*, 1991, **54**, 750).

10-Hydroxypericyclivine (317)

Voachalotinal (318)

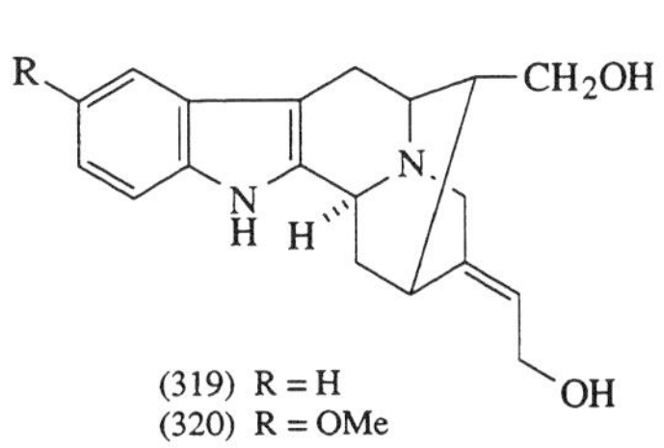

(319) R = H
(320) R = OMe

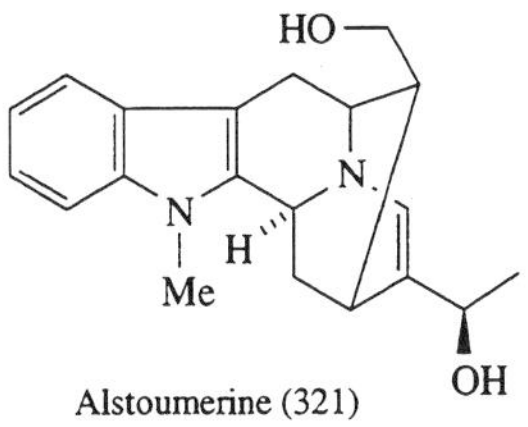

Alstoumerine (321)

Several new affinisine-type alkaloids are found in *Ervatamia hirta*, including (*E*)-16-*epi*-affinisine, *O*-acetyl-16-*epi*-affinisine, affinisine *N*(4) oxide, (*E*)-16-*epi*-normacusine B, and the ether-bridged dehydro-16-*epi*-affinisine (322) (P. Clivio, *et al.*, *Phytochemistry*, 1991, **30**, 3785). The plant *Gelsemium elegans* has yielded 19(*Z*)-akuammidine (S. Sakai, *et al.*, *Chem. Pharm. Bull.*, 1987, **35**, 4668; see also D. Ponglux, *et al.*, *Tetrahedron*, 1988, **44**, 5075, for a revision of akuammidine to the 19(*Z*) configuration). Trinervine (323), which is found in *Strychnos trinervis* (R. Mukherjee, *et al.*, *Heterocycles*, 1990, **31**, 1819), is a powerful muscle relaxant (M.F.F. Melo Diniz, *et al.*, *Phytomedicine*, 1994, **1**, 205). The *N*(4)-methyl quaternary alkaloid is venecurine and was isolated from Venezuelan curare (*Strychnos* sp.) (J. Quetin-Leclercq, *et al.*, *Phytochemistry*, 1989, **28**, 2221). Earlier, the plant *Catharanthus roseus* was found to contain 21α-hydroxy-10-methoxytrinervine (W. Kohl, *et al.*, *Planta Med.*, 1984, **50**, 242). Venezuelan curare (*Strychnos toxifera*) also produces panarine (324), one of a number of newly discovered *N*(4)-methosalts (J. Quetin-Leclercq, *et al.*, *Phytochemistry*, 1988, **27**, 4002), and 16-*epi*-panarine is found in *Stemmadenia minima* (H. Achenbach, *et al.*, *J. Nat. Prod.* 1991, **54**, 473). Others include 11-methoxymacusine A (*S. angolensis*) (R. Verpoorte, *et al.*, *ibid.*, 1983, **46**, 572), *N*(1)-methyl-11-hydroxymacusine A (*Stemmadenia obovata*) (A. Madinaveitia, *et al.*, *ibid.*, 1995, **58**, 250), 12-methoxy-*N*(4)-methylvoachalotine and the corresponding ethyl ester, and fuchsiaefoline (325) (*Peschiera fuchsiaefolia*) (R.M. Braga and F. de A.M. Reis, *Phytochemistry*, 1987, **26**, 833).

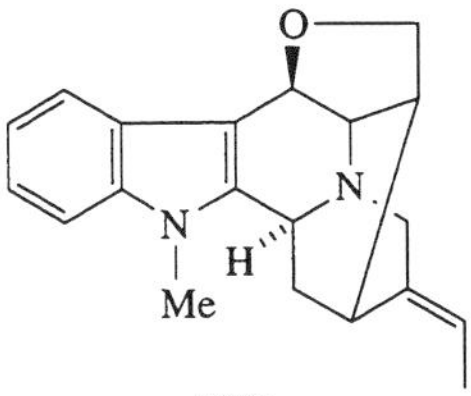

(322)

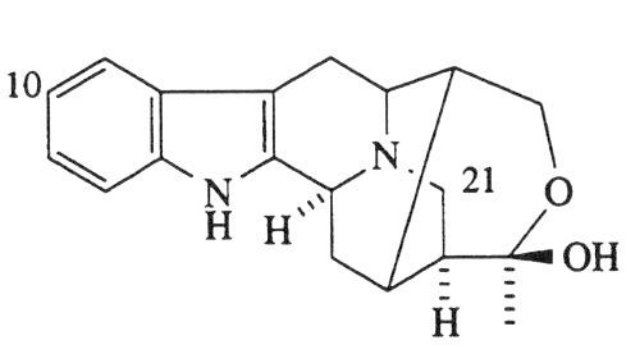

Trinervine (323)

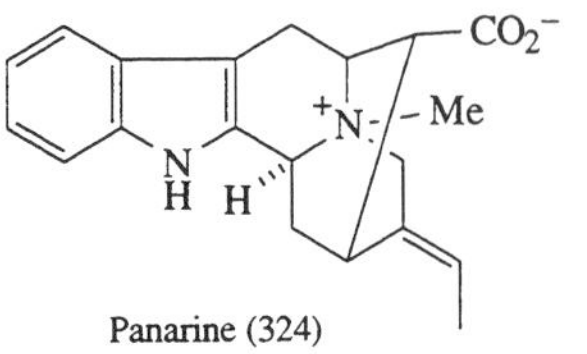

Panarine (324)

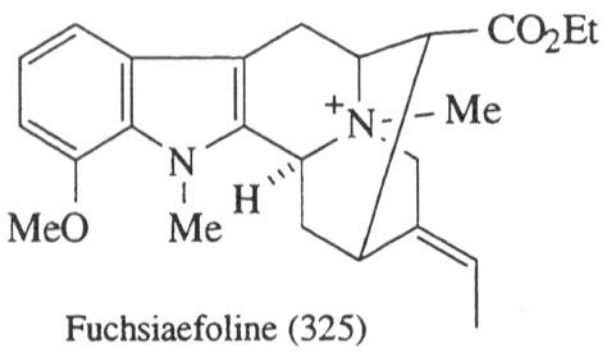

Fuchsiaefoline (325)

New alkaloids of the 2-acylindole vobasine type and compounds derived therefrom have been discovered in recent years. For example, 16-*epi*-voacarpine was discovered in *Gelsemium elegans* by two groups (S. Sakai, *et al.*, *Chem. Pharm. Bull.*, 1987, **35**, 4668; D. Ponglux, *et al.*, *Tetrahedron*, 1988, **44**, 5075), and pelirine (10-methyl-16-*epi*-affinine) (326) is found in *Rauwolfia perakensis* (A.S.C. Wan, *et al.*, *Heterocycles*, 1987, **26**, 1211). Pagicerine (327) is a novel ether-bridged alkaloid found in *Pagiantha cerifera* (M. Bert, *et al.*, *ibid.*, 1985, **23**, 2505), while erichsonine (*Strychnos erichsonii*) (328) represents the first report of a 2-acylindole-vobasine alkaloid from the Loganiaceae (P. Forgacs, *et al.*, *Phytochemistry*, 1986, **25**, 969). Vobasenal (329) and the 16-epimer are found in *Ervatamia polyneura* (P. Clivio, *et al.*, *ibid.*, 1990, **29**, 3007). Other novel 2-acylindole alkaloids are difforlemenine (330) from *Vinca difformis* (J. Garnier and J. Mahuteau, *Tetrahedron Lett.*, 1985, **26**, 1513), and the unusual

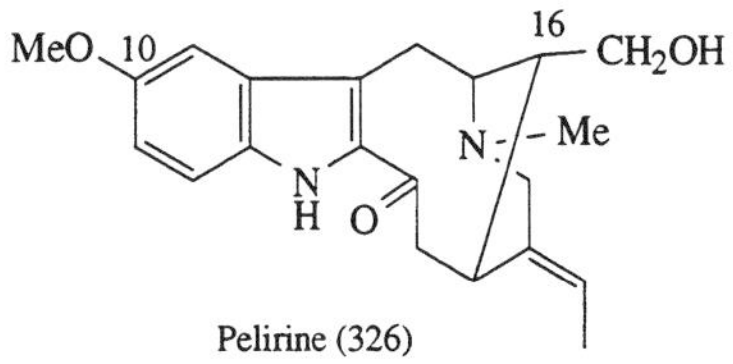

Pelirine (326)

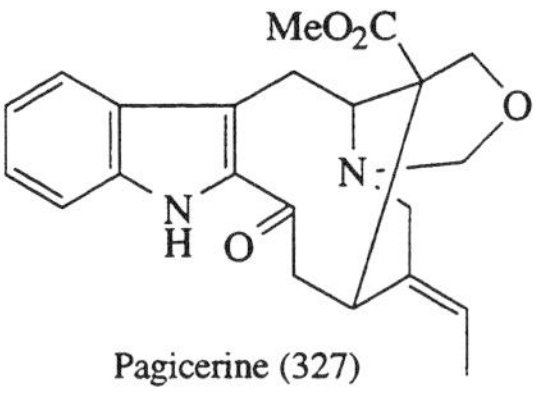

Pagicerine (327)

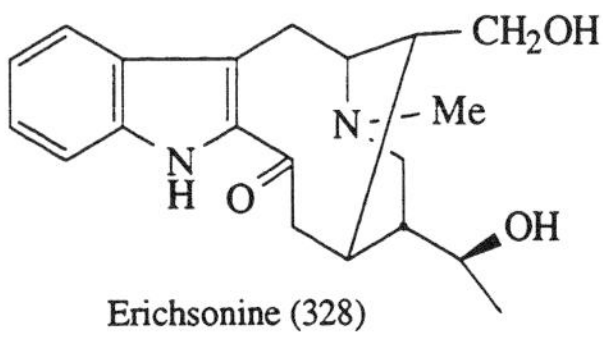

Erichsonine (328)

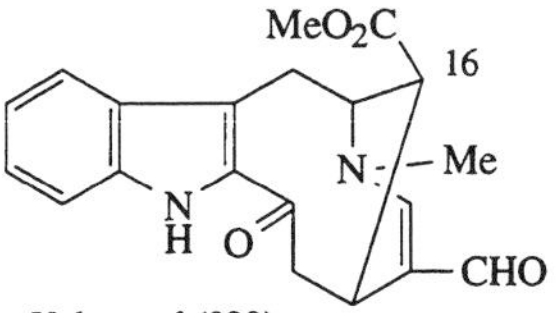

Vobasenal (329)

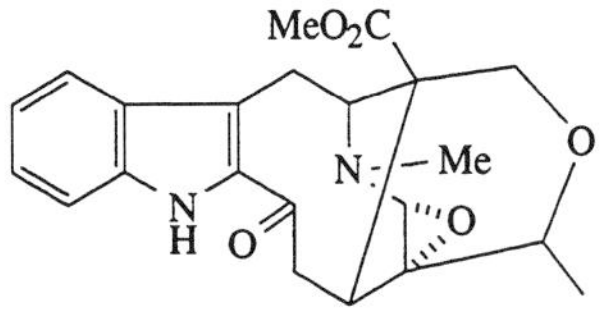

Difforlemenine (330)

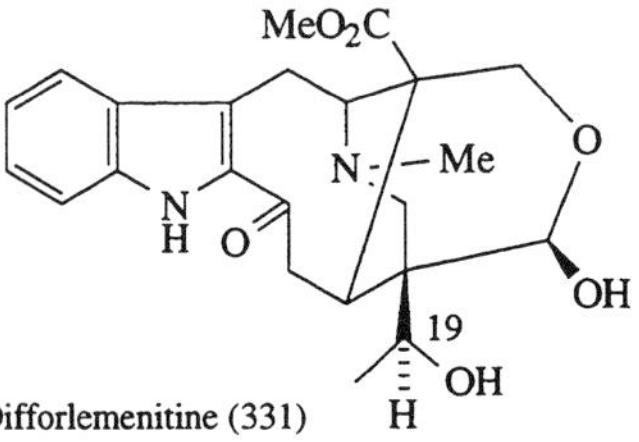

Difforlemenitine (331)

ring-contracted difforlemenitine (331) (and the 19-epimer) from *Tabernaemontana glandulosa* (H. Achenbach, *et al.*, *Phytochemistry*, 1994, **37**, 1737).

Several related vobasine-type alkaloids having lower level of oxidation at C-3 are known. New examples are the interesting sulfur-containing pagisulfine (332) from the New Caledonian plant *Pagiantha cerifera* (M. Bert, *et al.*, *Heterocycles*, 1986, **24**, 1567), and the 3-amino derivative (333) from *Hunteria zeylanica* (S. Subhadhirasakul, *et al.*, *ibid.*, 1995, **41**, 2049). The 19(*Z*) isomer of taberpsychine is found in *Gelsemium elegans* (D. Ponglux, *et al.*, *Tetrahedron*, 1988, **44**, 5075). The new alkaloids tabernaemontaninol (334), dregaminol (335), and dregaminol methyl ether have been characterized from *Tabernaemontana elegans* (R. van der Heijden, *et al.*, *Planta Med.*, 1986, 144). Macroxine-A is a novel oxindole that may arise from voachalotine oxindole found in *Alstonia macrophylla* (A. Rahman, *et al.*, *Tetrahedron*, 1991, **47**, 3129).

Pagisulfine (332)

(333)

Tabernaemontaninol (334)
R = β-Et
Dregaminol (335)
R = α-Et

Ajmaline alkaloids are characterized by a bond between C-7 and C-17, and several new examples have been uncovered in the interim since the last volume. For example, ajmalinimine (10-*C*, 17-*O*-diacetylajmaline) is present in *Rauwolfia serpentina* (S. Siddiqui, *et al.*, *Heterocycles*, 1987, **26**, 463), and 12-hydroxyajmaline has been discovered in the "hairy roots" of the same plant (H. Falkenhagen, *et al.*, *Can. J. Chem.*, 1993, **71**, 2201). Also found in this plant is the trimethoxybenzoate derivative ajmalimine (336) (S. Siddiqui, *et al.*, *Planta Med.*, 1987, 288). Likewise, several similar ester derivatives of vincamajine have been found in *Alstonia macrophylla*, such as (337) (F. Abe, *et al.*, *Phytochemistry*, 1994, **35**, 249), and in *A. angustifolia* (I.M. Said, *et al.*, *J. Nat. Prod.*, 1992, **55**, 1323). Both 11-methoxyvincamajine and its C-17 epimer are found in *Tonduzia pitteri* (A.-M. Morfaux, *et al.*, *Phytochemistry*, 1990, **29**, 3345), 19-hydroxy-19,20-dihydrovincamajine is present in *Alstonia macrophylla* (C.K. Ratnayake, *et al.*, *ibid.*, 1987, **26**, 868), 10-methoxyvincamedine and the corresponding *N*(4)-oxide have been identified from *A. sphaerocapitata* (C. Caron, *et al.*, *ibid.*, 1984, **23**, 2355), and 11-methoxyvincamedine is present in *T. pitteri* (A.-M. Morfaux, *et al.*, 1992, **31**, 1079). The novel 12-hydroxymauiensine (338) is found in *Rauwolfia media* (C. Kan,

122

et al., ibid., 1986, **25**, 1783). Raucaffricine (*R. caffra*) has been revised to vomilenine-β-D-glucoside by two groups (H. Schübel, *et al.*, *Helv. Chim. Acta*, 1984, **67**, 2078; M.A. Khan, *et al.*, *Z. Naturforsch.*, 1982, **B37**, 494). The novel ketone alkaloid leepacine (339) is found in *Rhazya stricta* (A. Rahman, *et al.*, *Phytochemistry*, 1991, **30**, 1285). Both 10-methoxy-perakine and vincawajine (340) are newly discovered alkaloids in *Vinca major* (A. Rahman, *et al.*, *ibid.*, 1995, **38**, 1057).

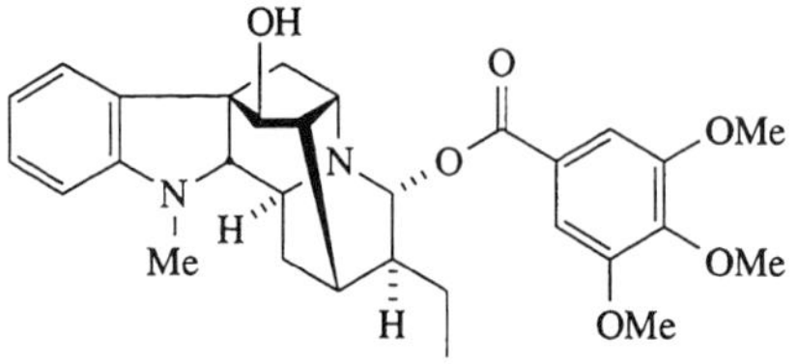

Ajmalimine (336)

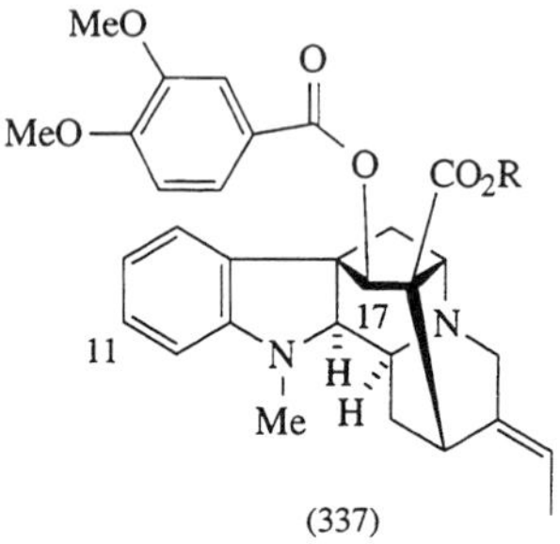

(337)

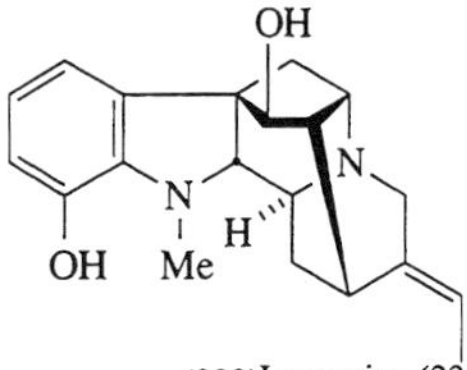

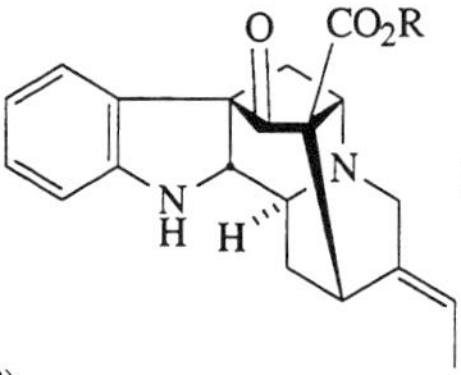

(338)Leepacine (339)

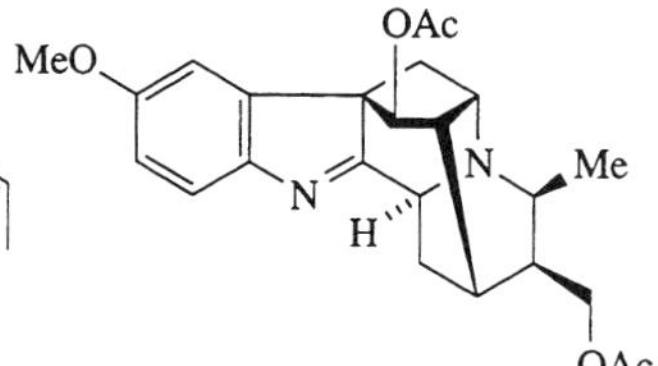

Vincawajine (340)

Several new seco ajmaline alkaloids have been characterized in recent years. For example, *Rhazya stricta* has yielded four such compounds: rhazimine (341) (A. Rahman and S. Khanum, *Tetrahedron Lett.*, 1984, **25**, 3913), rhazicine (2-hydroxy-1,2-dihydrorhazimine) (A. Rahman and S. Khanum, *Heterocycles*, 1984, **22**, 2183), 2-methoxy-1,2-dihydrorhazimine (A. Rahman and S. Khanum, *Phytochemistry*, 1985, **24**, 1625), and isohazicine (A. Rahman and S. Khanum, *Heterocycles*, 1987, **26**, 405). The alkaloid sandwicoline (342) is a different seco derivative of sandwicine (*Rauwolfia serpentina*) (S. Siddiqui, *et al.*, *ibid.*, 1985, **23**, 617), and

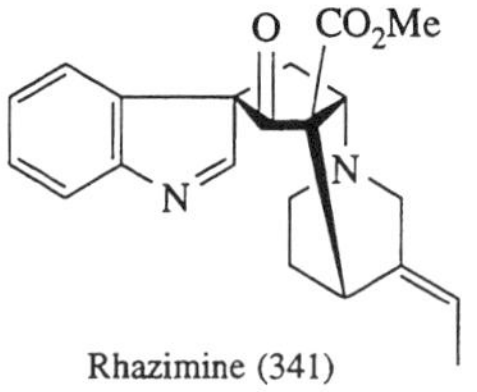

Rhazimine (341)

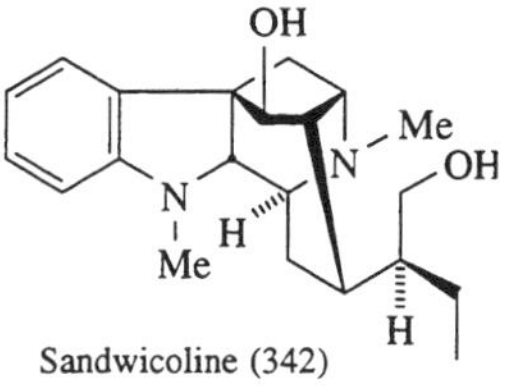

Sandwicoline (342)

sandwicolidine, from the same plant, is of biogenetic interest in that it represents the first alkaloid with the ethyl group next to N_b (S. Siddiqui, *et al.*, *Tetrahedron*, 1985, **41**, 4577).

Several new alkaloids that are biogenetically linked to the sarpagine-ajmaline class have been described. Thus, gardfloramine (343) and 18-demethoxygardfloramine are present in *Gardneria multiflora* (S. Sakai, *et al.*, *Chem. Pharm. Bull.*, 1987, **35**, 453). Some new macroline-related alkaloids have been discovered. *Rauwolfia serpentina* contains 6α-hydroxyraumacline (344), and several related raumaclines are produced when ajmaline is fed to cell cultures of this plant (S. Endress, *et al.*, *Planta Med.*, 1992, **58**, 410; *Phytochemistry*, 1993, **32**, 725). Alstomacrocine (345) is a related alkaloid from *Alstonia macrophylla* (A. Rahman, *et al.*, *Nat. Prod. Lett.*, 1994, **5**, 201). Sellowiine (346), which is *N*-demethyl-20-deethylsuaveoline, is found in *R. sellowii* (C.V.F. Batista, *et al.*, *Phytochemistry*, 1996, **41**, 969).

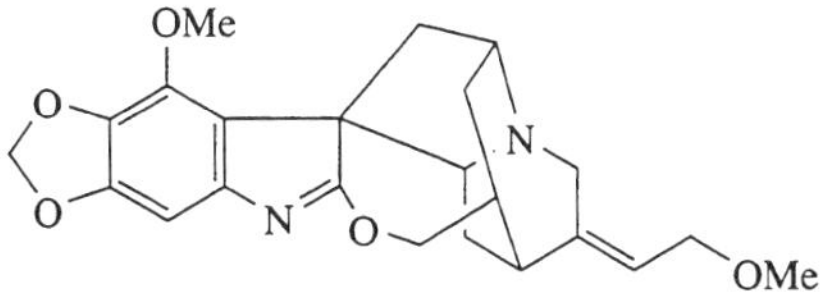

Gardfloramine (343)

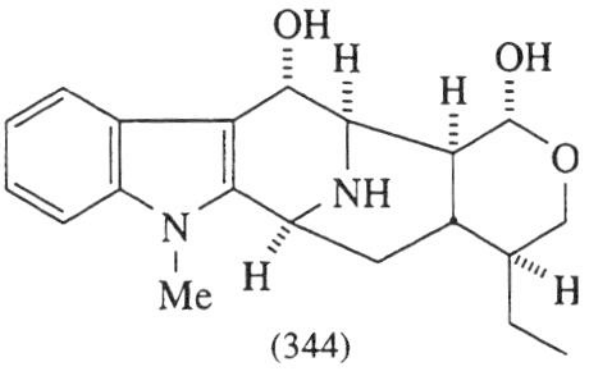

(344)

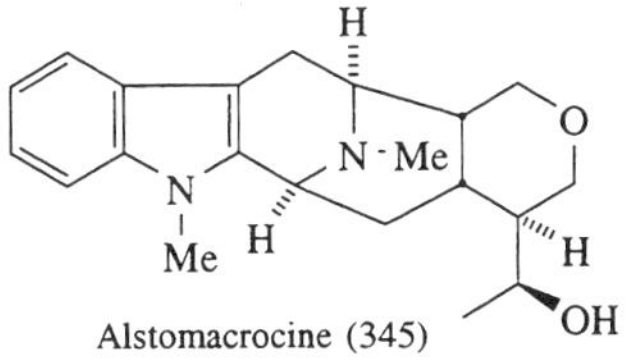

Alstomacrocine (345)

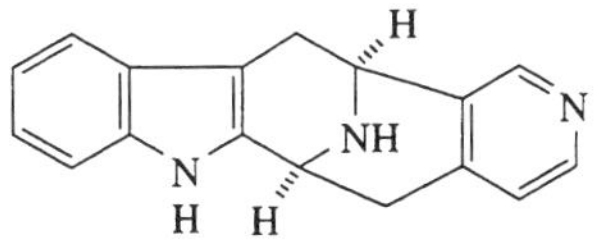

Sellowiine (346)

The 2-acylindole ervatamine family is related biogenetically to the vobasine class. Newly characterized ervatamine alkaloids are methuenine *N*-oxide from *Pterotaberna inconspieua* (P. Bakana, *et al.*, *Planta Med.*, 1984, **57**, 331), *N*(1)-methoxymethuenine and *N*(1)-methoxy-19,20-dehydroervatamine from *Ervatamia malaccensis* (P. Clivio, *et al.*, *Phytochemistry*, 1990, **29**, 2693), 5-oxo-19,20-dehydroervatamine (347) from *Tabernaemontana corymbosa* (T.-S. Kam and K.-Y. Loh, *ibid.*, 1993, **32**, 1357), and 16-*epi*-silicine (348) found in *Pandaca caducifolia* (P. Clivio, *et al.*, *ibid.*, 1995, **40**, 987).

124

(5) Carboline derivatives not containing a C-21 to N-4 bond

The prototypical members of this small group of alkaloids are akagerine and kribine, and several related compounds have been identified recently. Akagerinelactone (349) is present in *Strychnos decussata* along with akagerine and 10-hydroxyakagerine (A.A. Olaniyi and W.N.A. Rolfsen, *Lloydia*, 1980, **43**, 595). The structure of kribine (350) has been revised as shown, and the 10-hydroxy-17-*O*-methyl- and 10-hydroxy-*epi*-17-*O*-methylkribine derivatives have been found in several *Strychnos* species (R. Verpoorte, *et al.*, *J. Chem. Soc., Perkin Trans. 1*, 1984, 1455). Kribine is a mixture of anomers at C-17. The novel strychnohirsutine (351) and a tetradehydro derivative are found in *S. hirsuta* (C. Galeffi, *et al.*, *Tetrahedron*, 1981, **37**, 3167). The structure of correantoside (352), which was earlier isolated from *Psychotria correae*, was revised to that shown (H. Achenbach, *et al.*, *Phytochemistry*, 1995, **38**, 1537). Several other alkaloids found in this plant have a related structure (correantines A, B, and C, and 20-*epi*-correantine B).

Akagerinelactone (349)

Kribine (350)

Strychnohirsutine (351)

Correantoside (352)

The related pyridine alkaloid decussine (353), which was isolated from *Strychnos decussata*, has pronounced muscle relaxant activity *in vitro* and *in vivo* (W.N.A. Rolfsen, *et al.*, *Acta Pharm. Suec.*, 1980, **17**, 105). This plant is also the source of 3,14-dihydrodecussine and 10-hydroxy-3,14-dihydrodecussine (W.N.A. Rolfsen, *et al.*, *J. Nat. Prod.*, 1981, **44**, 415). The structure of mostueine (354) (*Mostuea brunonis*) has been revised to that shown (= 3,14-dihydrodecussine) (L.R. McGee, *et al.*, *Tetrahedron Lett.*, 1984, **25**, 2115). Tubulosine (355) has been isolated from *Pogonopus speciosus* (W.-W. Ma, *et al.*, *J. Nat. Prod.*, 1990, **53**, 1009) and 9-demethyltubulosine is found in *Alangium vitiense* (C. Kan-Fan, *et al.*, *Heterocycles*, 1985, **23**, 1089).

Decussine (353)

Tubulosine (355)

Mostueine (354)

(6) Strychnos alkaloids

Because of strychnine and Calabash-curare, no collection of indole alkaloids is more well known than the large *Strychnos* alkaloid family. There are some 200 *Strychnos* species known. Moreover, only 38 of the 75 known African *Strychnos* species have been investigated for their alkaloid content. Several excellent reviews of this area are available (G. Massiot and C. Delaude, in "The Alkaloids," Vol. 34, Chapter 5, 1988; N.G. Bisset, in "Alkaloids: Chemical and Biological Perspectives," Vol. 8, Chapter 1, 1992; J. Bosch and J. Bonjoch, in "Studies in Natural Products Chemistry," Vol. 1, p. 31, 1988; M. Lounasmaa and P. Somersalo, *Prog. Chem. Org. Nat. Prod.*, 1986, **50**, 27). Unfortunately, the very large number of newly discovered *Strychnos* alkaloids precludes presenting all of their structures herein.

Several simple derivatives of known *Strychnos*-type alkaloids have been discovered. These include 12-methoxyechitamidine (scholarine) and 12-hydroxyechitamidine (scholaricine) (*Alstonia scholaris*) (A. Banerji and A.K. Siddhanta, *Phytochemistry*, 1981, **20**, 540; A. Rahman, *et al.*, *ibid.*, 1985, **24**, 2771), N(1)-formylechitamidine (*A. boonei*) (J.U. Oguakwa, *ibid.*, 1984, **23**, 2708), 18-hydroxynorfluorocurarine (356) (strychnofluorine) (*Strychnos gossweileri*) (J. Quetin-Leclercq, *et al.*, *ibid.*, 1992, **31**, 4347) (*S. ngouniensis*) (G. Massiot, *et al.*, *Tetrahedron*, 1983, **39**, 3645), 10-methoxynorfluorocurarine (*Alstonia lanceolifera*) (T. Ravao, *et al.*, *Phytochemistry*, 1982, **21**, 2160), and norfluorocurarine *N*(4)-oxide (*Ervatamia hirta*) (P. Clivio, *et al.*, *ibid.*, 1991, **30**, 3785). The plant *Strychnos variabilis* contains 16-hydroxyisoretulinal (M. Tits, *et al.*, *ibid.*, 1980, **19**, 1531) and 11-methoxy-*O*-acetylisoretuline (P. Thepenier, *et al.*, *ibid.*, 1990, **29**, 686), while *S. kasengaensis* is reported to contain 11-methoxyisoretuline (P. Thepenier, *et al.*, *ibid.*, 1984, **23**, 2659), and *S. longicaudata* has yielded 23-hydroxy-2,16-dehydroretuline. This plant also contains 1,2-dehydrodesacetylretuline (357) and two C-18 oxygenated *N*(1)-C(2) dihydro analogues (e.g., *N*(1)-desacetyl-18-hydroxyisoretuline) (G. Massiot, *et al.*, *Tetrahedron*, 1983, **39**, 3645).

Strychnofluorine (356)

(357)

Several new akuammicine alkaloids are known. The previously synthesized 19α,20α-epoxyakuammicine has been isolated from *Rauwolfia sellowii* (C.V.F. Batista, *et al.*, *Phytochemistry*, 1996, **41**, 969), and the 12-methoxy derivative is present in *Alstonia lenormandii* (B. Legseir, *et al.*, *ibid.*, 1986, **25**, 1735), a paper that also revised the structure of 10-methoxycompactinervine to that of the 12-methoxy isomer. Both akuammicine *N*(4)-oxide (*A. angustifolia*) (W. Hu, *et al.*, *Planta Med.*, 1989, **55**, 463) and the 11-methoxy analogue (*A. macrophylla*) (F. Abe, *et al.*, *Phytochemistry*, 1994, **35**, 249) are new alkaloids. This plant also contains the novel 11-methoxy-19α,20α-epoxyakuammicine and 11-methoxy-19-oxo-20α-hydroxyakuammicine (F. Abe, *et al.*, *ibid.*, 1994, **35**, 253). The *N*(4)-oxide of demethylalstogustine as well as that of demethylalstogustine are present in *A. angustifolia* (W. Hu, *et al.*, *Planta Med.*, 1989, **55**, 463). Alstogustine (358) and 19-*epi*-alstogustine were also isolated from this plant (W.-L. Hu, *et al.*, *Phytochemistry*, 1989, **28**, 1963). The two diaboline derivatives, 3-hydroxydiaboline (*Strychnos castelnaeana*) (C. Galeffi, *et al.*, *ibid.*, 1982, **21**, 2393) and 12-hydroxy-11-methoxydiaboline (359) (*S. spinosa*) (F.C. Ohiri, *et al.*, *Planta Med.*, 1984, **50**, 446), have been described. Cultured cells of *Aspidosperma quebracho-blanco* have yielded 11-hydroxytubotaiwine (N. Aimi, *et al.*, *Heterocycles*, 1994, **38**, 2411), and stricticine (360) is a novel epoxide from *Rhazya stricta* (A. Rahman, *et al.*, *Tetrahedron Lett.*, 1987, **28**, 3609). This alkaloid is of the same enantiomeric series as (+)-20-*epi*-lochneridine. Strychnopivotine (361) is an unusual alkaloid from *Strychnos variabilis* because it is missing C-17 (M. Tits, *et al.*, *Phytochemistry*, 1980, **19**, 1531), and rosibiline (362) is also found in this plant. Isorosibiline has been isolated from *S. floribunda* (R. Verpoorte, *et al.*, *Planta Med.*, 1981, **42**, 32), and strychnozairine (363) is also found in *S. variabilis* (M. Tits, *et al.*, *Phytochemistry*, 1985, **24**, 205).

Alstogustine (358)

(359)

Stricticine (360)

Strychnopivotine (361)

Rosibiline (362)

Strychnozairine (363)

The plant *Strychnos henningsii*, which had yielded 39 alkaloids as of 1991, is the source of henningsamide (364), the first example of a *Strychnos* alkaloid with a broken C-3 to C-7 bond (G. Massiot, *et al.*, *Phytochemistry*, 1991, **30**, 3449). Three other related alkaloids are also present in this plant, henningsiine (365), 23-hydroxyspermostrychnine, and cyclostrychnine. Plant cell cultures of *Ochrosia elliptica* have furnished epchrosine (19*R*,20*R*-epoxyapparicine) (K.-H. Pawelka, *et al.*, *Plant Cell Rep.*, 1986, **5**, 147). A reinvestigation of tubotaiwine from *Pleiocarpa tubicina* and other plants has revealed that the configuration of C-20 is (*S*) (M. Lounasmaa, *et al.*, *Helv. Chim. Acta*, 1986, **69**, 1343; J. Schripsema, *et al.*, *J. Nat. Prod.*, 1987, **50**, 89). Lagunamine (19-hydroxytubotaiwine)

128

is found in *Alstonia scholaris* (T. Yamauchi, *et al.*, *Phytochemistry*, 1990, **29**, 3321). The related (20*S*)-19,20-dihydrocondylocarpine is found in both *Ervatamia coronaria* and *Alstonia scholaris* (A. Rahman, *et al.*, *Planta Med.*, 1986, 325). The plant *Strychnos soubrensis* has yielded 14-β-hydroxystrychnobrasiline (366) (F.C. Ohiri, *et al.*, *J. Nat. Prod.*, 1983, **46**, 369), and the novel ngouniensine (367), which represents a new structural type and may be formed by reductive opening of a dihydro-preakuammicine, is found in *S. ngouniensis* (G. Massiot, *et al.*, *J. Chem. Soc., Chem. Commun.*, 1982, 768). This alkaloid is the first example of C-3 linked to C-16. This plant also contains *epi*-ngouniensine along with tubotaiwinal (G. Massiot, *et al.*, *Tetrahedron*, 1983, **39**, 3645). Echitamidine N_b-oxide has been characterized in *Alstonia glaucenscens* (N. Keawpradub, *et al.*, *Phytochemistry*, 1994, **37**, 1745). Pericine is a new example of the stemmadenine type which has been isolated from *Picralima nitida* cell suspension cultures (H. Arens, *et al.*, *Planta Med.*, 1982, **46**, 210). This alkaloid has CNS activity.

Henningsamide (364)

Henningsiine (365)

(366)

Ngouniensine (367)

Several new members of the vallesamine-apparicine families have been identified. The plant *Alstonia scholaris* contains 19,20(*Z*)-vallesamine (A. Rahman, *et al.*, *Heterocycles*, 1987, **26**, 413), and 16(*S*)-hydroxy-16,22-dihydroapparicine is found in *Tabernaemontana dichotoma* (R. Perera, *et al.*, *J. Nat. Prod.*, 1984, **47**, 835). Brafouedine and isobrafouedine are dihydroxy and hydroxyether derivatives, respectively, of apparicine found in *Strychnos dinklagei* (S. Michel, *et al.*, *ibid.*, 1986, **49**, 452). Ervaticine is a 16-oxoapparicine alkaloid that may be derived from vallesamine and is found in *Ervatamia coronaria* (A. Rahman and A. Muzaffar, *Heterocycles*,

1985, **23**, 2975). Alstonamine and angustilobine B acid are cyclized analogues of vallesamine that are present in *Alstonia scholaris* (A. Rahman and K.A. Alvi, *Phytochemistry*, 1987, **26**, 2139; T. Yamauchi, *et al.*, *ibid.*, 1990, **29**, 3321). This latter study has also characterized losbanine (6,7-*seco*-6-norangustilobine).

(7) Gelsemine-type alkaloids

Unsurpassed in structural intricacy, new gelsemine alkaloids continue to be discovered. Two outstanding reviews of this field are available (Z. Liu and R. Lu, in "The Alkaloids," Vol. 33, Chapter 2, 1988; H. Takayama and S. Sakai, *Stud. Nat. Prod. Chem.*, 1995, **15**, 465). The latter review covers the literature through 1992.

The structure of koumine (368), which was first isolated in 1931 from *Gelsemium elegans*, was established independently by two groups (F. Khuong-Huu, *et al.*, *Tetrahedron Lett.*, 1981, **22**, 733; C. Liu, *et al.*, *J. Am. Chem. Soc.*, 1981, **103**, 4634).

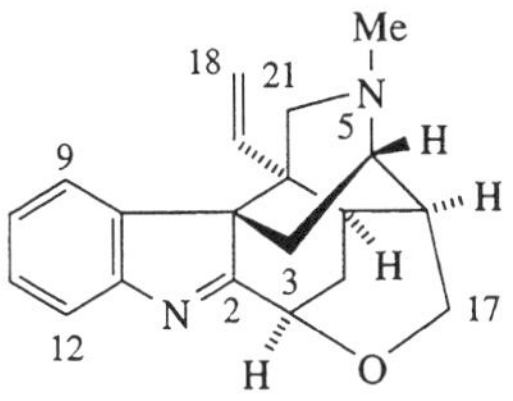

Koumine (368)

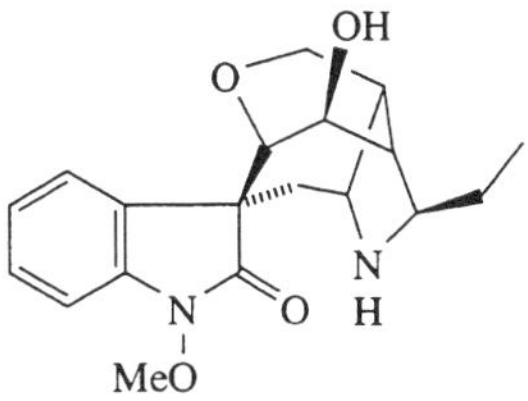

(369)

The new gelsemine-type alkaloids isolated since 1980 are listed below. The reader is referred to an excellent review (H. Takayama and S. Sakai, *Stud. Nat. Prod. Chem.*, 1995, **15**, 465) for the detailed discussion of these alkaloids. Both the 19(*R*)- and 19(*S*)-hydroxydihydrokoumines were found in *Gelsemium elegans* by two groups (L. Lin, *et al.*, *Phytochemistry*, 1990, **29**, 965; F. Sun, *et al.*, *J. Nat. Prod.*, 1989, **52**, 1180). Among the new alkaloids are several examples of the oxindole gelsemine type, such as 19-hydroxydihydrogelsevirine from *G. elegans* (S. Sakai, *et al.*, *Chem. Pharm. Bull.*, 1987, **35**, 4668; D. Ponglux, *et al.*, *Tetrahedron*, 1988, **44**, 5075), 20(*N*-4)-dehydrogelsemicine from *Mostuea brunonis* (M. Onanga and F. Khuong-Huu, *C. R. Acad. Paris C*, 1980, **291**, 191), 21-oxogelsevirine in *G. rankinii* (Y. Schun, *et al.*, *J. Nat. Prod.*, 1986, **49**, 483), 14β-hydroxygelsedine (369) from *G. sempervirens* (Y. Schun and G.A. Cordell, *ibid.*, 1985, **48**, 788). This latter plant has also yielded 19(*R*)-hydroxydihydrogelsevirine, the corresponding acetate, and 19(*R*)-hydroxydihydrogelsemine, compounds that are also found in *G. elegans* and *G. rankinii* (L.-Z. Lin, *et al.*, *Phytochemistry*, 1991, **30**, 679). Rankinidine is a related oxindole-gelsemine alkaloid found in *G. rankinii*, and the same investigation uncovered the new humantenirine and humantenine

130

in *G. elegans* (Y. Schun and G.A. Cordell, *J. Nat. Prod.*, 1986, **49**, 806; J.S. Yang and Y.W. Chen, *Acta Pharm. Sinica*, 1984, **19**, 399, 686). Similarly, three new humantenine types were found in this plant, 20-hydroxydihydrorankinidine, N_a-desmethoxyhumantenine, and 15-hydroxy-humantenine, along with the novel gelsemoxonine (L.-Z. Lin, *et al.*, *Phytochemistry*, 1991, **30**, 1311).

Gelsamydine is a novel alkaloid found in *G. elegans* consisting of a gelsenicine unit linked to a monoterpene (L.-Z. Lin, *et al.*, *J. Org. Chem.*, 1989, **54**, 3199). Gelselegine and 11-methoxy-19(*R*)-hydroxygelselegine are alkaloids also produced by this versatile plant (L.-Z. Lin, *et al.*, *Phytochemistry*, 1990, **29**, 3013).

(ii) Alkaloids with a seco-type unit

Only a few examples of alkaloids with a cleaved C-2 to C-3 bond have been described since the previous coverage. The potent anticholinesterase inhibitor crooksidine (370) is found in *Haplophyton crooksii* (M.A. Mroue, *et al.*, *Phytochemistry*, 1993, **33**, 217). The isomeric (371) has been isolated from *Aspidosperma marcgravianum* (G.M.T. Robert, *et al.*, *J. Nat. Prod.*, 1983, **46**, 694). The interesting salacin (372) and 3-oxo-7-hydroxy-3,7-secorhynchophylline (373) are found in *Uncaria salaccensis* (D. Ponglux, *et al.*, *Chem. Pharm. Bull.*, 1990, **38**, 573).

Crooksidine (370) R_1 = O, R_2 = H_2
(371) R_1 = H_2, R_2 = O

Salacin (372)

(373)

(iii) Alkaloids with the Aspidosperma unit
(1) Aspidospermine-aspidofractinine Alkaloids

Alkaloids of this category represent an enormous range of structural types. Without precise knowledge of the biogenetic origins, it is difficult to classify some of the more unusual examples into distinct alkaloid subcategories. Many new *Aspidosperma* alkaloids have been discovered since the last Chapter. Two reviews dealing with syntheses of *Aspidosperma* alkaloids and vindoline have appeared (L.E. Overman and M. Sworin, "Alkaloids: Chemical and Biological Perspectives," Vol. 3, Chapter 7,

1985; B. Danieli, *et al.*, *Stud. Nat. Prod. Chem.*, 1989, **4**, 29).

The new species *Vinca sardoa* has yielded the novel N_a-methyl-14,15-didehydroaspidospermidine (S. Crippa, *et al.*, *Heterocycles*, 1990, **31**, 1663), and 1,2-dehydroaspidospermidine N_b-oxide is present in *Rhazya stricta* (A. Rahman and K. Zaman, *Phytochemistry*, 1986, **25**, 1779). This plant has also provided the first alkaloid glucoside with an oxidized sugar, aspidospermidose (374) (A. Rahman, *et al.*, *J. Chem. Soc., Perkin Trans. 1*, 1987, 1701). Anomaline (375) and two related alkaloids are oxygenated A-ring compounds found in *Microplumeria anomala* (A.I. Reis Luz, *et al.*, *Phytochemistry*, 1983, **22**, 2301). Rosicine is a novel alkaloid from *Catharanthus roseus* that is the 14,15-β-epoxide of tabersonine but lacking the C-20 ethyl group (A. Rahman, *et al.*, *Tetrahedron Lett.*, 1984, **25**, 6051). Two new alkaloids of the pseudo-aspidospermidine class have been isolated from *Tabernaemontana eglandulosa* (T.A. van Beek, *et al.*, *Tetrahedron*, 1984, **40**, 737), and 11-hydroxyvincadifformine, which has significant antifertility activity, has been characterized from *Melodinus hemsleyanus* (L.-W. Guo and Y.-L. Zhou, *Phytochemistry*, 1993, **34**, 563). Modestanine (376) (11-demethoxyvandrikine) is found in *Hazunta modesta* along with 14,15-dihydroxyvincadifformine (A.-M. Bui, *et al.*, *ibid.*, 1980, **19**, 1473). The plant *Alstonia venenata* has yielded 19-*epi*-echitoveniline, which is the 3,4,5-trimethoxybenzoate ester of 19-hydroxyvincadifformine (minovincinine) (P.L. Majumder, *et al.*, *Tetrahedron*, 1981, **37**, 1243). Mehranine (377) is a novel epoxide found in *Tabernaemontana divaricata* (T.-S. Kam and S. Anuradha, *Phytochemistry*, 1995, **40**, 313), and *T. albiflora* has furnished a dihydroxylated alkaloid of the pseudo-tabersonine type (C. Kan, *et al.*, *Planta Med.*, 1981, **41**, 195).

Aspidospermidose (374)

Anomaline (375)

Modestanine (376)

Mehranine (377)

Strictanine, which is N_a-formyl-16-α-hydroxyaspidospermidine is found in *Rhazya stricta* (A. Rahman and S. Malik, *Phytochemistry*, 1987, **26**,

589), and *Petchia ceylanica* contains 19(*R*)- and 19(*S*)-*epi*-misiline (378) and the novel ether petchicine (379) (A. Rahman, *et al.*, *ibid.*, 1987, **26**, 543; 1989, **28**, 3221). The novel pyrrolidinone-vindoline derivative bannucine (380) is found in *Catharanthus roseus* (A. Rahman, *et al.*, *J. Chem. Soc., Perkin Trans. 1*, 1986, 923), and haplocidiphytine (381) and norisohaplophytine are two alkaloids from *Haplophyton cimicidum* that are comprised of a rearranged canthiphytine unit and a norisoaspidophytine unit (A.A. Adesomoju, *et al.*, *Heterocycles*, 1983, **20**, 1511).

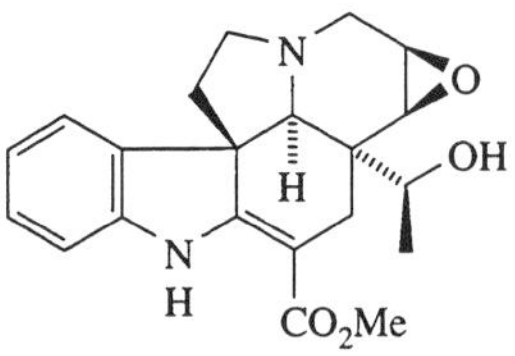

(378)

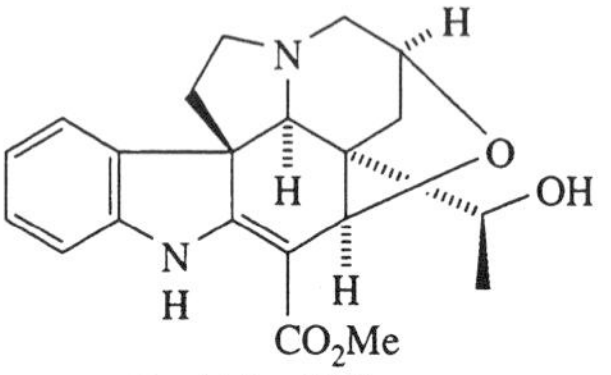

Petchicine (379)

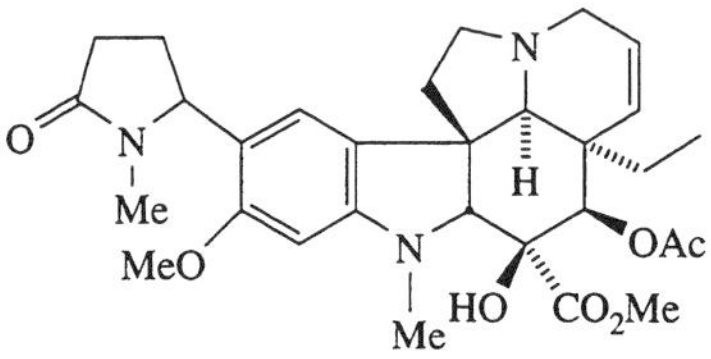

Bannucine (380)

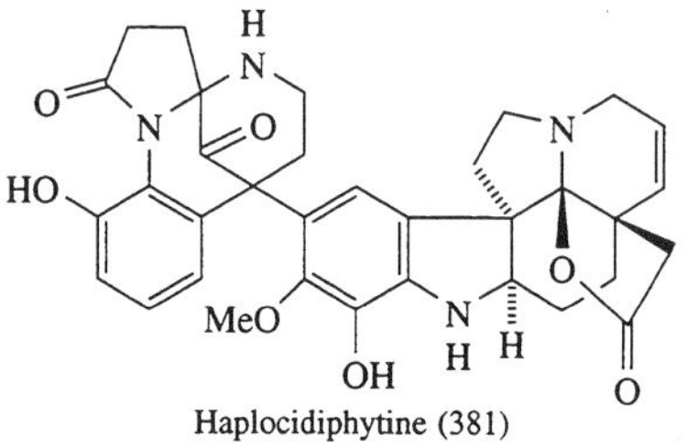

Haplocidiphytine (381)

Several new examples of cleavamine-velbanamine alkaloids have been described. For example, 14(*S*),20(*R*)-velbanamine is found in *Tabernae-montana eglandulosa* (T.A. van Beek, *et al.*, *Tetrahedron*, 1984, **40**, 737), and ervatinine (382) and stapfinine (383) are both alkaloids from *Ervatamia coronaria* (A. Rahman, *et al.*, *Phytochemistry*, 1985, **24**, 2473; 1986, **25**, 1781), a plant that also yielded the somewhat related hyderabadine (A. Rahman and N. Daulatabadi, *Z. Naturforsch.*, 1983, **38B**, 1310). Strictanol is the 3-hydroxyindolenine of the voaphylline type and is found in

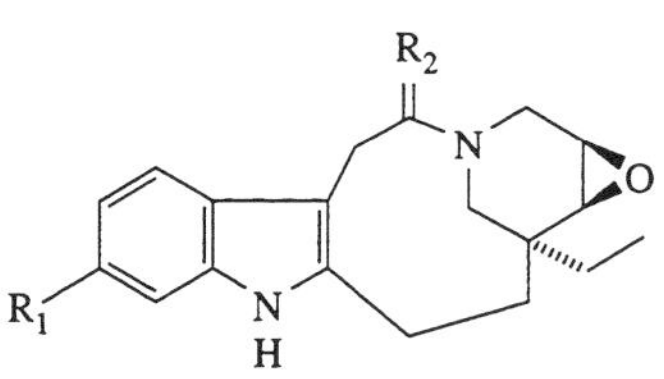

Ervatinine (382) R$_1$ = OH, R$_2$ = O
Stapfinine (383) R$_1$ = H, R$_2$ = H, OH

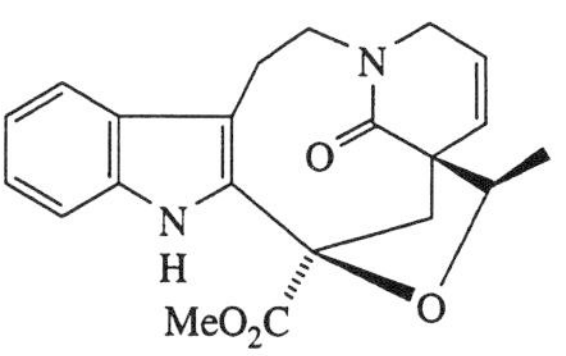

Suaveolenine (384)

Rhazya stricta (A. Rahman and S. Malik, *Phytochemistry*, 1987, **26**, 589).
The plant *Melodinus suaveolens* has yielded the new suaveolenine (384)
(J.H. Ye, *et al.*, *ibid.*, 1991, **30**, 3168).

The novel aspidofractinine *N*-methylindoline alkaloids (385) and the
14,15-dihydro derivative, and N_a-methyl-14,15-didehydrotuboxenine (386)
are found in the newly discovered *Vinca sardoa* (S. Crippa, *et al.*, *Heterocycles*, 1990, **31**, 1663). Other related alkaloids are 16β-hydroxy-19(*R*)-
and -19(*S*)-vindolinine from *Melodinus hemsleyanus* (L.-W. Guo and Y.-
L. Zhou, *Phytochemistry*, 1993, **34**, 563). The related 15α-hydroxy-
14,15-dihydrovindolinine and the 16-epimer are found in *M. morsei* (Y.-L.
He, *et al.*, *ibid.*, 1994, **37**, 1055), and the new 16-*epi*-19(*S*)-vindolinine
has been identified in *Catharanthus roseus* (A. Rahman, *et al.*, *ibid.*, 1983,
22, 1021). The plant *Tabernaemontana albiflora* produces four alkaloids of
the 21-nor-padolane-ibophyllidine type, such as 1-hydroxy-19-ibophillidine
(387) (C. Kan, *et al.*, *Tetrahedron Lett.*, 1980, **21**, 3363). Interestingly,
(+)-aspidospermine has been found in a plant for the first time (*Aspidosperma pyrifolium*) along with 6-demethoxypyrifoline (388) (A.A.
Craveiro, *et al.*, *Phytochemistry*, 1983, **22**, 1526).

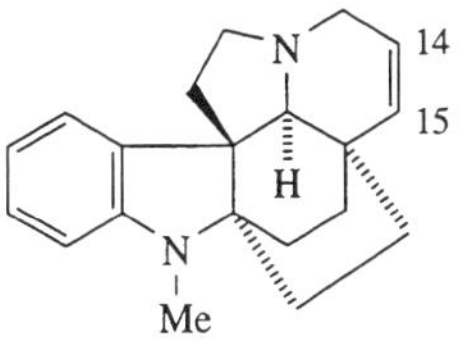

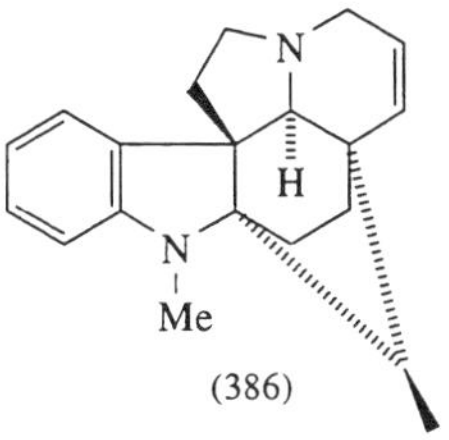

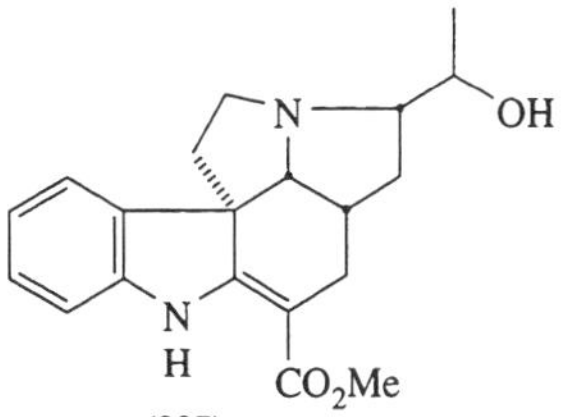

Several new examples of the aspidofractinine-type from *Kopsia* sp. have
been characterized over the past 15 years. For example, *Kopsia deverrei*
has yielded kopsinone (389) and two other new kopsinine alkaloids (C.
Kan-Fan, *et al.*, *J. Nat. Prod.*, 1988, **51**, 703), while *K. officinalis*
contains 12-methoxykopsinalidine and the 11,12-methylenedioxy analogue
(X.Z. Feng, *et al.*, *Planta Med.*, 1983, **48**, 280). This plant has also
yielded N_a-carbomethoxy-11-hydroxy-12-methoxykopsinaline, N_a-carbo-
methoxy-11-methoxy-12-hydroxykopsinaline, and kopsamine N_b-oxide
(J.-J. Zheng, *et al.*, *Acta Chim. Sin.*, 1989, 168), as well as three new

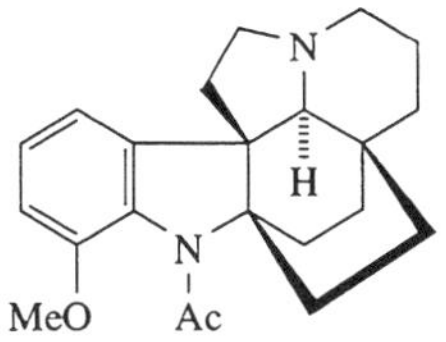

134

alkaloids with the two-carbon bridge linked between C-6 and C-20, e.g., (390) (C. Wei-shin, *et al.*, *Ann.*, 1981, 1886). *Kopsia dasyrachis* has afforded kopsidasine (391), the corresponding N_b-oxide, and a rearranged kopsidasinine (K. Homberger and M. Hesse, *Helv. Chim. Acta*, 1982, **65**, 2548), and five new alkaloids of the plumeran type are found in *K. profunda*, including N_a-carbomethoxy-11,12-methylenedioxy-$\Delta^{16,17}$-kopsinine, N_a-carbomethoxy-12-methoxy-$\Delta^{16,17}$-kopsinine, the N_b-oxides of these two alkaloids, and (392), which is 12-hydroxykopsijasmine (T.-S. Kam and P.-S. Tam, *Phytochemistry*, 1990, **29**, 2321; 1995, **39**, 469).

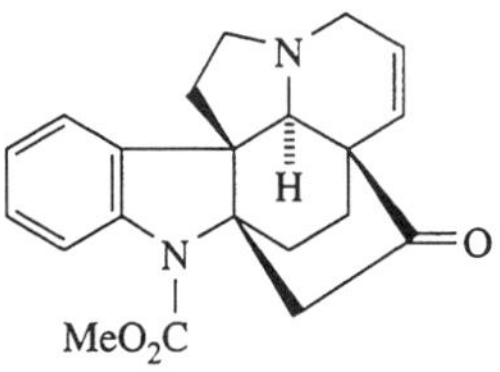

Kopsinone (389)

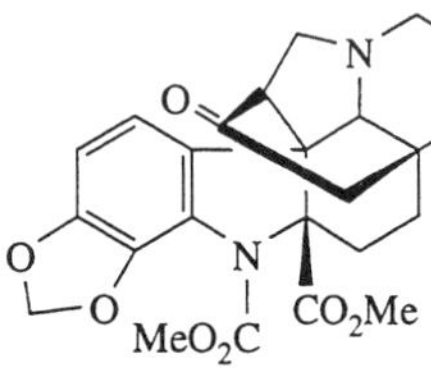

(390)

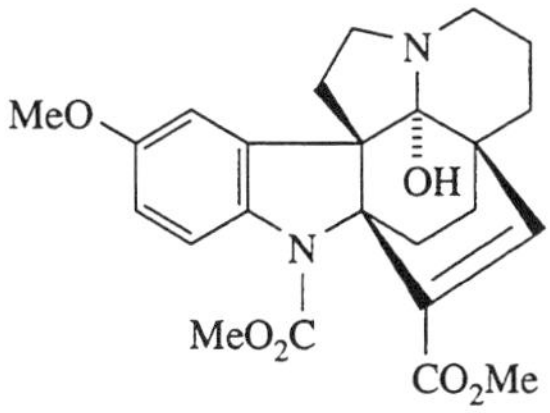

Kopsidasine (391)

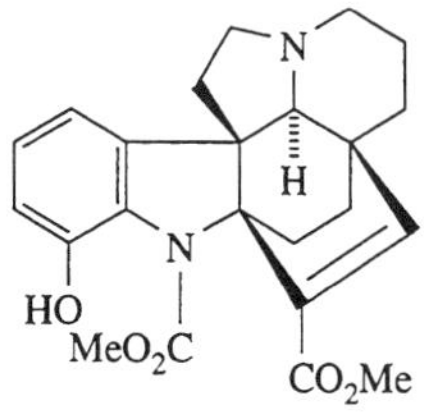

(392)

The Malaysian *Kopsia teoi* has provided several novel alkaloids in this family. For example, in addition to the new 17α-hydroxy-$\Delta^{14,15}$-kopsinine and its 16-epimer, the novel (393), and the related kopsinginine have been identified from this plant (T. Varea, *et al.*, *J. Nat. Prod.*, 1993, **56**, 2116; T.-S. Kam, *et al.*, *Phytochemistry*, 1993, **32**, 1343). This plant has also yielded both the novel kopsinitarines A-C, e.g., (394) (T.-S. Kam, *et al.*, *Tetrahedron Lett.*, 1994, **35**, 4457), and the kopsidines A (395) and B (396) (T.-S. Kam, *et al.*, *ibid.*, 1993, **34**, 1819), both groups of which contain a novel ring skeleton. The first example of a D/E ring seco *Kopsia* alkaloid is kopsijasminilam (397), which, along with two other related alkaloids, and kopsijasmine (398) and jasminiflorine (12-methoxyfruticosine) are present in *K. jasminiflora* (N. Ruangrungsi, *et al.*, *ibid.*, 1987, **28**, 3679). The plant *Melodinus reticulatus* has yielded the new 19-hydroxyvenalstonine, 19-hydroxyvenalstonidine, and 3-oxovenalstonidine (H. Mehri, *et al.*, *Planta Med.*, 1983, **48**, 72), while *M. guillauminii* has produced this new alkaloid, 14,15-*seco*-3-oxokopsinal (399) (M. Zeches, *et al.*, *Phytochemistry*, 1984, **23**, 171).

(393)

Kopsinitarine A (394)

Kopsidine A (395) R = Me
Kopsidine B (396) R = Et

Kopsijasminilam (397)

Kopsijasmine (398)

(399)

Other *Kopsia* species from Malaysia have revealed further structural virtuosity. For example, *K. lapidilecta* has yielded lapidilectines A (400), B (401), isolapidilectine, and two related alkaloids, all of which contain novel ring structures (K. Awang, *et al.*, *Tetrahedron Lett.*, 1992, **33**, 2493; *J. Nat. Prod.*, 1993, **56**, 1134). Similarly, the unusual lundurines A (402), B (403), and C, and (404) have been found in *K. tenuis* (T.-S. Kam, *et al.*, *Tetrahedron Lett.*, 1995, **36**, 759). The North Borneo *K. pauciflora* has yielded paucidactines A (405) and B (406), which contain a lactone unit (T.-S. Kam, *et al.*, *ibid.*, 1996, **37**, 3603). The unusual dichomine (407) is found in *Tabernaemontana dichotoma* (P. Perera, *et al.*, *Planta Med.*, 1983, **49**, 232).

Lapidilectine A (400)

Lapidilectine B (401)

Lundurine A (402) R = O
Lundurine B (403) R = H₂

(404)

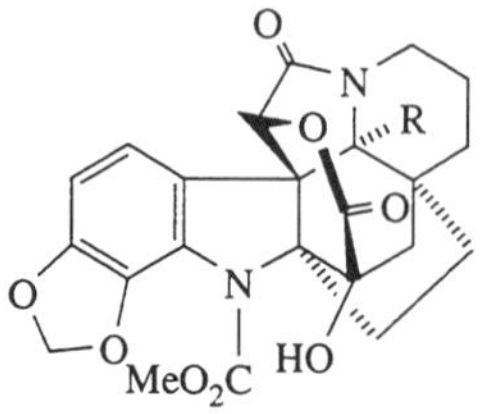

Paucidactine A (405) R = OH
Paucidactine B (406) R = H

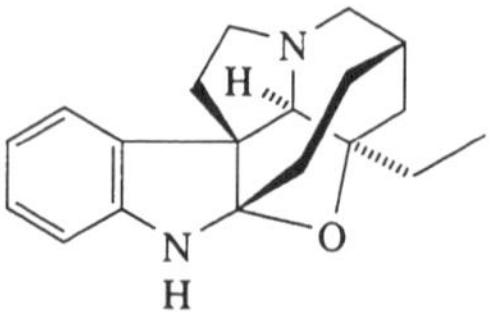

Dichomine (407)

Gilbertin (408) is a uleine-type alkaloid present in *Aspidosperma gilberti* (E.C. Miranda and S. Blechert, *Tetrahedron Lett.*, 1982, **23**, 5395), and undulifoline (409) is another uleine-type from *Alstonia undulifolia* (G. Massiot, *et al.*, *Phytochemistry*, 1992, **31**, 1078). Melonine (410) and the N_b-oxide are found in *Melodinus celastroides* (S. Baassou, *et al.*, *Tetrahedron Lett.*, 1983, **24**, 761). Upon being heated, this alkaloid, which has a new skeleton, rearranges to N_a-norvallesamidine and aspidospermidine. The ring cleaved trichophylline (410) from *Catharanthus trichophyllus* also has a novel skeleton that appears to be derived from a typical *Aspidosperma* alkaloid (S. Mukhopadhyay, *et al.*, *Tetrahedron*, 1983, **39**, 3639). Similarly, goniomitine (412) (*Gonioma malagasy*) is proposed to have a biogenesis from vincadifformine (L. Randriambola, *et al.*, *Tetrahedron Lett.*, 1987, **28**, 2123). A newly proposed biogenesis of vindoline from tabersonine has been advanced (J. Balsevich, *et al.*, *Heterocycles*, 1986, **24**, 2415).

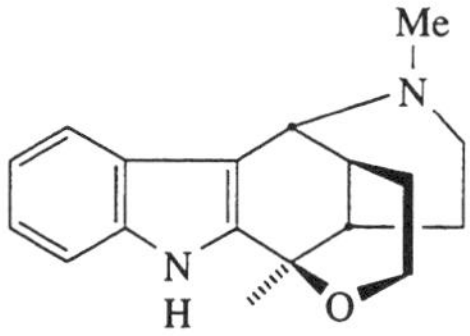

Gilbertin (408)

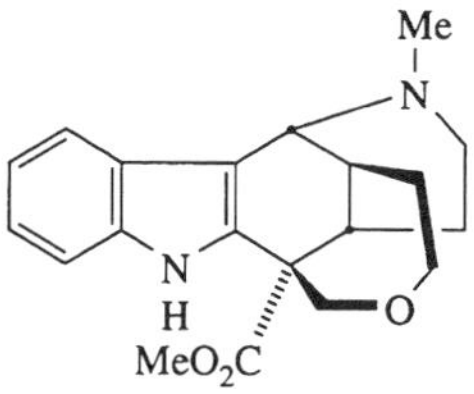

Undulifoline (409)

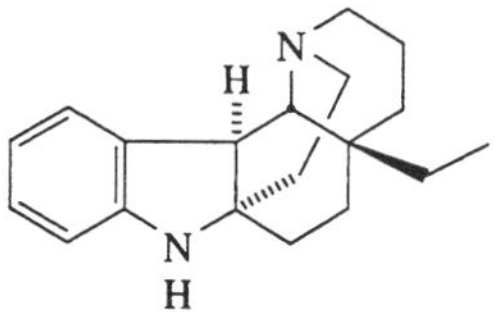

Melonine (410)

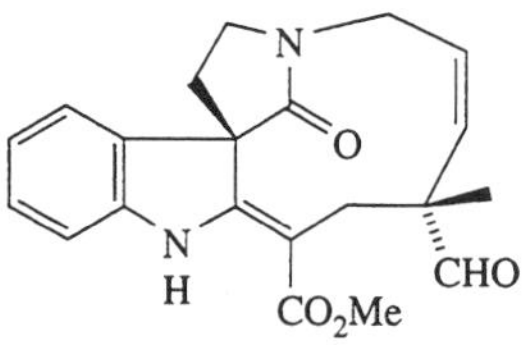

Trichophylline (411)

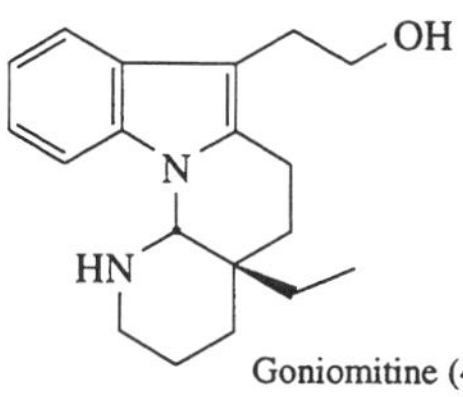

Goniomitine (412)

(2) Eburnamine group

The eburnamine subgroup of *Aspidosperma* alkaloids is represented by several new examples since 1981. Two reviews of this area are available

(W. Döpke, in "The Alkaloids," Vol. 20, Chapter 2, 1981; M. Lounasmaa and A. Tolvanen, in "The Alkaloids," Vol. 42, Chapter 1, 1992).

The previously known synthetic compound dihydroeburnamenine (413) is present in *Rhazya stricta* (A. Rahman, *et al.*, *Phytochemistry*, 1991, **30**, 1285). Likewise, the synthetic compounds *O*-ethyleburnamine, *O*-methyl-eburnamine, *O*-methylisoeburnamine, and isoeburnamine (414) are found in plants (*Hunteria elliotii*; J. Vercauteren, *et al.*, *ibid.*, 1980, **19**, 1959) (*H. zeylanica*; L.S.R. Arambewela and F. Khuong-Huu, *ibid.*, 1981, **20**, 349) (*Kopsia officinalis*; X.Z. Feng, *et al.*, *Planta Med.*, 1983, **48**, 280). It is possible that some of these eburnamine ethers are isolation artefacts. Eburnaminol (415) and larutensine (416) are produced by *K. larutensis* (K. Awang, *et al.*, *Phytochemistry*, 1991, **30**, 3164). This plant has also yielded eburnamonine N_b-oxide and larutenine (= larutensine) (T.-S. Kam, *et al.*, *ibid.*, 1992, **31**, 2936). *Melodinus guillauminii* is the source for the novel 11-methoxy-Δ^{14}-vincanol and 11-methoxy-$\Delta^{14,15}$-vincamenine (M. Zeches, *et al.*, *ibid.*, 1984, **23**, 171). The vincamine-type tacamine (417) is found in *Tabernaemontana eglandulosa* (T.A. van Beek, *Tetrahedron Lett.*, 1982, **23**, 4827), along with seven related new tacamine alkaloids (T.A. van Beek, *et al.*, *Tetrahedron*, 1984, **40**, 737).

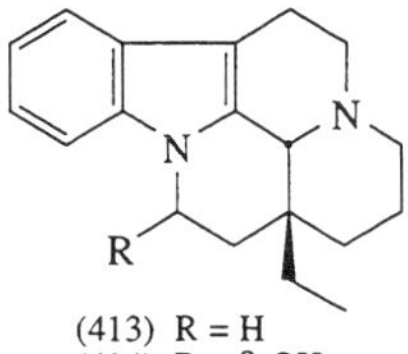

(413) R = H
(414) R = β-OH

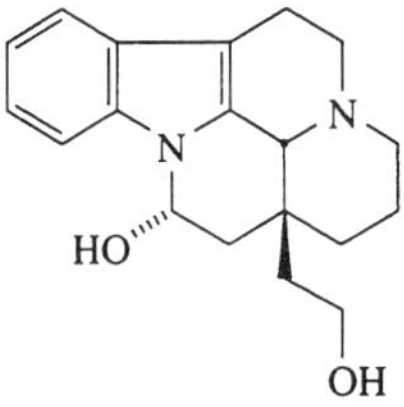

Eburnaminol (415)

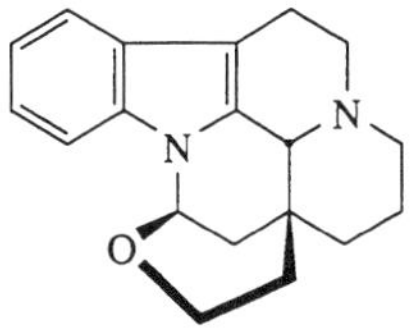

Larutensine (416)

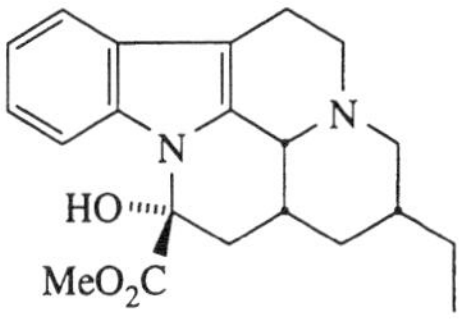

Tacamine (417)

Vincapusine (demethoxyvincarodine) (418) is an alkaloid found in *Vinca pusilla* (A.K. Mitra, *et al.*, *Phytochemistry*, 1981, **20**, 865). Strempelio-pine (419) is present in *Strempeliopsis strempeliodes* (A. Laguna, *et al.*, *Planta Med.*, 1984, **51**, 285).

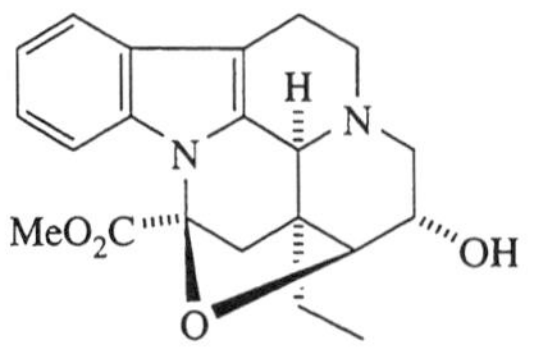

138

Vincapusine (418)

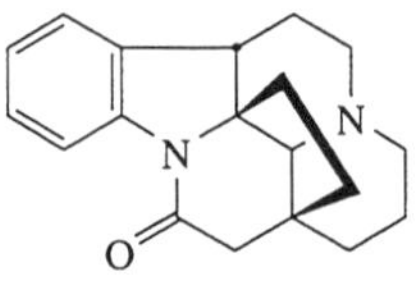

Strempeliopine (419)

(iv) Alkaloids containing the Iboga unit

Most of the newly discovered Iboga alkaloids are oxygenated derivatives of previously known alkaloids. For example, 19-hydroxycoronaridine was discovered in both *Tabernaemontana glandulosa* (H. Achenbach, *et al.*, *Phytochemistry*, 1980, **19**, 2185) and *T. divaricata* (K. Rastogi, *et al.*, *ibid.*, 1980, **19**, 1209). This latter plant contains the new 5- and 6-oxo-coronaridine and 5-hydroxy-6-oxocoronaridine. The plant *T. markgrafiana* has yielded 5,6-dehydrocoronaridine (420), 3(R)-methoxycoronaridine, 3(R)-methoxyvoacangine, and 10,11-demethoxychippiine (H.B. Nielsen, *et al.*, *ibid.*, 1994, **37**, 1729), and *Peschiera buchtieni* contains 18,19(R)-dihydroxycoronaridine (M. Azoug, *et al.*, *ibid.*, 1995, **39**, 1223). Albifloranine, which is found in *T. albiflora*, is 18-hydroxycoronaridine (C. Kan, *et al.*, *Planta Med.*, 1981, **41**, 72), and *Tabernanthe pubescens* has furnished 10-hydroxycoronaridine, 10-hydroxyheyneanine, 3,6-oxidoiboxygaine, and 3,6-oxidoibogaine (421) (T. Mulamba, *et al.*, *Lloydia*, 1981, **44**, 184). *Ervatamia heyneana* also contains 10-hydroxycoronaridine, along with the novel 10-methoxyglandine N_b-oxide (422), and heyneantine (19(S)-3,19-oxidovoacangine) (S.P. Gunasekera, *et al.*, *Phytochemistry*, 1980, **19**, 1213). The related 3-oxo-19-*epi*-heyneanine and 3-hydroxy-3,4-*seco*-coronaridine (423) are produced by *E. polyneura* (P. Clivio, *et al.*,

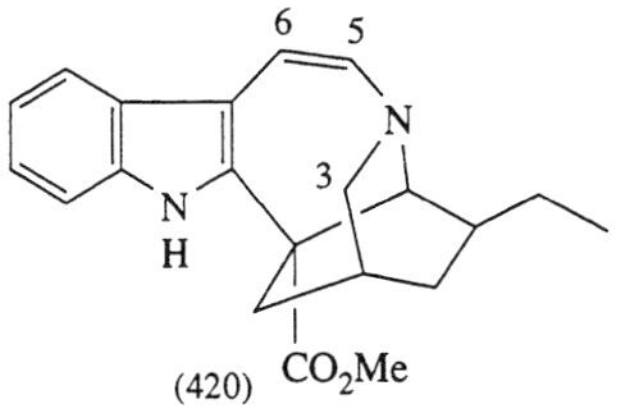

(420)

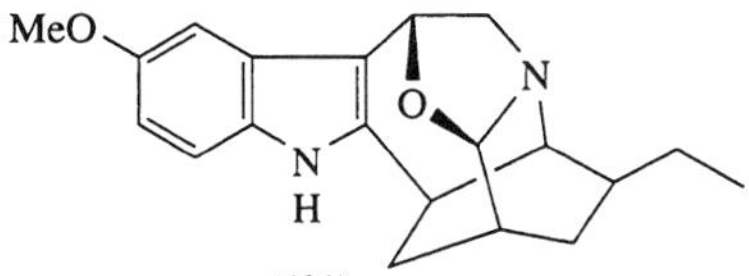

(421)

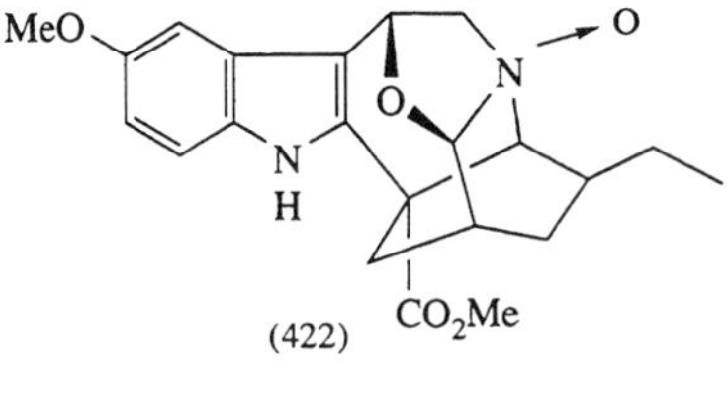

(422)

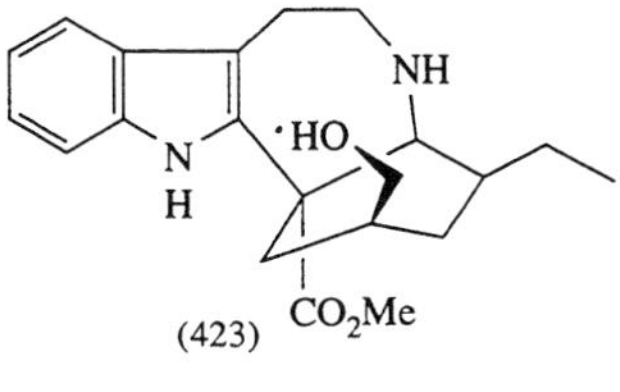

(423)

ibid., 1990, **29**, 3007). The new 10-methoxyglandine and 10-hydroxyhey-
neanine are found in *Peschiera echinata* (N. Ghorbel, *et al.*, *Lloydia*, 1981,
44, 717). The plant *Anartia* cf. *meyeri* has afforded 11-hydroxycoronari-
dine, and 10- and 11-hydroxyheyneanine (F. Ladhar, *et al.*, *ibid.*, 1981,
44, 459).

A callus culture of *Tabernaemontana elegans* has yielded the new 3-
oxoisovoacangine (R. van der Heijden, *et al.*, *Phytochemistry*, 1986, **25**,
843), and 19-oxavoacristine is present in *T. citrifolia* (J.P. Kutney and I.
Perez, *Helv. Chim. Acta*, 1982, **65**, 2242). The plant *Ervatamia
hainanensis* has yielded 10-hydroxyheyneanine and 3-(β-hydroxyethyl)-
coronaridine (S.Z. Feng, *et al.*, *Planta Med.*, 1982, **44**, 212). The
Thailand *E. coronaria* var. *plena* contains 19(*S*)-heyneanine hydroxyindole-
nine (P. Sharma and G.A. Cordell, *J. Nat. Prod.*, 1988, **51**, 528). The
plant *T. pachysiphon* is a source of both 3(*R*)- and 3(*S*)-hydroxycono-
pharyngine (424) (T.A. van Beek, *et al.*, *Phytochemistry*, 1984, **23**,
1771), and *T. dichotoma* has yielded 3,19(*R*)-oxidocoronaridine (P. Perera,
et al., *ibid*, 1985, **24**, 2097). The first natural *allo* iboga alkaloid is 16-
hydroxy-*allo*-ibogamine (425), isolated from *Strychnos ngouniensis* (G.
Massiot, *et al.*, *J. Chem. Soc., Chem. Commun.*, 1983, 1018).

(v) Novel types
This subsection in the previous volume included only the *Aristotelia*
alkaloids. The present discussion covers new examples of this type as well
as other alkaloids that do not obviously fit into other categories. Two excellent reviews of *Aristotelia* alkaloids have appeared since 1981 (I.R.C. Bick
and M.A. Hai, in "The Alkaloids," Vol. 24, Chapter 3, 1985; H.-J.
Borschberg, *Stud. Nat. Prod. Chem.*, 1992, **11**, 277). In view of the latter
review, only *Aristotelia* alkaloids isolated since 1991 will be depicted here.
Those alkaloids isolated during the period between 1980 and 1991 are
aristoserratine (*A. serrata* and *A. peduncularis*) (M.A. Hai, *et al.*, *Helv.
Chim. Acta*, 1980, **63**, 2130; I.R.C. Bick, *et al.*, *Aust. J. Chem.*, 1983,
36, 1037), tasmanine (*A. peduncularis*) (R. Kyburz, *et al.*, *Helv. Chim.
Acta*, 1981, **64**, 2555), makonine and makomakine (*A. serrata*) (I.R.C.
Bick and M.A. Hai, *Heterocycles*, 1981, **16**, 1301), aristomakine (*A.
serrata*) (I.R.C. Bick and M.A. Hai, *Tetrahedron Lett.*, 1981, **22**, 3275),
serratenone (*A. serrata*) (I.R.C. Bick, *et al.*, *Heterocycles*, 1983, **20**, 667),
peduncularistine, triabunnine, and aristolarine (*A. peduncularis*) (R.

140

Kyburz, *et al.*, *Helv. Chim. Acta*, 1984, **67**, 804), aristolasicone, aristo-lasicol, aristocarbinol, aristolasicolone, 11-*epi*-aristoteline, 9,10-dehydro-aristoteline, and 3-*epi*-aristoserratenine (*A. australasica*) (C. Kan-Fan, *et al.*, *Tetrahedron*, 1988, **44**, 1651), aristolasol and aristolasene (*A. australasica*) (J.C. Quirion, *et al.*, *Phytochemistry*, 1988, **27**, 3337), aristofruti-cosine (*A. fruticosa*) (I.R.C. Bick, *et al.*, *Tetrahedron Lett.*, 1988, **29**, 3355), 17-hydroxyhobartine and hobartidiol (*A. australasica*) (J.C. Quirion, *et al.*, *J. Nat. Prod.*, 1990, **53**, 713), and 8-oxo-9-dehydromakomakine and 8-oxo-9-dehydrohobartine (*A. chilensis*) (C. Cespedes, *et al.*, *Phyto-chemistry*, 1990, **29**, 1354). In addition, the structure of isopeduncularine, first isolated in 1975 (*A. serrata, A. fruticosa, A. peduncularis*) (B.F. Anderson, *et al.*, *J. Chem. Soc., Chem. Commun.*, 1975, 511), has been determined (I.R.C. Bick, *et al.*, *Tetrahedron*, 1985, **41**, 3127). An X-ray structure of aristone (*A. chilensis*) supports the novel structure originally proposed for this alkaloid (V. Zabel, *et al.*, *J. Chem. Soc., Perkin Trans. 1*, 1980, 2842). However, the structure originally suggested for serratoline (*A. serrata*) has been revised to that of a 3-hydroxyindolenine (426) (I.R.C. Bick, *et al.*, *Heterocycles*, 1983, **20**, 667). Likewise, the previously isolated aristolasicone and 11-*epi*-aristoteline, which is now renamed *allo*-aristoteline (*A. australasica*), have had their structures revised to (427) and (428), respectively (J.-C. Quirion, *et al.*, *J. Org. Chem.*, 1992, **57**, 5848). It is quite possible that the other alkaloid members of this family related to aristolasicone have such an "inverted" indole unit.

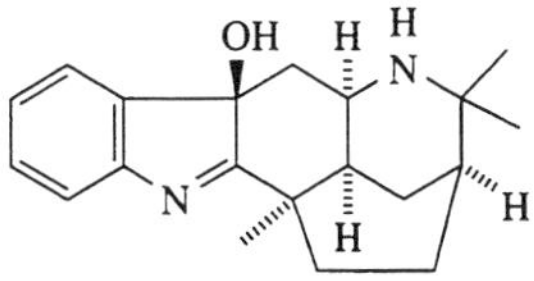

Serratoline (426)

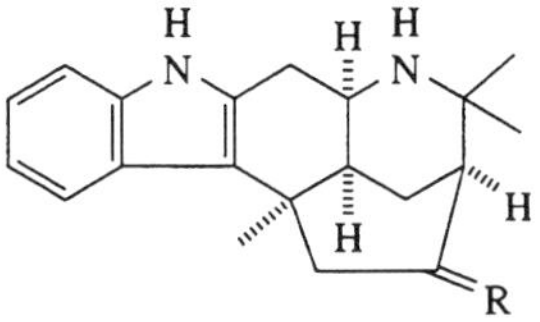

Aristolasicone (427)   R = O
allo-Aristoteline (428)   R = H$_2$

A few new members of the ellipticine (pyrido[4,3-*b*]carbazole) family of alkaloids have been isolated from *Strychnos dinklagei*, including 17-oxo-ellipticine (S. Michel, *et al.*, *Tetrahedron Lett.*, 1980, **21**, 4027), 10- and 18-hydroxyellipticine, and 17-oxoellipticine N_b-oxide (429) (S. Michel, *et al.*, *Lloydia*, 1982, **45**, 489). 3-Hydroxy-1,2,3,4-tetrahydroolivacine has been reported to be present in *Peschiera buchtieni* (M. Azoug, *et al.*, *Phyto-chemistry*, 1995, **39**, 1223). The novel alkaloid (430) has been found in *Aspidosperma gilbertii*, and the structure verified by synthesis (E.C. Miranda, *et al.*, *Chem. Ber.*, 1980, **113**, 3245). The biogenesis of this alkaloid is proposed to be similar to that of olivacine but diverges at a critical cyclization stage.

141

(429)

(430)

Over the past decade, an enormous number of naturally occurring organo-
halogen compounds have been discovered. Included in this array of novel
metabolites are a large group of indole-isonitrile alkaloids from blue-green
algae, most compounds of which contain chlorine. These unique isonitriles
appear to be derived from tryptophan and a monoterpene unit. The origin of
the isonitrile group in hapalindole A (431) involves a C_1 donor related to
tetrahydrofolate metabolism (V. Bornemann, *et al.*, *J. Am. Chem. Soc.*,
1988, **110**, 2339). This metabolite is the major isonitrile in the terrestrial
blue-green alga *Hapalosiphon fontinalis* (R.E. Moore, *et al.*, *ibid.*, 1984,
106, 6456). Subsequently, more than 20 related indoles have been isolated
from cultures of this alga (R.E. Moore, *et al.*, *J. Org. Chem.*, 1987, **52**,
1036, 3773; *Phytochemistry*, 1989, **28**, 1565), and one new hapalindole
has been found in *H. laingii*, in addition to several known hapalindoles (D.
Klein, *et al.*, *J. Nat. Prod.*, 1995, **58**, 1781). The absolute configuration
of these compounds ws established by synthesis (V. Vaillancourt and K.F.
Albizati, *J. Am. Chem. Soc.*, 1993, **115**, 3499). Related metabolites have
been found in other blue-green algae. Thus, hapalindolinone A (432) and
the non-chlorinated analogue are produced by *Fischerella* sp. (R.E.
Schwartz, *et al.*, *J. Org. Chem.*, 1987, **52**, 3704), and fischerindole L
(433) is a novel octahydroindeno[2,1-*b*]indole from *F. muscicola* (A. Park,
et al., *Tetrahedron Lett.*, 1992, **33**, 3257). The structurally related isonitriles, ambiguine isonitriles A-F, e.g., (434), are produced by the terrestrial
blue-green algae *F. ambigua*, *H. hibernicus*, and *Westiellopsis prolifica*
(T.A. Smitka, *et al.*, *J. Org. Chem.*, 1992, **57**, 857). The welwitindoli-
nones are another group of 15 related isonitrile indole alkaloids from the

Hapalindole A (431)

Hapalindolinone A (432)

142

Fischerindole L (433)

Ambiguine E Isonitrile (434)

Welwitindolinone A
Isonitrile (435)

blue-green algae *H. welwitschii* and *Westiella intricata* (K. Stratmann, *et al.*, *J. Am. Chem. Soc.*, 1994, **116**, 9935). One example is welwitindolinone A (435).

Cultured cells of *Aspidosperma quebracho blanco* have yielded aspidochibine having a completely novel structure (N. Aimi, *et al.*, *Tetrahedron Lett.*, 1991, **32**, 4949).

3. Bisindole Alkaloids

The extremely large number of diverse and novel bis-indole alkaloids that have been discovered over the last 15 years has necessitated the inclusion of some new categories. A review of noniridoid bisindole alkaloids has appeared (J. Sapi and G. Massiot, in "The Alkaloids," Vol. 47, Chapter 3, 1995).

(a) Compounds containing two identical "halves" linked symmetrically

(i) Calycanthaceous alkaloids
The Colombian poison-arrow frog *Phyllobates terribilis* secretes *d*-chimonanthine (436) the enantiomer of the plant alkaloid (T. Tokuyama and J.W. Daly, *Tetrahedron*, 1983, **39**, 41). Several new polymers involving 6-8 *N*-methyltryptamine units, similar to psychotridine, are found in *Calycodendron milnei* (Y. Adjibade, *et al.*, *Planta Med.*, 1990, **56**, 212), and the related compounds quadrigemine C and isopsychotridines A and B are found in *Psychotria oleoides* (F. Libot, *et al.*, *J. Nat. Prod.*, 1987, **50**, 468). This same plant has yielded psycholerine (437) (F. Guéritte-Voegelein, *et al., ibid.*, 1992, **55**, 923), while *Idiospermum australiense* contains (–)-idiospermuline (438) (R.K. Duke, *et al., ibid.*, 1995, **58**, 1200).

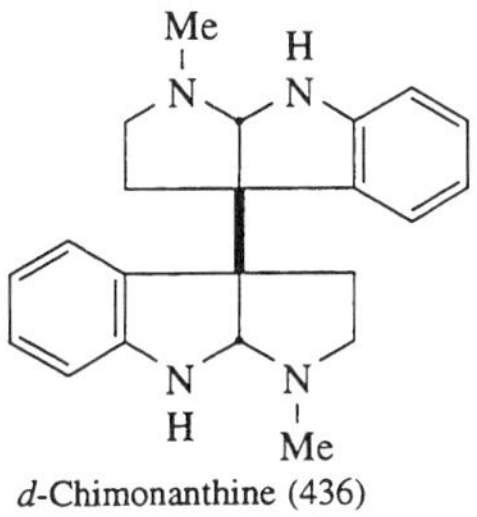

d-Chimonanthine (436)

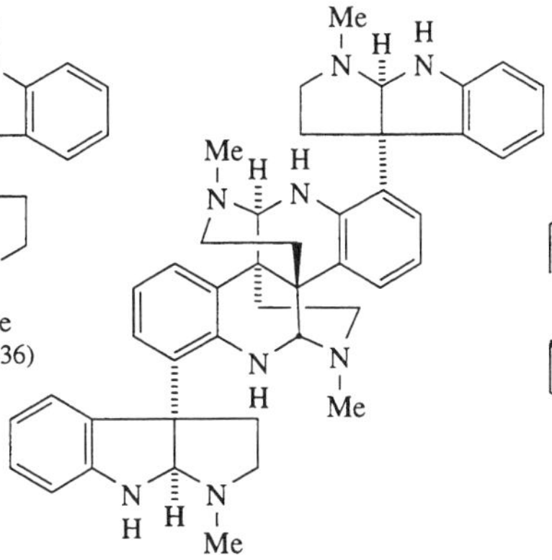

Psycholeine (437)

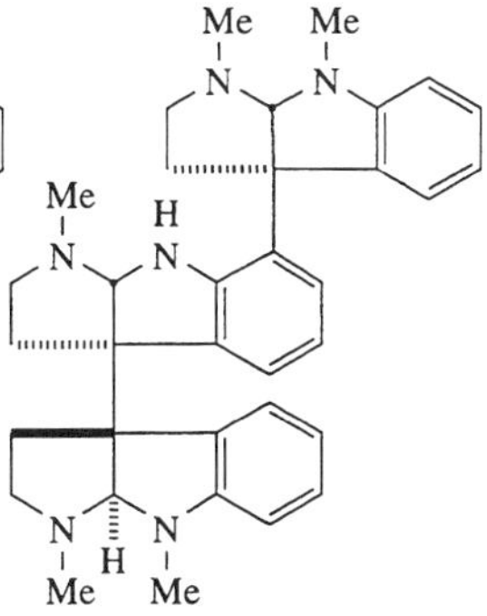

(−)-Idiospermuline (438)

(ii) Calabash-curare alkaloids

In addition to the well-known symmetrical alkaloids of this family, new related *Strychnos* bis-alkaloids that are not symmetrical are also listed here. Matopensine, which is 16(*S*),16'(*S*)-dihydro-17(*R*),17'(*R*)-oxobisnordihydrotoxiferine is found in *Strychnos matopensis* and *S. kasengaensis* (G. Massiot, *et al.*, *Heterocycles*, 1983, **20**, 2339). This latter plant has also yielded 16,17-dehydroisostrychnobiline and two related compounds (P. Thepenier, *et al.*, *Phytochemistry*, 1984, **23**, 2659). The isolation of 12'-hydroxystrychnobiline from *S. variabilis* has been reported (M. Tits, *et al.*, *J. Nat. Prod.*, 1983, **46**, 638). Several bis-indole alkaloids have been discovered which are comprised of a strychnine unit linked to a geissoschizine. For example, longicaudatine (439) and bisnor-C-alkaloid H are found in several *Strychnos* species (G. Massiot, *et al.*, *J. Org. Chem.*, 1983, **48**, 1869). The related longicaudatines F and Y are present in *S. ngouniensis* (G. Massiot, *et al.*, *Tetrahedron*, 1983, **39**, 3645), and two related dihydrolongicaudatines have been identified in *S. potatorum* (G. Massiot, *et al.*, *Phytochemistry*, 1992, **31**, 2873). Guianensine (440) is an interesting carboline bis-indole alkaloid in *S. guianensis* (J. Quetin-Leclercq, *et al.*, *ibid.*, 1995, **40**, 1557). Afrocurarine is a related carboline-strychnine hybrid alkaloid in *S. usambarensis* (M. Caprasse, *et al.*, *Planta Med.*, 1984, **50**, 131). The plant *S. divaricans* has afforded both divaricine (R. Mukherjee, *et al.*, *Heterocycles*, 1991, **32**, 985), a bis-alkaloid comprised of vellosimine and the 18-deoxy Wieland-Gumlich aldehyde N_b-oxide, and divarine (R. Mukherjee, *et al.*, *ibid.*, 1994, **38**, 1965), an alkaloid similar to 16-methoxyisomatopensine.

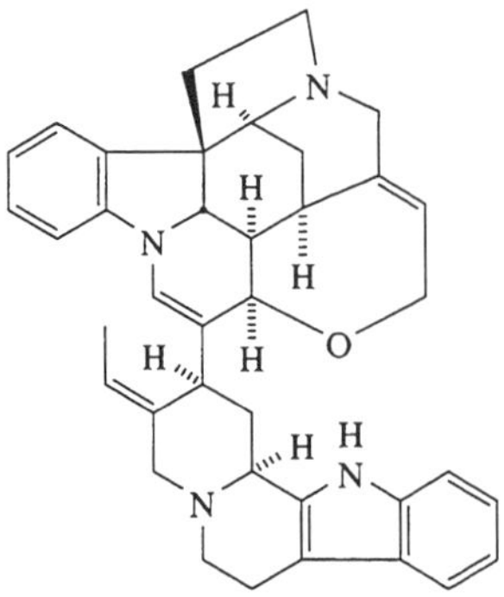

144

Longicaudatine (439)

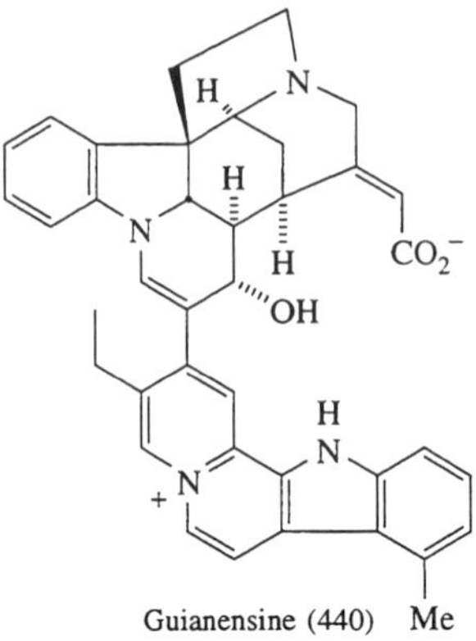

Guianensine (440)   Me

(iii) Diketopiperazine alkaloids

The very large number of tryptamine-derived diketopiperazine natural products suggests that they be included in a separate category. Most of these compounds are fungal metabolites, and several were mentioned in the last volume.

Bipolaramide (441) is a novel metabolite of *Bipolaris sorokiniana* (C.M. Maes, *et al.*, *J. Chem. Soc., Perkin Trans. 1*, 1985, 2489). Biosynthetic studies indicate that phenylalanine is a precursor to this dioxopiperazine. The absolute configuration of the previously isolated ditryptophenaline has been established (C.M. Maes, *et al.*, *ibid.*, 1986, 861). Two close analogues of ditryptophenaline, WIN 64821 and WIN 64745, which have different substituents on the piperazine ring, are produced by *Aspergillus* sp. and have activity as competitive antagonists to Substance P (D.M. Sedlock, *et al.*, *J. Antibiot.*, 1993, **47**, 391; C.J. Barrow, *et al.*, *J. Org. Chem.*, 1993, **58**, 6016). The 1'-(2-phenylethylene) derivative of ditryptophenaline has been isolated from the fungus *A. flavus* (C.J. Barrow and D.M. Sedlock, *J. Nat. Prod.*, 1994, **57**, 1239). Amauromine (442) from

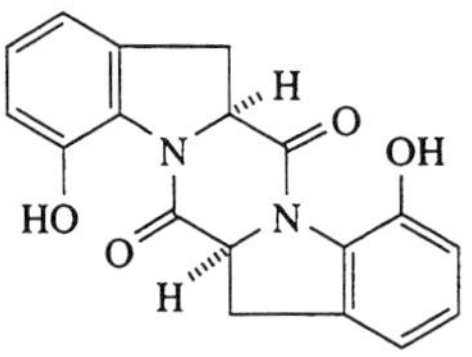

Bipolaramide (441)

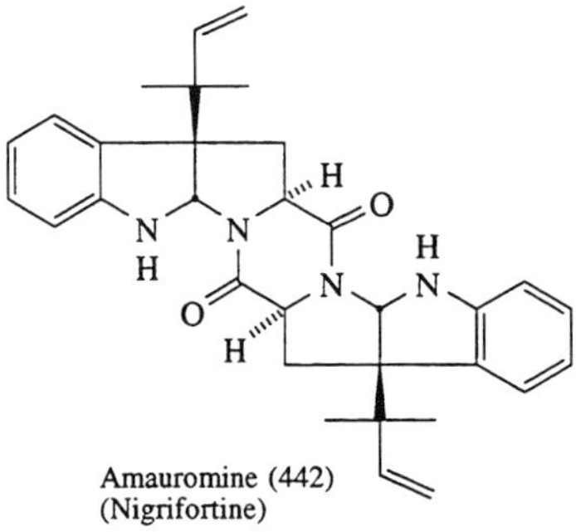

Amauromine (442)
(Nigrifortine)

Amauroascus sp. (= nigrifortine from *Penicillium nigricans*) (I. Laws and P.G. Mantle, *Phytochemistry*, 1985, **24**, 1395) is a hypotensive vasodilator (S. Takase, *et al.*, *Tetrahedron*, 1985, **41**, 3037). *Aspergillus ochraceus* has furnished *epi-* and *N*-methyl-*epi*-amauromine (F.S. de Guzman, *et al.*, *J. Nat. Prod.*, 1992, **55**, 931), which have some activity against the corn earworm.

Several new examples of the sulfur-containing *Chaetomium* family of alkaloids have been discovered. These include the highly cytotoxic chetracin A (443) from *C. abuense* and *C. retardatum* (T. Saito, *et al.*, *Tetrahedron Lett.*, 1985, **26**, 4731). This metabolite is the bis-tetrasulfide corresponding to melinacidin IV. Chetracin A and the related chaetocins B and C, which differ in the number of bridging sulfurs, are found in *C. nigricolor* and *C. virescens* var. *thielavioideum* (T. Saito, *et al.*, *Chem. Pharm. Bull.*, 1988, **36**, 1942). Dethiotetra(methylthio)chetomin is produced by *C. globosum* (T. Kikuchi, *et al.*, *ibid.*, 1982, **30**, 3846). The marine fungus *Leptosphaeria* sp., which grows on the alga *Sargassum tortile*, has been a source of the novel leptosins A-G, G_1, G_2, H-J (C. Takahashi, *et al.*, *J. Chem. Soc., Perkin Trans. 1*, 1994, 1859; *J. Antibiot.*, 1994, **47**, 1242; *Phytochemistry*, 1995, **38**, 155). Whereas leptosins A-C, G, G_1, G_2, H are related to chetracin A, leptosins D-F, e.g., (444), are linked to an indole ring, and leptosins I (445) and J are ether bridged.

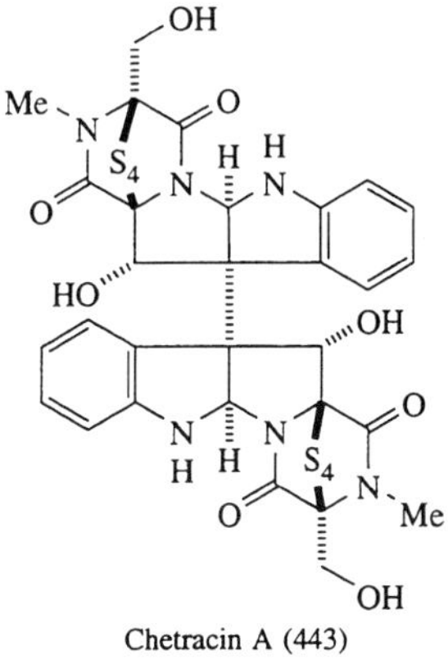

Chetracin A (443) 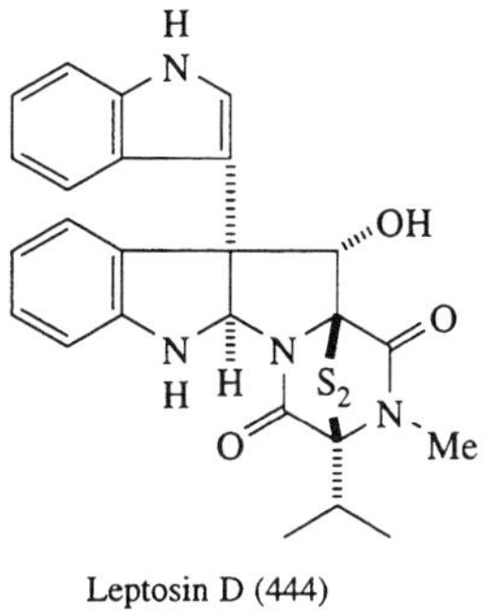   Leptosin D (444) 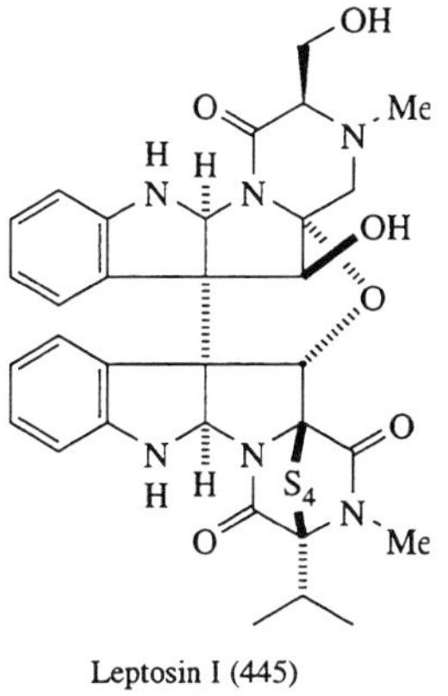   Leptosin I (445)

(iv) Bis-indole quinones

Several new fungal metabolites having two indole rings linked to a benzoquinone are known. Thus, *Aspergillus terreus* is the source of more than 20 such metabolites, such as the asterriquinones, e.g., (446), and asterridinone (447) (K. Arai, *et al.*, *Chem. Pharm. Bull.*, 1981, **29**, 961, 1005; A. Kaji, *ibid.*, 1994, **42**, 1682). The four prenylated ochrindoles A-D, e.g., (448), are produced by *A. ochraceus* (F.S. de Guzman, *J. Nat. Prod.*,

1994, **57**, 634). Related metabolites are isocochliodinol and neocochlio-dinol from *Chaetomium murorum* and *C. amygdalisporum* (S. Sekita, *Chem. Pharm. Bull.*, 1983, **31**, 2998), and hinnuliquinone (449) from *Nodulisporium hinnuleum* (M.A. O'Leary, *et al.*, *J. Chem. Soc.*, *Perkin Trans. 1*, 1984, 567).

Asterriquinone A-1 (446)

Asterridinone (447)

Ochrindole A (448)

Hinnuliquinone (449)

(v) Indolo[2,3-a]carbazoles and related alkaloids
This relatively new and growing group of alkaloids was inaugurated in 1977 with the isolation of staurosporine (450) from *Streptomyces staurosporeus* (S. Omura, *et al.*, *J. Antibiot.*, 1977, **30**, 275; A Furusaki, *et al.*, *Bull. Chem. Soc. Japan*, 1982, **55**, 3681). Since these novel alkaloids have been reviewed in depth (J. Bergman, *Stud. Nat. Prod. Chem.*, 1988, **1**, 3; G.W. Gribble and S.J. Berthel, *ibid.*, 1993, **12**, 365), only examples identified since 1990 will be explicitly shown herein.

Notable representative members of this group are rebeccamycin (*Saccharothrix aerocolonigenes*) (D.E. Nettleton, *et al.*, *Tetrahedron Lett.*, 1985, **26**, 4011; J.A. Bush, *et al.*, *J. Antibiot.*, 1987, **40**, 668), SF-2370 from *Actinomadura* sp. (M. Sezaki, *et al.*, *ibid.*, 1985, **38**, 1437), K-252 a-d from *Nocardiopsis* sp. (T. Yasuzawa, *et al.*, *ibid.*, 1986, **39**, 1072), UCN-01 and -02 from *Streptomyces* sp. (I. Takahashi, *et al.*, *ibid.*, 1989, **42**, 571), AT2433 A1, A2, B1, and B2 from *A. melliaura* (J.A.. Matson, *et al.*, *ibid.*, 1989, **42**, 1547, 1784), RK-286c (4'-deaminomethyl-4'-hydroxystaurosporine) from *Streptomyces* sp. (H. Takahashi, *et al.*, *ibid.*, 1990, **43**, 168), BE-13793c (K. Kojiri, *et al.*, *ibid.*, 1991, **44**, 723), and TAN-1030A from *Streptomyces* sp. (S. Tsubotani, *et al.*, *Tetrahedron*, 1991, **47**, 3565). Newer examples include RK-1409B (451) from *S. platensis* subsp. *malvinus* (H. Koshino, *et al.*, *J. Antibiot.*, 1992, **45**, 1428), and 7-oxostaurosporine is also produced by this microbe (H. Koshino, *et al.*, *ibid.*, 1992, **45**, 195). The previously known K252-c and arcyriaflavin are found in *Eudistoma* sp. (P.A. Horton, *et al.*, *Experientia*, 1994, **50**, 843).

Staurosporine (450)

RK-1409B (451)

The *Streptomyces* metabolite RK-286D (452) has the unusual digitoxose sugar attached to an indole nitrogen (H. Osada, *et al.*, *J. Antibiot.*, 1992, **45**, 278). A mutant of *S. longisporoflavus* has afforded 3'-demethoxy-3'-hydroxystaurosporine (P. Hoehn, *et al.*, *ibid.*, 1995, **48**, 300), and another investigation of this organism has revealed the presence of five new alkaloids, e.g., (453)-(456) (Y. Cai, *et al.*, *ibid.*, 1995, **48**, 143). The blue-green alga *Tolypothrix tjipanasensis* produces 15 indolo[2,3-*a*]carbazoles, most of which contain chlorine, e.g., (457) (R. Bonjouklian, *et al.*, *Tetrahedron*, 1991, **47**, 7739). These tjipanazoles, like many indolo[2,3-*a*]-carbazoles, are inhibitors of protein kinase C. The slime mould *Lycogala epidendrum* has yielded lycogarubins A-C, e.g., (458), and C shows anti-HSV-1 virus activity (T. Hashimoto, *et al.*, *Tetrahedron Lett.*, 1994, **35**, 2559). Chromopyrrolic acid from *Chromobacterium violaceum* is the didehydroxydicarboxylic acid corresponding to (458) (T. Hoshino, *et al.*, *Biosci. Biotechnol. Biochem.*, 1993, **57**, 775).

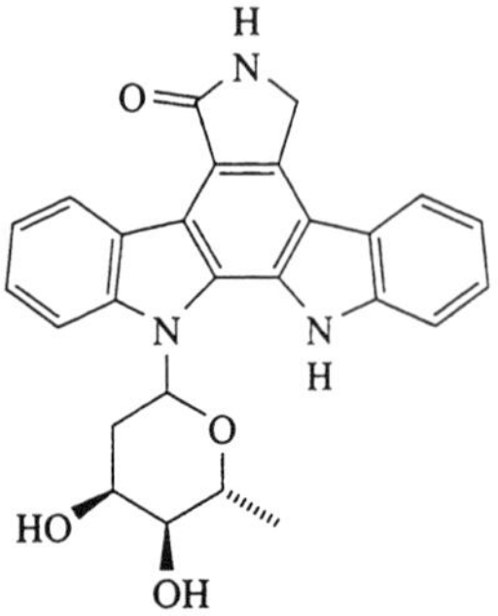

148

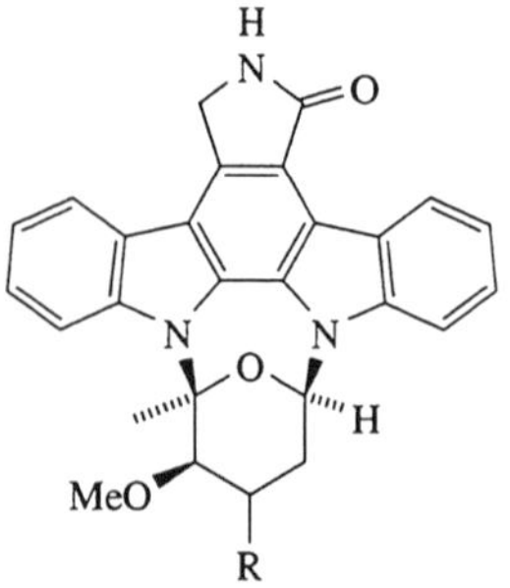

(453) R = α-NO$_2$
(454) R = β-NMeCHO
(455) R = β-NOHCHO
(456) R = β-NMeOCH$_2$OAc

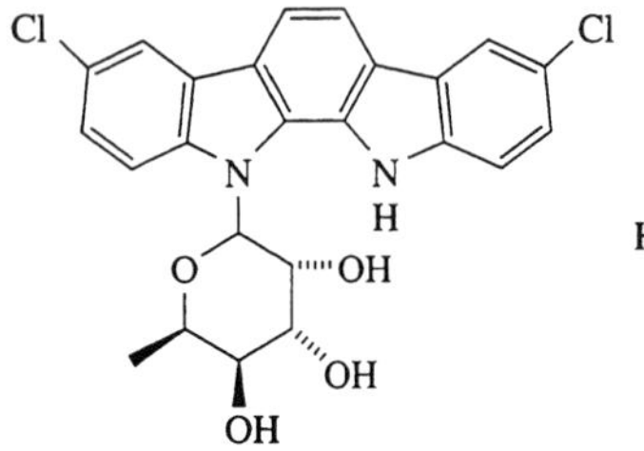

RK-286D (452)

Tjipanazole A1 (457)

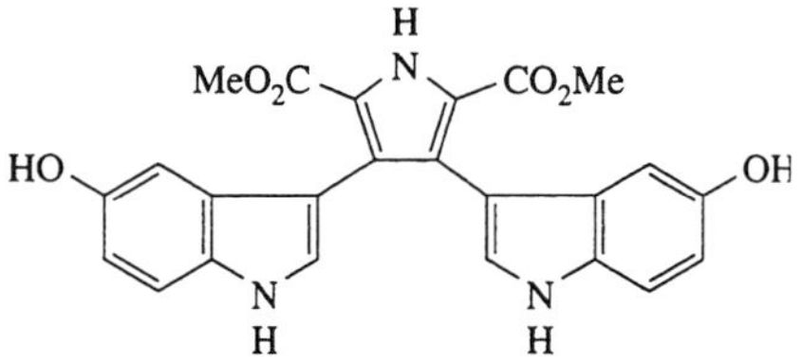

Lycogarubin A (458)

(vi) Marine bis-indoles

With the resurgence of marine natural products exploration has come the discovery of numerous bis- and tris-indoles from marine organisms, many alkaloids of which contain bromine. Some non-symmetrical marine bis-alkaloids are included here for convenience, when, for example, they occur with symmetrical compounds in the same organism.

The toxic mucus of the boxfish (*Ostracion cubicus*) contains the marine bacterium *Vibrio parahaemolyticus* which produces vibrindole A (459) and the previously known synthetic indole oxidation product (460) (R. Bell, *et al.*, *J. Nat. Prod.*, 1994, **57**, 1587). Hyrtiosin B (461) is found in the sponge *Hyrtios erecta* (J. Kobayashi, *et al.*, *Tetrahedron*, 1990, **46**, 7699), and the ascidian *Didemnum* sp. produces the novel bis-carboline (462) (P.S. Kearns, *et al.*, *J. Nat. Prod.*, 1995, **58**, 1075). The red alga *Chondria* sp. has yielded chondriamides A (463), B (464), and (465) (J.A. Palermo, *et al.*, *Tetrahedron Lett.*, 1992, **33**, 3097).

The blue-green alga *Rivularia firma* contains seven polybrominated bis-indoles, e.g., (466) and (467), six of which are optically active due to biphenyl chirality (restricted rotation) (R.S. Norton and R.J. Wells, *J. Am. Chem. Soc.*, 1982, **104**, 3628; A.R. Hodder and R.J. Capon, *J. Nat. Prod.*, 1991, **54**, 1661). The tunicate *Didemnum candidum* has afforded bis-indoles (468) and (469) (E. Fahy, *et al.*, *ibid.*, 1991, **54**, 564).

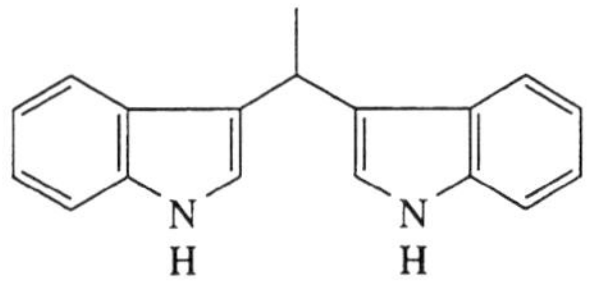

149

Vibrindole A (459)

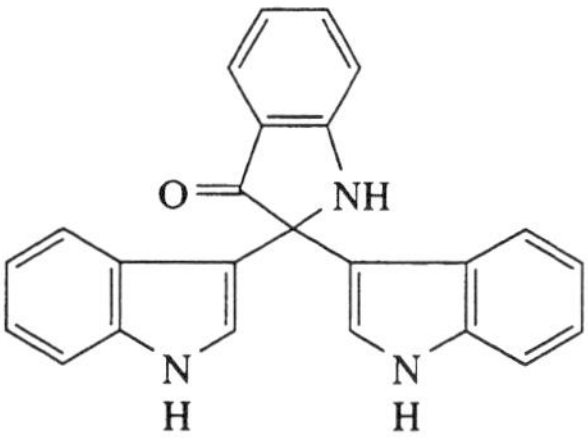

(460)

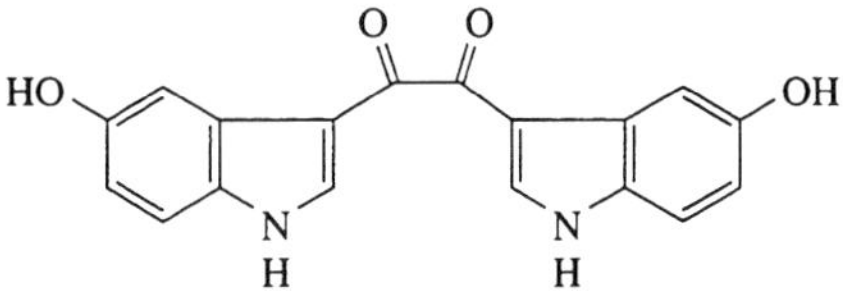

Hyrtiosin B (461)

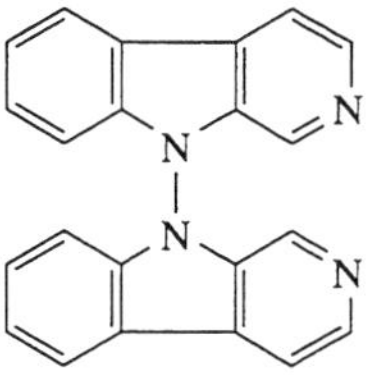

462

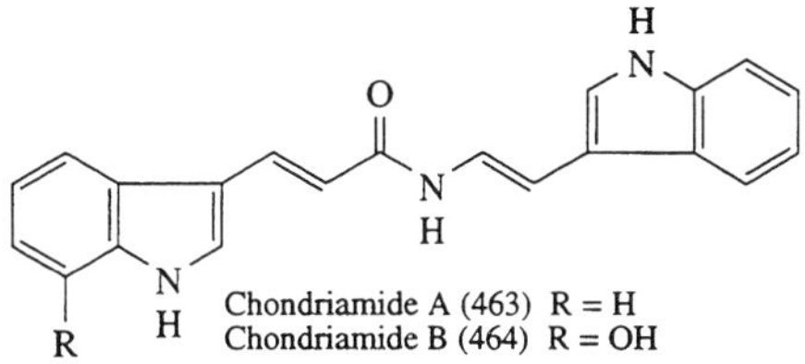

Chondriamide A (463) R = H
Chondriamide B (464) R = OH

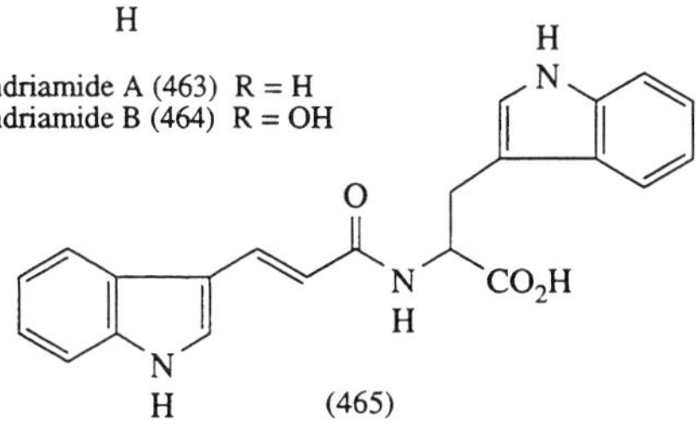

(465)

Dragmacidin is a related metabolite to (469) that is found in the deep water
sponge *Dragmacidon* sp. (S. Kohmoto, *et al.*, *J. Org. Chem.*, 1988, **53**,
3116), and the structurally similar dragmacidons A and B, which are
piperazine ring methylated analogues of (469), are produced by the sponge
Hexadella sp. along with the novel bis-indole imidazole alkaloid topsentin C
(470) (S.A. Morris and R.J. Andersen, *Tetrahedron*, 1990, **46**, 715).
Other related topsentins are found in sponges *Spongosorites* sp. (S. Tsujii,
et al., *J. Org. Chem.*, 1988, **53**, 5446; L.M. Murray, *et al.*, *Aust. J.
Chem.*, 1995, **48**, 2053), *S. ruetzleri* (S. Sakemi and H.H. Sun, *J. Org.
Chem.*, 1991, **56**, 4304), and *Topsentia genitrix* (K. Bartik, *et al.*, *Can. J.
Chem.*, 1987, **65**, 2118). This latter study provided evidence of the
chemical defensive role for these marine alkaloids. The pyrazinones
hamacanthins A and B are oxidized variants of (469) from the sponge
Hamacantha sp. (S.P. Gunasekera, *et al.*, *J. Nat. Prod.*, 1994, **57**, 1437),
and the novel dragmacidin d (471) is found in *Spongosorites* sp. (A.E.
Wright, *et al.*, *J. Org. Chem.*, 1992, **57**, 4772).

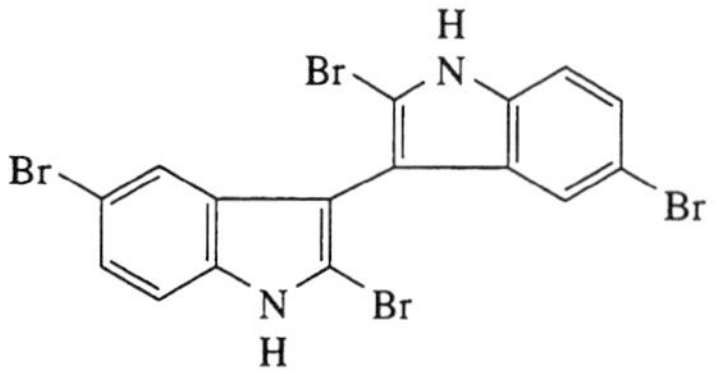

(466)

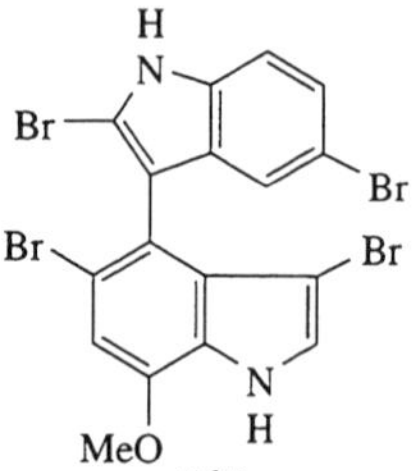

(467)

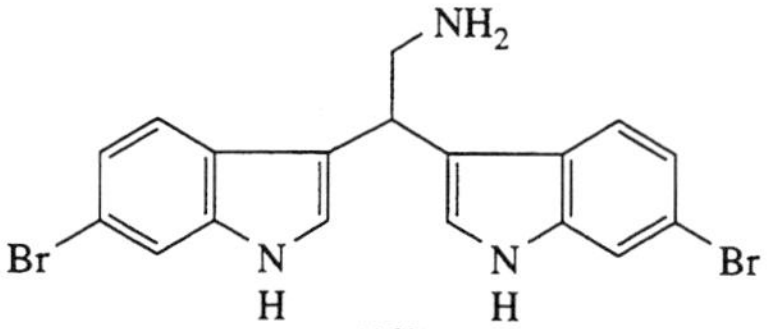

(468)

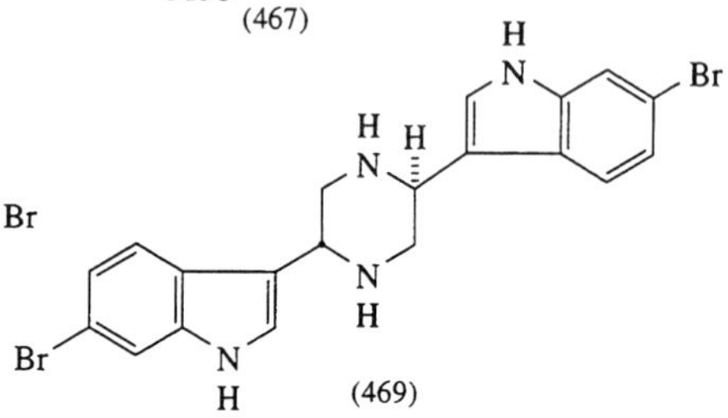

(469)

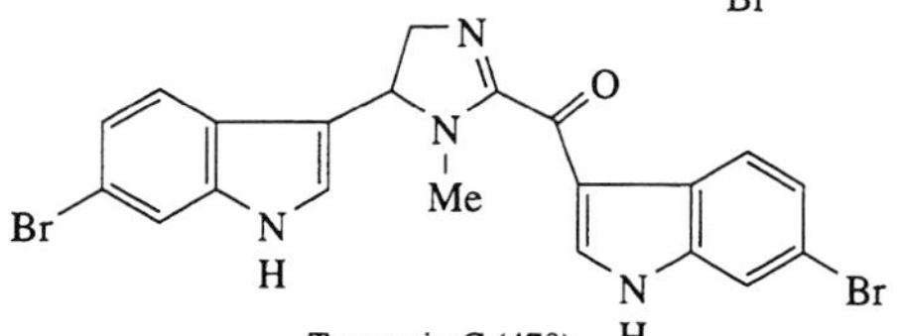

Topsentin C (470)

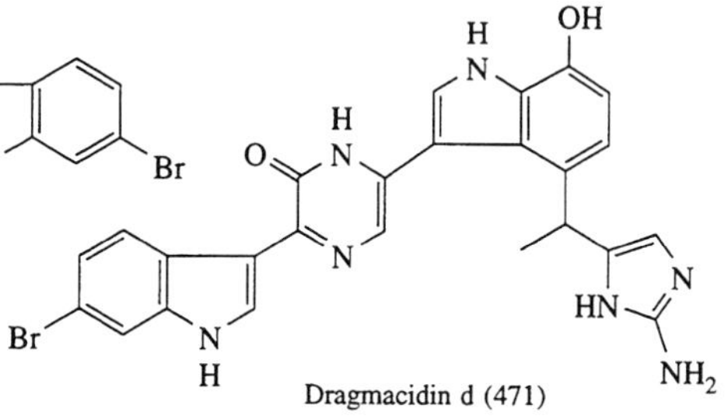

Dragmacidin d (471)

The gelliusines are a set of bis- and tris-indole alkaloids from a deep water New Caledonian sponge *Orina* sp. (G. Bifulco, *et al.*, *J. Nat. Prod.*, 1994, **57**, 1294; 1995, **58**, 1254). Two examples are gelliusines A and B (diastereomers) (472) and D (473). The ascidian *Eusynstyela misakiensis*

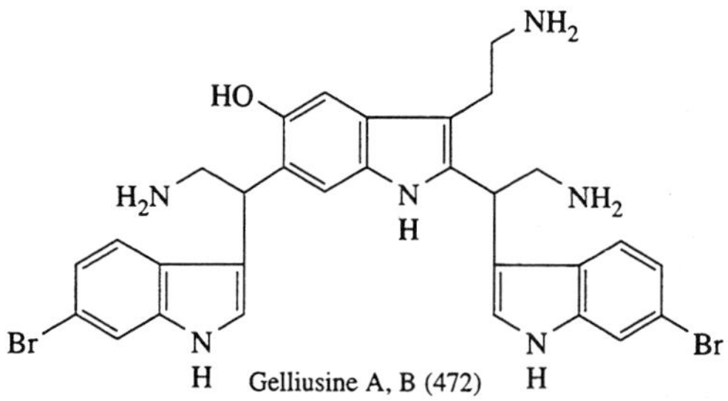

Gelliusine A, B (472)

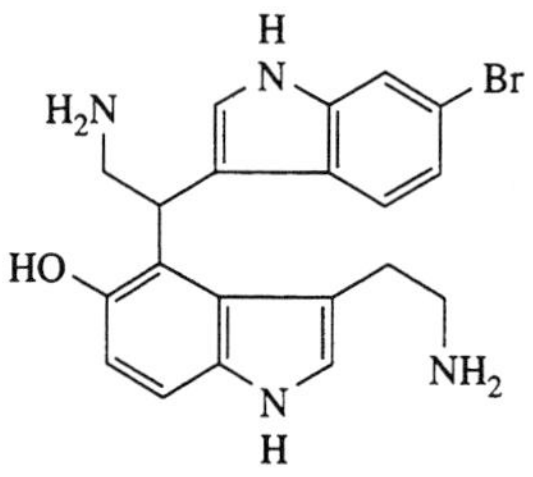

Gelliusine D (473)

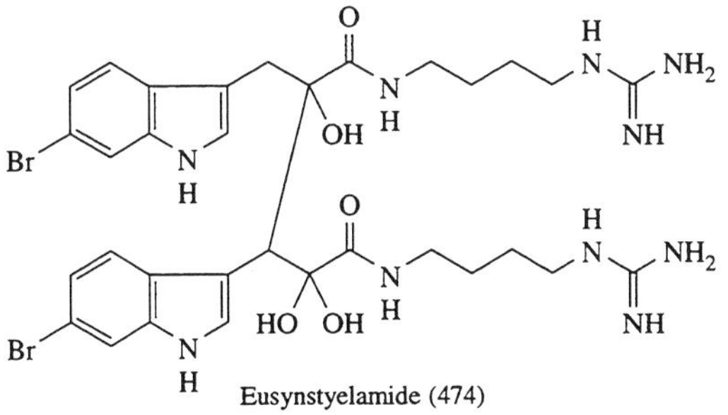

Eusynstyelamide (474)

produces eusynstyelamide (474), a novel bis-indole (J.C. Swersey, *et al.*, *ibid.*, 1994, **57**, 842). Kauluamine is a novel manzamine alkaloid dimer isolated from an Indonesian *Prianos* sp. sponge (I.I. Ohtani, *et al.*, *J. Am. Chem. Soc.*, 1995, **117**, 10743).

(vii) Other symmetrical bis-indoles

The simple bis-indole strepindole (475) is a genotoxic metabolite of human intestinal bacteria (*Streptococcus faecium* (T. Osawa and M. Namiki, *Tetrahedron Lett.*, 1983, **24**, 4719), and peronatins A (476) and the diastereomer B are found in the plant *Collybia peronata*, while two related ring-oxygenated compounds are produced by *Tricholoma scalpturatum* (Z. Pang and O. Sterner, *J. Nat. Prod.*, 1994, **57**, 852). Caulerpinic acid (477), along with the mono- and di-methyl ester (caulerpin), have been isolated from *Caulerpa racemosa* (A.S.R. Anjaneyulu, *et al.*, *Phytochemistry*, 1991, **30**, 3041). Picrasidine R (478) and an unsymmetrical dimer picrasidine H

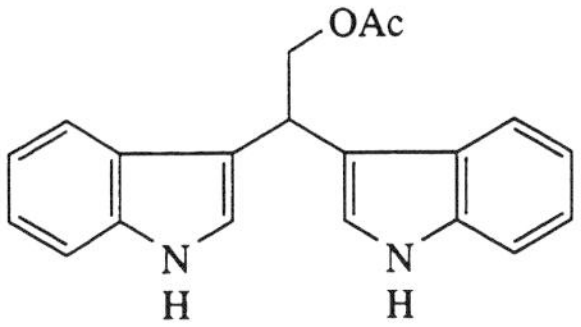

Strepindole (475)

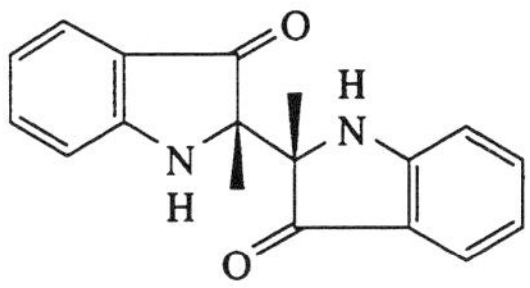

Peronatin A (476)

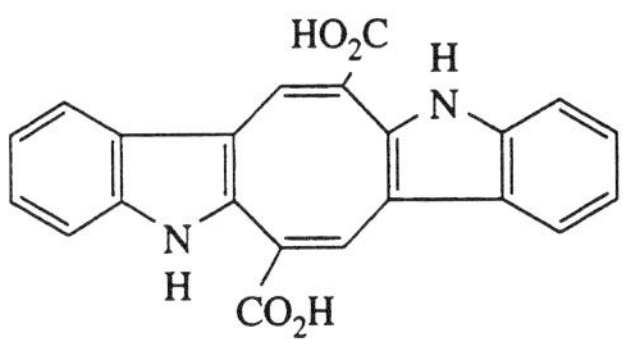

Caulerpinic Acid (477)

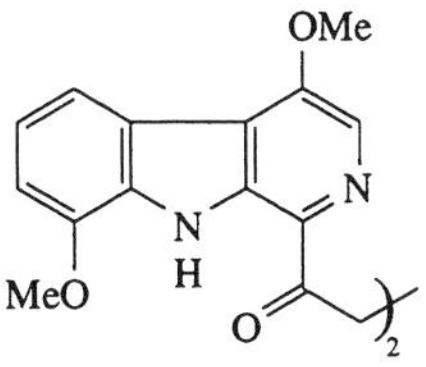

Picrasidine R (478)

were isolated from *Picrasma quassioides* (K. Koike and T. Ohmoto, *Chem. Pharm. Bull.*, 1986, **34**, 2090). Both symmetrical and unsymmetrical carbazole dimers are presented in the next section.

Blastmycetin A (479) is a novel dimer of the teleocidin family identified in cultures of *Streptoverticillium blastmyceticum* (K. Irie, *et al.*, *Agric. Biol. Chem.*, 1987, **51**, 285). The plant *Voacanga grandifolia* contains voacinol, which is 14,14'-methylene-bis-18-hydroxytabersonine (480) (T.R. Govindachari, *et al.*, *J. Chem. Soc., Chem. Commun.*, 1987, 1137), and *Hazunta modesta* var. *modesta* subvar. *divaricata* has yielded the novel hazuntiphylline (481) (A.-M. Bui, *et al.*, *J. Nat. Prod.*, 1986, **49**, 321). This alkaloid, which is comprised of two *Aspidosperma* units, is related to the *Strychnos* calabash-curare alkaloids. This plant has also afforded two related bis-alkaloids, hazuntiphyllidione and anhydrohazuntiphyllidine (A.-M. Bui, *et al.*, *ibid.*, 1991, **54**, 514). The structure of the yellow-brown pigment present in many species of blue-green algae, scytonemin (482) (F. Garcia-Pichel and R.W. Castenholz, *J. Phycol.*, 1991, **27**, 395), has finally been elucidated (P.J. Proteau, *et al.*, *Experientia*, 1993, **49**, 825). This compound appears to play an ultraviolet sunscreen role in these organisms (F. Garcia-Pichel, *et al.*, *Photochem. Photobiol.*, 1992, **56**, 17).

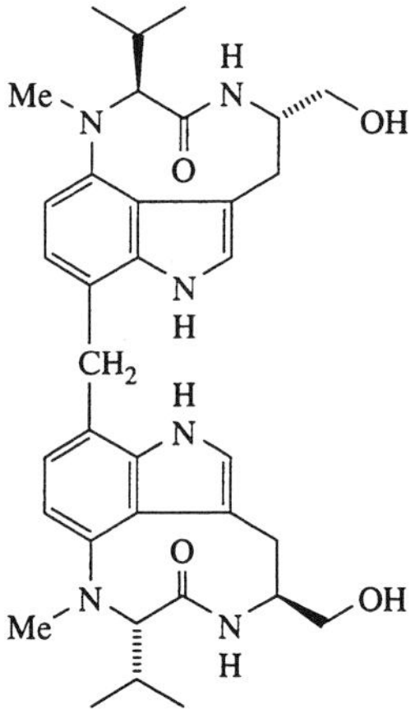

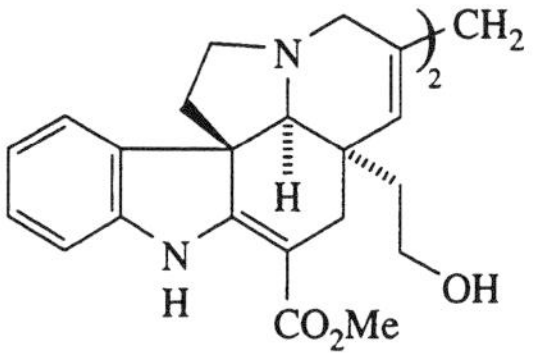

Voacinol (480)

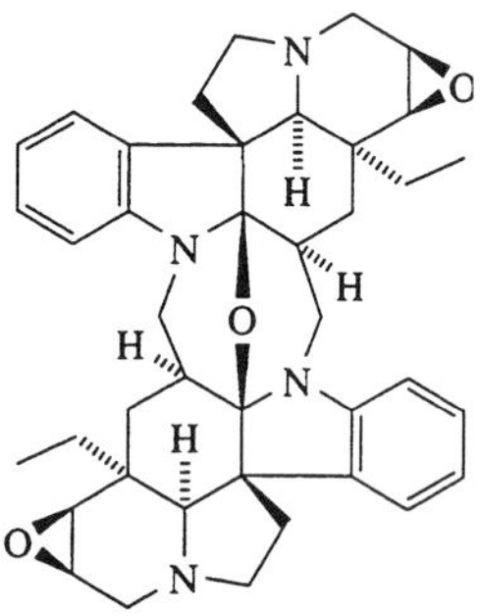

Hazuntiphylline (481)

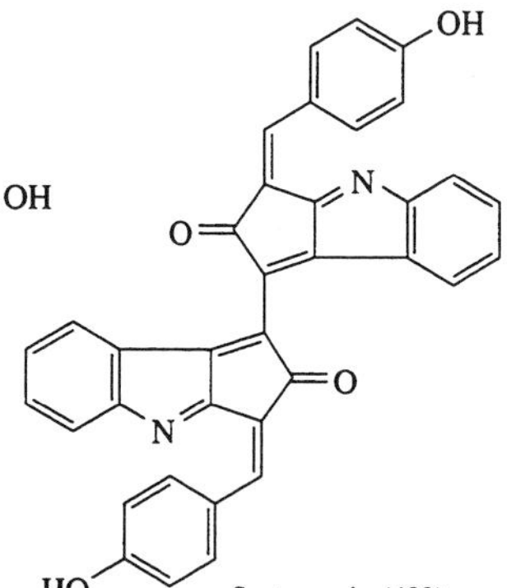

Blastmycetin A (479)

Scytonemin (482)

(b) Compounds not composed of identical halves nor linked symmetrically

(i) Sesquimeric compounds
Several new representative alkaloids of the tetrahydrousambarensine type
have been isolated, including five ochrolifuanines, e.g., (483), from *Dyera
costulata* (C. Mirand, *et al.*, *Phytochemistry*, 1983, **22**, 577), and two examples from *Strychnos dale* (R. Verpoorte, *et al.*, *Tetrahedron Lett.*, 1986,
27, 239). Buchtienine is a related bis-indole from *Peschiera buchtieni* (M.
Azoug, *et al.*, *Phytochemistry*, 1995, **39**, 1223). The novel geissoschizal
coupling product strychnofuranine (484) is found in *S. matopensis* (G.
Massiot, *et al.*, *Phytochemistry*, 1988, **27**, 3293). Uncaramine (485) from
Uncaria callophylla is a unique bis-alkaloid involving the union of pseudo-
yohimbine and gambirine (A. Arnone, *et al.*, *J. Chem. Soc., Perkin Trans.
1*, 1987, 571). The same alkaloid as callophylline and two isomers callo-
phyllines A and B were isolated independently from this plant (T.-S. Kam,
et al., *Phytochemistry*, 1991, **30**, 3441).

Ochrolifuanine E (483)

Strychnofuranine (484)

Uncaramine (485)

The two diastereomeric janussines A and B (486) (*Strychnos johnsonii*)
are biogenetically interesting bis-alkaloids of either the decussine-
mostuenine or the cathenamine types, depending on the depiction (G.
Massiot, *et al.*, *Tetrahedron Lett.*, 1985, **26**, 2441; *Phytochemistry*, 1987,
26, 2839). An interesting bis-ellipticine, strellidimine (487), formed
between 10-hydroxyellipticine and 3,14-dihydroellipticine, is found in *S.*

154

dinklagei (S. Michel, *et al., J. Chem. Soc., Chem. Commun.*, 1987, 229).
This is the first reported natural bis-ellipticine and the structure was con-
firmed by synthesis. The plant *Picrasma quassioides* has also yielded the
unsymmetrical bis-carboline alkaloids picrasidines F, G, and S (488) (K.
Koike, *et al., Chem. Pharm. Bull.*, 1986, **34**, 3228; K. Koike and T.
Ohmoto, *ibid.*, 1987, **35**, 3305). These alkaloids are racemic.

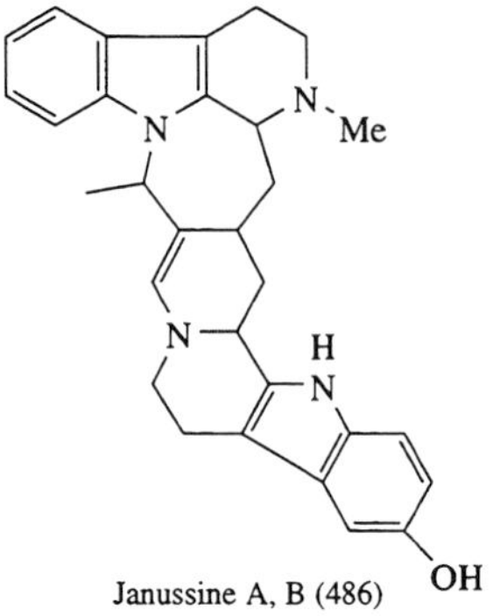

Janussine A, B (486)

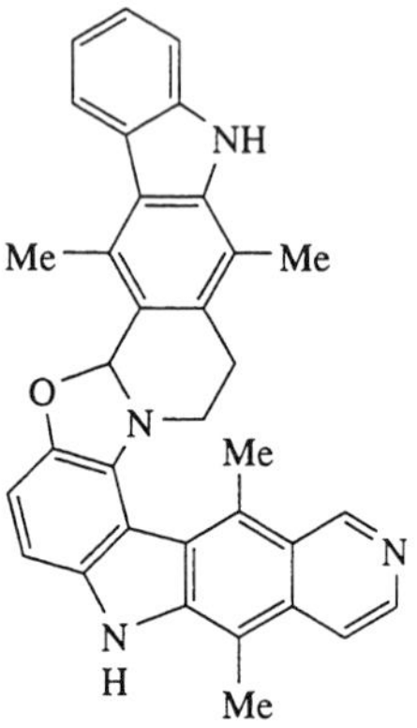

Strellidimine (487)

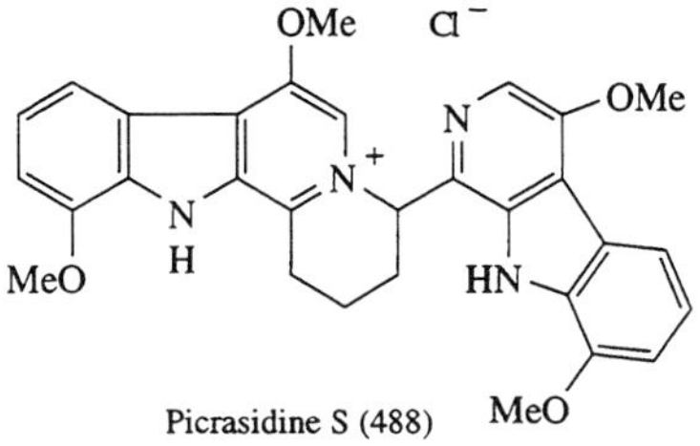

Picrasidine S (488)

Auricularine is the *N,N*-dimethyltryptamine analogue of borreverine and
is found in *Hedyotis auricularia* (K.K. Purushothaman and A. Sarada,
Phytochemistry, 1981, **20**, 351), and spermacoceine is the tryptophol
analogue of borreverine and was characterized from *Borreria verticillata*
(A.M. Baldé, *et al., ibid.*, 1991, **30**, 997). Yuehchukene (489) is a novel
racemic bis-indole from *Murraya paniculata* with potent anti-implantation
activity (Y.-C. Kong, *et al., J. Chem. Soc., Chem. Commun.*, 1985, 47).
The structure of this alkaloid has been supported by a biomimetic synthesis
(K.-F. Cheng, *et al., ibid.*, 1985, 48). A series of simple bis-indoles,
annonidines A-E, e.g., (490) and (491), are found in *Annonidium manni*
(H. Achenbach and C. Renner, *Heterocycles*, 1985, **23**, 2075).

Yuehchukene (489)

Annonidine B (490)

Annonidine E (491)

(ii) The secamines and presecamines
No new examples were discovered.

(iii) Bis-indoles from Vinca rosea
Although these bis-indole alkaloids remain an important component in the treatment of human cancer, only a few new alkaloids have been isolated from this plant since the last coverage. Some reviews are available (G.A. Cordell and J.E. Saxton, in "The Alkaloids," Vol. 20, Chapter 1, 1981; "The Alkaloids," Vol. 37 is devoted entirely to this topic; A. Rahman, *et al.*, *Stud. Nat. Prod. Chem.*, 1994, **14**, 805).

Catharanthus roseus has yielded both leurosidine (vinblastine) N'_b-oxide (S. Mukhopadhyay and G.A. Cordell, *Lloydia*, 1981, **44**, 611), and leurosinone, which has the novel structure (492), but appears not to be an

Leurosinone (492)

Roseadine (493) R = 10-vindolinyl

156

acetone artefact (A. Rahman, *et al.*, *J. Chem. Soc., Perkin Trans. 1*, 1988, 2175). The same plant has furnished catharanthamine, which is the first bis-indole with C-17 oxygenated in the velbanamine unit (A. El-Sayed and G.A. Cordell, *Lloydia*, 1981, **44**, 289), and rosedine (493) (A. El-Sayed, *et al.*, *J. Nat. Prod.*, 1983, **46**, 517).

(iv) Bis-carbazoles

The past decade has seen the discovery of numerous examples of bis-carbazoles, especially from genus *Murraya*. Thus, *M. euchrestifolia* has yielded more than 20 such compounds, the first of which was murrafoline A (494) (A.T. McPhail, *et al.*, *Tetrahedron Lett.*, 1983, **24**, 5377). This was followed by the isolation of bismurrayafolines A and B (H. Furukawa, *et al.*, *Chem. Pharm. Bull.*, 1983, **31**, 4202), murrafolines B and C (H. Furukawa, *et al.*, *ibid.*, 1985, **33**, 2611), murrastifolines A-D, e.g., (495), and chrestifolines A-C, (C. Ito, *et al.*, *ibid.*, 1990, **38**, 1143), murranim-bine (C. Ito and H. Furukawa, *ibid.*, 1991, **39**, 1355), bis-7-hydroxygiri-nimbines A and B (T.-S. Wu, *et al.*, *Phytochemistry*, 1991, **30**, 1052), and murrafolines D, G, H (H. Furukawa, *et al.*, *Chem. Pharm. Bull.*, 1993, **41**, 1249). The bis-methyl ether of bis-7-hydroxygirinimbine A is found in *M. exotica* (E.K. Desoky and D.W. Bishay, *Bull. Fac. Pharm. (Cairo)*, 1992, **30**, 231; *Chem. Abstr.*, 1993, **119**, 177557). *M. gleniei* contains the simple (496) (V. Kumar, *et al.*, *Tetrahedron Lett.*, 1990, **31**, 5217), and *M. koenigii* has yielded the new bis-carbazoles murrastifoline F, bismahanine, bikoeniquinone A, and others (C. Ito, *et al.*, *Chem. Pharm. Bull.*, 1993, **41**, 2096).

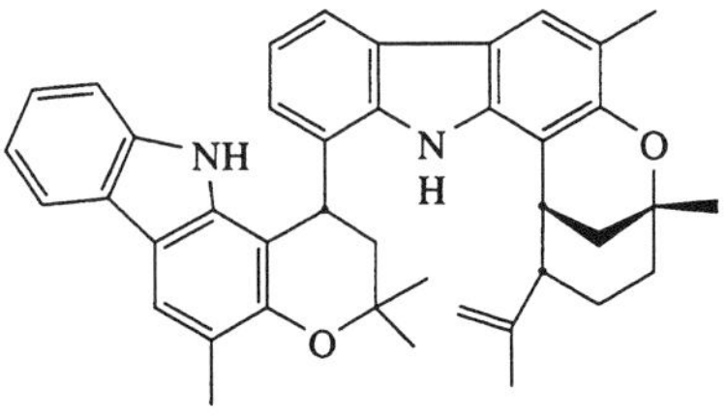

Murrafoline A (494)

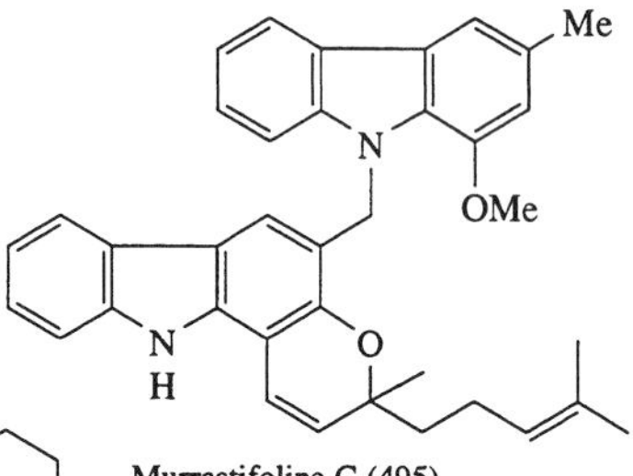

Murrastifoline C (495)

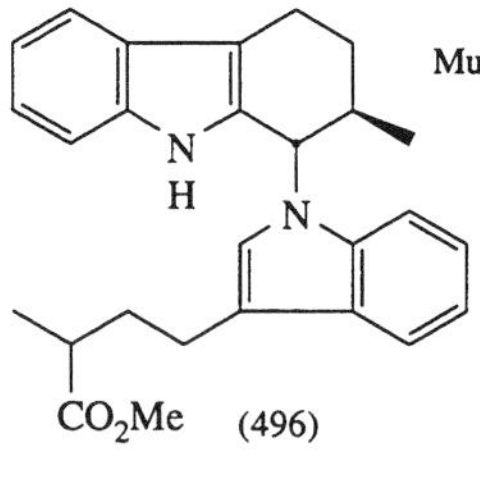

(496)

(v) Duocarmycins and related alkaloids
The earlier discovery of CC-1065 opened the door to a remarkable group of *Streptomyces* metabolites possessing enormous antitumor activity. More recently, the duocarmycins A, C1, C2, D have been identified as members of this family. These bis-indoles, which have potent DNA affinity (A. Asai, *et al., J. Am. Chem. Soc.*, 1994, **116**, 4171), can be represented by duocarmycin A (497) (I. Takahashi, *et al., J. Antibiot.*, 1988, **41**, 1915; T. Yasuzawa, *et al., Chem. Pharm. Bull.*, 1988, **36**, 3728; 1995, **43**, 378). An independent study isolated these compounds as pyrindamycins (K. Ohba, *et al., J. Antibiot.*, 1988, **41**, 1515; S. Ishii, *et al., ibid.*, 1989, **42**, 1713). Subsequent research identified the related duocarmycin SA (M. Ichimura, *et al., ibid.*, 1990, **43**, 1037; 1991, **44**, 1045; T. Yasuzawa, *et al., ibid.*, 1991, **44**, 445), which is one of the most potent cytotoxic compounds ever discovered. An enormous amount of synthetic work has been performed in this important area of drug design.

Duocarmycin A (497)

(vi) Simple bis-indoles
A review of several types of bis-indole alkaloids has appeared (P.G. Waterman, in "Alkaloids: Chemical and Biological Perspectives," Vol. 4, Chapter 3, 1986).
Sciodole (498) is produced by *Tricholoma sciodes* (O. Sterner, *Nat. Prod. Lett.*, 1994, **4**, 9), and both eudistomin U (499) and isoeudistomin U

Sciodole (498)

Eudistomin U (499)

Isoeudistomin U (500)

Grossularine-1 (501)

(500) are unusual indole-carboline hybrid metabolites of the ascidian *Lisso-clinum fragile* (A. Badre, *et al.*, *J. Nat. Prod.*, 1994, **57**, 528). The marine tunicate *Dendrodoa grossularia* produces grossularine-1 (501) (C. Moquin-Pattey and M. Guyot, *Tetrahedron*, 1989, **45**, 3445). This structure was revised from that earlier reported (C. Moquin and M. Guyot, *Tetrahedron Lett.*, 1984, **25**, 5047).

Alangiobussine (502) and alangiobussinine, with a fully-oxidized carboline ring, are found in *Alangium bussyanum* (A.O. Diallo, *et al.*, *Phytochemistry*, 1995, **40**, 975). Synthesis confirmed these structures. Although candidine (503) was first isolated in 1922 from the yeast *Candida lipolytica*, the structure of this novel bis-alkaloid was only determined in 1985 (J. Bergman and U. Tilstam, *Tetrahedron*, 1985, **41**, 2883). Several new violacein analogues have been isolated from the bacterial species *Chromobacterium violaceum*, namely, prodeoxyviolacein (504), provio-lacein (505), and pseudoviolacein (T. Hoshino, *et al.*, *J. Chem. Soc.*, *Perkin Trans. 1*, 1995, 1565). Cytoblastin (506) is a novel unsymmetrical dimer of indolactam V found in *Streptoverticillium eurocidicum* (H. Kumagai, *et al.*, *J. Antibiot.*, 1991, **44**, 1029), and the marine sponge *Cribrochalina olemda* has furnished kapakahine B (507) (Y. Nakao, *et al.*, *J. Am. Chem. Soc.*, 1995, **117**, 8271).

Alangiobussine (502)

Candidine (503)

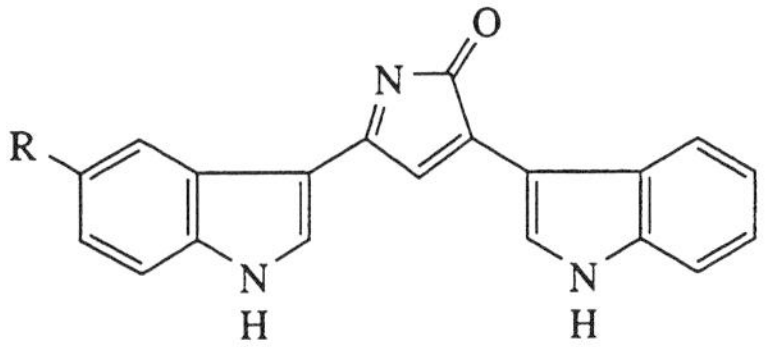

Prodeoxyviolacein (504) R = H
Proviolacein (505)   R = OH

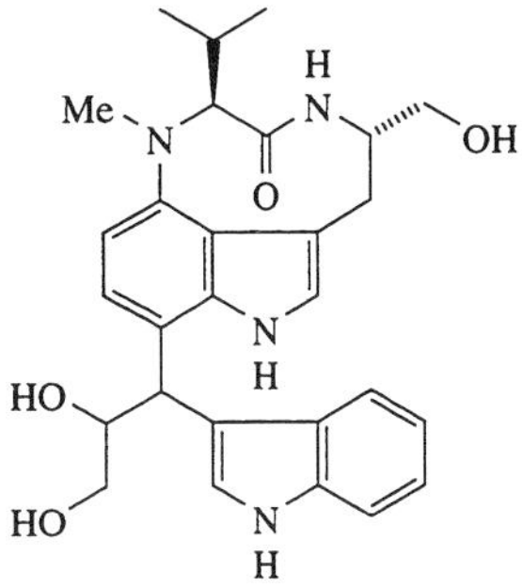

Cytoblastin (506)

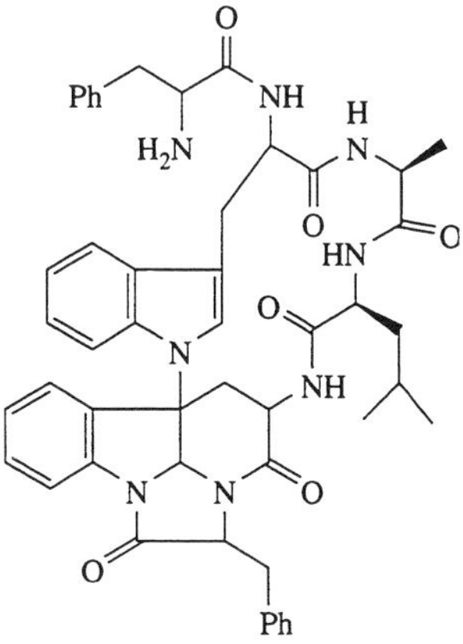

Kapakahine B (507)

Wakayin (508) represents the first pyrroloiminoquinone from an ascidian (*Clavelina* sp.) (B.R. Copp, *et al.*, *J. Org. Chem.*, 1991, **56**, 4596), and pseudophrynamine A (509) is secreted by the Australian frog *Pseudophryne coriacea* (T.F. Spande, *et al.*, *ibid.*, 1988, **53**, 1222). The roots of *Antirhea lucida* contain the novel alkaloid (510) (B. Weniger, *et al.*, *Planta Med.*, 1995, **61**, 569).

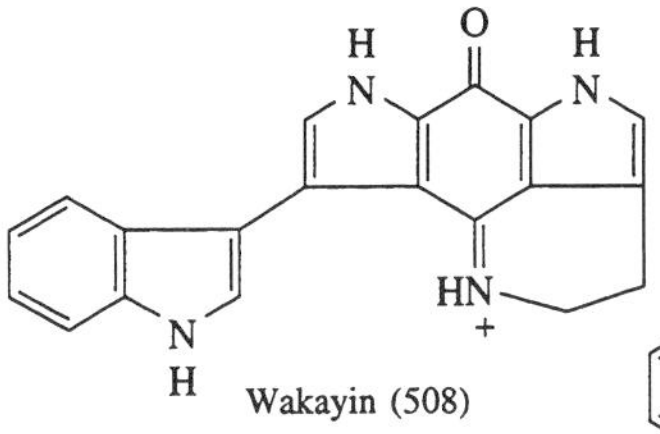

Wakayin (508)

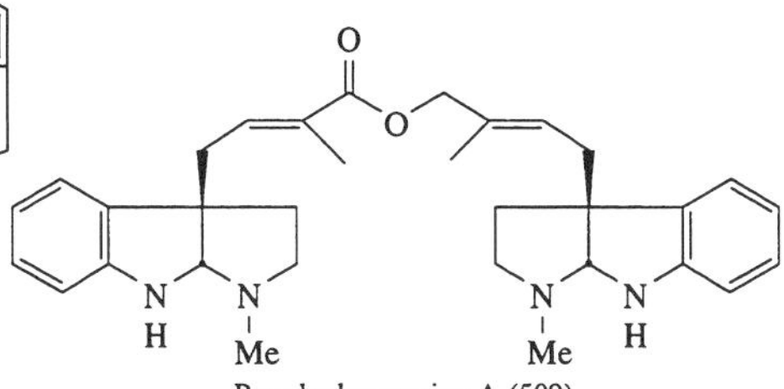

Pseudophrynamine A (509)

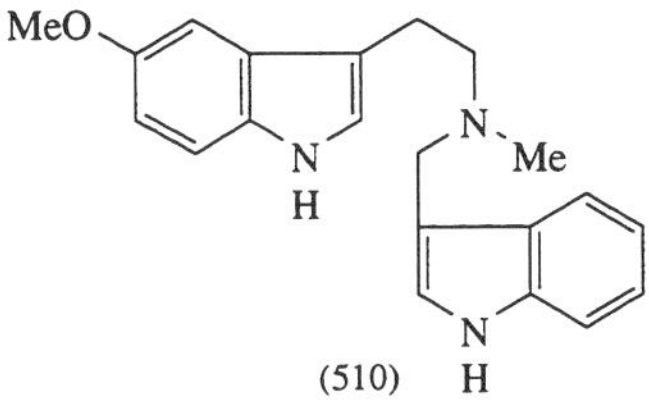

(510)

160

(vii) Other complex bis-indole alkaloids

A very large number of new bis-indole alkaloids have been discovered over the past 15 years that are comprised of various combinations of sarpagine-vobasine-*Iboga-Aspidosperma* units. The sheer number of these complex structures precludes the illustration of all but a few examples, but each new one is cited herein.

Most of the new alkaloids are found in genus *Tabernaemontana* and, in fact, over 300 alkaloids have been characterized from these plants as of 1994. The 19'(R)-hydroxy derivatives of conodurine and conoduramine have been identified in *T. subglobosa* (H. Takayama, *et al.*, *Chem. Pharm. Bull.*, 1994, **42**, 280), and 11-demethylconoduramine is found in *T. pachysiphon* (T.A. van Beek, *et al.*, *Phytochemistry*, 1984, **23**, 1771). Both *T. citrifolia* and *Peschiera echinata* yield 14-dehydrotetrastachyne (511) (J.Abaul, *et al.*, *C. R. Acad. Sci.*, 1984, **298**, 627). Eight new oxygenated bis-alkaloids of the conoduramine, conodurine, and voacamine families have been discovered in *T. chippii* (T.A. van Beek, *et al.*, *J. Nat. Prod.*, 1985, **48**, 400), and monogagaine is a novel bis-alkaloid formed from vobasinol and apparicine units isolated from both *T. chippii* and *T. dichotoma* (T.A. van Beek, *et al.*, *Z. Naturforsch.*, 1985, **40B**, 693). This latter plant has afforded five new bis-indole alkaloids of the tabernamine and ervahanine types (P. Perera, *et al.*, *Phytochemistry*, 1985, **24**, 2097). Vobparicine (512) is a novel bis-indole from *T. chippii* (T.A. van Beek, *et al.*, *Tetrahedron Lett.*, 1984, **25**, 2057). The isomeric pseudovobparicine is found in *T. divaricata* (T.A. van Beek, *et al.*, *Planta Med.*, 1985, 277). Here the attachment of the vobasinyl unit is to C-10 rather than to C-22 in the apparicine ring system. This plant has also afforded these novel bis-*Aspidosperma* alkaloids, conophylline (T.-S. Kam, *et al.*, *Tetrahedron*

(511)

Vobparicine (512)

Lett., 1992, **33**, 969), conophyllidine (T.-S. Kam, *et al.*, *J. Nat. Prod.*, 1993, **56**, 1865), and conofoline (T.-S. Kam, *et al.*, *Phytochemistry*, 1995, **40**, 313).

Voafrines A (513) and B (3'-epimer) are also bis-*Aspidosperma* alkaloids from *Voacanga africana* cell suspension cultures (J. Stöckigt, *et al.*, *Helv. Chim. Acta*, 1983, **66**, 2525). The first spiro-*Aspidosperma* eburnamine bis-alkaloid, vobtusamine, was isolated from *V. chalotiana* (B. Danieli, *et al.*, *J. Org. Chem.*, 1983, **48**, 381), and strempeliopidine is a related bis-alkaloid from *Strempeliopsis strempeliodes* (A. Laguna, *et al.*, *Planta Med.*, 1984, **51**, 285). Obovatine is a bis-*Iboga* alkaloid found in *Stemmadenia obovata* (E. Valencia, *et al.*, *J. Nat. Prod.*, 1995, **58**, 134). The plant *Stenosolen heterophyllus* has yielded eight new bis-alkaloids of the ervafoline and ervafolidine types (A. Henriques, *et al.*, *J. Org. Chem.*, 1982, **47**, 803). Callichiline, which was first isolated from *Callichilia subsessilis* in 1959, has now been found to be a bis-alkaloid consisting of 11-demethoxy-vandrikine and beninine units (A.T. McPhail, *et al.*, *Tetrahedron*, 1983, **39**, 3629). Cimilophytine (514) is a new type of bis-indole derived by coupling an unrearranged canthiphytine with a dehydrocimicidine unit. This alkaloid is found in *Haplophyton cimicidum* (A.A. Adesomoju, *et al.*, *J. Org. Chem.*, 1983, **48**, 3015). This plant has also furnished the similar cimiciduphytine (A.A. Adesomoju, *et al.*, *Heterocycles*, 1991, **32**, 1461), and crooksiine from *H. Crooksii* is another *Aspidosperma*-canthinone bis-alkaloid closely related to cimilophytine (M. Mroue and M. Alam, *Phytochemistry*, 1991, **30**, 1741).

Voafrine A (513)

Cimilophytine (514)

Kopsoffine (515) is a 1-norpleiomutine type from *Kopsia officinalis* (X.Z. Feng, *et al.*, *J. Nat. Prod.*, 1984, **47**, 117). Scandomelidine, which evidently results from the union of venalstonine and pachysiphine, is found in *Melodinus scandens* (H. Mehri and M. Plat, *ibid.*, 1992, **55**, 241).

162

Melomorsine is another bis-*Aspidosperma* alkaloid from *M. morsei*, a plant used in Chinese folk medicine (Y.-L. He, *et al.*, *ibid.*, 1994, **57**, 411). The plant *M. celastroides* contains several dimeric and quasi-dimeric alkaloids comprised of melonine and eburnamine units (H. Mehri, *et al.*, *ibid.*, 1991, **54**, 372). Some of these, however, may be dichloromethane-derived artefacts. The structure of gardmultine (*Gardneria multiflora*), proposed in 1975 (S. Sakai, *et al.*, *Tetrahedron Lett.*, 1975, 719), has now been confirmed by X-ray crystallography (J.V. Silverton and T. Akiyama, *J. Chem. Soc., Perkin Trans. 1*, 1982, 1263). Pandicine (516) is a bis-alkaloid involving a unique highly oxygenated tabersonine linked to macro-line that is found in *Pandacastrum saccharatum* (C. Kan-Fan, *et al.*, *J. Org. Chem.*, 1981, **46**, 1481).

Kopsoffine (515)

Pandicine (516)

N-Demethylpleiomutine (norpleiomutine) has been isolated from *Hunteria zeylanica* (C. Lavaud, *et al.*, *Phytochemistry*, 1982, **21**, 445), a plant that has also yielded the sarpagine-echitamine bis-alkaloids coryzeylamine and deformylcoryzeylamine (H. Takayama, *et al.*, *Chem. Pharm. Bull.*, 1994, **42**, 1957). This plant has also furnished hunteriatryptamine (517), a novel sarpagine-tryptamine bis-alkaloid (S. Subhadhirasakul, *et al.*, *Heterocycles*, 1995, **41**, 2049). The related ceridimine (minus hydroxymethyl) is found in *Pagiantha cerifera* (G. Baudouin, *et al.*, *J. Chem. Soc., Chem. Commun.*, 1986, 3). Ceridimine is the first example of a secologanin-derived bis-indole monoterpene alkaloid in which the extra tryptamine unit has a free side chain and is coupled to its benzene ring. Reaction of a mixture of vobasinol and tryptamine with acid gives ceridimine. Demethylceridimine is present in *Peschiera buchtieni*, along with demethylaccedinisine (M. Azoug, *et al.*, *Phytochemistry*, 1995, **39**, 1223). The 19(*R*)-hydroxyl derivative of the well known tabernaelegantine A as well as 19',20'(*S*)-dihydrotaberna-mine have been isolated from *Hazunta* sp. (M. Urrea, *et al.*, *Bull. Soc. Chim. Fr. II*, 1981, 147). The bis-alkaloid hazuntamine, from *H. modesta* var. *methuenii* subvar. *methuenii*, has a novel bis-dregamine structure (A.-M. Bui, *et al.*, *Heterocycles*, 1994, **38**, 1025). Vincarubine, which is the first bis-alkaloid to be identified from *Vinca minor*, is an *Aspidosperma*-echitamine type (B. Proksa, *et al.*, *Tetrahedron Lett.*, 1986, **27**, 5413). Cabufiline (518) and two related alkaloids (desoxycabufiline and norcabufiline) are found in *Alstonia plumosa* and *Cabucala caudata* (G. Massiot, *et al.*, *C. R. Acad. Sci. Ser. II*, 1982, **294**, 579), and plumocraline and nordesoxycabufiline are found in *Alstonia plumosa* (M.J. Jacquier, *et al.*, *Phytochemistry*, 1982, **21**, 2973).

Hunteriatryptamine (517)

Cabufiline (518)

Pedunculine and peduncularidine are two bis-*Aspidosperma* alkaloids from *Ervatamia peduncularis* (M. Zéches-Hanrot, *et al.*, *Phytochemistry*, 1995, **40**, 587), while *E. hainanensis* has yielded seven bis-alkaloids including ervahaimine A (519) and the related ervahanines A-C, all of which are of the voacamine type (X.-Z. Feng, *et al.*, *Lloydia*, 1981, **44**, 670; *J. Nat. Prod.*, 1989, **52**, 928). *E. hirta* has afforded 16-decarbomethoxyvoacaminepseudoindoxyl (P. Clivio, *et al.*, *Phytochemistry*, 1991, **30**, 3785), and *E. polyneura* contains the bis-*Aspidosperma* alkaloids polyervine and polyervinine (P. Clivio, *et al.*, *ibid.*, 1995, **40**, 953). The plant *Tonduzia pittieri* contains four bis-alkaloids of the ajmaline-cathafoline (or vincorine) type (A.-M. Morfaux, *et al.*, *Phytochemistry*, 1990, **29**, 3345; 1992, **31**, 1079). Rausutrine (520) and rausutranine are novel bis-alkaloids from *Rauwolfia sumatrana* (S. Subhadhirasakul, *et al.*, *Chem. Pharm. Bull.*, 1994, **42**, 1427). Flexicorine is a related quinoneimine bis-alkaloid from *R. reflexa* (A. Chatterjee, *et al.*, *J. Org. Chem.*, 1982, **47**, 1732). *Cinchona ledgeriana* contains two novel quasi-dimer alkaloids (M. Zeches, *et al.*, *Phytochemistry*, 1980, **19**, 2451), and *Aristotelia australasica* has yielded three novel dimers, bis-aristones A and B, and aristoaristone (J.-C. Quirion, *et al.*, *J. Org. Chem.*, 1987, **52**, 4527; *Nat. Prod. Lett.*, 1993, **2**, 41).

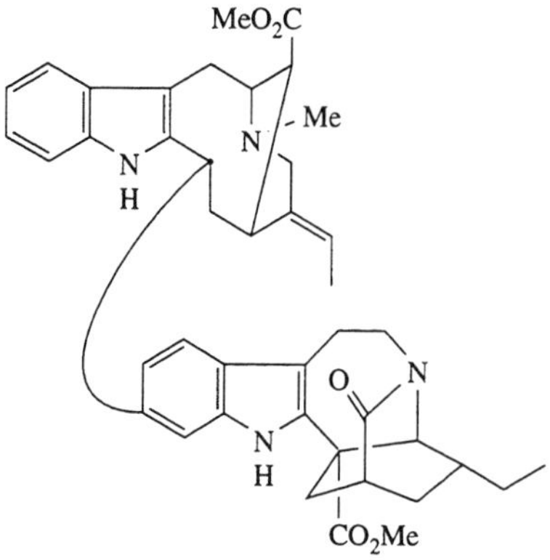

Ervahaimine A (519)

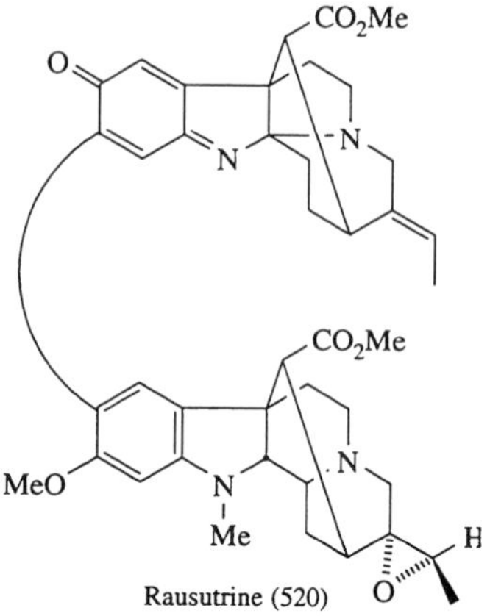

Rausutrine (520)

Cryptospirolepine (521) is a novel alkaloid from *Cryptolepis sanguino-lenta* related to cryptolepine from this same plant (A.N. Tackie, *et al.*, *J. Nat. Prod.*, 1993, **56**, 653). The novel cofactor TTQ (tryptophan trypto-phylquinone) is the redox center of bacterial methylamine dehydrogenase that catalyzes the oxidation of methylamine to formaldehyde and ammonia (S. Itoh, *et al.*, *J. Am. Chem. Soc.*, 1995, **117**, 1485; L. Chen, *et al.*, *Proteins*, 1992, **14**, 288). The structure of TTQ comprises the bis-indole unit (522) (W.S. McIntire, *et al.*, *Science*, 1991, **252**, 817).

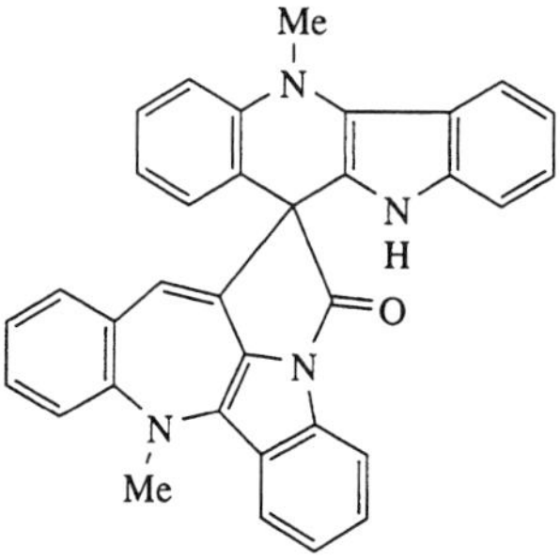

Cryptospirolepine (521)

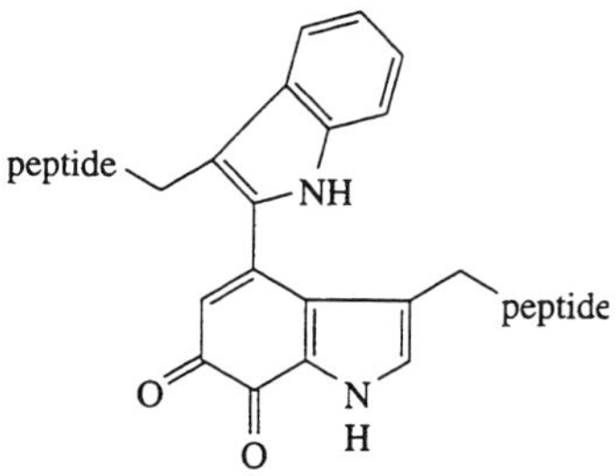

TTQ (522)

A fitting conclusion is the report describing the constitution of ancient Mayan blue paint. This beautiful, nanostructured material, which is resistant to acids and biocorrosion, is a composite of metal particles and the indole oxidized dimer, indigo (M. José-Yacamán, *et al.*, *Science*, 1996, **273**, 223).

Second Supplements to the 2nd Edition of Rodd's Chemistry of Carbon Compounds, Vol.IV B, edited by M. Sainsbury
© 1997 Elsevier Science B.V. All rights reserved.

Chapter 10

FIVE-MEMBERED MONOHETEROCYCLIC COMPOUNDS, AMARYLLIDACEAE ALKALOIDS

J.R. LEWIS

1. Introduction

The diversity of structures associated with alkaloids produced by Amaryllidaceae plants using only phenylalanine as a precursor exemplifies the armoury of synthetic procedures that Nature has under its control. The original reviews in this series covered the literature up to 1980 and annual reviews have appeared firstly in 'The Alkaloids', Volumes 1 to 13 (Special Periodical Reports, Royal Society of Chemistry) which have been superseded by 'Natural Product Reports', Volumes 1-12-, also published by the Royal Society of Chemistry.   A comprehensive review of these alkaloids covered the period 1972 to 1987 (S.F. Martin in The Alkaloids. Volume 30, Ed. A. Brossi, Academic Press, 1987) and this review covers the literature from 1987 primarily highlighting new alkaloids, new synthetic methods, structural determinations and biological activities.

In a few cases the isolation of unusual alkaloidal products has suggested artefact formation.   Thus of two chlorinated alkaloids, N-chloromethyl-narcissidinium chloride (1) was thought to be an artefact of its isolation procedure (R. Suau, A.I. Gome and R.R. Rico, *An. Quim.*, 1990, *86*, 672). The formation of N-chloromethylgalanthaminium chloride (2) by crystallising galanthamine (11) from dichloromethane confirmed this suggestion.   X-ray studies (see X-ray section) clearly established the quaternary nature of this last product.

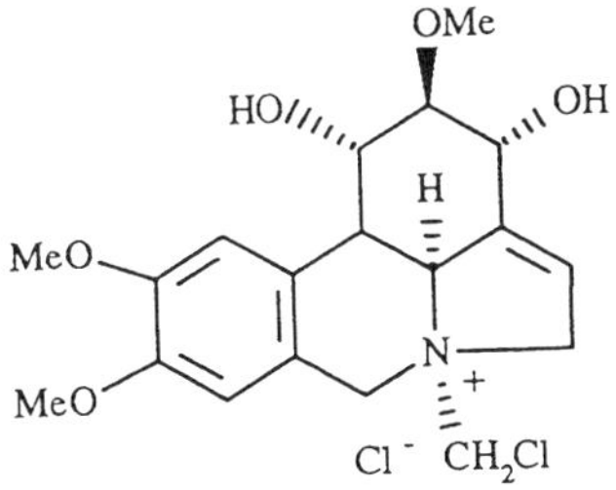

(1)

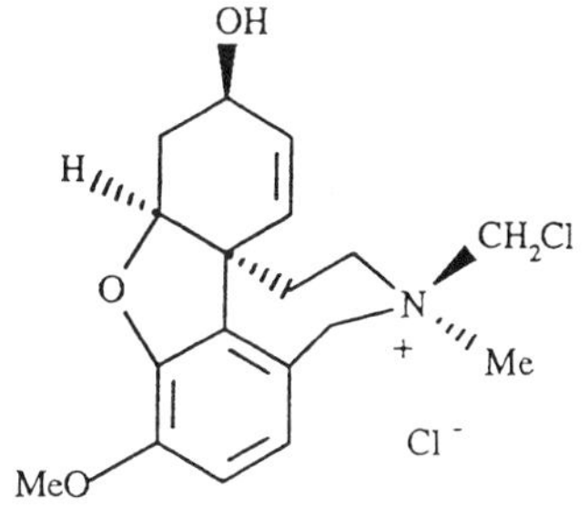

(2)

N-Oxides of known Amaryllidaceae alkaloids homolycorine (73) and O-methyllycorenine (74) were first reported in 1988 (Table 1) closely followed by ungiminorine N-oxide (107; Table 1). Later studies established that these N-oxides were not artefacts (Table 1, ref. 27).

Since enzyme hydroxylation can be induced *in vitro* (see Biogenesis Section) imprecise plant pretreatments could lead to cell rupture with the subsequent release of oxidative, hydrolytic and other enzymes. This could result in alkaloid modification. Now that purification and structural identification can be carried out on micro-molecular quantities the case arises for detailed descriptions of plant material collections, storage and pretreatment so as to create reproducible extraction procedures. The isolation of more glucosides (Table 1) stems from isolation procedures which avoid enzymatic hydrolysis due to mechanical cell injury and subsequent (prolonged) contact time with alkaloid conjugates.

Most of the alkaloids of this family can be classed into seven principle, skeletally homogeneous subgroups, although there are several other alkaloids having structures derived from these main molecular frameworks. Examples of alkaloids from each of these classes include lycorine (3), lycorenine* (4) crinine (5), narciclasine (6), tazettine (7), latisodine (8), montanine (9), mesembrine (10), galanthamine (11), crinafolidine (12), phenanthridine trisphaeridine (13) followed by miscellaneous types of alkaloid such as ismine (14) and isocraugsodine (15).

(3)

(4)

*This group of alkaloids are also called homolycorines which use a different numbering system.   Both are used in Section 7.

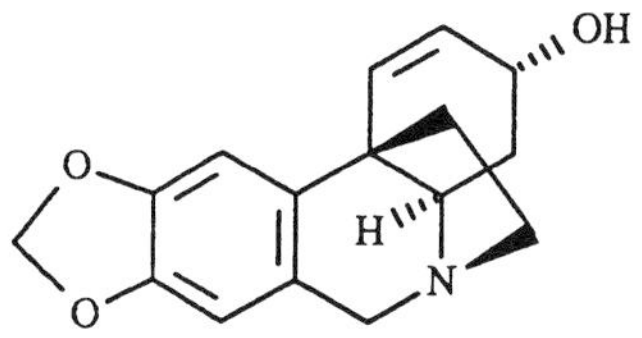

(5)

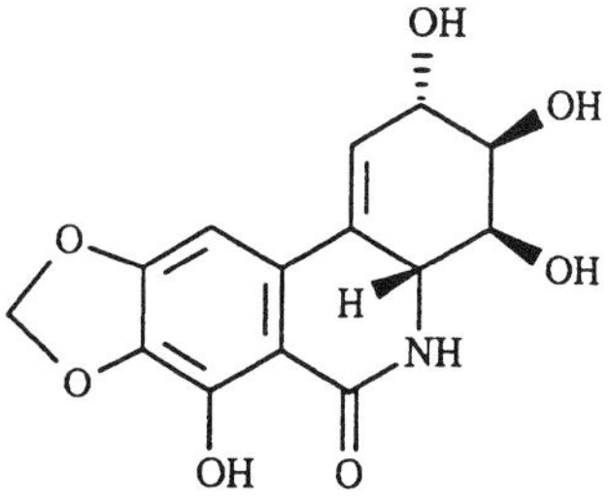

(6)

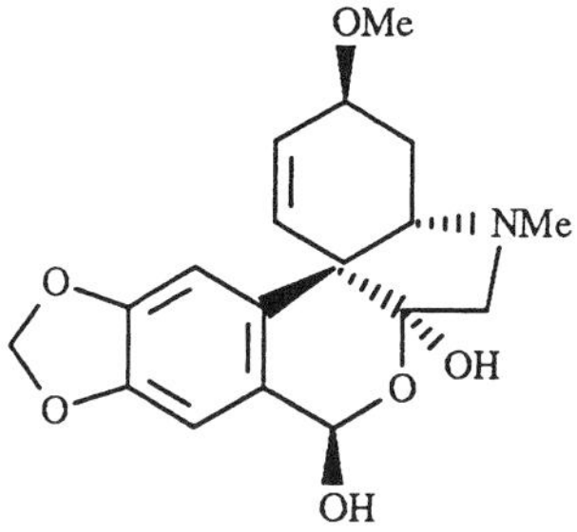

(7)

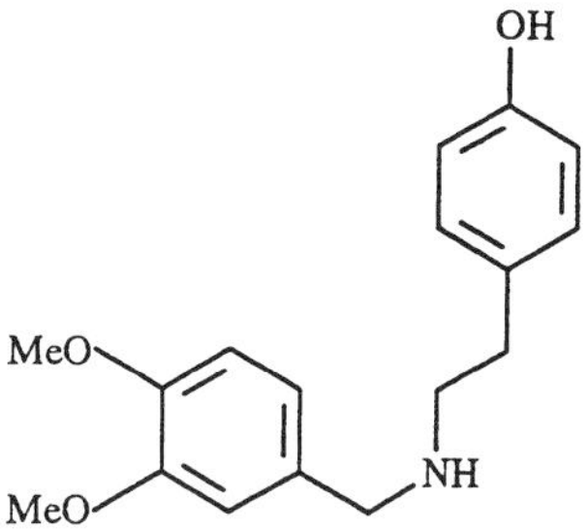

(8)

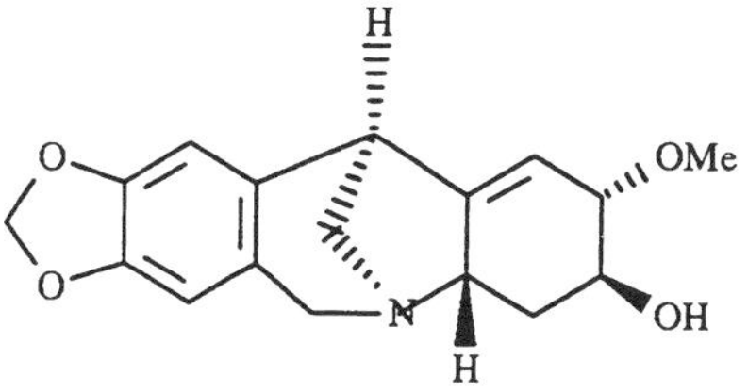

(9)

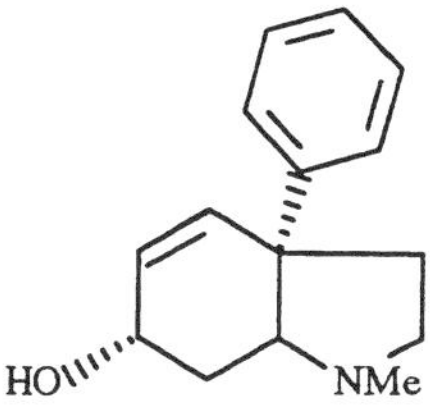

(10)

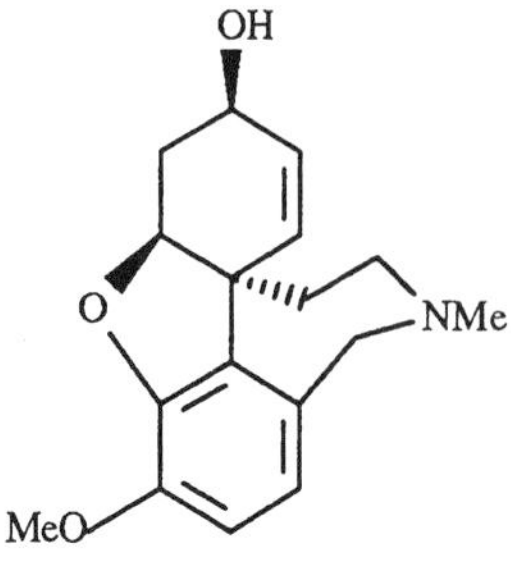

168

(11)

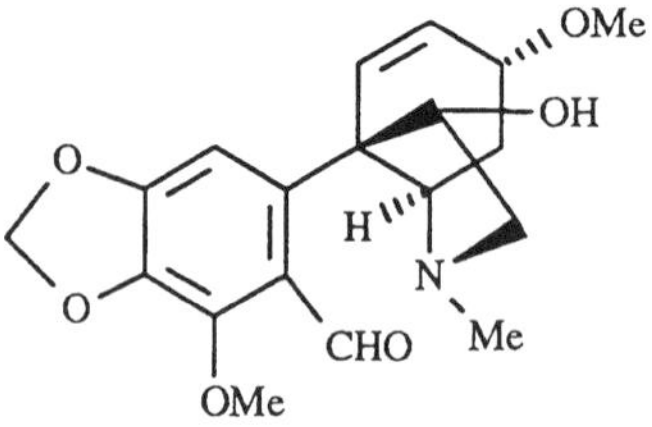

(12)

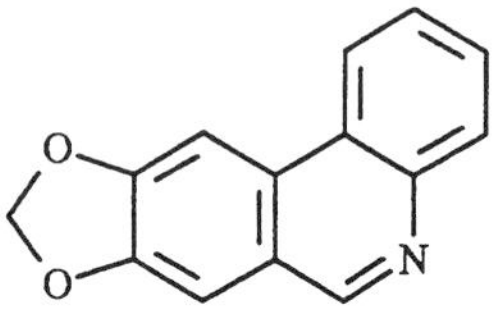

(13)

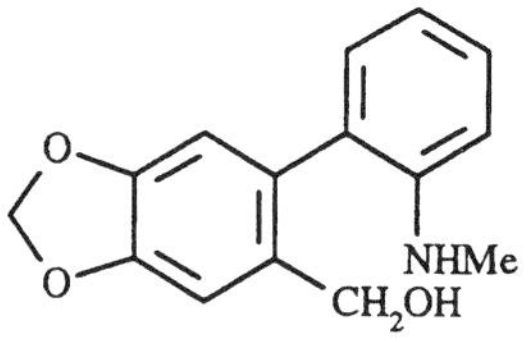

(14)

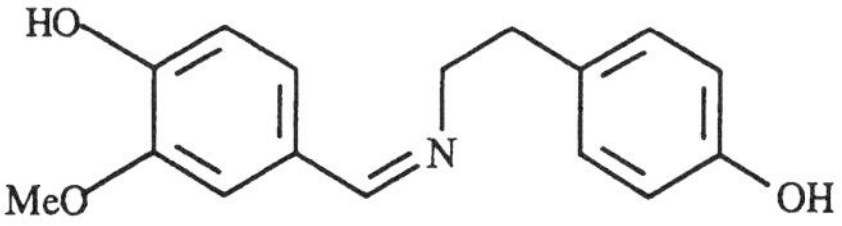

(15)

The most complex alkaloid reported in the period reviewed is pallidiflorine (16). It was obtained from fresh bulbs and aerial parts of *Narcissus pallidiflorus*. Galanthamine (11) and tazettine (7) are obvious contributors to this heterodimeric alkaloid (Table 1, ref. 36).

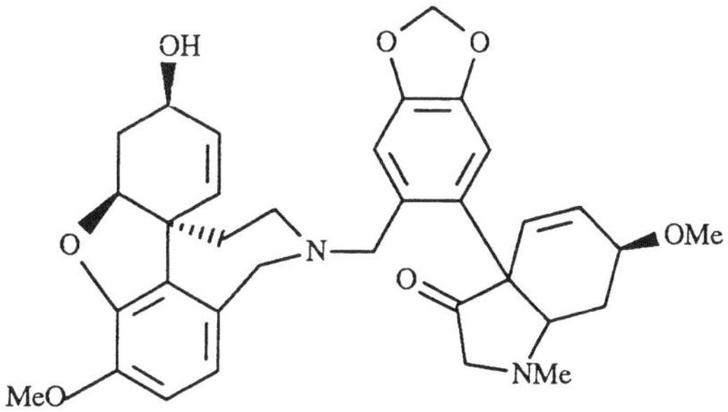

(16)

170

2. Biogenesis

In this review period most biogenetic studies were targeted at the late stages in Amaryllidaceae alkaloid production. Well established as are phenylalanine and tyrosine their subsequent modification is still under scrutiny. In earlier studies doubly labelled benzylamines when introduced into King Alfred daffodils were shown to be incorporated into galanthamine (11), haemanthamine (18) and oduline (19). N-Methyl-vanillamine was considered an intermediate which underwent benzylic hydrogen removal (C. Fuganti, D. Chiringhelli, P. Grasselli and M. Mazza, *Tetrahedron Letters*, 1974,*26*, 2261). It has now been shown that the benzylidine derivative craugsodine (17) is present in the flower-stem fluid of *Crinum augustum* and that it can be easily converted into the maritidine ring (20) (S. Ghosal, Y. Kummar, S.K. Singh, and A. Shanthy, *J. Chem. Res. Synop.*, 1986, 28).

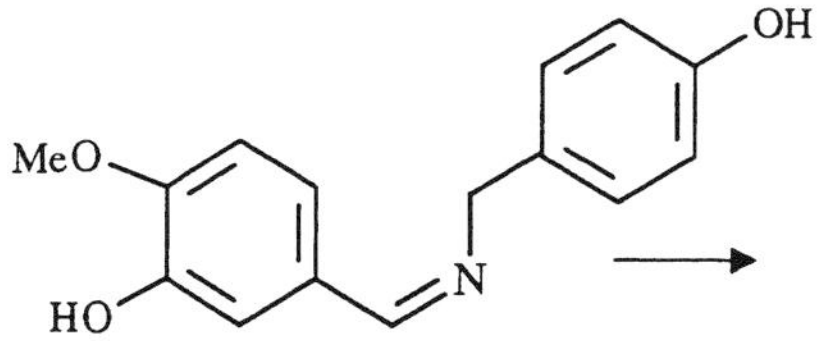

CRAUGOSIDINE (17)

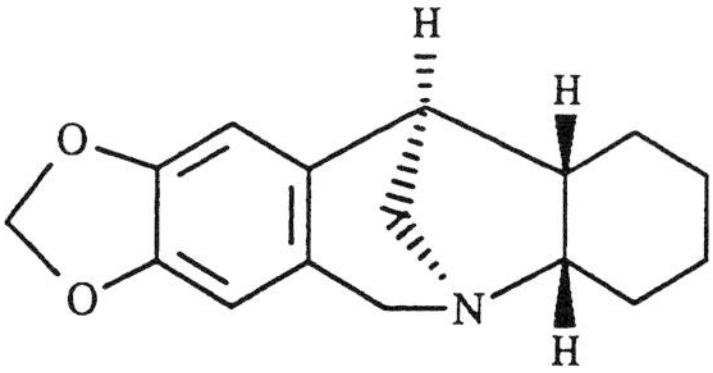

NOR-OXOMARTIDINE (20)

N-METHYLVANILLAMINE

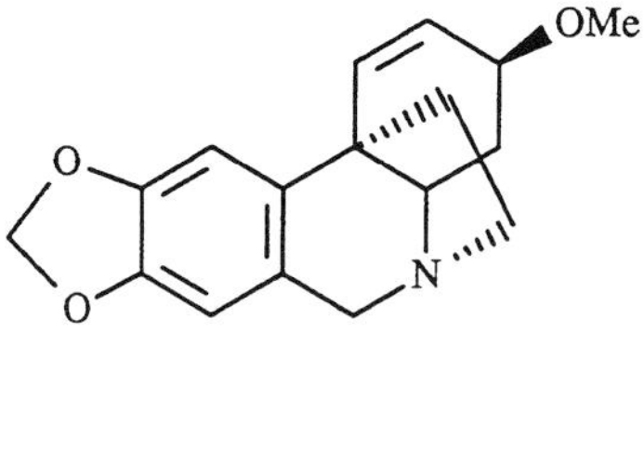

HAEMANTHAMINE (18)

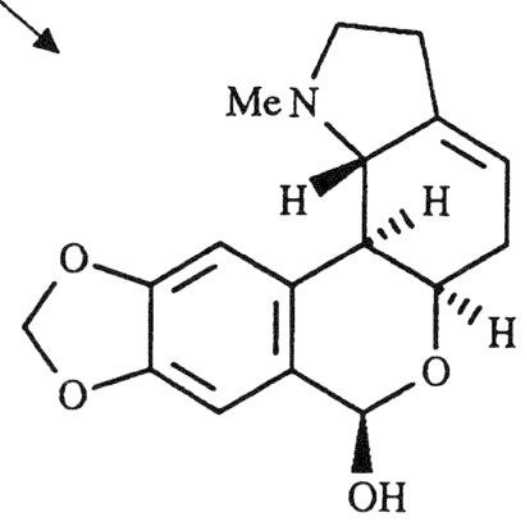

ODULINE (19)

Tritium labelled crinine (5) and oxovittatine (21) are not interconverted in *Nerine bowdenii* but vittatine (22) is converted into haemanthamine (18) and montanine (9). This latter route involves a ring rearrangement which is reflected by a lower specific incorporation (A.I. Feinstein and W.C. Wildman, *J. Org. Chem.*, 1976, *41*, 2447)

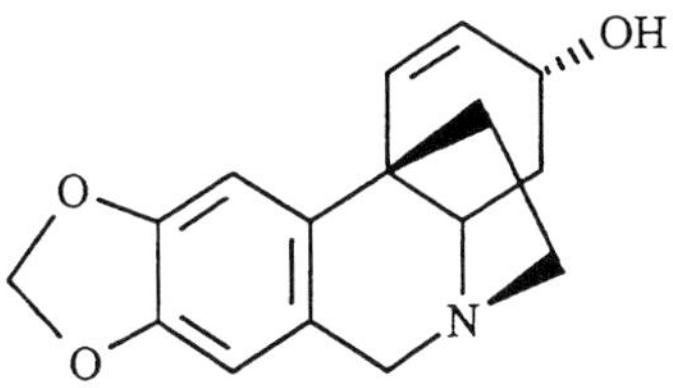

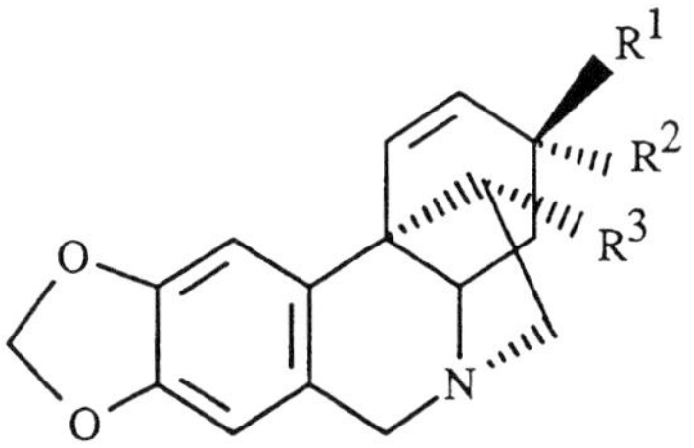

(21: $R^1 + R^2 = O, R^3 = H$)

(22: $R^1 = OH, R^2 = R^3 = H$)

(18: $R^1 = OMe, R^2 = H, R^3 = OH$)

(5)

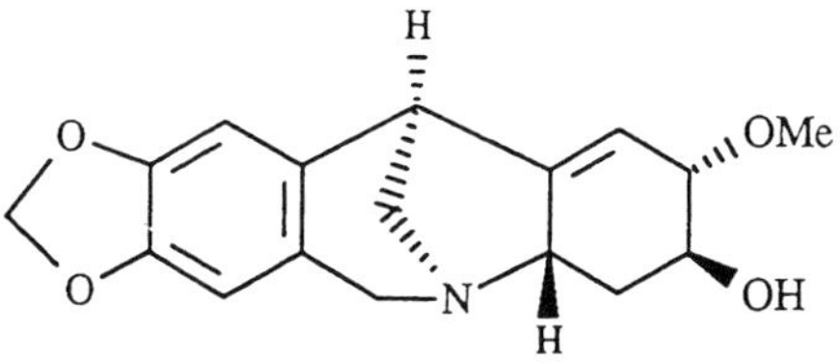

(9)

172

Late stages in the biosynthesis of mesembrine alkaloids such as the
joubertiamines (24; R=H or Me, *ab sat*) have been identified by feeding,
respectively, H-labelled-(4-hydroxyphenyl)propanal and phenylethyl-
phenylpropylamine (23) into intact plants of *Sceletium subvelutinum*. In
this study six structurally related joubertiamine like alkaloids (24; where
ab is saturated or unsaturated and R is or Me) were isolated. These
isolates also supported the biogenetic pathways' interrelationship (R.B.
Herbert and A.E. Kattah, *Tetrahedron*, 1990, *46*, 7105).

Phenylalanine
Tyrosine

(23)   (24)

The isolation of new alkaloids biogenetically related to established Amaryllidaceae alkaloids lends additional, but not experimental, support to the pathways used by plants to biosynthesise the more complex structures. Craugsodine (17) found in the bulbs of *Crinum angustum* is such an intermediate (S. Ghosal, Y. Kummar, I.K. Singh and A. Shanthy, *J. Chem. Res. S*, 1986, 28) as is isocraugsodine (15) (S. Ghosal, A. Shanthy and S.K. Singh, *Phytochemistry*, 1988, 27, 1849).

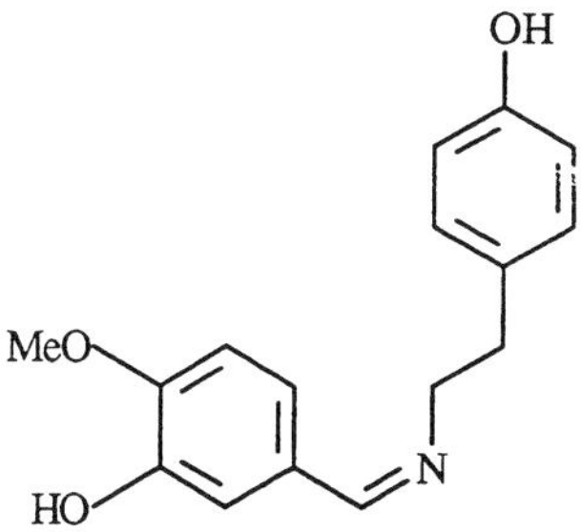

(17) 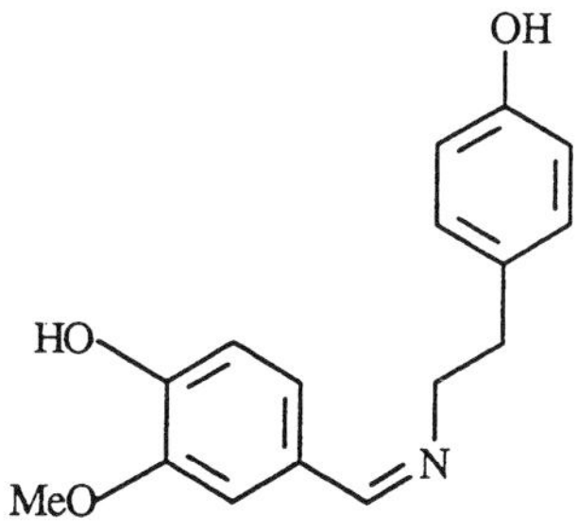   (15)

Ismine (14) co-occurs with three phenanthridine alkaloids, 8,9-methylenedioxyphenanthridine (13), N-methyl-8,9-methylenedioxy-6-phenanthridone (25), N-methyl-8,9-methylenedioxyphenthridinium chloride (86), in *Lapiedra martinezii* thus indicating a common biosynthetic relationship (R. Suau, A.I. Gomez and R. Rico, *Phytochemistry*, 1990, *29*, 1710). Related alkaloids are 8,9-methylenedioxy-6-phenanthridone (27) found in *Crinum asiaticum* (S. Ghosal, S.K. Saini, S. Razdan and Y. Kumar, *J. Chem. Res. S*, 1985, 100) and N-methyl-5,6-dihydro-8,9-methylenedioxyphenanthridine (87) from *Narcissus bicolor* (Table 1, ref. 30).

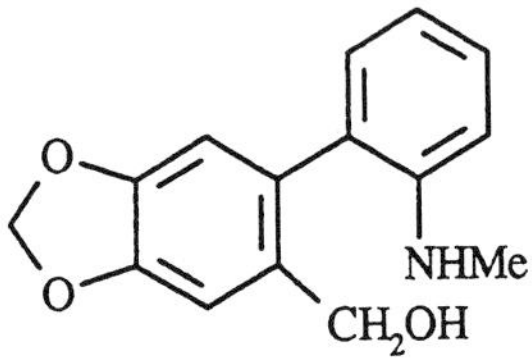

(14)

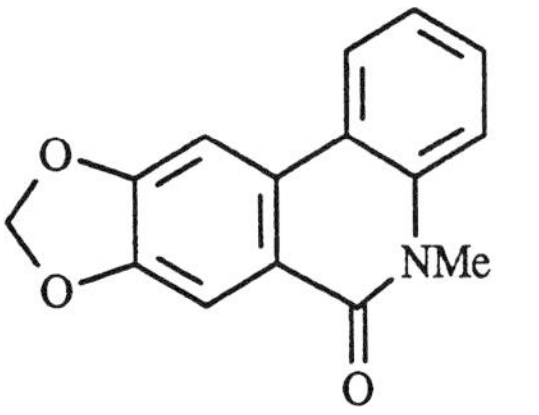

(25)

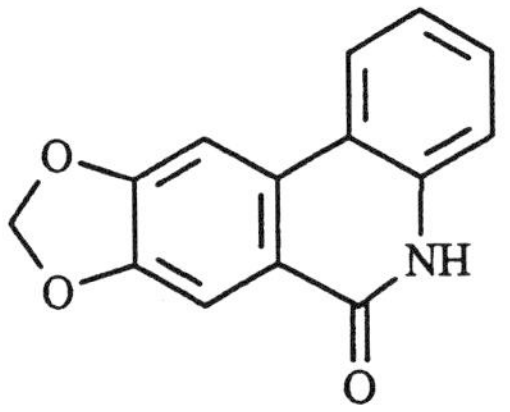

(27)

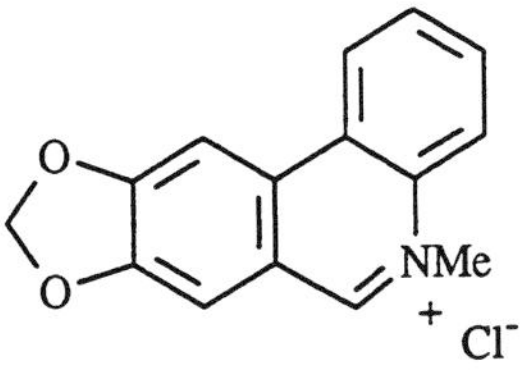

(86)

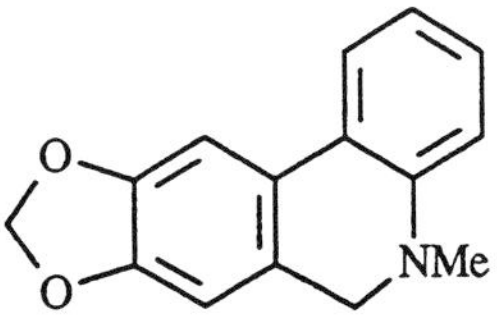

(87)

Additional modification of the phenanthridine nucleus can occur by enzymatic hydroxylation. Thus benzo[c]phenanthridines and their precursor protopine (28), can be hydroxylated in a stereospecific manner to (29) by P_{450} type cytochromes, then ring opened to (30), and recyclised to dihydrosanguinarine (31) Scheme 1 (T. Tanahashi and M.H. Zenk, *J. Natural Products*, 1990, *53*, 579).

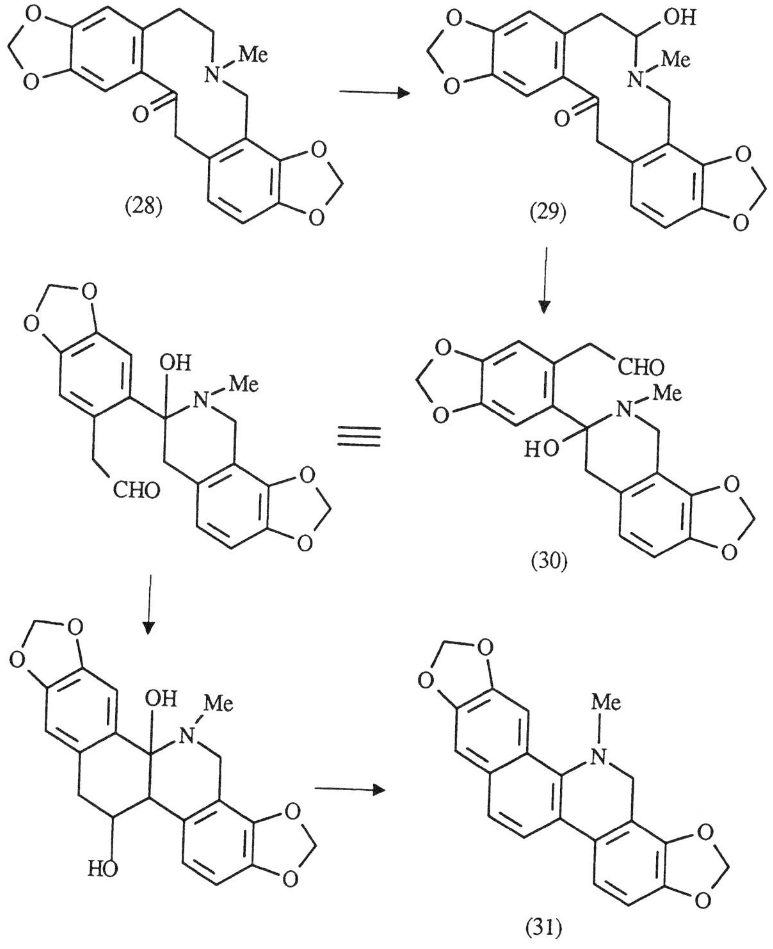

(28)

(29)

(30)

(31)

SCHEME 1

Infection of a callus culture of *Eschscholtzia californica* by a *Penicillium* fungus brought about extensive hydroxylation of the benzo[c]phenanthridine nucleus; in addition to the six known alkaloids (see Table 1) five new ones were isolated, where hydroxylation had occurred at positions 10 and/or 12, frequently followed by O-methylation.

3.   **Biological activity**

Perhaps the most significant and certainly the most well publicised event on the biological profile of any Amaryllidaceae alkaloid, is that associated with galanthamine (11) and its possible use as a treatment of Aldzheimer's disease.   Particularly important are the long known biological properties of galanthamine, its hydrobromide NIVALIN has been used for some time to treat neuromuscular, neurological and glaucoma problems (A. Barak and S.J. Harik, *J. Amer. Med. Assoc.*, 1977, *238*, 2293) and as a result it is known to have a large therapeutic margin, good tolerance and a reliable reaction.   Its anticholinesterase activity, as well as its muscarinic receptor blocking ability (J.E. Sweeney, S.K. Han, M.M. Jouille and J.T. Coyle, *Soc. Neurosci. Abstr.*, 1989, *15*, 731) which that has promoted further studies (S.Y. Han, J.E. Sweeney, E.S. Bachman, E.J. Schweiger, G. Forloni, J.T. Coyle, B.M. Davies and M.M. Joullie, *Eur. Med. J.*, 1992, *27*, 673).   The kinetics of the disappearance of the acetylcholinesterase inhibition in the rat intestine suggests reasonable stability and therefore absorption into the gut [I. Yamboliev, V. Mutafova-Yambolieva and D. Milailova, *Eur. J. Drug Metab. Pharmacokinet.*, 1993, 50 (Special Issue)].   Improved methods of screening have pinpointed biological activities over a wide range: anticancer properties have been found for lycorine (3), 6α-hydroxybuphanisine (32), narciclasine (6), acetylcaranine (33), anhydrolycorinium chloride (34), kalbretorine (35), pretazettine (36), unigerimine (37), criasbetaine (38), galanthamine (11), pancratistatin (39), tazettine (7), haemanthidine (40) and hippeastrine (41) (M.D. Antoun, M.J. Mendoza, Y.R. Rios, G.R. Proctor, D.B. Wickramaratne, J.M. Pezzuto and A.D. Kinghorn, *J. Nat. Prod.*, 1993, *56*, 1423).

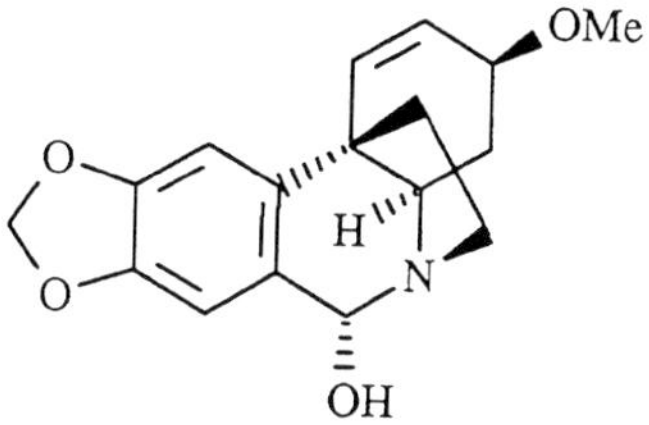

(32)

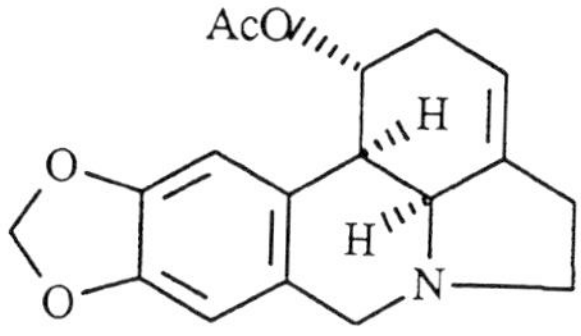

(33)

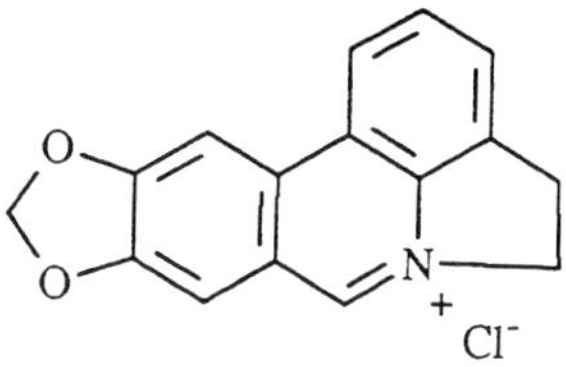

(34)

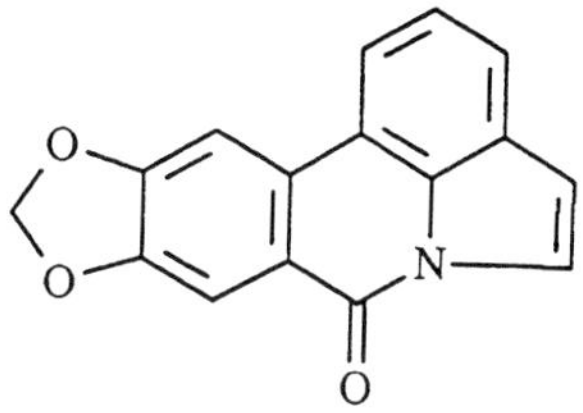

(35)

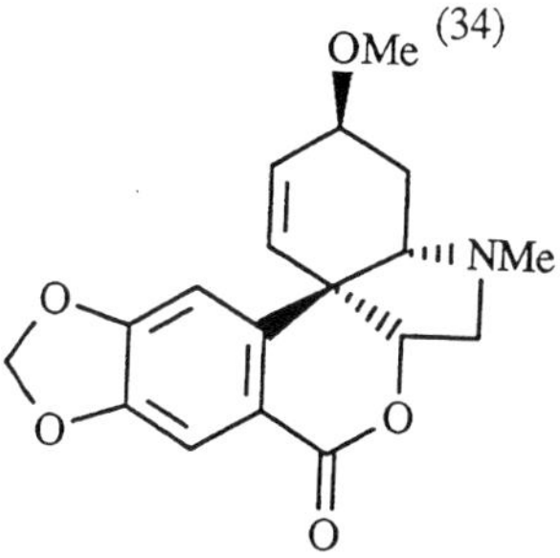

(36)

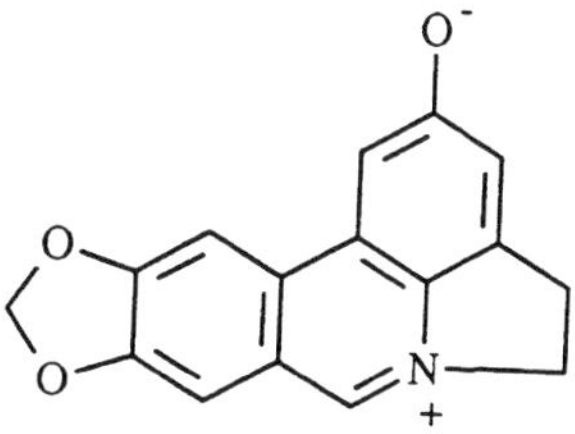

(37)

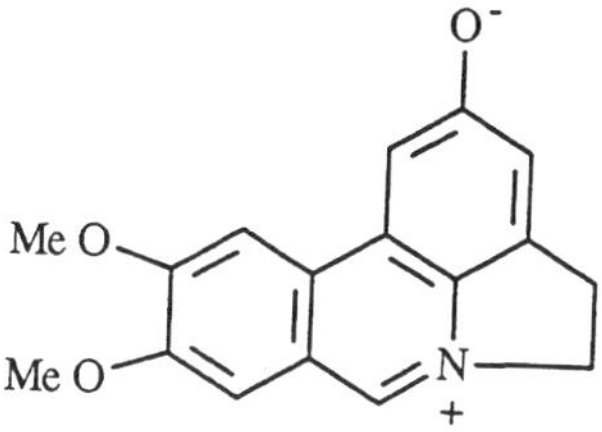

(38)

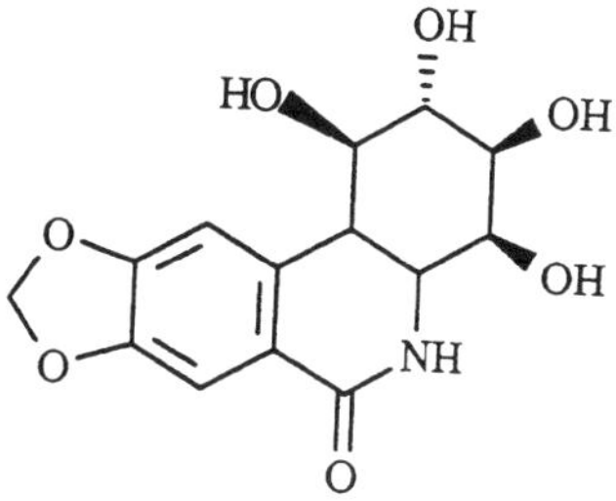

(39)

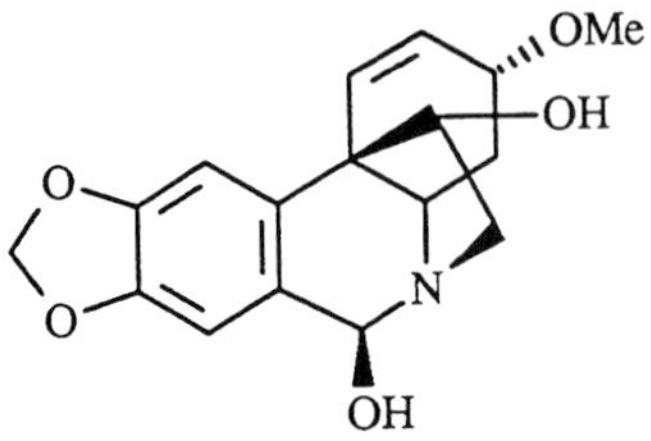

178

(40)

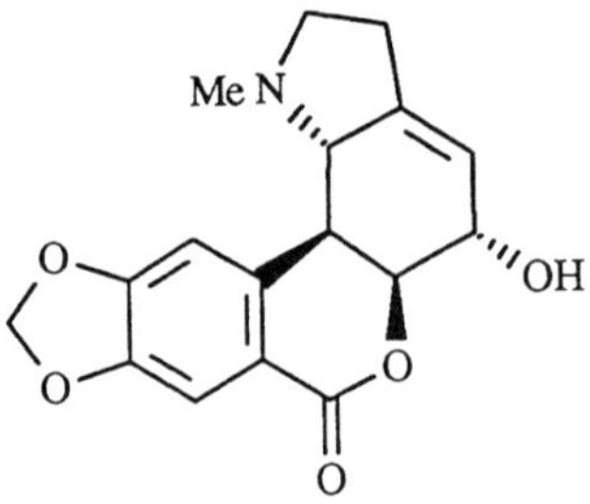

(41)

Antifeedant activity has been found for alkaloids: 9,O-demethylhomolycorine (42), lycoricidinol (43), and lycoridine (44), all present in the bulbs of *Lycoris radiata*. Narciclasine (6) and galanthamine (11) have similar properties, while lycorine (3) was active against the locust but tazettine (7) was not (R.P. Singh and N.C. Pant, *Experientia*, 1980, *36*, 552 and see reviews by Grundon, Lewis and Martin *loc. cit.*).

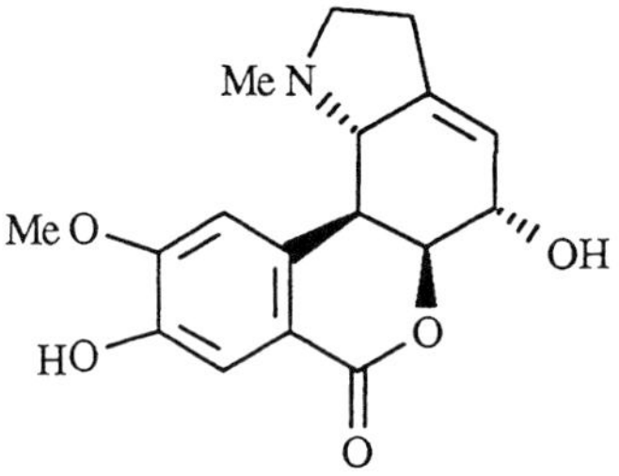

(42)

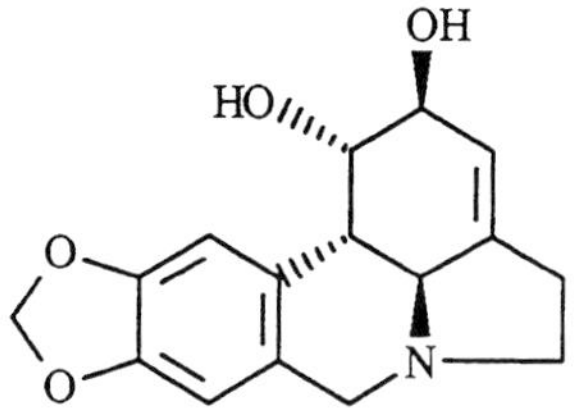

(43)

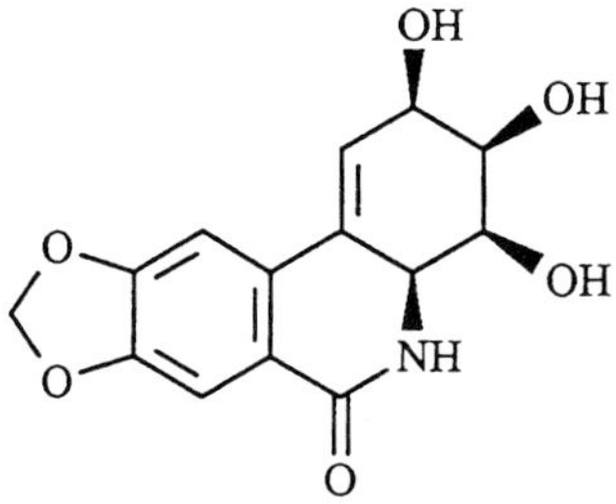

(44)

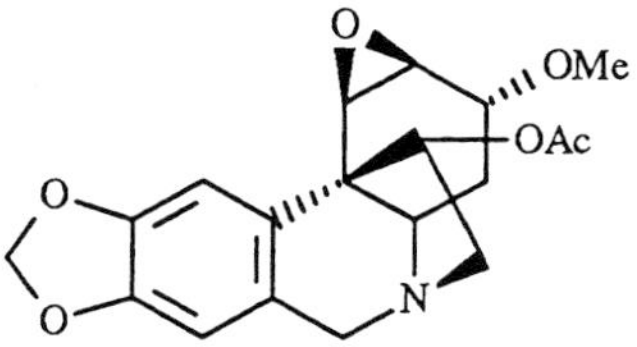

(45)

Immunoregulatory behaviour has been found for 11-O-acetyl-1,2–β-epoxyambelline (45) and 11-O-acetylambelline (46) (S. Ghosal, P.H. Rao and S.K. Saini, *Pharm. Res.*, 1985, 251) while lycorenine (4) depressed the blood pressure in rats.   Activity against animal viruses has been found for lycorine (3) and galanthamine (11).

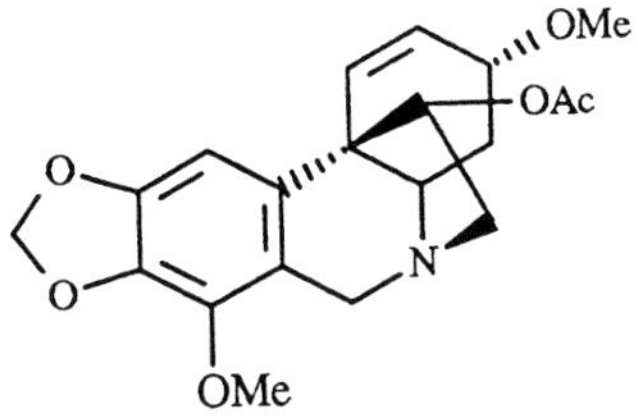

(46)

Table 1.   New Amaryllidaceae Alkaloids (1987-1994) and their Botanical Sources

Plant	Alkaloid	References
Brunsviga josephinae	Josephinine (47)	1
Clavia nobilis	Nobilisine (48)	2
Crinum amabile	Amabiline (49)	3
Crinum americanum	Oxocrinine (50)	4
	Crinan type alkaloid (unnamed) (51)	5
Crinum asiaticum	Criasbetaine (38)	6
	Lycoriside (52)	7
	Palmilycorine (53)	7
	Isocraugsodine (15)	8
	Lycorine-1,2-diglucoside (54)	9
Crinum asiaticum var sinicum	Crinisin (55)	10
Crinum latifolium	Crinafolidine (12)	11
	Ambelline β-epoxide (56)	11
	Crinafoline (57)	11
	2-Epilycorine (58)	12
	2-Epipancrassidine (59)	12
Crinum zeylanicum	Zeylamine (60)	13
Eschscholtzia califonica	11-Hydroxysanguinarine (61)	14
	12-Hydroxychelirubine (62)	
	10-Hydroxychelerythrine (63)	
	10-Hydroxy-5,6-dihydros-anguinarine (64)	
	12-Hydroxy-5,6-dihydro-chelirubin (65)	
Haemanthus albiflos	Albiflomanthine (66)	15
Haemanthus kalbreyeri	7-Deoxypancratistatine (67)	16
	Pancratiside (68)	16
Haemanthus multiflorus	3-O-Acetylchlidanthine (69)	17
Haemocallus caymanensis	4-Hydroxyanhydrolycorine (70)	18
Haemocallus rotata	8-O-Demethylmaritidine (71)	19
	N-Demethyllycoramine (72)	19
	6-Hydroxyvittatine (73)	19

Hippeastrum puniceum	3-O-Acetylnarcissidine (74)	20
Lipiedra martinezii	Homolycorine N-oxide (75)	21
	O-Methyllycorenine N-oxide (76)	22
	N-Methylcrinasidine (25)	22
	8,9-Methylenedioxyphenthridine (Trisphaeridine)(13)	22
	N-Methyl-8,9-methylenedioxy-phenanthridine (26)	22
	N-Methyl-8,9-methylenedioxy-6-phenanthridone (27)	22
	N-Methyl-8,9-methylenedioxy-phenthridinum chloride (28)	22
	N-Methylassoaninum chloride (77)	23
Lycoris incarnata	Incartine (78)	24
Lycoris radiata	Acetyl-O-methyllycorine (79)	25
	Hippeastrine N-oxide (80)	26
Lycoris sanguinea	Galanthamine N-oxide (81)	27
	Sanguinine N-oxide (82)	27
	Norsanguine (83)	28
	Norbutsanguine (84)	28
Narcissus assoanus	Oxoassoanine (85)	29
Narcissus bicolor	Bicolorine (86)	30
	5,6-dihydrobicolorine (87)	30
Narcissus confusus	N-formylgalanthamine (88)	31
	9-O-Demethylhomolycorine (42)	32
Narcissus dubius	Dubiusine (89)	32,45
Narcissus leonensis	Epinorgalanthamine (90)	33
	Epinorlycoramine (91)	34
Narcissus obesus	Obesine (92)	35
Narcissus pallidiflorus	Pallidiflorine (16)	36
	Roserine (93)	37
Narcissus papyraceus	O-Methylpapyramine (94)	38
	O-Methylmaritidine (95)	38
	9-O-Demethylhomolycorine N-oxide (96)	38

Narcissus pseudonarcissus	Narcidine (97)	39
	O-Methyloduline (98)	40
	N-Demethylmasonine (99)	40
Narcissus radinganorum	9-O-Demethylmaritidine (100)	41
Narcissus sp 'Fortune'	Fortucine (101)	42
Narcissus tazetta	Narcisine (102)	43
Narcissus tazetta var *chinensis*	6α-Hydroxy-3-O-methyl epimaritidine (103)	44
	6β-Hydroxy-3-O-methyl-epimaritidine (104)	45
Narcissus tortifolius	9-O-Demethylhomolycorine (40)	45,32
Narcissus vasconicus	8-O-Acetylhomolycorine (105)	46
	Vasconine (106)	46
Pancratium biflorum	Telastaside (107)	47
Pancratium maritimum	Ungiminorine N-oxide (108)	48
	4-O-β-D-glucosylnarciclasine (109)	49
	3β,11α-Dihydroxy-1,2-dehydro-crinane (110)	50
	8-Hydroxy-9-methoxycrinine (siculine) (111)	50
Sternbergia lutea	Epimaritinamine (112)	51
	Maritinamine (113)	51
	Deacetylbutessine (unigiminorine) (114)	51
	Buphanisine (115)	51
Sternbergia sicula	Siculinine (116)	52
Zephyranthes flava	Lycorine deriv. (117)	53
	Criasbetaine (118)	53,6
	Zefbetaine (119)	53
	Zeflabetaine (120)	53

References

1. F. Vildomat, J. Bastida, C. Codina, W.G. Campbell and S. Mathee, *Phytochemistry*, 1994, *35*, 809.

2. P.W. Jeffs, L. Mueller, A.H. Abou-Donia, A.D.S. El-Din and D. Campau, *J. Nat. Products*, 1988, *51*, 549.

3. K. Likhitwitayawuid, C.V. Angerhofer, H. Chai, G.M. Pezzuto and G.A. Cordell, *J. Nat. Products*, 1993, *120*, 1331.

4. A.A. Ali, H.M.El Sayad, O.M. Abdallan and W. Steglich, *Phytochemistry*, 1986, *25*, 2399.

5. Z. Trimino, C. Iglesias and I. Spengler, *Rev. Cubana Quim.*, 1987, *3*, 67 (*Chem. Abstr.*, 1988, *109*, 1872510).

6. S. Ghosal, Y. Kummar, S.K. Singh and A. Kumar, *J. Chem. Res. (S)*, 1986, 112.

7. S. Ghosal, A. Shanthy, A. Kumar and U. Kumar, *Phytochemistry*, 1985, *24*, 2703.

8. S. Ghosal, A. Shanthy and S.K. Singh, *Phytochemistry*, 1988, *27*, 1849.

9. S. Ghosal, S.K. Singh and S.G. Unnikrishnan, *Phytochemistry*, 1989, *28*, 2535.

10. R.J. Tang, N.J. Bi and G.E. Ha, *Chim. Chem. Letters*, 1994, *5*, 855 (*Chem. Abstr.*, 1995, *122*, 128545).

11. S. Ghosal and S.K. Singh, *J. Chem. Res. S*, 1986, 312.

12. S. Ghosal, S.G. Unnikrishnan and S.K. Singh, *Phytochemistry*, 1989, *28*, 2535

13. W. Doepke, E. Reich, E. Sewerin, R. Donau, D. Ferras, C. Iglesias, I. Spengler and Z. Trimino, *Z. Chem.*, 1986, *26*, 438.

14. T. Tanahashi and M.H. Zenk, *J. Nat. Products*, 1990, *53*, 579.

15. G. Bandouin, F. Tellequin and M. Koch, *Heterocycles*, 1994, *38*, 965.

16. S. Ghosal, S. Singh, Y. Kumar and R.S. Srivastava, *Phytochemistry,*. 1989, *28*, 611.

17. O.M. Abdallah, A.A. Ali and H. Itokawa, *Phytochemistry*, 1989, *28*, 3248.

18. Z. Trimino, I. Spengler, C. Iglesias and J. Borrego, *Rev. Cubana Quim.*, 1989, *5*, 55 (*Chem. Abstr.*, 1991, *115*, 203303).

19. M. Kihara, T. Koike, Y. Imakura, K. Kida, T. Shingu and S. Kobayashi, *Chem. Pharm. Bull.*, 1987, *35*, 1070.

20. J.C. Quirion, H.P. Hussan, H. Phillippe, B. Weniger, F. Jiminez and T.A. Zanone, *J. Nat. Products*, 1991, *54*, 1112.

21. R. Suau, A.I. Gomez, R. Rico, M.P.V. Tato, L. Castedo and R. Riguera, *Phytochemistry*, 1988, *27*, 3285.

22. R. Suau, A.I. Gomez and R. Rico, *Phytochemistry*, 1990, *29*, 1710.

23. R. Suau, A.I. Gomez and R. Rico, *An. Quim.*, 1990, *86*, 672 (*Chem. Abstr.*, 1991, *114*, 98281).

24. M. Kihara, L. Xu, K. Konishi, Y. Nagao, S. Kobayashi and T. Shingu, *Heterocycles*, 1992, *34*, 1299.

25. A. Numata, T. Takemura, H. Ohbayashi, T. Katsuno, K. Yamamoto, K. Sato and S. Kobayashi, *Chem. Pharm. Bull.*, 1983, *31*, 2146.

26. M. Kihara, K. Konoshi, L. Xu and S. Kobayashi, *Chem. Pharm. Bull.*, 1991, *39*, 1849.

27. S. Kobayashi, K. Satoh, A. Numata, T. Shingu and M. Kihara, *Phytochemistry*, 1991, *30*, 675.

28. O.M. Abdallah, *Phytochemistry*, 1995, *39*, 477.

29. J.M. Llabres, F. Vildomat, J. Bastida, C. Codina and M. Rubiralta, *Phytochemistry*, 1986, *25*, 2637.

30. F. Wildomat, J. Bastida, G. Tribo, C. Codina and M. Rubiralta, *Phytochemistry*, 1990, *29*, 1307.

31. J. Bastida, F. Viladomat, J.M. Llabres, C. Codina, M. Feliz and M. Rubiralta, *Phytochemistry*, 1987, *26*, 1519.

32. J. Bastida, J.M. Llabres, F. Viladomat, C. Codina, M. Rubiralta and M. Feliz, *Phytochemistry*, 1988, *27*, 3657.

33. J. Bastida, F. Viladomat, S. Bergonon, J.M. Fernandez, C. Codina, M. Rubiralta and J.C. Quiron, *Phytochemstry*, 1993, *34*, 1656.

34. S.A. Adesanya, T.A. Olugbade, O.O. Odebiyi and J.A. Aladesanami, *Int. J. Pharmacogn.*, 1992, *30*, 303 (*Chem. Abstr.*, 1993, *119*, 113351).

35. F. Viladomat, J. Bastida, C.Codina, M. Rubirala and J.C. Quirion, *J. Nat. Products*, 1992, *55*, 804.

36. C. Codina, F. Viladomat, J. Bastida, M. Rubiralta and J.C. Quirion, *Phytochemistry*, 1990, *29*, 2685.

37. J. Bashida, C. Codina, F. Viladomat, M. Rubiralta, J.C. Quirion and B. Weniger, *J. Nat. Products*, 1992, *55*, 122.

38. R. Suau, R, Rico, A.I. Garcia and A.I. Gomez, *Heterocycles*, 1990, *31*, 517.

39. E. Tojo, *J. Nat. Products*, 1991, *54*, 1387.

40. M. Kreh and R. Matusch, *Phytochemistry*, 1995, *38*, 1533.

41. J. Bastida, J.M. Llabres, F. Viladomat, C. Cordina, M. Rubiralta and M. Felix, *Planta Medica*, 1988, *54*, 524.

42. G.M. Tokhtabaeva, V.I. Sheichenko, I.V. Yartseva and O.N. Tolkachev, *Khim. Prir. Soedin.*, 1987, 872 (*Chem. Abstr.*, 1988, *109*, 89687).

43. O.M. Abdullah, *Phytochemistry*, 1993, *34*, 1447.

44. G.E. Ma, H.Y. Li, C.E. Lu, X.M. Yang and S.H. Hong, *Heterocycles*, 1986, *24*, 2089.

45. J. Bastida, C. Codina, F. Vildomat, M. Rubiralta, J.C. Quirion and H.P. Husson *Phytochemistry*, 1990, *29*, 2683.

46. J. Bastida, C. Codina, F. Vildomat, M. Rubiralta, J.C. Quirion and B. Weniger, *J. Nat. Products.*, 1992, *55*, 122.

47. S. Ghosal, K. Dalta, S.K. Singh and Y. Kumar, *J. Chem. Res. (S)*, 1990, 334.

48. R. Suau, A.I. Gomez, R. Rico, M.P.V. Tato, L. Castedo and R. Riguera, *Phytochemistry*, 1988, *27*, 3285.

49. A.H. Abou-Donia, A.De Gulio, A. Evidente, M. Gaber, A.A. Habib, R. Lanzetta and A.A.S. El Din, *Phytochemistry*, 1991, *30*, 3445.

50. B. Sener, S. Konukol, C. Kruk, P. Cornelis and K. Upendra, *Nat. Products Letters*, 1993, 287.

51. V.P. Rabuccuoglu, P. Richomme, T. Gozler, B. Kivcak, A.J. Freyer and M. Shamma, *J. Nat. Products*, 1989, *52*, 785.

52. P. Richomme, V. Rabuccuoglu, T. Gozler, A.J. Freyer and M. Shamma, *J. Nat. Products*, 1989, *52*, 1150.

53. S. Ghosal, S.K. Singh and S.G. Unnikrishan, *Phytochemistry*, 1987, *26*, 823.

54. S. Ghosal, S.K. Singh and R.S. Srivastava, *Phytochemistry*, 1986, *25*, 1975.

186

5. X-ray Structure Determination

The multiplicity of chiral centres in the Amaryllidaceae alkaloids demands assignment of uniquivocal stereochemistry. This has been achieved by X-ray measurements on six alkaloids, belonging to the four main alkaloidal classifications, namely hemanthamine and hippeastidine-crinine group; lycorenine-homolycorine group, pratorinine with pratorimine in the narchiclasine group and brunsvigine of the montanine group. Hemanthamine (18) was obtained as opaque crystals from an extract of *Hippeastrum bicolor* following chromatography. X-ray measurements (W.H. Watson, J. Galloy and M. Silva, *Acta Crystallographica*, 1984, *C40*, 156) showed it to have the cyclohexene ring and the nitrogen containing ring to be 1,2-diplanar in conformation while the five membered methylene-dioxy ring is a flattened envelope. The five membered nitrogen containing ring exhibits a slightly flattened half-chair conformation.

(18)

A previous structural determination on hemanthamine p-bromobenzoate also described the same stereochemistry (J. Clardy, F.M. Hauser, D. Dahm, R.A. Jacobson and W.C. Wildman, *J. Amer. Chem. Soc.*, 1970, *92*, 6337). Hippeastidine (121) and epihomolycorine (122) were obtained from an extract of *Hippeastrum ananuea* and their structures determined unequivocally both as their picrates (G.M. Gopalakrishna, W.H. Watson, M. Silva and P. Pacheco, *Cryst. Struct. Commun.*, 1978, *7*, 41) or in the case of hippeastidine (121) as the free base (W.J. Watson, V. Zabel, M. Silva and P. Pacheco, *Cryst. Struct. Commun.*, 1982, *11*, 157).

(121)

(122)

Pratorimine (123) and pratorinine (124) are isomeric phenanthridine alkaloids differing in the location of the methoxy grouping. Previous to this X-ray determination pratorinine was thought to be (123) on the basis of chemical shifts of the aromatic protons 8 and 11 adjacent to the hydroxy group after addition of D_2O, NaOD. However, the X-ray measurement of alkaloid A, identical to pratorinine by ir comparison, obtained by extensive column and vacuum liquid chromatography and crystallised from methanol indicated that the methoxy group was located at C-10 and thus structural revision was necessary (J.A. Maddry, B.S. Joshi, A.A. Ali, M.G. Newton and S.W. Pelletier, *Tetrahedron Letters*, 01985, *26*, 4301).

(123; R^1 = H, R^2 = Me)
(123; R^1 = Me, R^2 = H)

Latifine (125) and cheryline (126) are both found in the leaves of *Crinum latifolium*, are isomeric and have similar mass spectral fragmentation patterns. The structure of latifine has now been confirmed by a direct X-ray crystallographic study unambiguously locating the position of the phenolic and methoxy groups as indicated in structure (125) (S. Kobayashi, T. Tokumoto and Z. Taira, *J. Chem. Soc., Chem. Commun.*, 1984, 1043).

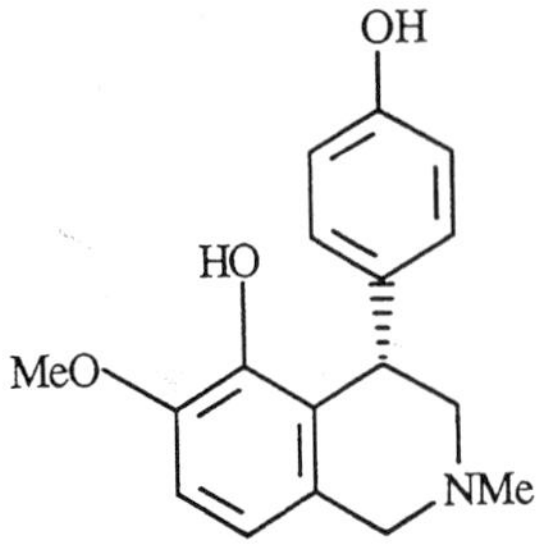

(125)

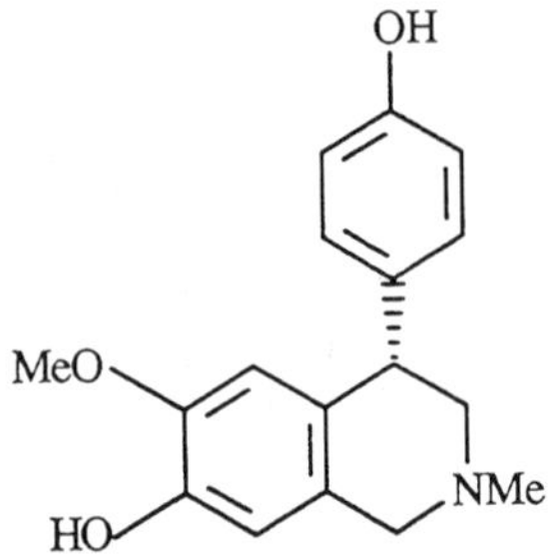

(126)

Brunsvigine (127) was originally thought to have a lycorine type structure but the X-ray measurements of its O,O'-di-*para*-bromobenzoate (128) clearly assigned it to the montanine group (M. Laing and R.C. Clark, *Tetrahedron Letters*, 1974, *15*, 583).

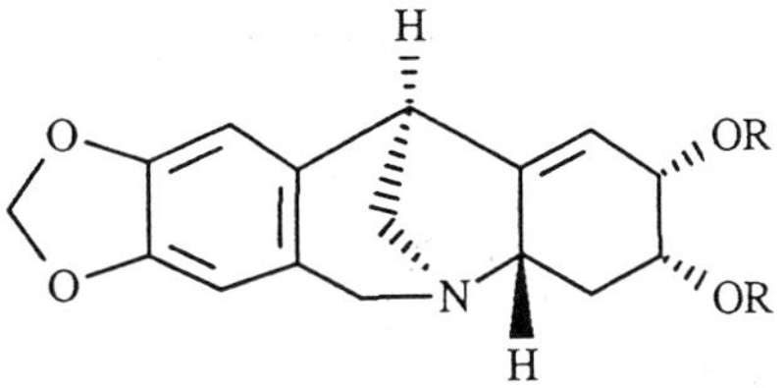

(127; R =H)
(128; R = COC$_6$H$_4$Br)

Galanthamine (11), the main alkaloid of several *Narcissus* species, and a candidate for screening for the treatment of Aldzheimer's disease has, both anticholinesterase and antimuscarinic activity. Its structure is drawn so that its N-methyl group is β (11) as determined by X-ray measurements on galanthamine itself (P. Carroll, G.T. Furst, S.Y. Han and M. Jouille, *Bull. Chem. Soc. Fr.*, 1990, *127*, 769). However, if the alkaloid is crystallised from dichloromethane a quaternary salt is formed which has been identified as (-)-N-(chloromethyl)galanthaminium chloride (2), where the CH$_2$Cl group is attached to the quaternary nitrogen in a stereospecific (R)-configuration (R. Matusch, M. Kreh and U. Muller, *Helv. Chim. Acta*, 1994, *77*, 1611).

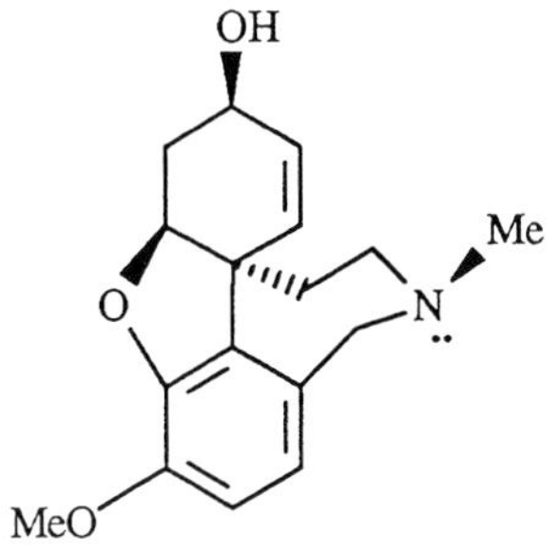

(11)

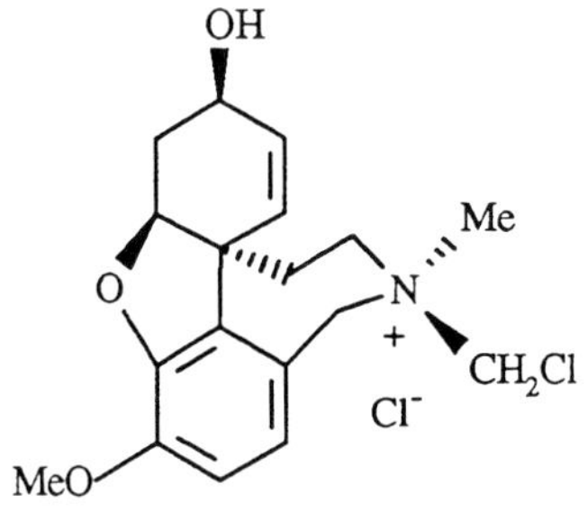

(2)

Eugenine (129) is a new alkaloid, which contains an unusual ethoxygroup: its structure was confirmed by X-ray measurements. The possibility of it being an 'artefact' has not yet been resolved (J. Via, M.I. Arriortura, L.F. Ocehando, M.M. Reventos, J.M. Amigo and J. Bastida, *Acta Crystallogr., Sect. C*, 1989, *45*, 2020).

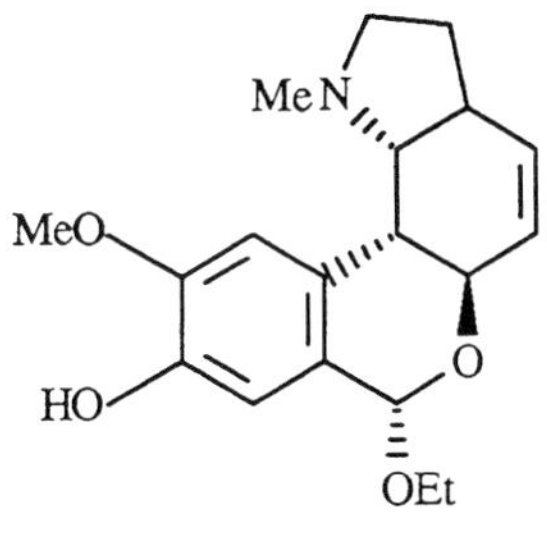

(129)

6.   Lycorine Type Alkaloids

(i)   Isolation and structural studies

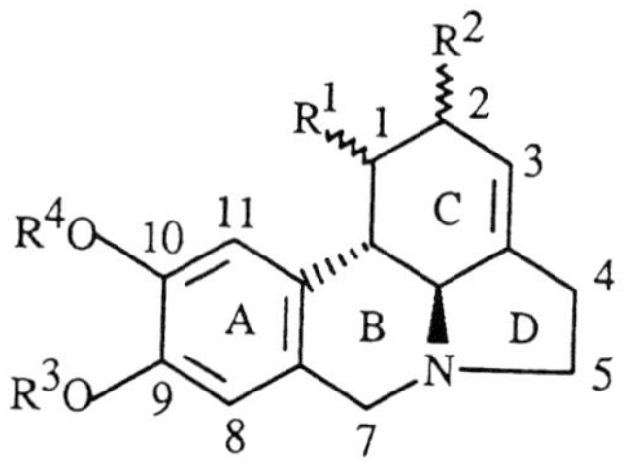

(3) R^1=α-OH; R^2=β-OH, R^3+R^4=Me
Lycorine
(58) R^1=R^2= α-OH; R^3+R^4=CH$_2$
Epilycorine
(54) R^1=α-1–β-D-glucosyl; R^2=β-1-β-D-glucosyl;
R^3=H, R^4=Me
1β-D-glucosyl-2β-D-glucosylycorine

In an investigation into the effect of stress (mechanical or insect injury) on secondary metabolites the fruits of *Crinum asiaticum* under 'normal' extraction conditions were shown to contain epilycorine (58) but not under protected conditions (extraction in the presence of ether or lidocaine lycorine (3) being isolated).   Likewise alkaloidal conjugates such as 1-O-palmitoyl-2-O(1'-O-palmitoyl-2'-O-stearoyl)glycerophos-phoryl   lycorine were absent after mechanical injury had allowed access to hydrolytic enzymes.   Oxidative modification was also observed with lycorine N-oxide (130), this compound being isolated after an unprotected work up procedure (Table 1, ref. 9).   A bisglycoside of lycorine (54) has been obtained from the bulbs of *Hymenocallis caymanensis* while the leaves contained 4-hydroxyanhydrolycorine (70) (Table 1, ref. 18).   The leaves of the 'Fortune' species of *Narcissus* yield the new lycorine derivative fortucine (101); this has the unusual 'cis-fusion' of rings B and C.

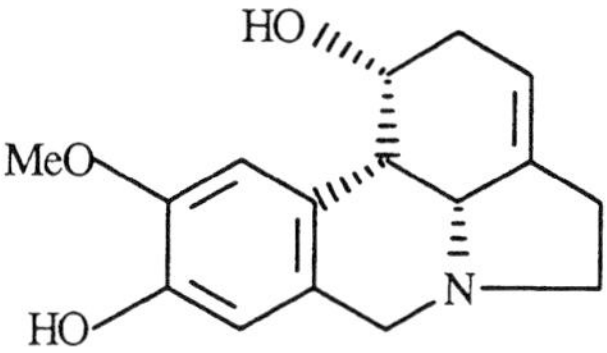

(101)

Incartine (78) has been suggested as a biogenetic precursor of galanthine (131) and narcissidine (132). 3-O-Acetyl-narcissidine (74) has, surprisingly, not suffered enzymatic hydrolysis as it was obtained from the dried bulbs of *Hippeastrum puniceum* (Table 1, ref. 20). An investigation of Turkish *Sternbergia sicula* has identified a new lycorine alkaloid siculinine (116), also present is its diastereosisomer ungiminorine (133). Both alkaloids have had their ^{1}H chemical shifts critically assessed using inter locking decoupling and NOE measurements; CD_3CN was found to be a good solvent for these spectral studies. (Table 1, ref. 52).

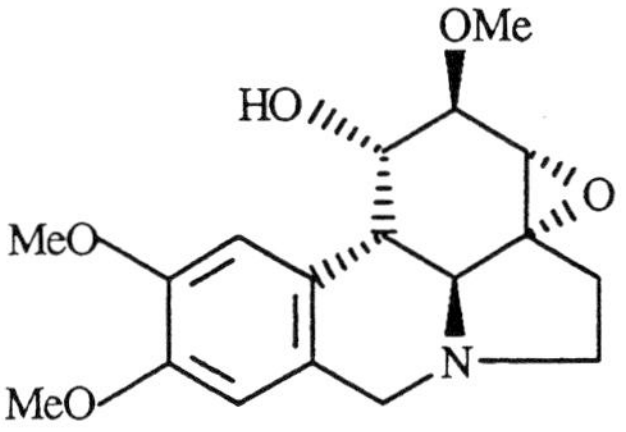

(78)

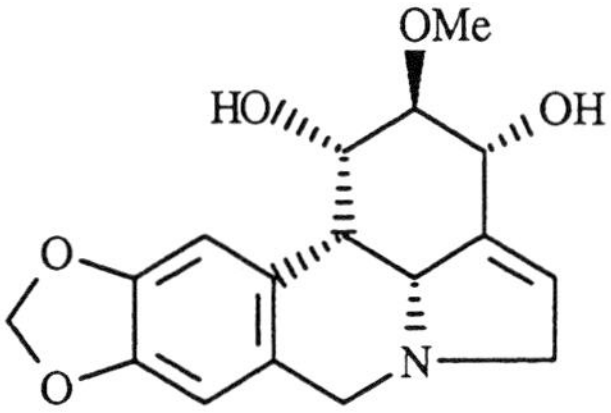

(116)

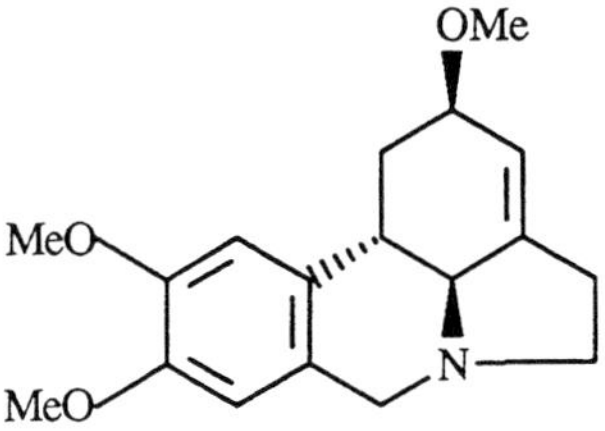

(131)

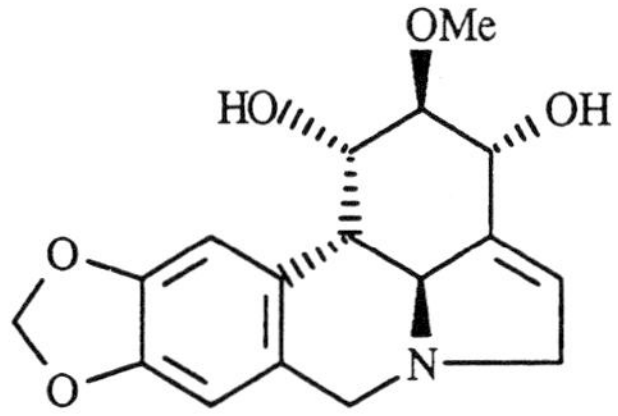

(133)

192

The N-oxide of ungiminorine (108) has been found in the aerial parts of *Pancratium maritimum* (Table 1, ref. 48) and it has been synthesised by treatment of ungiminorine (133) with MCPBA.

(59) $R^1=R^2=R^3=\alpha$-OH, $R^4+R^5=CH_2$, $R^6=$ lone pair
Epipancrassidine
(74) $R^1=\alpha$-OH; $R^2=\beta$-OMe; $R^3=\alpha$-OAc; $R^4=R^5=$Me,
$R^6=$lone pair
3-O-Acetyl-narcissidine
(132) $R^1=R^3=\alpha$-OH; $R^2=\beta$-OMe; $R^4=R^5=$Me, $R^6=$lone pair
Narcissidine
(133) $R^1=R^3=\alpha$-OH; $R^2=\beta$-OMe; $R^4+R^5=CH_2$; $R^6=$lone pair
Ungiminorine
(108) $R^1=R^3=\alpha$-OH; $R^2=\beta$-OMe; $R^4+R^5=CH_2$; $R^6=$O
Ungiminorine N-oxide

A number of lycorine alkaloids with ring C aromatised have been characterised. These include oxoassoanine (85), assoanine (134) from *Narcissus assoanus* (Table 1, ref. 29), vasconine (105) from *Narcissus vasconicus* (Table 1, ref. 46) and pratosine (135) from *Crinum americanum* (Table 1, ref. 4).

(85) $R^1+R^2=O-R^3=H$; $R^4=R^5=CH_3$
    Oxoassoanine
(68) $R^1=R^2=H$, $R^3=OH$; $R^4=R^5=Me$
    4-Hydroxyanhydrolycorine
(134) $R^1=R^2=R^3=H$; $R^4=R^5=Me$
    Assoanine

(105) Vasconine

Known betaines include criasbetaine (38), zefbetaine (119) and zeflabetaine (120), the latter two are obtained from mature seeds of *Zephyranthes flava* (Table 1, ref. 53) and the former from *Crinium asiaticum* (Table 1, ref. 6).

(38) $R^1=R^2=OMe$; $R^3=H$
    Criasbetaine
(119) $R^1+R^2=OCH_2O$; $R^3=H$
    Zefbetaine
(120) $R^1+R^2=OCH_2O$; $R^3=OMe$
    Zeflabetaine

(135) Pratosine

194

Uusually Amaryllidaceae plants propagate through offsets of their bulbs, the physiological purpose of their flowers is not clear, but they are a rich source of free and conjugated alkaloids. *Zephyranthes flava* flowers contain phosphatidyl-lycorines, pseudolycorine (136) and a lycorinium methocation alkaloid conjugate. Typical of these conjugates are the fatty esters of 2-O-glycerophosphoryllycorine (137), (138) and the lycorinium methocation (117) (Table 1, ref. 53)

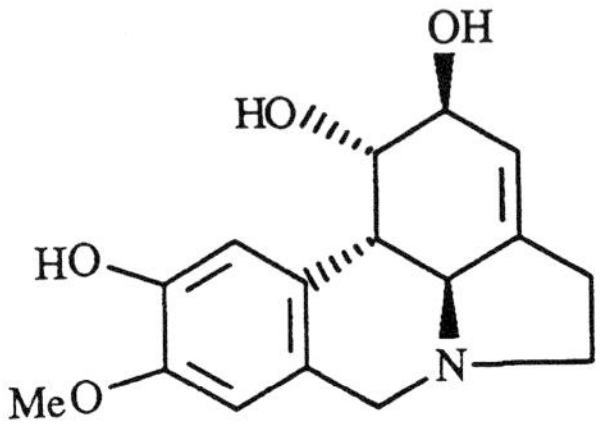

(136)

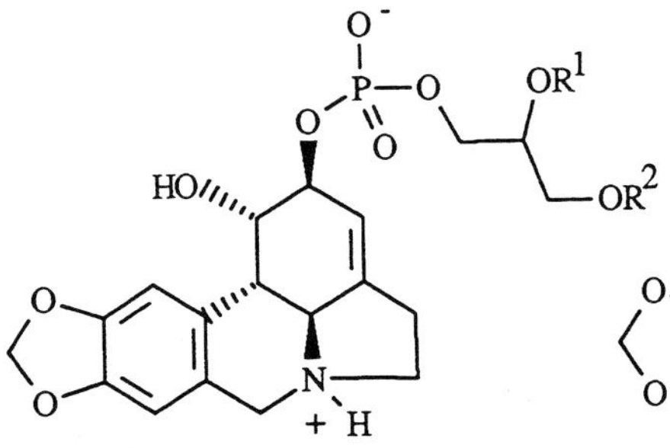

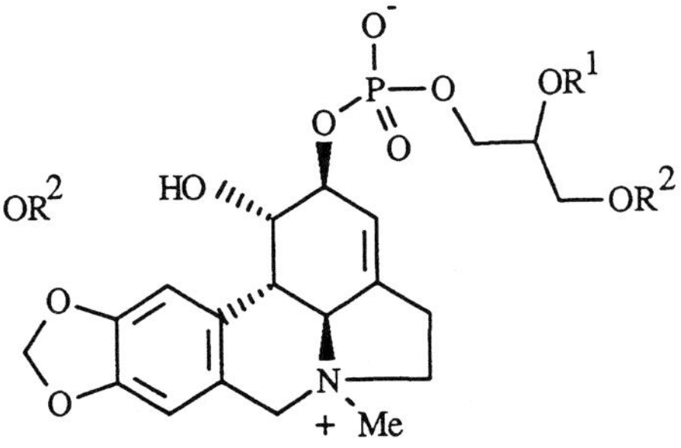

(137; R^1 = palmitoyl, R^2 = stearoyl)
(138; R^1 = palmitoyl, R^2 = oleoyl)

(117; R^1 = palmitoyl, R^2 = stearoyl)

(ii) Synthesis

As mentioned previously (Section 5) an uniquivocal structure assignment of pratorinine (124) was established by X-ray measurement. This has been confirmed by synthesis (B.S. Joshi, H.K. Desai and S.W. Pelletier, *J. Nat. Products*, 1986, *49*, 445) (Scheme 1)

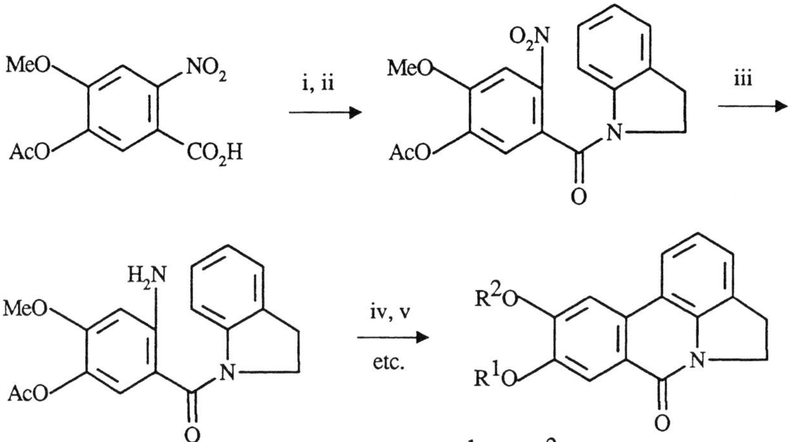

(123; R^1 = H, R^2 = Me) PRATORIMINE

(124; R^1 = Me, R^2 = H) PRATORININE

SCHEME 1

Reagents: (i) ClC(O)C(O)Cl, reflux; (ii) indoline. Et_2O; (iii) H_2.Pd/C.EtOH; (iv) $NaNO_2$, H_2SO_4, AcOH at 10°C, then at 100°C; (v) 5% KOH in MeOH at 50°C

Lycorine alkaloids are coming increasingly under scrutiny because of their antifertility and antitumor properties. A new approach to the synthesis of the lycorine ring system has been reported (R.K. Boeckman Jr., J.P. Sabuticci, S.W. Goldstein, D.M. Springer and P.F. Jackson, *J. Org. Chem.*, 1986, *51*, 3740). The masked dienamine and cinnamate residues within the spiro-cyclic salt (139) were liberated to give the triene (140); thermolysis then resulted in an intramolecular (4+2) cycloaddition to provide compound (141), which possesses the skeleton and stereochemistry present in lycorine (3) (Scheme 2).

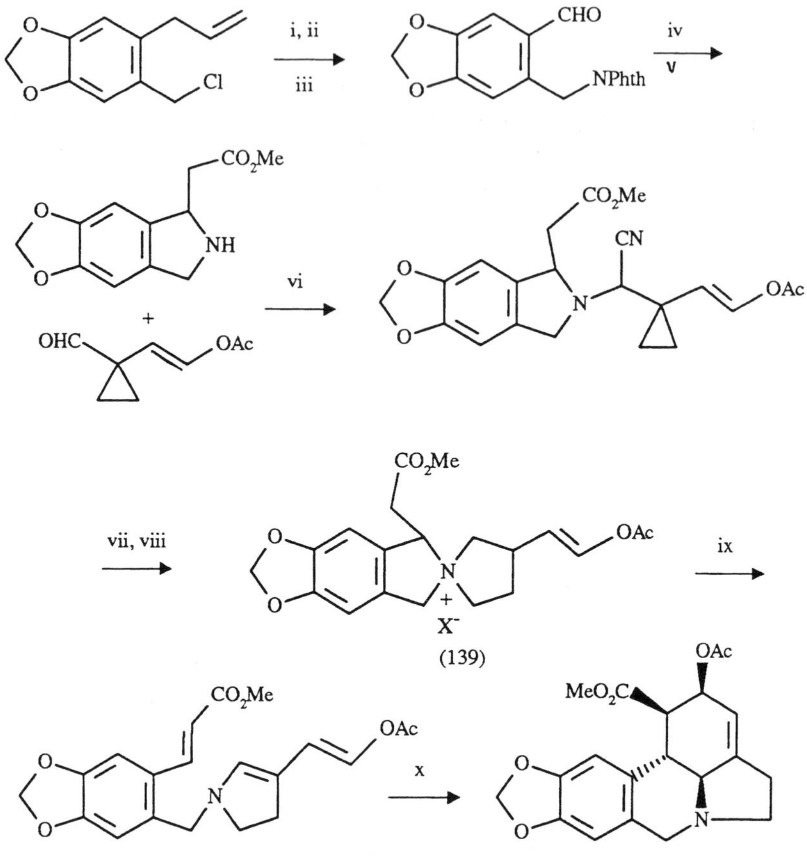

(139)

(140)

(141)

Scheme 2

Reagents: (i) potassium phthalimide, DMF; (ii) $Fe(CO)_5$, xylenes, heat; (iii) OsO_4, N-methylmorpholine N-oxide, Me_2CO, H_2O then $NaIO_4$; (iv) $(AcO)_2CCH_2CO_2H$, piperidine, $PhNH_2$, pyridine, heat; (v) H_2NNH_2, H_2O, MeOH, heat; (vi) $HClO_4$, EtOH, CH_2Cl_2, then the salt added to (55), KCN, $CaCl_2$, THF; (vii) $AgBF_4$, DME; (viii) LiBr,MeCN; (ix) 1,8-diazabicyclo[5,4,0]undec-7-ene, CH_2Cl_2 at 0°C; (x) PhMe, heat

The use of a 1-aza-1'-oxa[3,3]sigmatropic rearrangement of the hydroxamic acid (142) creates the lactam alkaloid hippadine (143) (Scheme 3) (S. Prebhakar, A.M. Lobo and M.M. Marques, *J. Chem. Res.*, 1987, 167).

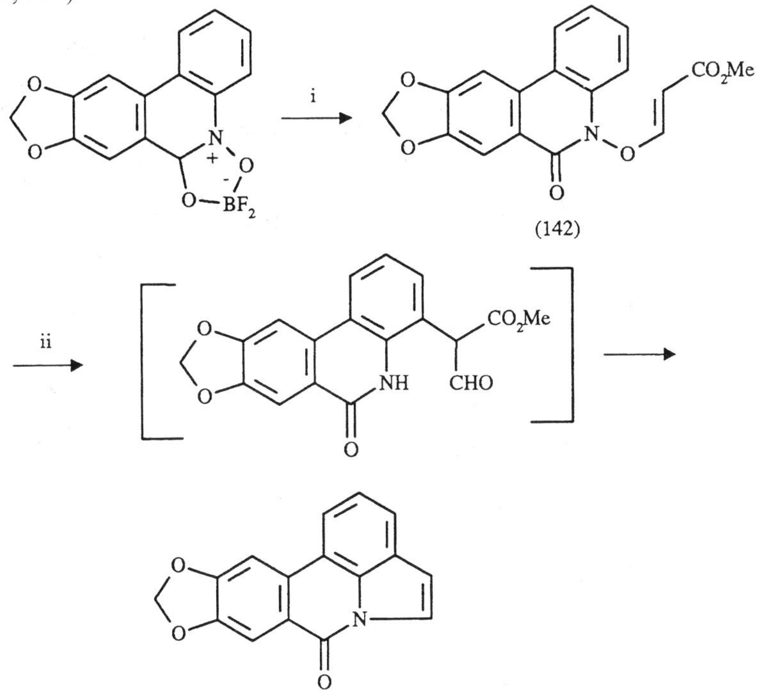

(142)

(143)

Scheme 3

Reagents: (i) HC CCO$_2$Me, Et$_3$N; (ii) DMSO, H$_2$O (trace), reflux

An elegant, simple, high yield, synthesis of the lycorine ring system (147) has been achieved by treating the bromoamide (146) with lithium diethylamide in tetrahydrofuran (Scheme 4). Since the bromoamide is easily prepared from veratroyl chloride (144) and bromohomo-eratrylamine (145) this procedure offers a route especially suitable for the active anti-tumour alkaloids of the Amaryllidaceae family, e.g. lycorine (3), ungeremine (37) and O-acetylcaranine (149). In one step (146) >

(148) via aryne (147), intramolecular cycloaddition occurred in 74% yield (D.P. Meiras, E. Guitian and L. Castedo, *Tetrahedron Letters*, 1980, *31*, 2331).

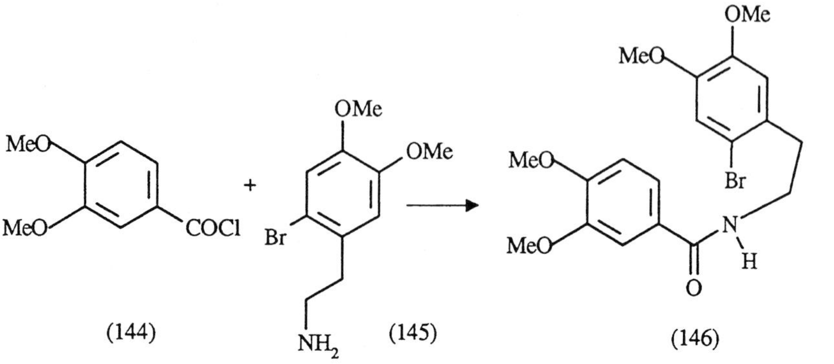

(144)

(145)

(146)

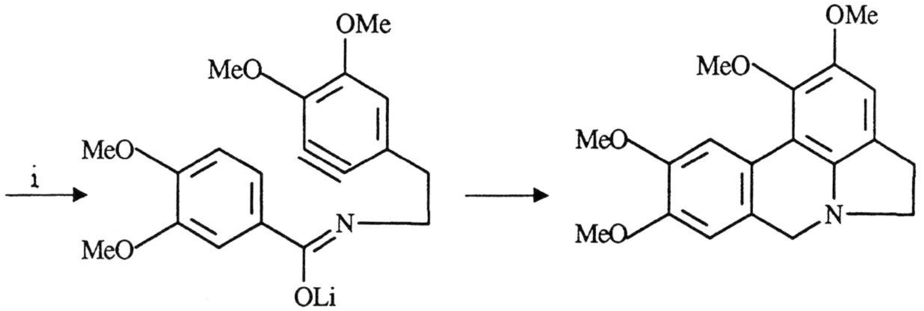

i

(147)

(148)

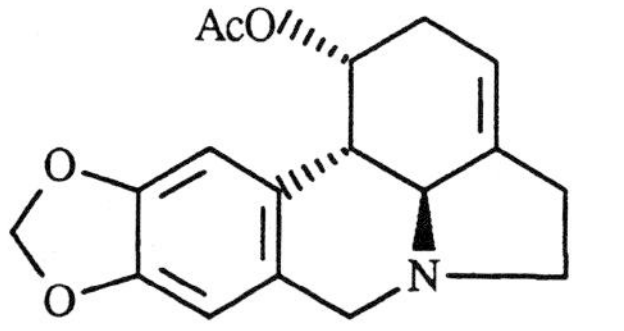

(149) O-ACETYLCARANINE

Scheme 4

Reagents: LDA/THF/-70°C

A concise synthesis of ungeremine (37) and hippadine (143) has been reported using the Suzuki aryl-aryl cross coupling reaction (Scheme 5). 5-Hydroxyindole (150) was mesylated, reduced, and brominated in the 7-position to give the indoline (151; R^1=OMS, R^2=H, X=Br). Cross coupling with the *ortho* formyl boronic acid (152) under modified Suzuki conditions gave concommitant cyclisation and aerial oxidation leading to the lactam (153; R=OMs). Reduction with Red-Al gave ungeremine (37). In a similar sequence 1-acetylindoline (151; R^1=X=H, R^2=Ac) was first iodinated (151; R=H, R^2=Ac, X=1) then cross Suzuki coupled to give (153; R=H) which upon DDQ oxidation gave hippadine (143) (M.A. Siddique and V. Snieckus, *Tetrahedron Letters*, 1990, *31*, 1523).

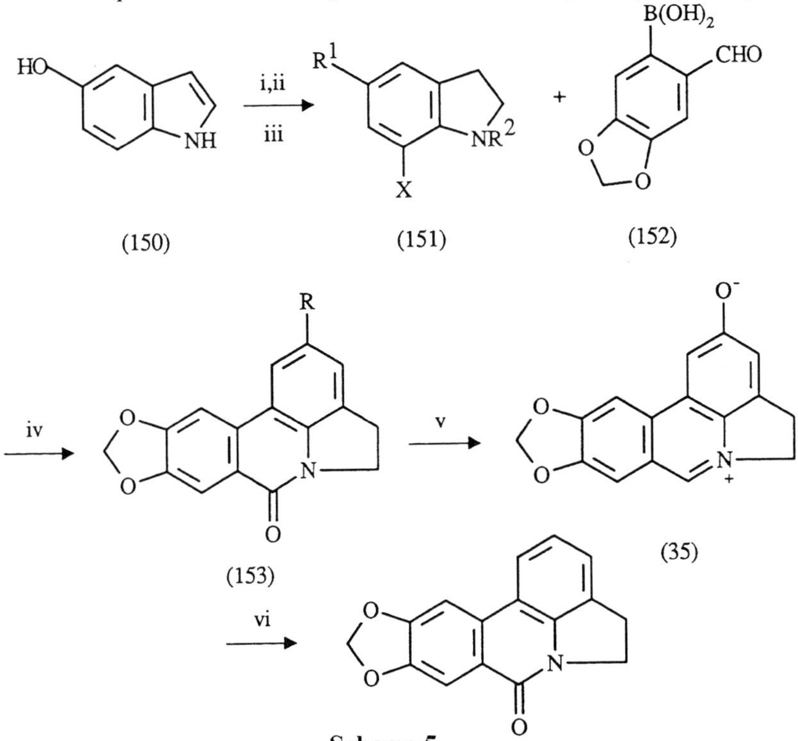

Scheme 5

Reagents:(i) $MsCl,Et_3N,CH_2Cl_2$; (ii) $NaBH_3CN,HOAc$; (iii) Br_2,CH_2Cl_2; (iv) $Pd(PPh_3)_4,Na_2CO_3,DMF$,reflux; (v) Red-Al,toluene,reflux (vi) DDQ,dioxane,reflux

200

A new stereoselective synthesis of (±)-lycorine (3) has been achieved
(Scheme 6). Starting with the triene lactone (154) an intramolecular
Diels-Alder reaction was obtained, by heating it in a sealed tube at 235ºC
with o-dichlorobenzene, giving a mixture of stereoisomers (155; R=α, or
β-H), the cis-isomer (155; R=α-H) predominating (86%). Reduction of
(155; R=α-H), followed by oxidation afforded γ-lactone (156) which
could be converted into the γ-iodolactone (157) by water-iodine
treatment. Dehydroiodination of (157) occurred only if prior protection
(tetrahydropyranyl) of the hydroxymethyl functionality was carried out.
Conversion of the hydroxymethyl unsaturated lactone (158) into lactam
(160) was accomplished by oxidation first to acid (159; R=OH), followed
by Curtis rearrangement of its amide (159; R=NH$_2$). Ring opening of the
lactone resulted in concomitant ring closure to give the lactam (160;
R=H). Introduction of the epoxy group 'anti' to the hydroxyl functionality
required that this group first be siloxylated (161; R=SiMe$_3$Bu^t) prior to
epoxidation. This allowed epimeric transfer of the epoxy group in (161)
to give (162) which upon phenylselenation gave the key intermediate
(163). Conversion to (±)-lycorine (3) was obtained by reduction,
followed by cyclization to (164) and finally elimination of phenylselenide
(O. Hoshuno, M. Ishizaki, K. Kamei, M. Taguchi, T. Nagao, K. Owaoka,
S. Sawaki, B. Yinezawa and Y. Iitaka, *Chem. Letters*, 1991, 1365).

(Ar = 3,4-methylenedioxyphenyl)

(154)   (155)   (156)

(157)   (158)

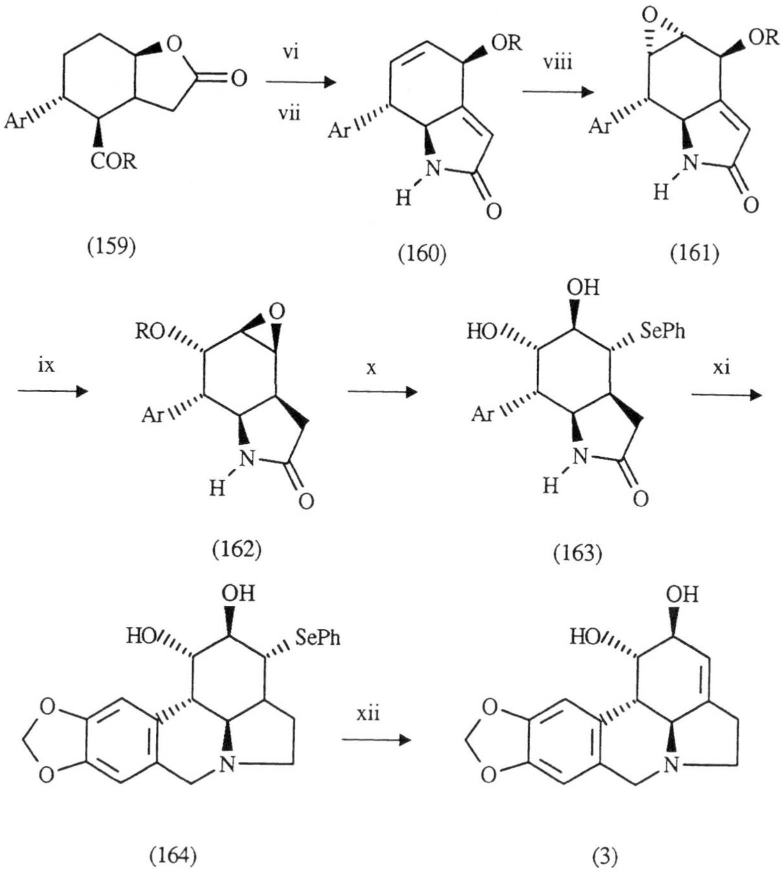

Scheme 6

Reagents: (i) o-Cl$_2$C$_6$H$_4$, 235°C, sealed tube; (ii) LiAlH$_4$, THF, reflux; Ag$_2$CO$_3$-celite, C$_6$H$_6$ reflux; (iii) K$_2$CO$_3$, MeOH, H$_2$O reflux; I$_2$ aq. K$_2$CO$_3$, MeOH, r.t.; (iv) DHP, CH$_2$Cl$_2$, H$^+$, r.t.; DBU, C$_6$H$_6$ reflux; MeOH, CH$_2$Cl$_2$, H$^+$, rt.; (v) CrO$_3$, H$_2$O, acetone, 0°C; (vi) DPPA, Et$_3$N, Bu^{t}OH, reflux; (vii) Bu^{t}Me$_2$SiCl, imidazole, DMF, r.t.; (viii) MCPBA, CH$_2$Cl$_2$, r.t.; (ix) 5% aq. K$_2$CO$_3$, MeOH, r.t.; (x) Ph$_2$Se$_2$, NaBH$_4$, EtOH, reflux; (xi) Na(MeOCH$_2$CH$_2$O)$_2$AlH$_2$. toluene, reflux; Me$_2$N$^+$=CH$_2$I$^-$, THF, reflux; (xii) NaIO$_4$, THF, MeOH, H$_2$O, 40°C

Anhydrolycorinium chloride (34) has been synthesised by an elegant intramolecular aryne cycloaddition procedure (P.D. Meiras, E. Guitian and L. Castedo, *Tetrahedron Letters*, 1990, *31*, 2331) and another, minor, variation of the method of ring closure has appeared (Scheme 7). Not only does the amide (165) cyclise to amide (166), but so does imine (167). It was suggested that in this latter cyclisation oxygen was introduced during the work-up of the reaction mixture as it was not present prior to it (B. Gomez, C. Gonzalez, D. Perez, E. Guitian and L. Castedo, *Planta Medica*, 1990, *56*, 516).

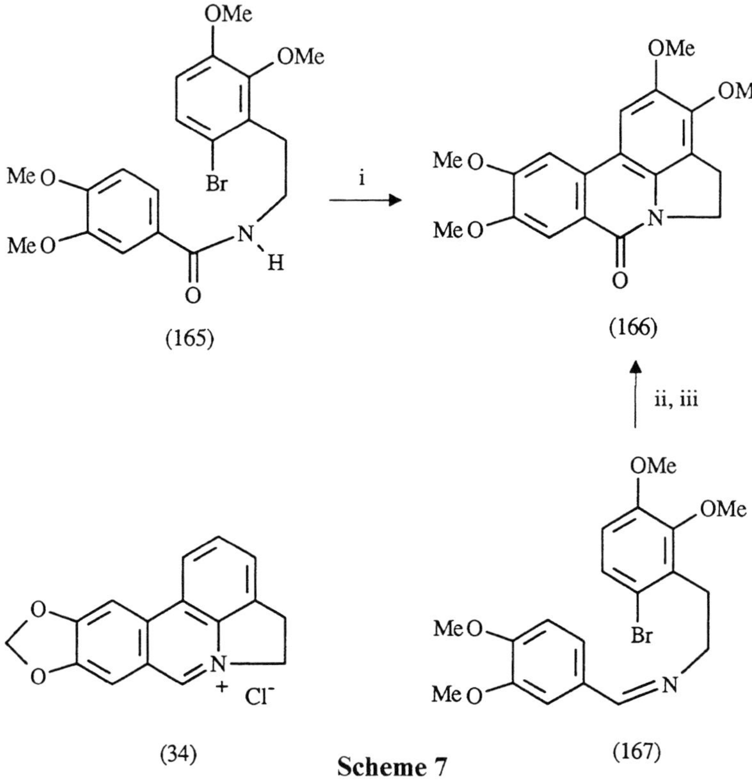

Scheme 7

Reagents: (i) LDA(2 eq), THF, -78° to 20°

A slightly different approach to this alkaloid ring system used a radical cyclization process (Scheme 8). The technique has been employed to make ungeremine (37), anhydrolycorinone (172) and hippadine (143). Bromonitrohydrindole (168) was condensed with 3,4-methylenedioxybenzoyl chloride to give amide (169), which was converted into the pyrrolophenanthridone system (170; $R=NO_2$) by heating to 155° in dimethylsulfoxide in the presence of potassium carbonate and benzyltriethylammonium chloride (BTAC). The nitro group (170; $R=NO_2$) was converted through conventional procedures into the O-benzyl ether (171), which upon deprotection and dehydrogenation gave ungeremine (37) in gram quantities Reductive removal of the diazo-group in (170; $R = N_2^+ BF_4^-$), enabled anhydrolycorinone (172) and hippadine (143) to be synthesised in respectable yields (U. Lauk, D. Durst and W. Fischer, *Tetrahedron Letters*, 1991, *32*, 65).

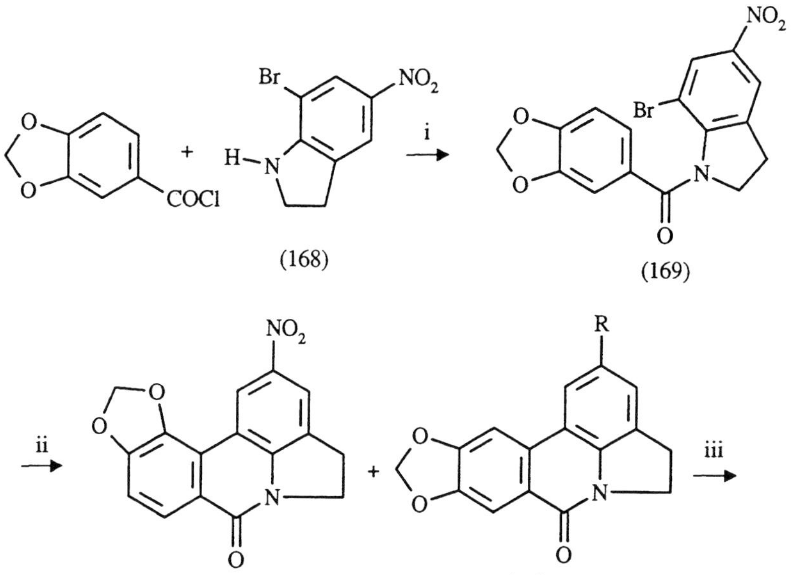

Scheme 8

continued overleaf

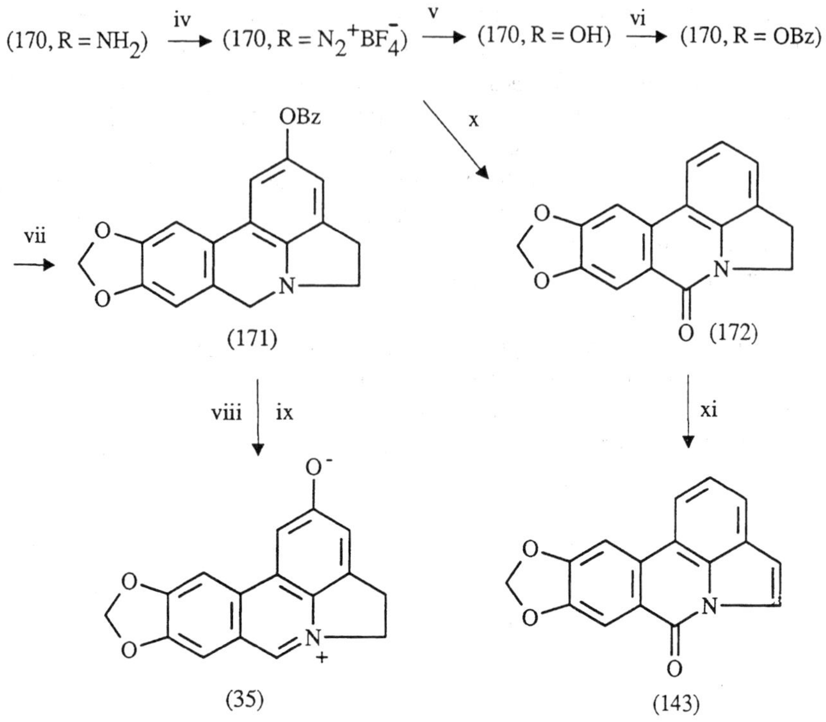

204

$(170, R = NH_2) \xrightarrow{iv} (170, R = N_2^+BF_4^-) \xrightarrow{v} (170, R = OH) \xrightarrow{vi} (170, R = OBz)$

(171)

(172)

(35)

(143)

Scheme 8

Reagents: (i) pyridine, 100°C, 13 h; (ii) K$_2$CO$_3$, DMSO, BTAC, 155°C; (iii) TFA, Pd/C, H$_2$, 25°C, 0.5 h; (iv) H$_2$SO$_4$, ONOSO$_3$H, NaBH, 0°C, 1 h; (v) TFA, reflux, 40 h; (vi) DMF, NaH, BzBr, 70°C, 0.75 h; (vii) LiAlH$_4$, THF, reflux, 1.5 h; (viii) EtOH, Pd/C, H$_2$, 70°C, 2 h; (ix) H$_2$O$_2$, AcOH, MnO$_2$, 25°C, 2 h;   (x) Cl(CH$_2$)$_2$Cl, H$_3$PO$_4$, Cu$_2$O, 70°C, 0.5 h; (xi) PhCN, DDQ, 100°C, 2 h.

A new approach to the lycorine ring system has been developed utilizing the reactivity of α-pyrones with multiple bonds whereby the Diels-Alder adduct loses CO_2 to give a benzene derivative (Schemes 9/10).

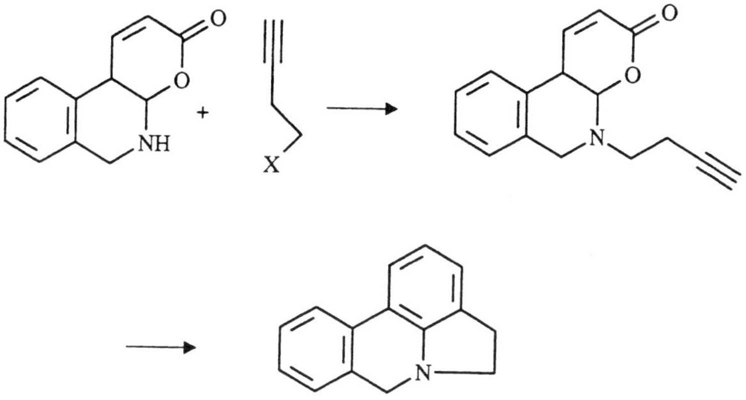

Scheme 9

Pyrone (175) was prepared from homophthalimide (173) by reaction with trimethylorthoformate to give (174) which through reaction with ethyl cyanoacetate and strong base followed by an acidic work up yielded the pyrone (175). Alkylation of (175) was unspecific, both N- and O-alkylated products resulted from treament with 3-butyn-1-tosylate. Separation gave (176) which on heating at 210° in nitrobenzene produced tetracyclic (177). The overall yield was 32% and further studies are underway to minimize the O-alkylation in step (iii) (D. Perez, E. Guitan and L. Castedo, *Tetrahedron Letters*, 1992, *33*, 2407).

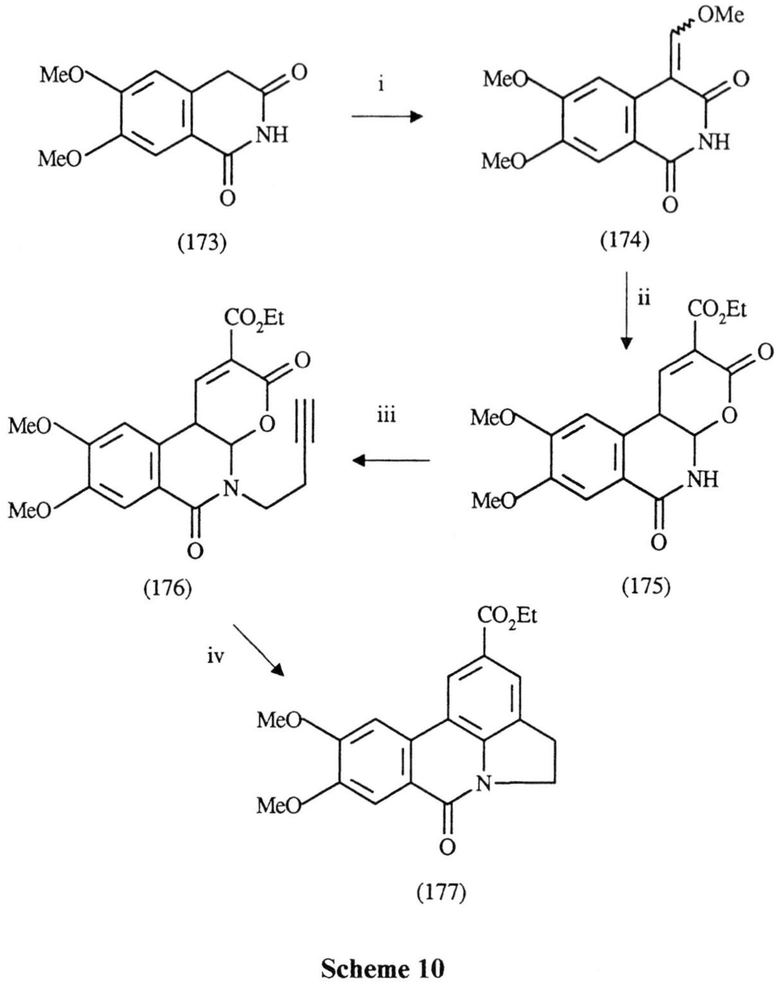

Scheme 10

Reagents: (i) HC(OMe)$_3$, Ac$_2$O, DMF, reflux;   (ii) NCCH$_2$CO$_2$Et, NaOMe; (iii) HCCCH$_2$CH$_2$OTs, Bu^{t}OK, DMF; (iv) C$_6$H$_5$NO$_2$, 210°

Alkaloids vasconine (106), assoanine (134) and oxoassoanine (85) have been synthesised concisely using the Zeigler procedure (room temperature) for an Ullmann reaction. Thus, N-BOC-indoline (178; X=H) reacted with sec. butyllithium followed by the Cu-1-P(OET)$_3$ complex to give the copper complex (178; X=Cu) which condensed with iodo-enzaldoxime (179; R=C$_6$H$_{11}$) to give aldoxime (180; R=C$_6$H$_{11}$). Thus upon hydrolysis gave the aldehyde (181). Intramolecular cyclisation was achieved by removal of the pyrrolidine protecting group creating the quaternary salt vasconine (106). Sodium borohydride reaction gave assoanine. Alternatively, oxidation of (135) with alkaline permanganate gave oxoassoanine (85) (Scheme 11) (J.S. Barnes, D.S. Carter, L.J. Kurtz and L.A. Flippin, *J. Org. Chem.*, 1994, *59*, 3497).

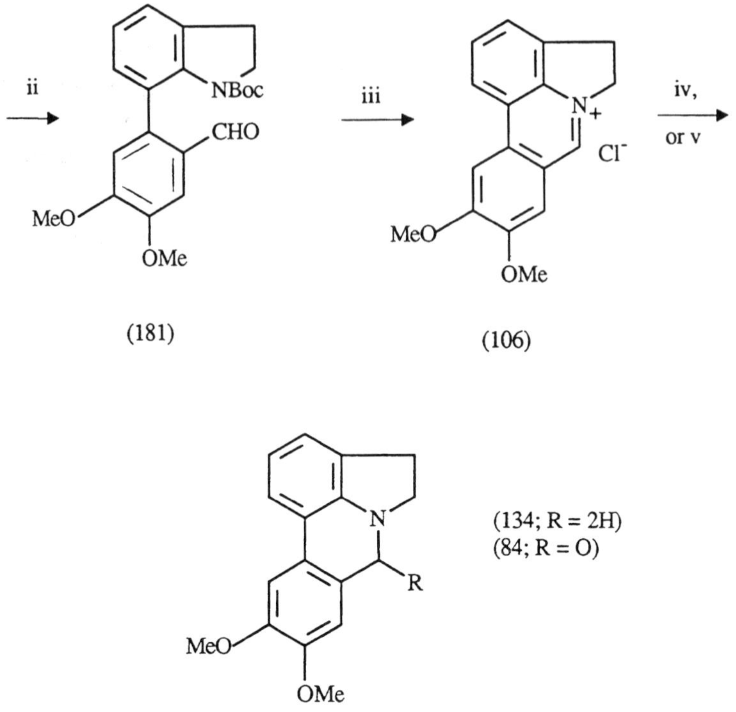

(181)

(106)

(134; R = 2H)
(84; R = O)

Scheme 11

Reagents: (i) Indoline (178; X=H), N,N,N',N'-tetramethylethylene diamine, N_2, Et_2O, -40° then BuLi, cyclohexane, -40° to -50° then CuI-P(OEt)$_2$ complex, stir -45°, 30 min, then aldoxime (189; R=C$_6$H$_{11}$) to r.t., 1 h; (ii) 1MHCl, THF, reflux, 30 min; (iii) HCl (g), CHCl$_3$, r.t., 15 min; (iv) EtOH, NaBH$_4$, r.t., 30 min; (v) H_2O, N_2, NaOH, KMnO$_4$, CH_2Cl_2, 20°, stir.

7. **Lycorenan Type alkaloids (Lycorenines and Homolycorines)**

(i) Isolation and Structural Studies

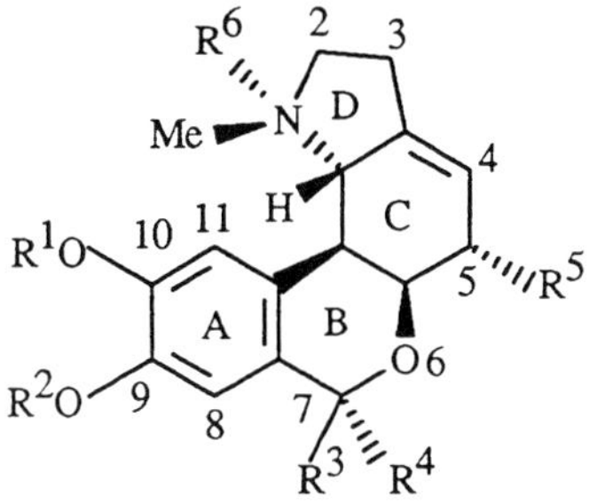

LYCORENINE

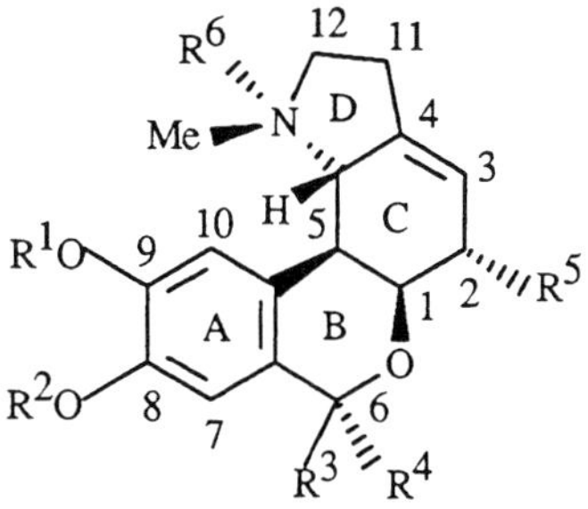

HOMOLYCORINE

(4) $R^1=R^2=Me$, $R^3=OH$, $R^4=R^5=H$, $R^6=$lone pair
Lycorenine

(182) $R^1=R^2=Me$, $R^3+R^4=O$, $R^5=H$, $R^6=$lone pair
Homolycorine

(75) $R^1=R^2=Me$, $R^3+R^4=R^6=O$, $R^5=H$
Homolycorine N-oxide

(76) $R^1=R^2=Me$, $R^3=OMe$, $R^4=R^5=H$, $R^6=O$
O-Methyllycorenine N-oxide

(80) $R^1+R^2=CH_2$, $R^3+R^4=O$, $R^5=OH$, $R^6=O$
Hippeastrine N-oxide

(183) $R^1=H$, $R^2=Me$, $R^3+R^4=O$, $R^5=OH$, $R^6=$lone pair
9-O-Demethyl-2-α-hydroxyhomolycorine

(184) $R^1=Me$, $R^2=R^5=H$, $R^3+R^4=O$, $R^6=$lone pair
8-O-Demethylhomolycorine

(42) $R^1=R^5=H$, $R^2=Me$, $R^3+R^4=O$, $R^6=$lone pair
9-O-Demethylhomolycorine

(96) $R^1=R^5=H$, $R^2=Me$, $R^3+R^4=R^6=O$
9-O-Demethylhomolycorine N-oxide

(89) $R^1=COCH_2CH(OH)CH_3$, $R^2=Me$, $R^3+R^4=O$, $R^5=OAc$,
$R^6=$lone pair

210

(98) $R^1+R^2=CH_2$, $R^4=OMe$, $R^3=R^5=H$, R^6=lone pair
O-Methyloduline

(105) $R^1=Me$, $R^2=Ac$, $R^3+R^4=O$, $R^5=H$, R^6=lone pair
8-O-Acetylhomolycorine

Lycorenine and homolycorine alkaloids differ only in that the latter has a keto-group at position 6 (homolycorine numbering). Although lycorenine alkaloids have a differing numbering system (*loc cit*) these two groups of alkaloids warrant single classification. Original nomenclature is used however in this section to avoid confusion. Lycorenan is the skeletal name used by Chemical Abstracts to cover both groups of alkaloids. *Narcissi* are the predominant plant species producing both lycorenine and homolycorine alkaloids. *Narcissus confusus* produces both homolycorine (182) and its 9-O-demethylated counterpart (40) both being found in the bulbs (Table 1, ref. 32). In *Narcissus tortifolius*, a rare endemic species, obained from the Iberian peninsula, 9-O-demethyl-2α-hydroxyhomolycorine (183) was found (Table 1, ref. 45) together with homolycorine (182), 8-O-demethylhomolycorine (184) and dubuisine (89) which is a homolycorine having the unusual hydroxybutyl ester substituent at position 2. Dubuisine having been previously isolated from *Narcissus dubius* (Table 1, ref. 32). *Narcissus confusus* bulbs contain homolycorine (182) and 9-O-demethylhomolycorine (40) (Table 1, ref. 32). Nobilisine (46) from bulbs of *Clivia nobilis* is 5α-hydroxy-3,4-dihydrohomolycorine having the stereochemical classification IV; with rings B/C and C/D trans-fused (Table 1, ref. 2). A number of N-oxides of both series of alkaloids have been reported (see Table 1) and careful examinations of the extraction conditions have precluded artificial production. N-Demethylmansonine (98) is unusual as only the second N-demethylhomolycorine alkaloid to be identified, the first being N-demethylhomolycorine (185) itself.

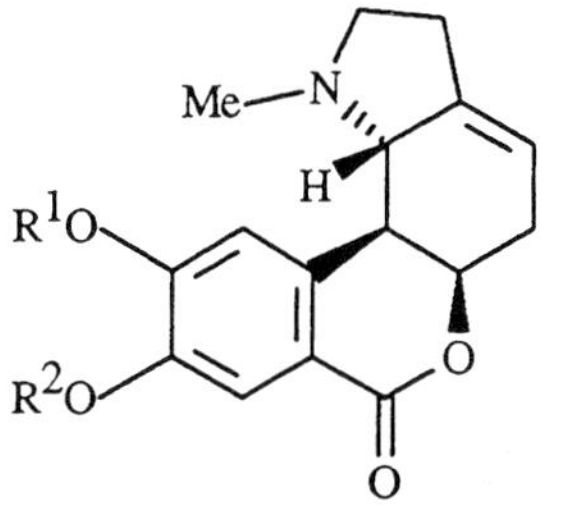

$$(98; R^1 + R^2 = CH_2)$$

$$(185; R^1 = R^2 = Me)$$

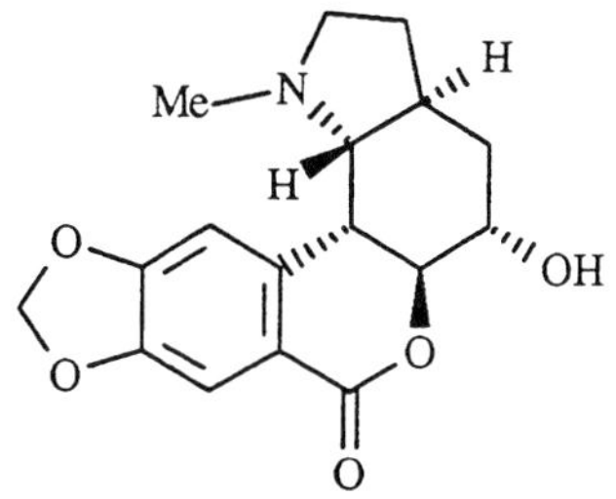

$$(46)$$

Structural assignments to these alkaloids much depended upon uniquivocal assignments for ^{1}H and ^{13}C data. Several of the publications quoted contain detailed interpretations of chemical shifts and their mutual interactions.

212

8. Crinine Type alkaloids

(i) Isolation and structural studies

(5) $R^1=\alpha\text{-OH}$, $R^2=R^3=R^4=H$, $R^5+R^6=CH_2$
Crinine

(187) $R=\alpha\text{-OAc}$, $R^2=R^3=R^4=H$, $R^5+R^6=CH_2$
O-Acetylcrinine

(50) $R^1=O$, $R^2=R^3=R^4=H$, $R^5+R^6=CH_2$
Oxocrinine

(57) $R^1=\alpha\text{-OMe}$, $R^2=OH$, $R^3=H$, $R^4=\alpha\text{-OH}$, $R^5+R^6=CH_2$
Crinafoline

(189) $R^1=\alpha\text{-OMe}$, $R^2=OH$, $R^3=H$, $R^4=OMe$, $R^5+R^6=CH_2$
Ambelline

(188) $R^1=\beta\text{-OH}$, $R^2=R^3=R^4=H$, $R^5=R^6=Me$
Maritidine

(100) $R^1=\beta\text{-OH}$, $R^2=R^3=R^4=R^5=H$, $R^6=Me$
9-O-Demethylmaritidine

(109) $R^1=\beta\text{-OH}$, $R^2=\alpha\text{-OH}$, $R^3=R^4=H$, $R^5+R^6=CH_2$
3β,11α-Dihydroxy-1,2-dehydrocrinane

(110) $R^1=\alpha\text{-OH}$, $R^2=R^3=R^4=R^5=H$, $R^6=Me$
8-Hydroxy-9-methoxycrinine

(115) $R^1=\alpha\text{-OMe}$, $R^2=R^3=R^4=H$, $R^5+R^6=CH_2$
(-)-Buphanisine

(194) $R^1=\beta\text{-OMe}$, $R^2=R^3=R^4=H$, $R^5+R^6=CH_2$
(+)-Buphanisine

An original investigation into the alkaloids present in *Crinum americanum* (1957) identified lycorine (3) and crinine (5); now a reinvestigation (Table 1, ref. 4) has characterised ten alkaloids in the air dried bulbs. The main four, crinine (5), flexinine (186) O-acetylcrinine (187) and oxocrinine (50) belong to the crinine group. Since 40 kg of dried bulbs were used in this study minor akaloids, ungermine (37), trisphaeridine (13), hippadine (143), protorinine (122), pratrimine (123) and pratosine (135) were also isolated. In contrast *Narcisus radinganorum* is considered quite rare and only 4.1 kg of fresh plant material was available for study. Nevertheless, four alkaloids were isolated, homolycorine (4), 8-O-demethylhomolycorine (184), (Table 1, ref. 41) together with crinans, maritidine (188) and 9-O-demethylmaritidine (100). Crinafoline (57), from the bulbs of *Crinum latifolium* is the 6-hydroxy derivative of ambelline (189), into which it can be converted by first diacetylation, partial hydrolysis to the 3-monoacetate, conversion of the 6-hydroxy group to chloride (thionyl chloride) and reduction with lithium aluminium hydride (Table 1, ref. 11). In the same plant is the tricyclic aldehyde crinafolidine (12) and the β-epoxide (56).

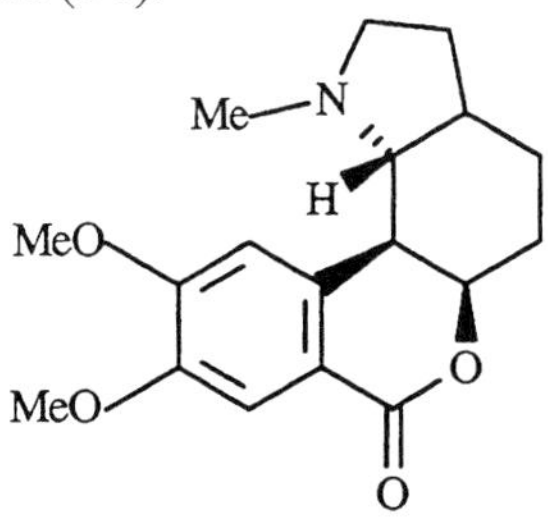

(184)

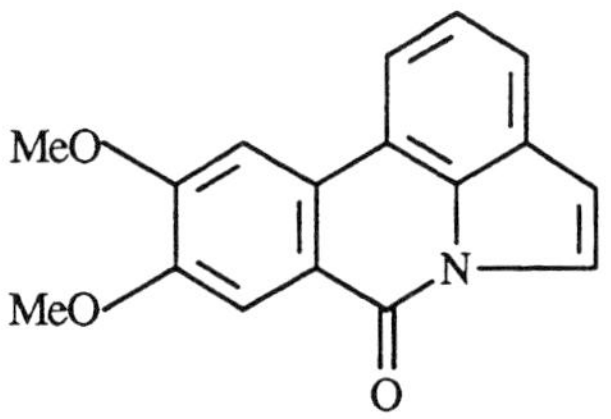

(135)

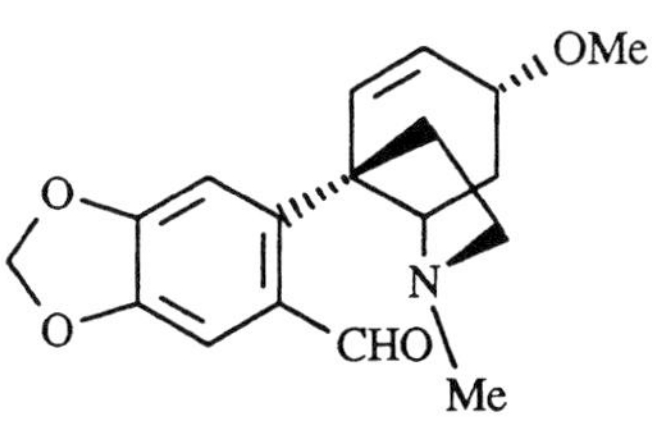

(12)

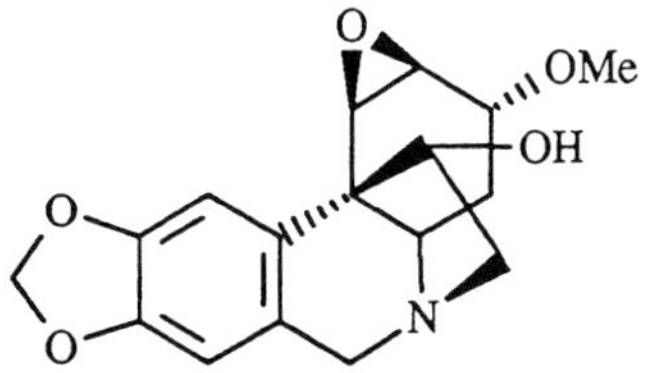

(56)

214

The methiodide of crinafoline (57) can be converted into crinafolidine (56) by treatment with sodium hydroxide solution.

In the new publication Natural Product Letters the isolation of 3β,11α-dihydroxy-1,2-dehydrocrinane (siculine) (110) and 8-hydroxy-9-methoxy-crinine (111) from *Pancratium maritimum* was reported (Table 1, ref. 50).

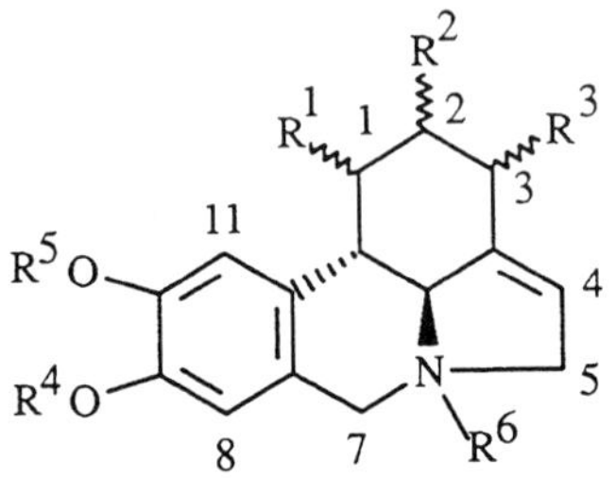

(47) R^1=α-OH, R^2=β-OAc, R^3=R^4=R^5=H
          Josephinine
(49) R^1=R^2=α-OH, R^3=R^4=R^5=H
          Amabiline
(55) R^1=R^2=R^4=α-OH, R^3=H, R^5=OMe
          Crinisin
(186) R^1+R^2=β-O, R^3=α-OH, R^4=R^5=H
          Flexinine
(193) R^1+R^2=β-O, R^3=α-OMe, R^4=R^5=H
          Augustine (augustisine)

Aqueous extracts of the bulbs of the South African *Brunsvigia josephinae* are used in traditional medicine. However, an ethanol extract of macerated bulbs produced six alkaloids. One, josephinine (45) is new and co-occurs with four other crinine alkaloids, 3-O-acetylhamayne (190), hamayne (191), crinamine (192) and ambelline (189). Comprehensive nmr measurements enabled all chemical shifts to be assigned (Table 1, ref. 1). Amabiline (49) was characterised as a new alkaloid present in the bulbs of *Crinum amabile* (Table 1, ref. 3) and was accompanied by augustine (193), lycorine (5), crinamine (192) and (-)-buphanisine (115). An alkaloid isolated from *Sternbergia lutea* (Table 1, Ref. 51) has the same chemical properties as (115) but differs only in having the opposite C.D. curve; it is thus named (+)-buphanisine (194).

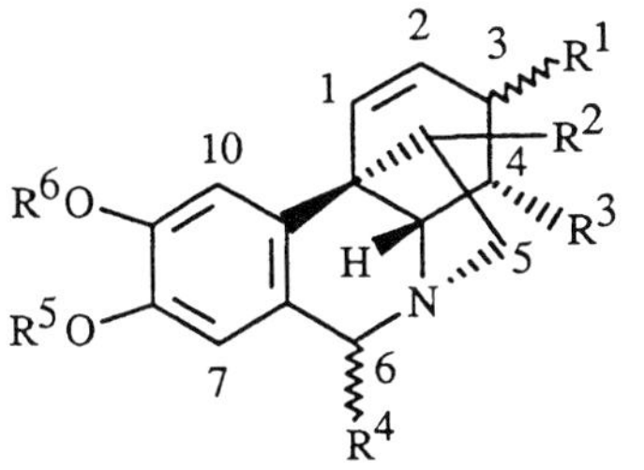

(66) R^1=β-OMe, R^2=R^3=OH, R^4=H, R^5+R^6=CH_2
Albiflomanthine

(97) R^1=β-OMe, R^2=OH, R^3=R^4=R^5=H, R^6=Me
Narcidine

(93) R^1=R^4=β-OMe, R^2=R^3=H, R^5=R^6=Me
O-Methylpapyramine

(94) R^1=β-OH, R^2=R^3=R^4=H, R^5=R^6=Me
Maritidine

(95) R^1=β-OMe, R^2=R^3=R^4=H, R^5=R^6=Me
O-Methylmaritidine

(100) R^1=β-OH, R^2=R^3=R^4=R^6=H, R^5=Me
9-O-Demethylmaritidine

(102) R^1=α-OMe, R^2=R^3=H, R^4=α-OMe, R^5=R^6=Me
6α-Hydroxyl-3-O-methylepimaritidine

(191) R^1=α-OH, R^2=OH, R^3=R^4=H, R^5+R^6=CH_2
Hamayne

(190) R^1=α-OAc, R^2=OH, R^3=R^4=H, R^5+R^6=CH_2
3-O-Acetylhamayne

(192) R^1=α-OMe, R^2=OH, R^3=R^4=H, R^5+R^6=CH_2
Crinamine

(110) R^1=α-OMe, R^2=R^3=R^4=R^5=H, R^6=Me
Siculine

A reinvestigation of the alkaloidal constituents in King Alfred daffodil bulbs (*Narcissus pseudonarcissus*) reconfirmed the presence of crinine (5), lycorenine (4), galanthine (131) and lycorine-type alkaloids

216

haemanthamine (18), hippeastrine (41) and narcissidine (132); new for this source was 8-O-demethylhomolycorine (184) while narcidine (97) is completely novel (Table 1, ref. 39). O-Methylated and 9-O-demethylated derivatives of maritidine (188) i.e. (95) and (100) have been identified in *Narcissus papyraceus* (Table 1, ref. 38) and in *Narcissus radinganorum* (Table 1, ref. 41). Albiflomanthine (66) possesses the unusual feature of having an oxygen substituent at C-4 (Table 1, ref. 15). Spectroscopic evaluation of one of the alkaloid fractions obtained from *Narcissus tazetta* var *chinensis* (Table 1, ref. 44) indicated that it was a mixture of 6α- and 6β-hydroxy-3-O-methylepimaritidine, i.e. (103) and (104). Eight species of *Sternbergia* are known, the *lutea* variety being the most common and investigated. Turkish *Sternbergia lutea* (Table 1, Ref. 51) contains epimaritinamine (112) and maritinamine (113) while *Sternbergia sicula* contains siculine (110) and previously quoted (+)-buphanisine (194). Later investigations into this latter plant yielded the lycorine type alkaloids - lycorine (3), hippadine (143), unigiminorine (133) and a new one, siculinine (126) (Table 1, ref. 52). These two publications give comprehensive nmr assignments for all the alkaloids quoted.

(112)

(113)

(ii) Synthesis

Martin's synthesis of (±)-haemanthidine (40), which involved the efficient preparation of compounds that contain a quaternary carbon atom (bearing functionalised groups) at a carbonyl centre (S.F. Martin, S.K. Davidson and J.A. Puckette, *J. Org. Chem.*, 1987, *52*, 1962), has been applied to synthesise (+)-crinine (5) and (+)-buphanisine (115/194).   (S.F. Martin and C.L. Campbell, *Tetrahedron Letters*, 1989, *28*, 503).   The key intermediate (196) was prepared from ketone (195) in 60% yield and then was converted into the indoline derivatives (197) and then to (198) and thence into (199); compounds (197) and (199) gave (±)-crinine (5) and (±)-buphanisine (115/194), respectively (Scheme 1).

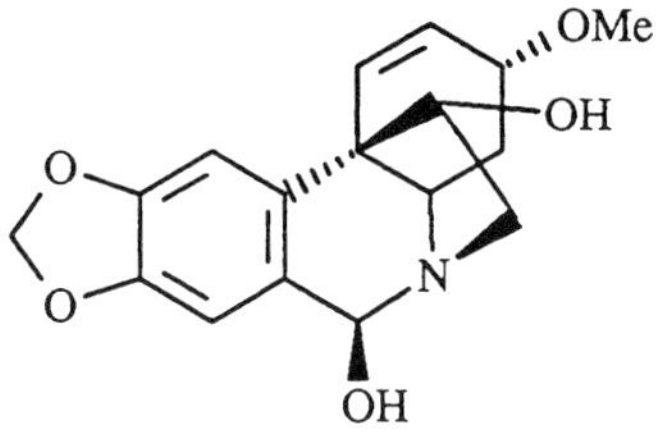

(40)

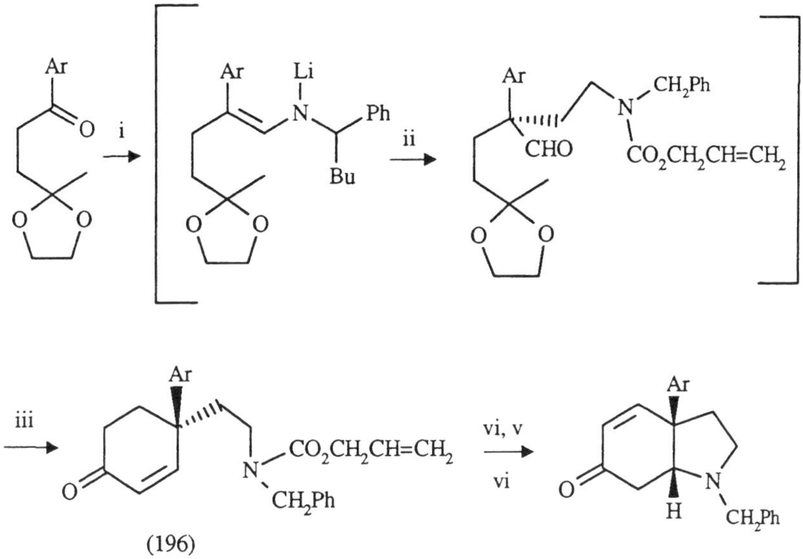

Scheme 1

218

Ar = 3,4-methylenedioxyphenyl

(197; X = OH, Y = H)
(198; X = H, Y = OH)
(199; X = OMe, Y = H)

(5; R = H, CRININE)
(194; R = Me, BUPHANISINE)

Reagents: (i) PhCH=NCH(Li)P(O)(OEt)$_2$, THF at -78°, then reflux, then ^{n}BuLi at -78°; (ii) Br(CH$_2$)$_2$N(CO$_2$CH$_2$CH=CH$_2$)CH$_2$Ph, then H$_3$O$^+$; (iii) pyrrolidine-33% aq. AcOH-MeOH (1:3:30); (iv) PhNMe$_3$Br$_3$, EtOAc, catalytic H$_2$SO$_4$; (v) 1,8-diazabicyclo-[5:4:0]undec-7-ene, benzene, reflux; (vi) Pd(Ph$_3$P)$_4$, Ph$_3$Bu(Et)CHCO$_2$H, CH$_2$Cl$_2$; (vii) AlH$_3$, THF; (viii) Ms$_2$O, Et$_3$N, CH$_2$Cl$_2$; (ix) CsOAc, DMF; (x) MeOH, K$_2$CO; (xi) MeOH, then MeCH(Cl)OC(O)Cl, ClCH$_2$CH$_2$Cl, Proton Sponge [1,8-bis(dimethylamino)naphthalene], reflux; (xii) MeOH, reflux; (xiii) HCHO, MeOH, then 6M HCl.

A Diels-Alder addition of benzoquinone to an alkoxycyclohexa-1,3-diene, when followed by an acid catalysed rearrangement of its adduct, can yield a crinine ring system (Scheme 2) (C.W. Bird, A.L. Brown, C.C. Chan and A. Lewis, *Tetrahedron Letters*, 1989, *30*, 6223).

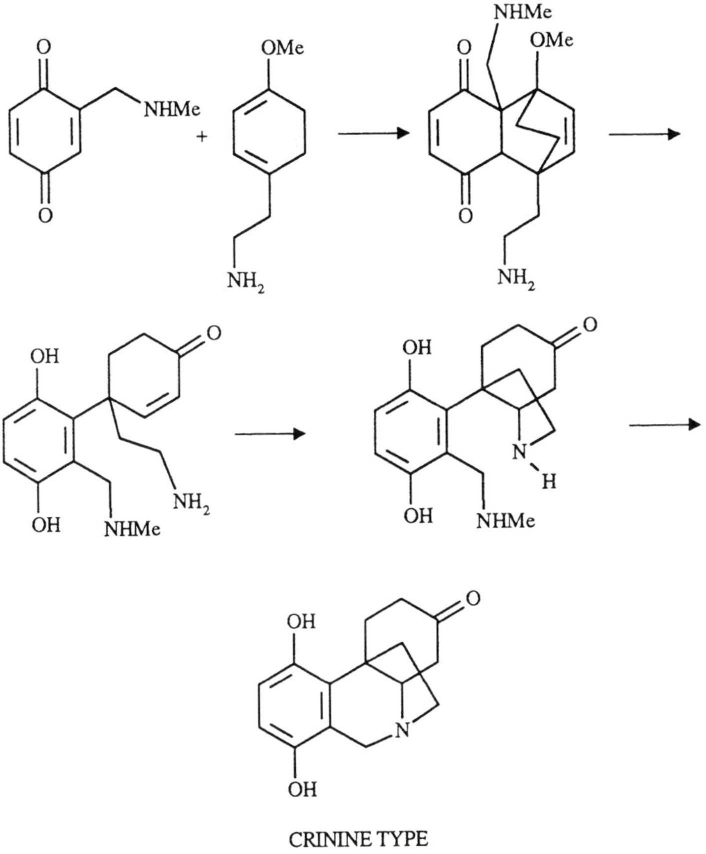

CRININE TYPE

Scheme 2

9. Galanthamine Type Alkaloids

(i) Isolation and structural studies

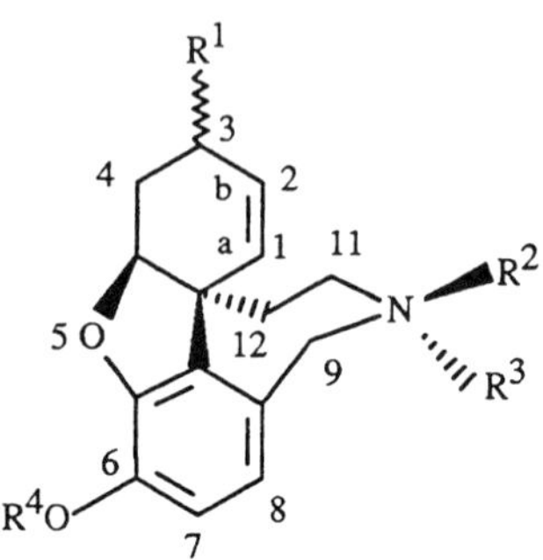

(11)	$R^1=\beta$-OH, $R^2=R^4$=Me, R^3=lone pair Galanthamine
(69)	$R^1=\beta$-OAc, $R^2=R^4$=H, R^3=lone pair 3-O-Acetylchlidanthine
(72)	$R^1=\beta$-OH, R^2=H, R^3=lone pair, R^4=Me, ab sat N-Demethyllycoramine
(81)	$R^1=\beta$-OH, $R^2=R^4$=Me, R^3=O Galanthamine N-oxide
(82)	$R^1=\beta$-OH, R^2=Me, R^3=O, R^4=H Sanguinine N-oxide
(83)	$R^1=\beta$-OH, $R^2=R^4$=H, R^3=lone pair Norsanguinine
(84)	$R^1=\beta$-OCOCH$_2$CH(OH)Me, $R^2=R^4$=H, R^3=lone pair Norbutsanguinine
(88)	$R^1=\beta$-OH, R^2=CHO, R^3=lone pair, R^4=Me N-Formylgalanthamine
(90)	$R^1=\alpha$-OH, R^2=H, R^3=lone pair, R^4=Me Epinorgalanthamine
(91)	$R^1=\alpha$-OH, R^2=H, R^3=lone pair, R^4=Me, ab sat Epinorlycoramine

(107) R^1=α-OH, R^2=Ac, R^3=lone pair, R^4=Me
       Narcisine
(200) R^1=β-OH, R^2=R^4=Me, R^3=O, ab sat
       Lycoramine N-oxide
(201) R^1=β-OH, R^2=H, R^3=lone pair, R^4=Me
       N-Demethylgalanthamine

It has been found that one of the alkaloids of the galanthine group when extracted or crystallised from methylene chloride produces a quaternary salt (2) (see Introduction). There is no reason why other members of this group should not do the same. N-oxides were one time suspected of being artifacts but this has now been discounted; careful extraction procedures have established their authenticity. Three have been identified in *Lycoris sanguinea* (Table 1, ref. 27) namely galanthamine N-oxide (81), sanguinine N-oxide (82) and lycoramine N-oxide (200); also in these bulbs were found the parent alkaloids. Present in *Hymonocallus rotata* bulbs are N-demethyllycoramine (72) and N-demethyl-galanthamine (201) as well as their N-methylated counterparts (Table 1, ref. 19). Epinorgalanthamine (90) and epinorlycoramine (91) have been found in *Narcissus leonensis* (Table 1, refs. 33,34). An additional modification of the functionality associated with the amino-group is the isolation of N-formylgalanthamine (88) which was found in fresh plants of *Narcissus confusus* (Table 1. ref. 31). This publication also highlights the use of nmr techniques NOE, DQF, COSY and 2D.

The 2-position is normally associated with hydroxy or methoxy substituents but interestingly 3-O-acetylchlidanthine (69) from *Haemanthus multiflorus* (Table 1, ref. 11) indicate that variation in O-substitution can also occur. The alkaloid pallidiflorine (16) isolated from whole plants of *Narcissus pallidiflorus* is the first heterodimeric Amaryllidaceae alkaloid isolated to date. It can be considered formed from N-demethyllycoramine (72) and the reverse form of the hemiacetal at C11 of tazettine (7). Only pretazettine (36) was found as a constituent of the extract however.

(ii) Synthesis

Under the title 'Synthesis of biologically active natural substances' the synthesis of galanthamine precursors and steroids has been reviewed (R. Vlakhov, D. Krikoryan, V. Tarpanov, G. Spasov, G. Snatzke, H. Duddeck, H, Schaefer and K. Kislich, *Izv. Khim.*, 1987, *20*, 59; *Chem. Abstr.*, 1998, *108*, 150799). The 'lycorine' ring (208) system can be created from intermediate (207); itself produced by a Diels-Alder addition procedure which is an adaptation of the previously described synthesis of a 'crinine' ring system (C.W. Bird *et al. loc. cit*). Starting from benzoquinone (202) and acetate (203) its adduct (204) rearranged to dibenzofuran (205; R=OAc) which after conversion to the iodide (206) was followed by reduction of the carbonyl group with sodium borohydride to give diol (207; $R^1=R^2=H$, $R^3=I$). The iodo-group (R^3) was converted through azide into amine then N-methylamine (208) which could be cyclised to the 'lycorine' system (209) by treatment with bis(dimethylamino)methane in refluxing dioxane (Scheme 1).

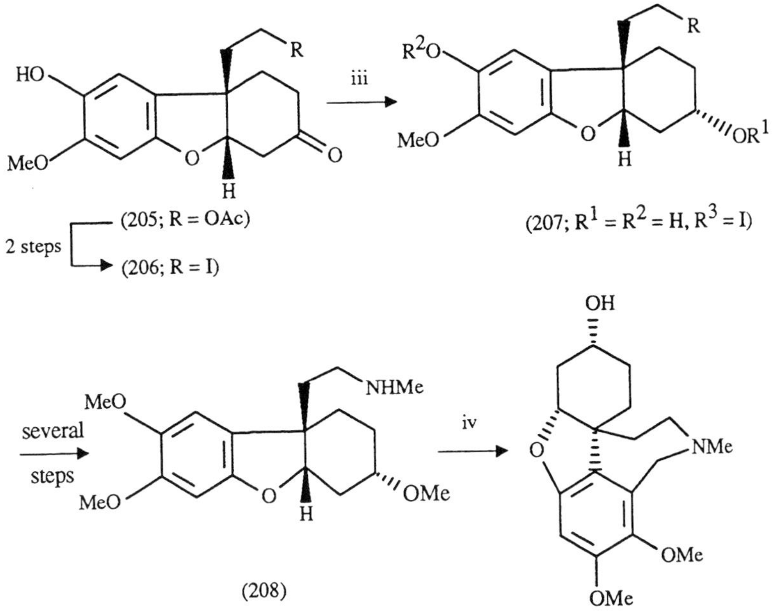

Scheme 1

(205; R = OAc)
2 steps
(206; R = I)

(207; R^1 = R^2 = H, R^3 = I)

(208)

(209)

Reagents: (i) Benzene, 80°C; (ii) 2M HCl, EtOH, r.t; (iii) NaBH$_4$, EtOH; (iv) CH$_2$(NMe$_2$)$_2$, dioxane, reflux

Racemic lycoramine (216) has been synthesised by two routes, both methods producing the quaternary centre (charactertic of all galanthamine alkaloids) by a radical cyclisation reaction induced by Bu$_3$SnH. The first (Scheme 2) starts with a condensation of protected bromophenyl-ethylamine (210) with bromocyclohexanone (211) to give ether (212) whence a Wittig reaction introduced the allylester (213), which upon treatment with tributyltin hydride cyclised to afford the cis-fused dihydrobenzofuran (214). Treatment with aqueous trifluoroacetic acid followed by neutralisation with sodium hydroxide solution gave the lactam (215), which upon LiAlH$_4$ reduction afforded (+)-lycoramine (216) (K.A. Parker and Ho Jin Kim, *J. Org. Chem.*, 1992, 57, 752).

(210) (211) (212) OSiMe$_2$(t-Bu)

(213) (214)

(215) **Scheme 2** (216)

Reagents: (i) Bu$_4$NBr, aq NaOH, CH$_2$Cl$_2$, 0°C, 12 h; (ii) Ph$_3$P=CHCO$_2$Me, THF, 24 h, reflux; (iii) Bu$_3$SnH, AIBN, benzene, Ar, reflux, 2 h; (iv) CF$_3$CO$_2$H, H$_2$O(1:1), r.t. 1 h; (v) LiAlH$_4$, THF, reflux, 24 h.

Another synthetic procedure also uses this radical cyclisation step to create the quaternary centre and hence to (±)-lycoramine (216) in a 13% overall yield (Scheme 3). Starting with bromoguiacol ether (217) condensation with the protected cyclohexanone (218) gave cyclohexanol (219), which upon dehydration allowed radical cyclisation to take place (219 > 220 > 221). Boiling o-xylene was used to optimise the yield (91%). Cleavage of the α-phenoxy group (221) was accomplished using samarium iodide (produced *in situ*) in THF/HMPA at room temperature (1.5 hours) the resulting lactone mixture (unprotected and protected ketone (222) and (223) was completely converted into ketone (223) by acid treatment. Oxidation with benzene selenenic anhydride in boiling toluene gave enone (224) in better yield than did a reaction with DDQ. Reaction with methylamine gave a Michael addition creating ketoamide (225) as the sole product, which with para-formaldehyde and TFA in a modified Pictet-Spengler reaction produced lactam (226). Reduction with $LiAlH_4$/THF first at -78°C then to 0°C then to reflux yielded (±)-lycoramine (216) (M. Ishizaki, K. Ozaki, A. Kanematsu, T. Isoda and O. Hoshino, *J. Org. Chem.*, 1993, *58*, 3877).

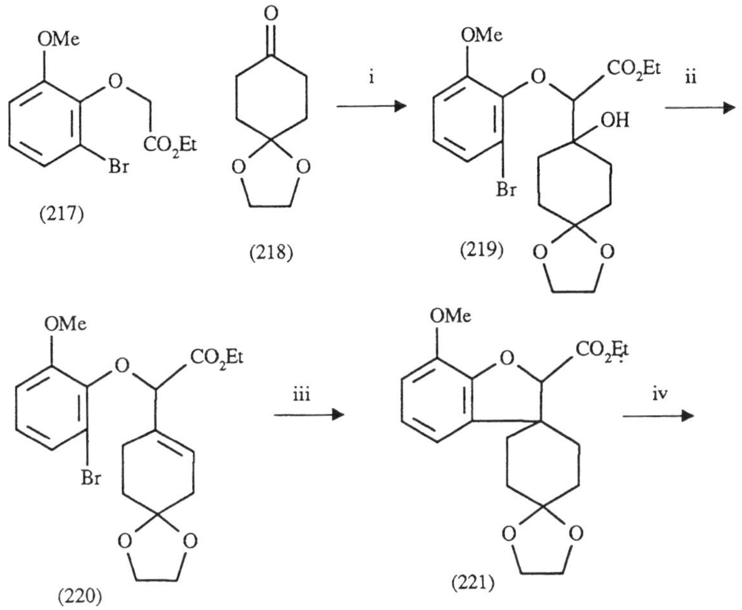

(217)   (218)   (219)

(220)   (221)

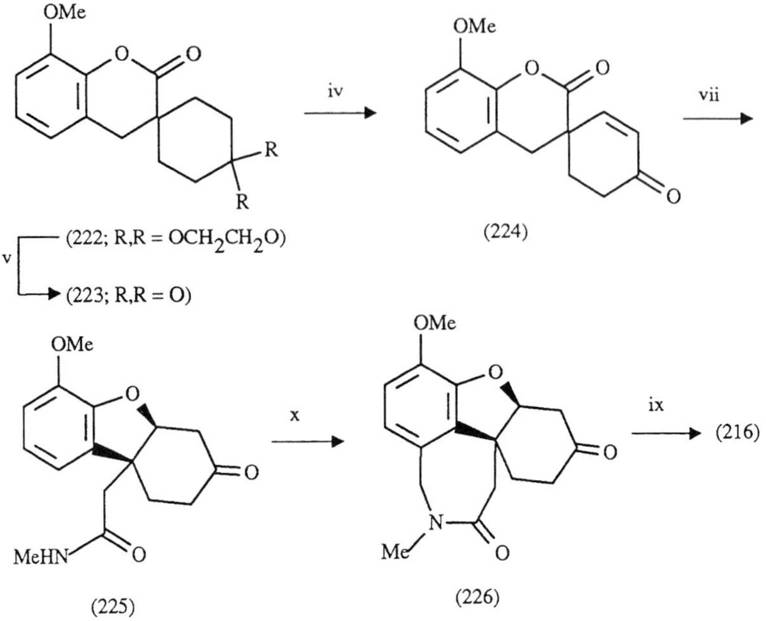

Scheme 3

Reagents: (i) LDA, THF, Ar, -78°C then (218), 15 min; (ii) POCl$_3$, py. 90°C, 1 h; (iii) AIBN, Bu$_3$SnH, o-xylene, reflux, 1.5 h; (iv) Sm(Powder), 1.2-diiodomethane, THF, Ar, r.t., 0.5 h then HMPA 15 min (224) r.t. 25 min; (v) 3M HCl, r.t. 1 h; (vi) benzene selenenic anhydride, toluene, reflux, 0.5 h; (vii) MeNH$_2$ (40% aq), THF, r.t. 20 min; (viii) HCHO, TFA, (ClCH$_2$)$_2$ r.t., 1 h; (ix) LiAlH$_4$, THF, Ar, -78°C 1 h then reflux 2 h.

The biomimetic synthesis of (±)-narwedine (230) reported in 1988 (R.A. Holton, M.P. Sibi and W.S. Murphy, *J. Am. Chem. Soc.*, 1988, *110*, 314) uses an activating functionality which is replaced by a carbon unit. Quite simply methylthiomethyl-norbelladine (227) reacted with lithium tetrachloropalladate and the product gave on oxidation with thallic trifluoroacetate (±)-narwedine in 51% yield (Scheme 4). A palladium complex (228) is first formed, which on oxidation gives a radical and then dienone (229). This cyclises to give racemic narwedine (230).

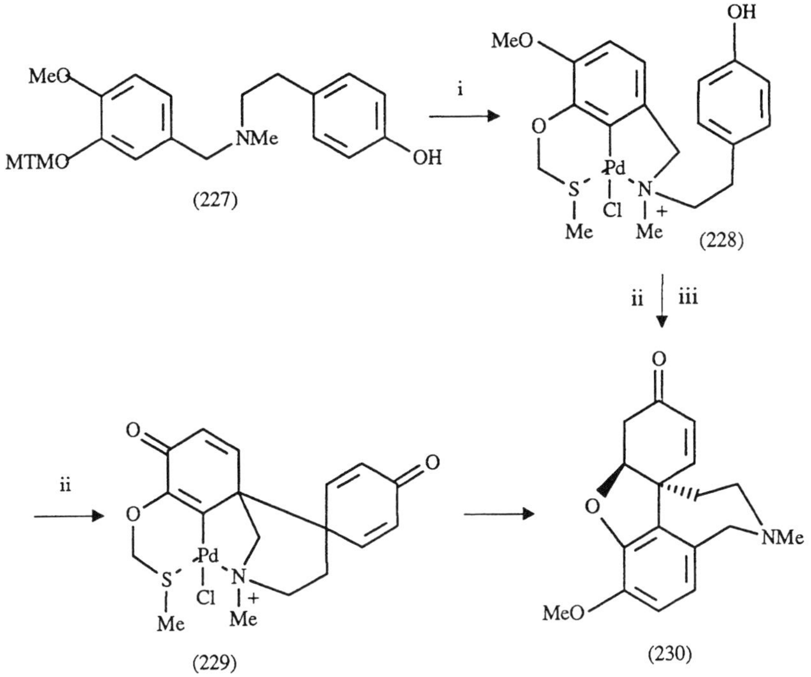

Scheme 4

Reagents: (i) Lithium tetrachloropalladate, MeOH, (^{i}Pr)$_2$NH, -78°; (ii) Thallic trifluoroacetate (2 eq), CH$_2$Cl$_2$/TFA (2:1, -10°C, 1.5 h.; (iii) then r.t. several hours.

An improved route to (±)-galanthamine (Scheme 5) follows: the Barton procedure for intramolecular phenol oxidative coupling to generate the quaternary centre. This was examined first in 1962. But gave only a 1.4% yield due to para-para- rather than para-ortho- coupling. Now a modification using a bromine substituent to block the para-para-coupling process has resolved this difficulty (Scheme 5). Norbelladine (231; R=H) was N-formylated (232; R=CHO), prior to bromination at C6 (233). Oxidative coupling with potassium ferricyanide in sodium bicarbonate buffer gave (±)-2-bromo-N-desmethyl-N-formylnarwedine (234) in 21% yield. Reduction with lithium aluminium hydride concomitantly removed the bromo group and reduced the aldehyde group, to produce a mixture of (±)-galanthamine (11) and (±)-epigalanthamine (235) (J. Szewczk, A.H. Lewin and F.I. Carroll, *J. Het. Chem.*, 1988, *25*, 1809).

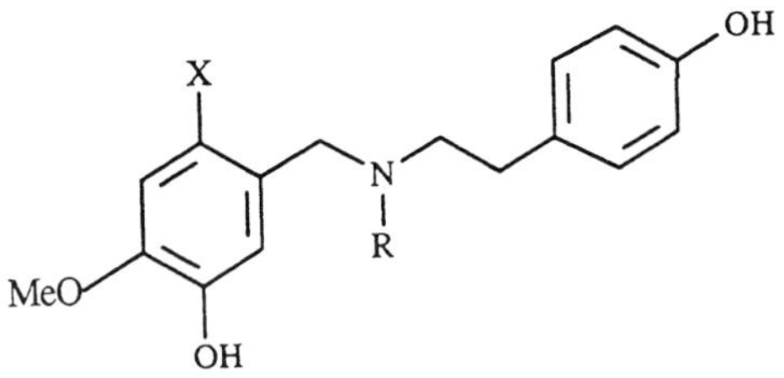

(231; R= X = H)

(232; R = CHO, X = H)

(233; R = CHO, X = Br)

i

ii

Scheme 5

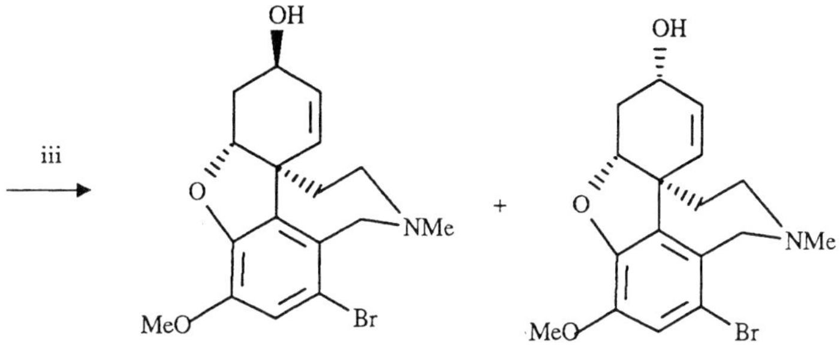

SCHEME 5

Reagents: (i) Br$_2$ 1 eq, CHCl$_3$/MeOH(4:1), -65°C, 2 h; (ii) K$_3$Fe(CN)$_6$, 5% aq NaHCO$_3$, CHCl$_3$, 60°C, 1.5 h; (iii) LiAlH$_4$, THF, 0° then reflux12 h.

Using (±)-narwedine (230) an asymmetic transformation can take place *via* a total spontaneous resolution process, either enantiomer will transform but (-)-narwedine is the required synthon. Shieh and Carlson have been able to convert (+)-narwedine (230) into (-)-narwedine by dissolving the racemate in 95% ethanol:triethylamine (9:1) at 80°C to produce a saturated solution. Reducing the temperature to 68°C supersaturates the solution which upon seeding with (-)-narwedine produces a precipitate containing 84% of the (-)-enantiomer. Likewise, using (+)-narwedine 85% of the (+)- product was obtained (Scheme 6). Deuteration studies indicated that a retro-Michael reaction, followed by reclosure takes place. A diastereoisomeric reduction of (-)-narwedine (230) to (-)-galanthamine (11) was achieved using L-selectide in THF at -78°C (Wen-Chung Shieh and J.A. Carlson, *J. Org. Chem.*, 1994, *59*, 5463).

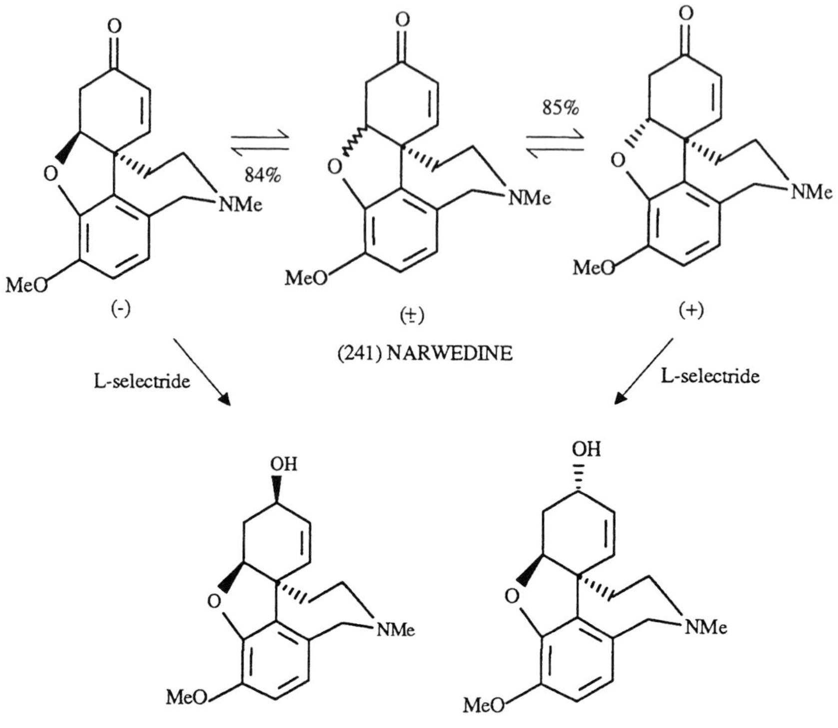

O
O
O
85%
84%
NMe
NMe
NMe
MeO
MeO
MeO
(-)
(±)
(+)
(241) NARWEDINE
L-selectride
L-selectride
OH
OH
NMe
NMe
MeO
MeO
(11) (-)-GALANTHAMINE
(11) (+)-GALANTHAMINE

Scheme 6

10. **Pancracine Type alkaloids**

(i)   Isolation and structural studies

(9) R^1=α-OMe, R^2=H, R^3=OH
           Montanine
(127) R^1=α-OH, R^2=H, R^3=OH
           Brunsvigine
(132) R^1=α-OH, R^2=OH, R^3=H
           Pancracine (hippagine)
(236) R^1=β-OMe, R^2=OH, R^3=H
           Coccinine
(237) R^1=β-OMe, R^2=OAc, R^3=H
           O-Acetylcoccinine

All these alkaloids were reported in previous reviews and no new ones have been obtained during this review period, but montanine (9) and pancracine (132) have now been obtained from the bulbs of *Hippeastrum* hybrides, which hitherto has not been a normal source for this type of alkaloid (C. Muegge, B. Schablirski, K. Obst and W. Doppe, *Pharmazie*, 1994, *49*, 444).   Comprehensive nmr studies including HOHAHA, ROESY and ^{13}C-HMBC techniques helped to assign ^{1}H and ^{13}C shift values and are reported in this publication.

(ii) Synthesis

A preliminary investigation into the synthesis of the pancracine ring system appeared in 1990 which described the production of both the α- and β-isomers of the methanomorphanthridine ring system (O. Hoshimo, M. Ishizaki, K. Saito and K. Yumoto, *J. Chem. Soc., Chem. Commun.*,

232

1990, 420). In 1991 this procedure was used to synthesise racemic coccinine (236), O-acetylcoccinine (237), mantanine (9) and pancracine (132) (M. Ishizaki, O. Hoshino and Y. Iitaka, *Tetrahedron Letters*, 1991, *32*, 7079). Starting with cis-cyclohex-4-enedicarboxylic anhydride (238), its bromide (239) gave the keto acid (240) which was converted into the tosylamide (241). This key compound was 'cis' hydroxylated to give mainly (93%) the β-diol (242; R=H) which was purified *via* its diacetate (243; R=Ac). Wittig reaction gave the exo-methylene derivative (244), which could be epoxidised and ring opened to give diacetate (245; R=H) and thence triacetate (246; R=Ac). Formylation introduced the functionality necessary to ring close *via* a modified Pictet-Spengler reaction to give methanomorphanthridine (247). The 'cis' triacetate was hydrolysed to the triol (248) and reprotection of the 'cis' diol as its benzylidine acetal allowed an intramolecular cyclisation to give the pancracine ring synthon (249). Reduction of the 'cis' protecting group gave the benzyl ether (250; R^1=H, R^2=CH$_2$Ph) which on oxidation, dehydrogenation,dimethylacetal formation followed by reduction gave the benzyl ether of (+)-coccinine whence debenzylation gave the racemic natural product (136). Alternatively (250; R^1=H, R^2=CH$_2$Ph) could be converted into its mesylate (251; R^1=Ms, R^2=CH$_2$Ph), this upon elimination, gave (252) and thence to β-epoxide (253) which could be converted into (±)-pancracine (142) by hydrolysis; or to (±)-montanine (9) by methanolic ring opening or to (±)-brungsvigine (127) by elimination to diene (254) followed by 'cis' hydroxylation, acetylation (255) and methanolysis (Scheme 1).

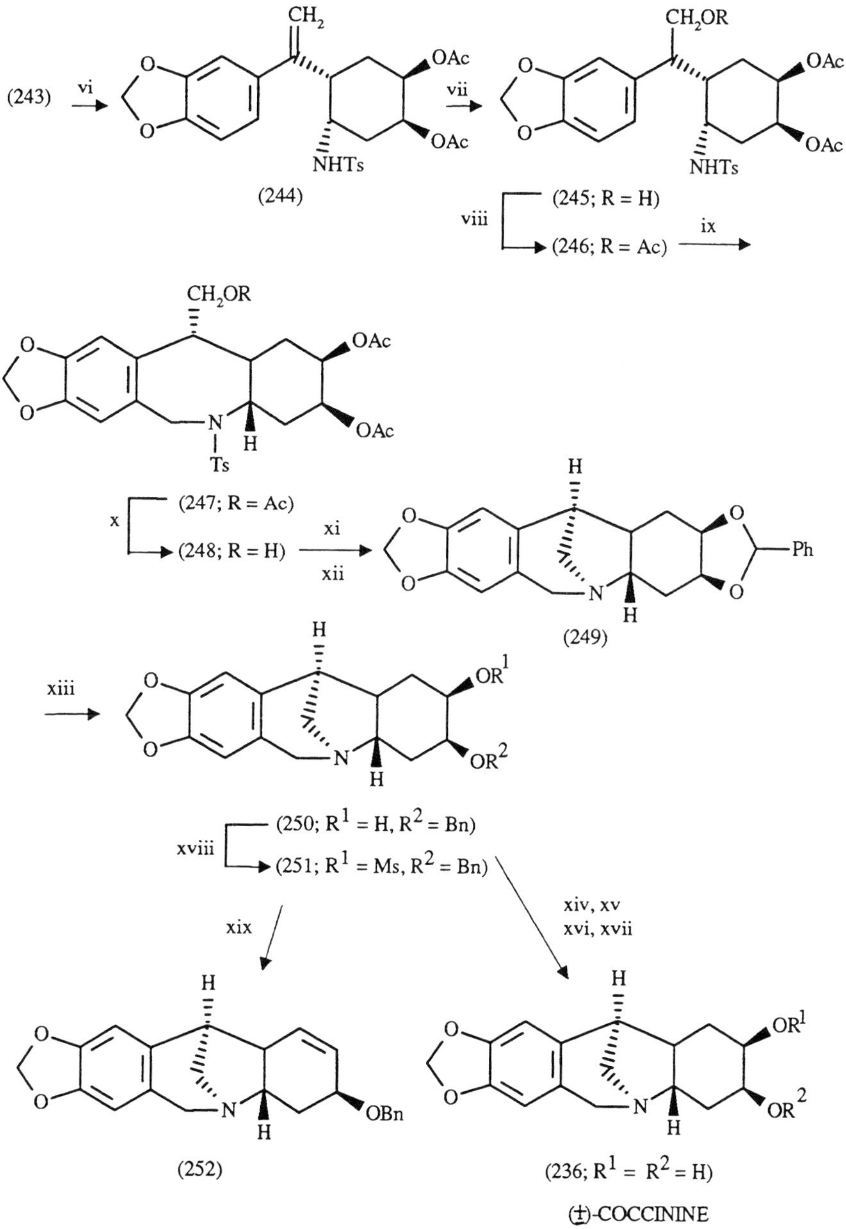

(243) →[vi] (244) →[vii]

viii ⌐ (245; R = H)
     └→ (246; R = Ac) →[ix]

x ⌐ (247; R = Ac)
  └→ (248; R = H) →[xi][xii]

(249)

→[xiii]

xviii ⌐ (250; R^1 = H, R^2 = Bn)
      └→ (251; R^1 = Ms, R^2 = Bn)

→[xix]

xiv, xv
xvi, xvii

(252)

(236; R^1 = R^2 = H)

(±)-COCCININE

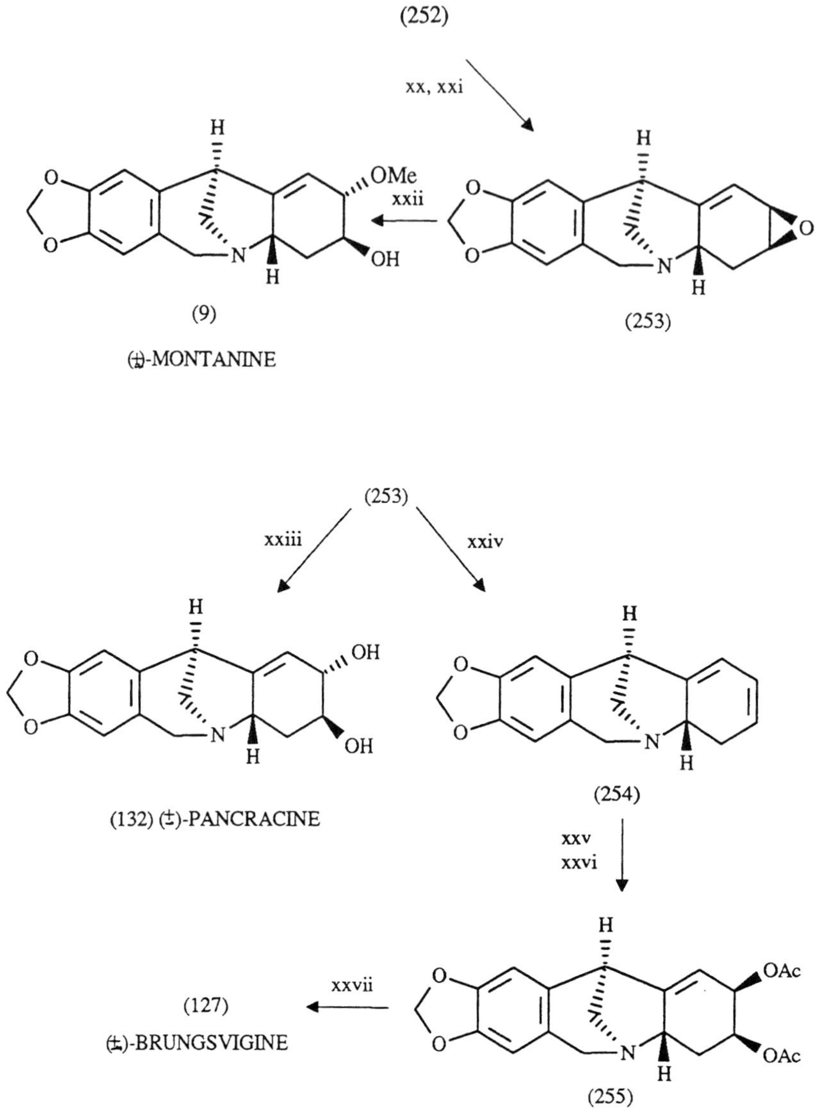

234

(252)

xx, xxi

xxii

(9)

(±)-MONTANINE

(253)

(253)

xxiii

xxiv

(132) (±)-PANCRACINE

(254)

xxv
xxvi

xxvii

(127)

(±)-BRUNGSVIGINE

(255)

Scheme 1

Reagents: (i) THF, Ar O° 1.5 h; (ii) ClCO$_2$Et, acetone, Et$_3$N, 0°C, 1.5 h then NaN$_3$, 0°C, 1 h, then ^{t}BuOH reflux 16 h; (iii) TFA, CH$_2$Cl$_2$, r.t. 1 h then Et$_3$N, pTsCl, CHCl$_3$ r.t. 1 h; (iv) N-methylmorpholine N-oxide, OsO$_4$ (trace), dioxane/H$_2$O (4:1), r.t. 2 h; (v) Ac$_2$O, DMAP, py, r.t. 2 h; (vi) PPh$_3$MeBr, ^{t}BuOK, THF, Ar, r.t. 0.5 h; (vii) BH$_3$ THF (1 eq), THF, 0°, 4 h, then H$_2$O, 3MNaOH, H$_2$O$_2$ (30%) r.t. 40 min; (viii) Ac$_2$O, py, r.t. 1 h; (ix) MeSO$_3$H, (CH$_2$)$_2$Cl$_2$, (HCHO)$_n$, r.t. 2 h; (x) NaOMe, MeOH, r.t. 20 min; (xi) benzaldehydedimethylacetal, pTsOH, CHCl$_3$, r.t. 4 h; (xii) SMEATH, o-xylene reflux, 35 min; (xiii) DIBAH, toluene, Ar, r.t. 2 h; (xiv) Jones reagent, acetone, 0°, 10 min; (xv) DDQ, dioxane, Na$_2$HPO$_4$, reflux, 1 h; (xvi) trimethylorthoformate, pTsOH, MeOH, r.t. 2 h; (xvii) DIBAH, toluene, Ar, r.t. 2 h; (xviii) MsCl, Et$_3$N, CH$_2$Cl$_2$, 0°, 10 min; (xix) ^{t}Bu,OK. DMSO, rt, 2 h; (xx) BF$_3$Et$_2$O, Me$_2$S, CH$_2$Cl$_2$, r.t. 5 h; (xxi) PhSeCl, MeOH, r.t. sonication, 2 h, then NaIO$_4$, 10 min; (xxii) BF$_3$Et$_2$O, MeOH, 0°, 15 min; (xxiii) 3N H$_2$SO$_4$, THF, reflux 5 h; (xxiv) TMSCl, NaI, MeCN, r.t. 0.5 h; (xxv) OsO$_4$ (cat), N-methylmorpholine N-oxide, dioxane/water (4:1), r.t. 3 h; (xxvi) Ac$_2$O, DMAP, py, r.t. 19 h; (xxvii) NaOMe, MeOH, r.t. 20 min.

Earlier in 1991 a different approach to the synthesis of ($\pm$)-pancracine (132) was reported (L.E. Overnan and J. Shim, *J. Org. Chem.*, 1991, *56*, 5005) which used a tandem azo-Cope rearrangement coupled with a Mannich cyclisation as a central step in the creation of the tetracyclic ring system. Starting with the cyclopentanone nitrile (256) condensation with alkyne (257) gave alkyne (258; R=CH$_2$CN) which, after deprotection to (259; R=H) was easily converted to allylic alcohol (260). This upon formylation in the presence of camphosulphonic acid produced oxazole (261). This key intermediate was rearranged by Lewis acid (BF$_3$Et$_2$O) in 97% yield to perhydroindolone (262; R=Bn) which, after deprotection, underwent formylation, followed by Pictet-Spengler cyclisation to the pancracine ring system (264). Stereoselective β-face reduction of the carbonyl group was achieved using lithium sec-butylborohydride to give 1-α-alcohol (265), which on dehydration (266) allowed introduction of a hydroxy group at C-2 (267). Oxidation to the enone (268) was followed by enol silylation to produce 3β-hydroxy-2-ketopancracine (269) which on stereoselective acyloxyborohydride reduction gave the racemic natural product (132) (Scheme 2).

236

(256)
+
(257)
i
(258; R = CH₂CN)
(259, R = H)
ii
iii
OH
NRBn
(260)
OH
NHBn
iv
(261)
O
NBn
v
(262; R = Bn)
(263; R = H)
vi
vii
viii
(264)

Scheme 2

Reagents: (i) THF, -78°C; (ii) AgNO$_3$ (1.1 eq), EtOH, 30°C, sonication; (iii) LiAlH$_4$ (3.5 aq), Et$_2$O, -20°C to reflux; (iv) formalin (2 eq) Na$_2$SO$_4$ (4 eq), camphorsulphonic acid (0.2 eq), CH$_2$Cl$_2$, 23°C; (v) BF$_3$OEt$_2$ (2.4 eq), CH$_2$Cl$_2$, -20 to 23°C; (vi) H$_2$.PdC, HCl, MeOH; (vii) formalin (50 eq), Et$_3$N (2 eq), MeOH, 23°C; HCl-MeOH, 23°; (viii) Bu$_3$LiH (2 eq), THF, -78°C; (ix) SOCl$_2$ (13 eq), CHCl$_3$, -30°C to 23°C; (x) SeO$_2$ (4.3 eq), dioxane, 85°C, 6 h; (xi) Swern oxid.; (xii) Me$_3$SiOSO$_2$CF$_3$, (10 eq), Et$_2$N (30 eq), Et$_2$O, -60°C to 0°, then OsO$_4$ (cat), N-methylmorpholine N-oxide (3 eq), ^{t}BuOH, H$_2$O, py, -5° to 23°C, 6 h; (xiii) NaBH$_4$ (40 eq), HOAc/MeCN (1:1), -35°C, 19 h

238

Using a radical-mediated synthetic procedure involving a phenyl sulphide it has been shown that the 1-phenyltetrahydroisoquinoline sulphide (275), prepared from 1-hydroxytetrahydroisoquinoline (270) *via* 4-bromocyclohex-2-enone (274), was conveniently rearranged using SnBu$_3$H and AIBN (Barton procedure) (M. Ishizaki, Ken-ichi Kurthara, E. Tanazawa and O. Hoshino, *J. Chem. Soc. Perkin Trans. 1*, 1993, 101). Scheme 3 details the relevant steps to methanophenanthridine-2-one (276) which by standard procedures has been converted into alkaloids pancracine (132) coccinine, (236), O-acetylcoccinine (237) and montanine (9).

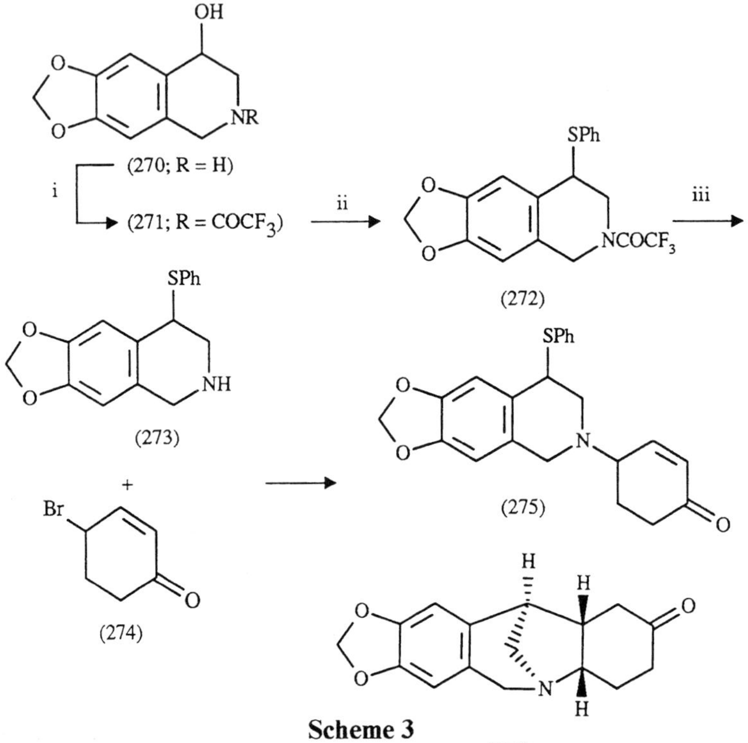

(270; R = H)

(271; R = COCF$_3$)

(272)

(273)

(274)

(275)

Scheme 3

(276)

Reagents: (i) (CF$_3$CO)$_2$O, CHCl$_3$, K$_2$CO$_3$, r.t., 0.5 h; (ii) PhSH, ZnI$_2$, (CH$_2$)$_2$Cl$_2$, r.t., 1 h; (iii) K$_2$CO$_3$ (5% aq), MeOH, reflux, 0.5 h; (iv) 4-bromocyclohex-2-enone, Et$_3$N, Et$_4$Ni, CH$_3$CN, r.t. then reflux, 3 h; (v) Bu$_3$SnH, AIBN, o-xylene, reflux, 1 h + 1 h.

11. Phenanthridine Alkaloids

(i) Isolation and structural studies

$R^1=R^2=R^3=H$, $R^4+R^5=CH_2$, ab unsat.
8,9-Methylenedioxyphenanthridine
(trisphaeridine) (13)
$R^1=R^3=H$, $R^2=Me$, $R^4+R^5=CH_2$, ab unsat.
N-Methyl-8,9-methylenedioxyphenanthridine (25)
$R^1=H$, $R^2=Me$, $R^3=O$, $R^4+R^5=CH_2$, ab sat.
N-Methyl-8,9-methylenedioxy-6-phenanthridone (26)
$R^1=R^3=H$, $R^2=Me$, $R^4+R^5=CH_2$, ab unsat
N-Methyl-8,9-methylenedioxyphenanthridinium chloride
(bicolorine) (86)
$R^1=R^3=H$, $R^2=Me$, $R^4+R^5=CH_2$, ab sat.
5,6-Dihydrobicolorine (87)

240

An investigation into the alkaloids present in fresh bulbs of *Narcissus bicolor* (Table 1, ref. 30) has identified six alkaloids, pretazettane (36), 9-O-demethylhomolycorine (42), epimacronine (277), oxoassonine N-oxide (278) and two phenanthridine alkaloids, namely bicolorine (86) and 5,6-dihydrobicolorine (87).

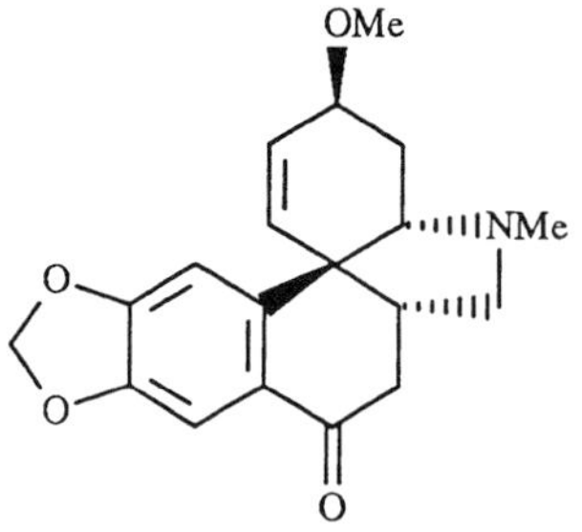

(277) 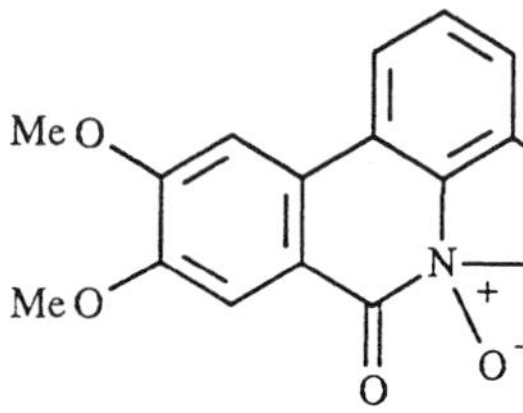   (278)

The parent alkaloid, 8,9-methylenedioxyphenanthridine (13), which is also called trisphaeridine, has been isolated from an extract of fresh leaves and bulbs of *Lapiedra martinezii* (Table 1, ref. 22) also present in this extract were N-methyl-8,9-methylenedioxy-6-phenanthridone (25) and N-methyl-8,9-methylenedioxyphenanthridinium chloride (86). Since ismine (14) was also found, the biogenesis can be easily inferred.

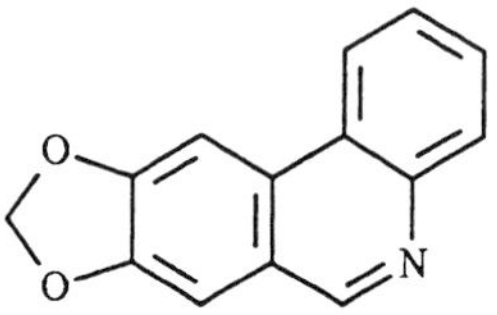

(13) 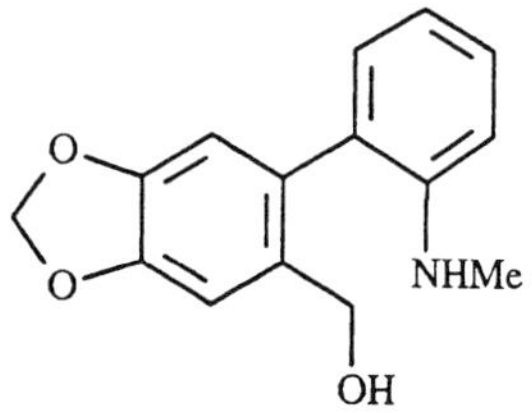   (14)

These alkaloids have been referred to as phenanthridines (Table 1, ref. 28,29).

12. **Miscellaneous Alkaloids**

(i) Benzylidene-phenylethylamines and belladines

The isolation of benzylidene-phenylethylamines has filled a gap in the biosynthetic pathway from tyrosine to the cyclic members of the Amaryllidaceae alkaloids. Craugsodine (17) and isocraugsodine (15) were mentioned earlier (Biogenesis Section) and the first four oxygenated member of this group rylliston (284) has been obtained from the flower fluid of *Amaryllis vittata* (S. Ghosal and S. Rzadan, *J. Chem. Res. (S)*, 1984, 412). Confirmation of its structure was obtained by its synthesis from veratraldehyde (279) and homoveratrylamine (280), followed by reduction of the Schiff's base (281) with sodium borohydride. Reduction of this type of Schiff's base, with or without N-methylation, leads to the belladines e.g. N-norbelladine (282).

(279)   (280)

heat

(281; $R^1 = R^3 = R^4 = Me, R^2 = H$)

NaBH$_4$

(282; $R^1 = H, R^2 = Me$)

(283; $R^1 = R^2 = H$)

(284; $R^1 = OMe, R^2 = H$)

242

(ii) 4-Phenyltetrahydroisoquinolines

The two main alkaloids in this group are cherylline (126) and latifine (125), both are found in *Crinum* species and both have been synthesised by routes based on their biogenesis hence O-demethylbelladines were involved. Both alkaloids have been synthesised from (±)-6-hydroxy-O,O-demethylnorbelladines (283) and (284). In the former case base induced the cyclisation with ring closure occurring para- to the hydroxyl group giving (±)-cherylline (126) if however the para- position was blocked (by bromine) ring closure could be obtained using acid conditions, subsequent debromination giving (±)-latifine (125) (S. Kobayashi, T. Tokumoto, S. Iguchi, M. Kihara, Y. Imakura and Z. Taira, *J. Chem. Res. S*, 1986, 280).

(iii) Mesembrinanes

Unexpectedly mesembrenone (286) has been isolated from the aerial parts of *Narcissus pallidulus*; normally it is found in Sceletium species and very occasionally in the Amaryllidaceae (J. Bastida, F. Viladomat, J.M. Lhabres, G. Ramidez, C. Codina and M. Rubiralta, *J. Nat. Products*, 1989, *52*, 479). Only one other instance of the isolation of a mesembrane from the Amaryllidaceae has been reported with mesembrenol (287) being found in *Crinum oliganthum* (W. Dopke, E. Sewerin, Z. Trimino and G. Gutierrez, *Z. Chem.*, 1981, *21*, 358). Two other alkaloids, related to mesembrinanes, are amisine (288) (W. Dopke, E. Swerin and Z. Trimino, *Z. Chem.*, 1980, *20*, 298) from *Hymenocallis arenicola* and joubertiamine (26) from *Sceletium subvelutinum* (R.B. Herbert and A.E. Kattah, *Tetrahedron*, 1970, *46*, 7105).

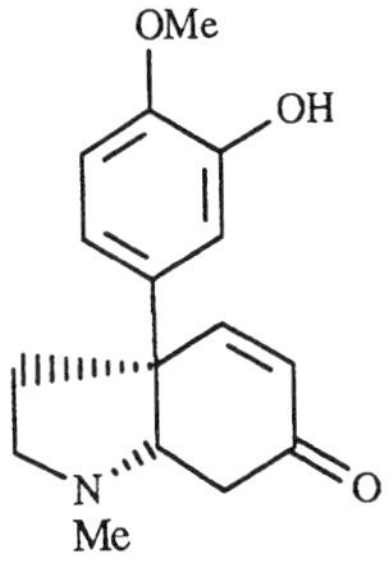

(286)

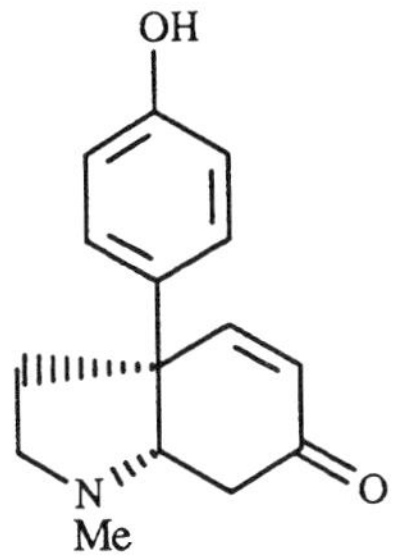

(287)

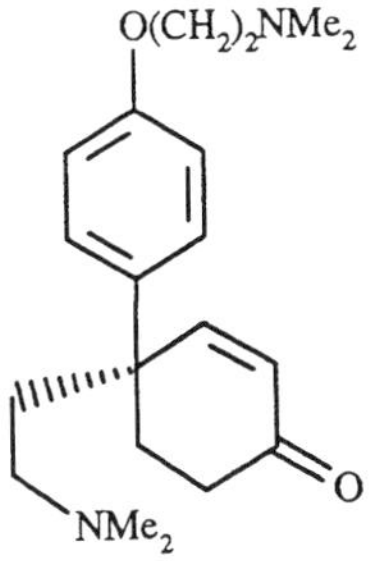

(288)

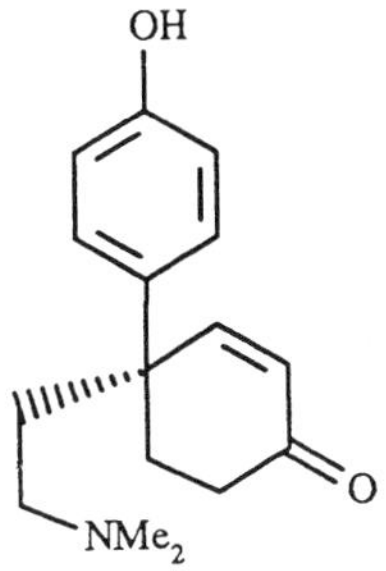

(26)

244

A possible synthetic procedure to the mesembrine ring system was discussed in 1990 (D.A. Owen, G.R. Stephenson, H. Finch and S. Swanson, *Tetrahedron Letters*, 1990, *31*, 3401).   Using the carbonyl-cyclohexadiene complex (289) prepared by Birch in 1968, it was possible to introduce successively a phenyl group (290) and a malonyl group into the complex (291) and thus to enone (292).   By a similar sequence it was proposed that an intermediate such as (293) would be suitable for ring closure to the mesembranone system (294) (Scheme 1).

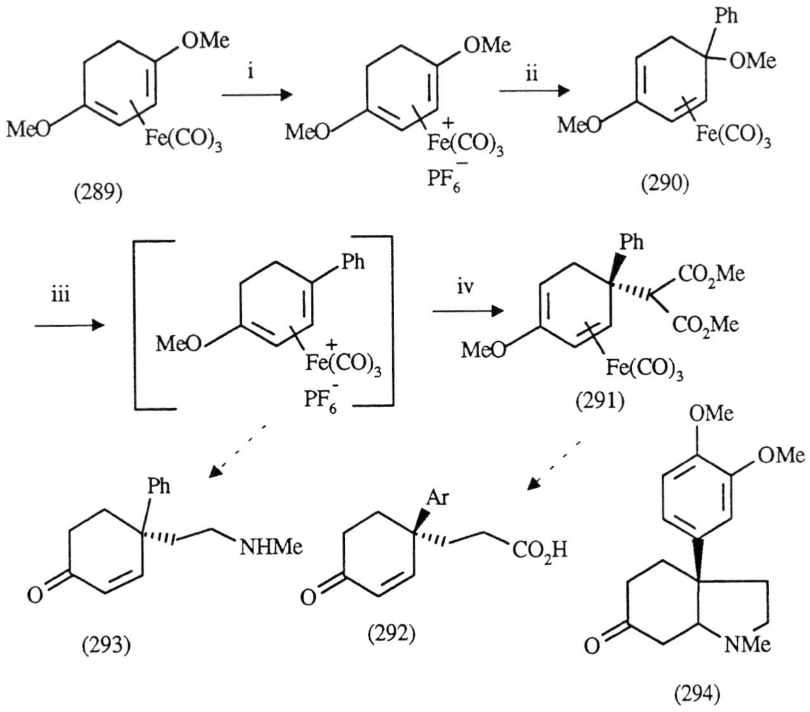

Scheme 1

Reagents: (i) $Ph_3C.PF_6$; (i) PhLi, CH_2Cl_2, -78°; (iii) CF_3CO_2H, 0°C, 1 hr, then NH_4PF_6, -78° to r.t.; (iv) $NaCH(CO_2Me)_2$, CH_3CN, CH_2Cl_2, Ar, reflux 6 h.

An alternative route to synthesise mesembrine (10) uses a cyclic enamine such as (298) (Y. Matsumura, J. Terauchi, T. Yamamoto, T. Konno and T. Shono, *Tetrahedron*, 1993, *49*, 8503). This was prepared by an anodic α-methoxylation procedure starting with pyrrolidine (295) (Scheme 2). Firstly the N-formate ester (296) was prepared, which was followed by its anodic oxidation in methanol whereby creating enamine (296) that upon bromo-methoxylation gave (297). Displacement of the α-methoxy group by a 3,4-bimethoxyphenyl anion easily led to (298), which underwent an aryl migration under the influence of Ag^+ ions, concomitant loss of HBr gave the aryl substituted enamine (299) which gave the N-methyl derivative (300) by reduction with $LiAlH_4$. Finally reaction with methylvinyl ketone gave mesembrine (10).

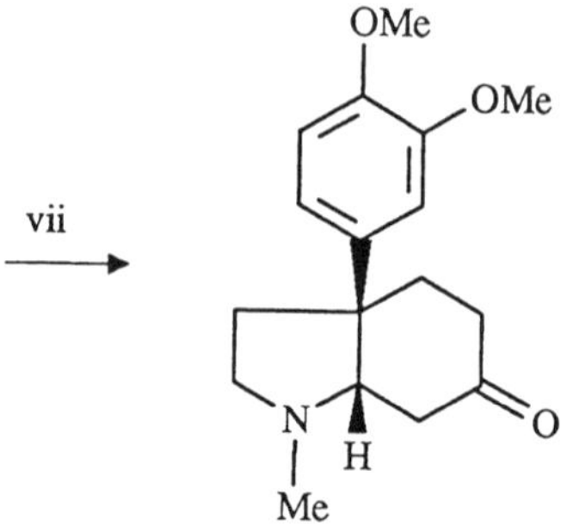

(10)

Scheme 2

Reagents: (i) ClCO$_2$Me, Na$_2$CO$_3$, CH$_2$Cl$_2$, reflux; (ii) anodic cell, C anode, Et$_4$Np-Tos$^-$, MeOH, 10-15°C; (iii) Br$_2$, MeOH, NaOMe, 1 h; (iv) TiCl$_4$, CH$_2$Cl$_2$N$_2$, -78°, 5 min then veratrole, -78° to r.t. 3 h; (v) AgNO$_3$, MeOH, reflux 30 min; (vi) LiAlH$_4$, ether, N$_2$, reflux; (vii) methylvinyl-ketone, ethylene glycol, N$_2$, 57°C, 2.5 h then 119°C, 1 h.

A stereospecific synthesis of (-)-mesembranol (317; R=H) has been described starting from D-glucose using the established Ferrier rearrangement to produce methyl-4,6-O-benzylidine-α,β-altopyranoside (301; R=H). This was protected as its bismethoxymethyl ether (302; R=MOM) before being transformed into bromo-ether (303) by treatment with NBS in carbon tetrachloride containing BaCO$_3$. Dehydro-bromination gave exomethylene (304), which upon Ferrier carbocyclisation produced cyclohexanone (305). This without purification was converted into enone (306). Reaction with 3,4-dimethoxyphenyl lithium at -78° gave enol (307), which was reduced catalytically to (308), prior to being converted into the thiocarbonate (309), elimination then giving alkene (310). After several attempts the perhydroindole (311; R=CHO) was obtained which upon formic acid deprotection gave a number of products all of which were hydrolysed and separated to give mostly alcohol (312; R=H). Triethyl orthoacetate acidified with a trace of propionic acid in the presence of a 3A molecular sieve gave ester (313; R=CO$_2$Et) which was converted into N-methyl-methylamine derivative (315; CH$_2$NHMe) via aldehyde (314). Intra-molecular cyclisation with mercuric acetate gave the saturated protected mesembranol (316; R=MOM) which upon deprotection gave the natural product (-)-mesembranol (317; R=H) (Scheme 3) (N. Chida, K. Sugihara and S. Ogawa, *J. Chem. Soc., Chem. Commun.*, 1994, 901).

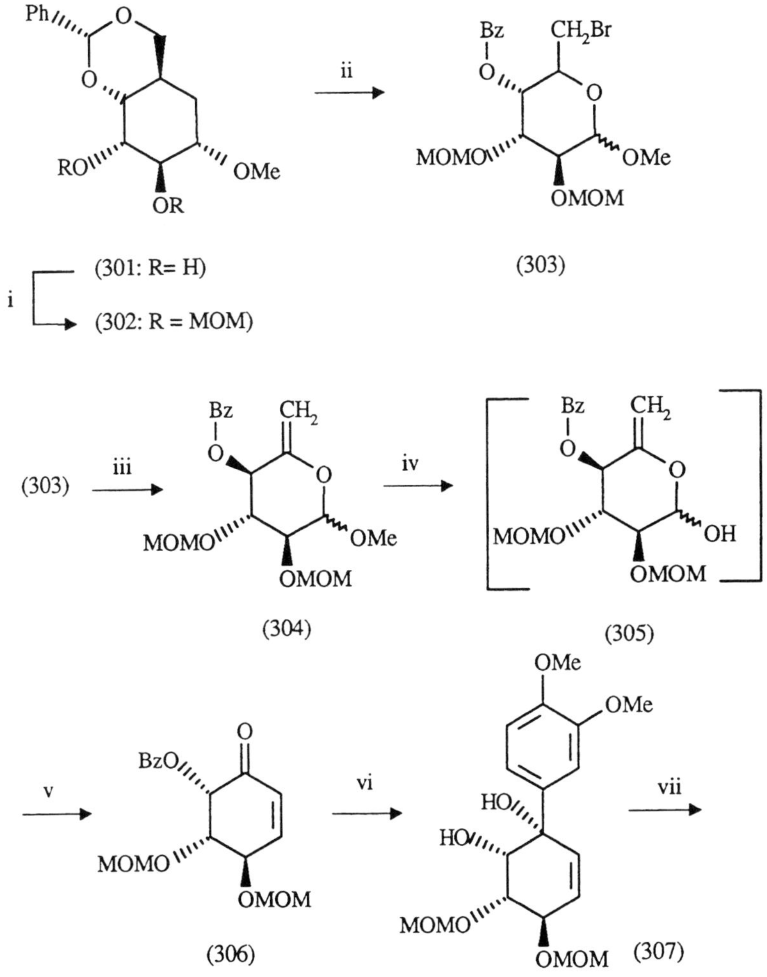

OMe
OMe
S
O
O
HO
HO
MOMO
OMOM
MOMO
OMOM
viii
(308)
(309)
OMe
OMe
(309)
ix
MOMO
OMOM
x
OMe
OMe
RO
OMOM
(311; R = H)
+
(312; R = CHO)
xi
OMe
OMe
R
OMOM
MeO
OMe
NMe
OR
(313; R = CO₂Et)
(314; R = CHO)
(315; R = CH₂NHMe)
xii
xiii
xiv
(316; R = MOM)
(317; R = H)
xv

Scheme 3

Reagents: (i) chloromethyl ether, Pr^i_2NEt, CH_2Cl_2, reflux 18 h; (ii) NBS, $BaCO_3$, CCl_4, 1,1,2,2-tetrachloroethane, reflux, 20 h; (iii) DBU, toluene, 75°, 20 h; (iv) $Hg(OCOCF_3)_2$ (0.5 mol %), acetone/H_2O (2:1), r.t., 72 h; (v) MsCl, Et_3N, CH_2Cl_2, r.t., 30 min; (vi) 3,4-dimethoxyphenyl lithium, Et_2O, -78° then MeONa:MeOH; (vii) H_2, $Pd(OH)_2$, EtOAc; (viii) 1,1'-thiocarbonyl diimidazole, acetone, reflux, 48 h; (ix) $P(OMe)_3$, reflux, 72 h; (x) 30% aq HCO_2H, 30°, 72 h, then K_2CO_3, MeOH, r.t.; (xi) triethyl orthoacetate, 2% propionic acid v/v, molecular sieves 3 A, 135°, 48 h; (xii) DIBAL, toluene, -78°; (xiii) $MeNH_2$ (30% MeOH soln), $NaBH_3CN$, MeOH, r.t., 24 h; (xiv) $Hg(OAc)_2$, THF, r.t. then $NaBH_4$, THF, aq NaOH, r.t.; (xv) 6 mol dm^{-3}, HCl/THF (1:2 v/v), r.t.

*Second Supplements to the 2nd Edition of Rodd's Chemistry
of Carbon Compounds, Vol.IV B*, edited by M. Sainsbury
© 1997 Elsevier Science B.V. All rights reserved.

Chapter 11

TROPANE ALKALOIDS

G. Fodor

Introduction

Alkaloids, that contain the 8-azabicyclo[3.2.1^{1,5}]tropane skeleton, named by
Holmes "nortropane" (*L. Holmes*, Alkaloids (NY) 1950, **1**, 273) are found in
plants of the Convolvulaceae, Dioscoraceae, Erythroxylaceae, Proteaceae
and Solanaceae. In 1987, 151 such alkaloids were known (*M. Lounasmaa*,
Alkaloids (NY) 1988, **33**, 1), but by 1993, this number had increased to 200
(*M. Lounasmaa, and T. Tamminen*, Alkaloids (NY), 1993, **44**, 1).
As for other natural products, NMR spectroscopy and mass spectrometry are
used to identify the alkaloids and as early as 1964 a MS map of the tropane
skeleton was presented (*H. Budzikiewicz, C. Djerassi, G. Fodor, E. Blossey
and M. Ohashi*, Tetrahedron 1964, **20**, 585) and this work has been
amplified by many authors since then. Frequently, however, the tropane
alkaloid was not hydrolyzed, therefore its absolute configuration was not
correlated to the known absolute configuration of the alkamine. Tables I, II
and III contain many, though not all, of these compounds, named by trivial
or rational names. The relative configurations of the commoner alkaloids
were determined in the fifties when the alpha-beta convention was introduced
(*G. Fodor and K. Nádor*, J. Chem. Soc., 1953, 7121). For a survey of this
and later work see (*G. Fodor,* Alkaloids 1968, **6**, 149; 1966, **9**, 269; 1971,
13, 352).
This current chapter updates Chapter 11 in C.C.C. 2nd Ed., Vol. IVB, by
H. C. S. Wood and *R. Wigglesworth* and deals with the literature up to
December 1995. The author does not intend to record all newly-discovered
alkaloids, rather, a few new structural types will be emphasized. A modern
and complete biosynthesis of tropanes, based mostly on brilliant work by (the
late) *Edward Leete*, will also be outlined.
Many of the tropane alkaloids are esters of tropan-3α-ol and 3β-ol, named
tropeines, of tropane-3α,6β-diol, of tropane-3α,6β,7β-triol and of
3-hydroxy-2-tropanecarboxylic acid. A few are derived from 2β-

hydroxytropane. Fortunately, although many tropanes occur as racemates, most of the major chiral alkaloids have been correlated with simpler compounds of known absolute configuration. For example, cocaine was correlated with L-(+)-glutamic acid (*E. Hardegger and H. Ott*, Helv., 1955, **38**, 312). Similarly 6β-methoxytropinone derived from (-)-3α,6β-dihydroxytropane was degraded to *S*-(-)methoxysuccinic acid; therefore the parent alkamine was the (3*S*:6*S*) enantiomer, the opposite (3*R*:6*R*) configuration was ascribed, by inference, to the esters of (+)-3,6-tropanediol (*G. Fodor and F. Sóti*, J. Chem. Soc., 1965, 1830). Unfortunately, there is some unwarranted confusion in describing a tropanediol as a 3α,6β- or a 3α,7β-dihydroxylated compound: 3α,6β-tropanediol is the antipode of the 3α,7β-tropanediol (SCHEME I. However, it is clearer to designate the stereocentres in terms of the *R/S* convention than to present the enantiomers as regioisomers.

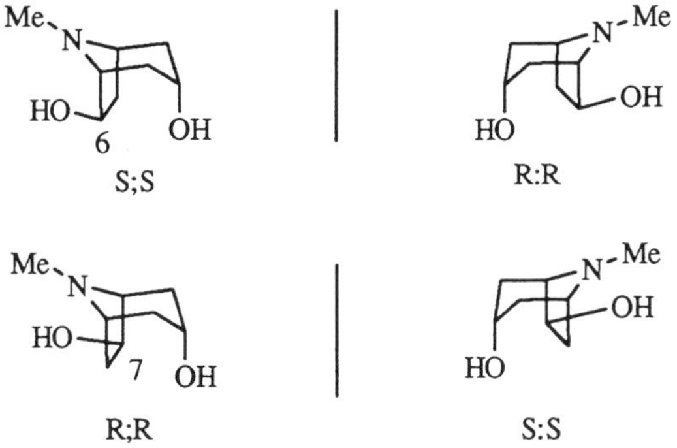

SCHEME I

Teloidine, 3α,6β,7β-tropanetriol is a mesoid molecule; most of its natural ester derivatives are also optically inactive mesoid or, racemic modifications. Here again it is confusing to refer to the racemate as 3α,6β and 3α,7β-substituted tropane esters, as if they were regiomers. The simple rule applies to the diols, that the sign of rotation of the chiral alkamine obtained on hydrolysis determines the absolute configuration of the natural ester. The 2-hydroxylated tropanes pose another problem, but again the absolute configuration of the alkamine is decisive.

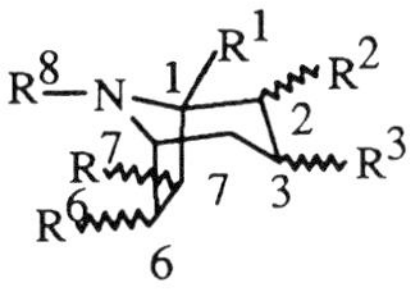

TABLE I. TROPEINES

Alkaloid name	R3	R8
Nortropane	H	H
Tropane	H	Me
Tropine	α-OH	Me
Pseudotropine	β-OH	Me
Atropine	α-(±)tropoyloxy	Me
Hyoscyamine	α-*S*-tropoyloxy	Me
Apoatropine	α-atropoyloxy	Me
Butropine	α-isobutyryloxy	Me
Brugine	α-2,4-dithiolbutyryloxydisulfide	Me
Convolamine	α-veratroyloxy	Me
Convolvine	α-veratroyloxy	H
Littorine	α-(*S*)-2-hydroxyphenylpropionoxy	Me
Phyllalbine	α-vanilloyloxy	H
Isoporoidine	α-methylbutyryloxy	H
Valtropine	α-methylbutyryloxy	Me
Poroidine	α-isovaleryloxy	H
Tigloidine	β-tigloyloxy	Me
Tropacocaine	β-benzoyloxy	Me
Tropine trimethoxy cinnamate	α-(3,4,5-trimethoxycinnamoyloxy)	Me
Subhirsine	α-bis(veratroyloxy)	CHO
Tropine cinnamate	α-cinnamoyloxy	Me
Nortropanyl cinnamate	α-phenylacetoxy	H
Tropane-3α-yl phenylacetate	α-phenylacetoxy	Me
Tropane-3α-yl hydroxyphenylacetate	α-(3-hydroxyphenylacetoxy)	Me

TABLE II. 1-Hydroxy and 2-Hydroxytropanes; 2-Benzyltropanes

Name	R1	R2	R3	R6	R8
Physoperuvine	OH	H	H	H	H
Calystegine B1	OH	α-OH	β-OH	β-OH	H
Calystegine B2	OH	α-OH	β-OH	α-OH	H
Calystegine A2	OH	α-OH	β-OH	H	H
1-Hydroxytropacocaine	OH	H	β-OBz	H	Me
Baogongteng A	H	β-OH	H	β-OAc	H
(±) 2-Benzyltropine	H	α-Bz	α-OH	H	Me
acetate	H	α-Bz	α-OAc	H	Me
benzoate	H	α-Bz	α-OBz	H	Me
Knightinol	H	α-(α-hydroxybenzyl)	α-OH	H	Me
Acetylknightinol	H	α-(α-acetoxybenzyl)	OAc	H	Me

Modern methods including capillary gas chromatographic-electron capture detection and capillary gas chromatography-mass spectrometry are now routinely applied to the isolation and the structural elucidation of the tropane alkaloids (*J. M. Moore, et al.*, J. Chromatography, 1987, **410**, 297). In addition NMR methods are essential in deducing structural details and relative stereochemistry. The foundations of this work are described in papers published some 30 years ago (*R. J. Bishop, G. Fodor et al.*, J. Chem. Soc., C1, 1966, 74: *A. Sinnema, L. Maat, A. J. van der Gugten, and H. C. Beyerman*, Rec. Trav. Chim., 1968, **87**, 1027).

Syntheses of Tropanes

There are numerous synthetic approaches to tropanes. However, it must be emphasized that they can all be summarized in three main pathways (SCHEME II).

TABLE III. Tropanediol and Tropanetriol Esters

Name	R3	R6	R7	R8
(-)3α,6β-Ditigloyloxytropane	3α-tigloyloxy	tigloyloxy	H	Me
6β-Acetoxy-3α-tigloyloxytropane	3α-tigloyloxy	β-acetoxy	H	Me
3α-Hydroxy-6β-tigloyloxytropane	3α-OH	tigloyloxy	H	Me
(+)2α-Benzoyloxy-3β-hydroxynortropane	α-benzoxy 3β-OH	H	H	H
Scopolamine	(±)3α-tropoyloxy		6β,7β-epoxy	Me
Hyoscine	3α-(S-tropoyloxy)		6β,7β-epoxy	Me
Scopine	3α-OH		6β,7β-epoxy	Me
Oscine (Scopoline)	3α,6α-oxydo	6 or 7β-OH		Me
Norscopolamine	3α(±)-tropoyloxy		6β,7β-epoxy	H
Norhyoscine	3α(S)-tropoyloxy)		6β,7β-epoxy	H
Apohyoscine	3α-atropoyloxy		6β,7β-epoxy	Me
Meteloidine	3α-trigloyloxy	β-OH	β-OH	Me
3α,6β-Ditigloyloxy-7β-hydroxytropane	3α-tigloyloxy	β-tigloyloxy	β-OH	Me
6β-Isovaleroxy-3α-tigloyloxytropan-7β-ol	3α-tigloyloxy	β-isovaleroxy	OH	Me
Schizantine F	Rα-mesaconoyloxy	R-tigloyloxy	H	Me
Schizantine G	Rα-itaconate	R-methyltigloyloxy	H	Me
Schizanthine H	Rα-itaconate	R-methylagelicoyloxy	H	Me
Schizanthine I	R-mesaconate	R-angelicoyloxy	H	Me
Schizanthine K	Rα-mesaconate	R-tigloyloxy	H	Me
Schizanthine L	R-itaconate	R-angelicoyloxy	H	Me
Schizanthine M	α-itaconate	R-tigloyloxy	H	Me
Valeroidine	α-isovaleroxy	β-OH	H	Me
6β-Hydroxyhyoscyamine	α-S-tropyloxy	β-OH	H	Me
α-Acetoxy-6β-hydroxytropane	3α-acetoxy	β-OH	H	Me
Catuabine A	α-(3,4,5-trimethoxybenzoyloxy)		β-(2-N-methylpyrroloyloxy)	
Catuabine B	α-(3,4,5-trimethoxybenzoyloxy)		β-benzoyloxy	Me
Catuabine C	α-(2-N-methylpyrroloyloxy)	β-(2-N-methylpyrroloyloxy)		
(3R:6R)-tropanediol-3-phenylacetate	α-phenylacetoxy	β-OH		Me

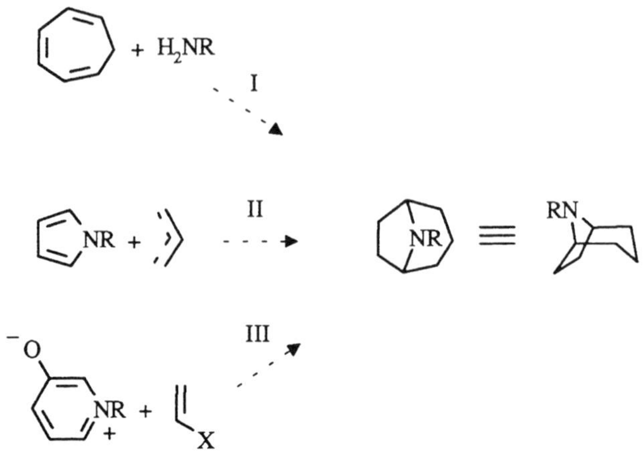

SCHEME II

I. Type Syntheses of this type are based upon the classical work of *Willstätter* in which a nitrogen bridge is inserted across cycloheptatriene, or a derivative. Some modern variants of this process are described below:
(i) Dianions derived from butane-1,4-diphenyl sulfone react smoothly with 2-methylene-1,3-dichloropropane to give a 1,4-diphenylsulfonyl-6-methylenecycloheptane which, upon ozonolysis followed by S_N2-type displacement by methylamine, cyclizes to tropan-3-one (*P. Lansbury et al.*, Tetrahedron, 1990, **13**, 3965) (SCHEME III).

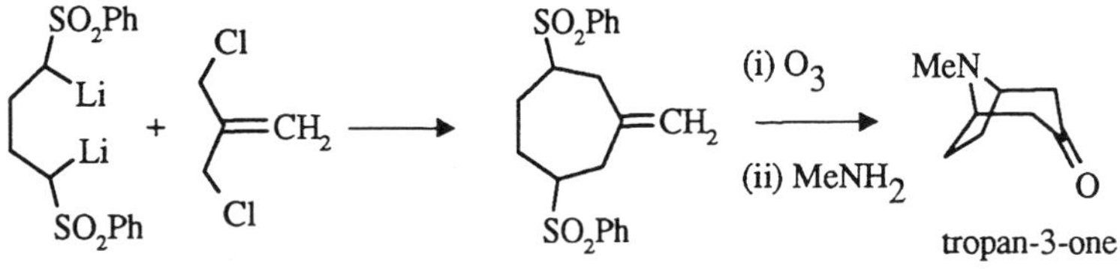

SCHEME III

(ii) Physoperuvine, the first known 1-hydroxytropane, was synthesized from 1,3-cycloheptadiene and the benzyl ester of N-hydroxycarbamic acid; the latter was oxidized *in situ*, by tetramethylammonium perchlorate, to the corresponding nitroso compound, and this reacted *via* a hetero-Diels-Alder process to give a bridged hydroxylamine. Reduction by diimide gave the dihydro analogue, which was ring opened by treatment with sodium amalgam. This gave *cis*-4-(benzoxycarbonylamino)cycloheptanol, which upon hydrogenolysis with LAH afforded *cis*-(4-methylamino)cycloheptanol. Finally Jones oxidation led to 4-(methylamino)cycloheptanone, which exists preferably in the tautomeric carbinolamine form, *i.e.* physoperuvine (*D. E. Justice and J. R. Malpass*, J. Chem. Soc., Perkin I, 1994, 2559) (SCHEME IV).

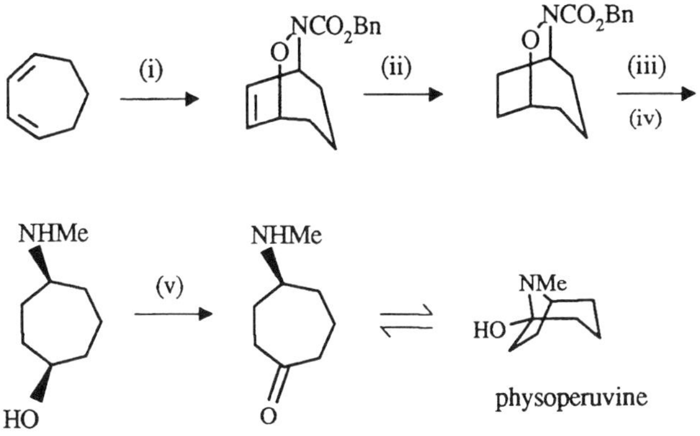

Reagents: (i) BnO$_2$CNO (ii) dimide (iii)Na/Hg (iv) LAH (v) Jones oxidation

SCHEME IV

(iii) 1-Chloro-1-nitrosocyclohexane served as a nitrosyl group donor to 6-benzoxy-1,3-cycloheptadiene *(H. Iida, Y. Watanabe and C. Kibayashi, J. Org. Chem., 1985, **50**, 1818)* to give a bicyclic hydroxylamine, which upon hydrogenolysis led to *cis*-1-benzoxy-3-hydroxy-6-aminocycloheptane. After protection of the amino group and the conversion of the hydroxyl group into a *trans* chloro substituent, cyclization was achieved by using potassium-*t*-butoxide to afford N-(ethoxycarbonyl)pseudotropine benzoate and this, by reduction gave benzoylpseudotropine, *i.e.*, tropacocaine. Hydrolysis gave pseudotropine (SCHEME V).

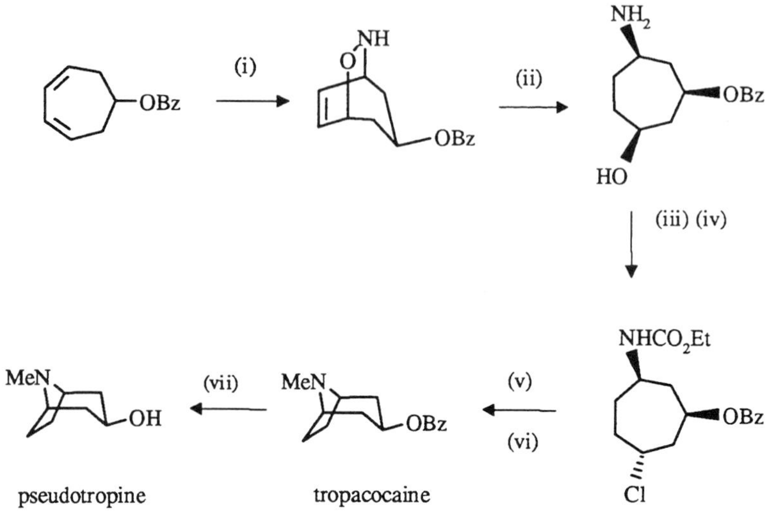

258

Reagents: (i) 1-chloro-1-nitrosocyclohexane (ii) H_2/Pd (iii) $ClCO_2Et$ (iv) $SOCl_2$ (v) ^{t}BuOK (vi) LAH (vii) acid hydrolysis

SCHEME V

(iv) Scopolamine was synthesized in low yield by the epoxidation of 6-tropene-3α-yl acetate *via* scopine (*G. Fodor et al.,* Chem. & Ind., 1956, 764); later, epoxidation by pertungstic acid of (*S*)-tropoyl-6-tropenol gave overall yields of hyoscine up to 15%, based on furan (*G. Fodor, S. Kiss and J. Rákóczi,* Chim. & Ind. (Paris), 1963, **90**, 225; *G. Fodor,* Magy. Tud. Akad. Kém. Tud. Közl., 1963, **20**, 163). Since then several attempts have been made to obtain scopine and its esters by starting from a cycloheptadiene. First, *Bäckwall et al.* (Tetrahedron Letters, 1987, **28**, 4199) succeeded in introducing an acetoxy group and a chlorine atom into 6-benzyl-oxy-1,3-cycloheptadiene by *cis*-1,4-addition, exchanging stereoselectively (with inversion)* the chlorine atom by a tosylamino group and hydrogenating the carbon-carbon double bond. The *trans*-acetoxy-tosylamino product was then cyclised by an intramolecular S_N2 reaction to N-tosyl-3β-nortropanol benzoate. The all-*cis*-adduct from the original cycloaddition reaction was similarly converted into N-tosylnortropine benzyl ether taking advantage of a Mitsunobu cyclisation procedure to form the bicyclic ring system (SCHEME VI). [* Pd-catalyst - retention].

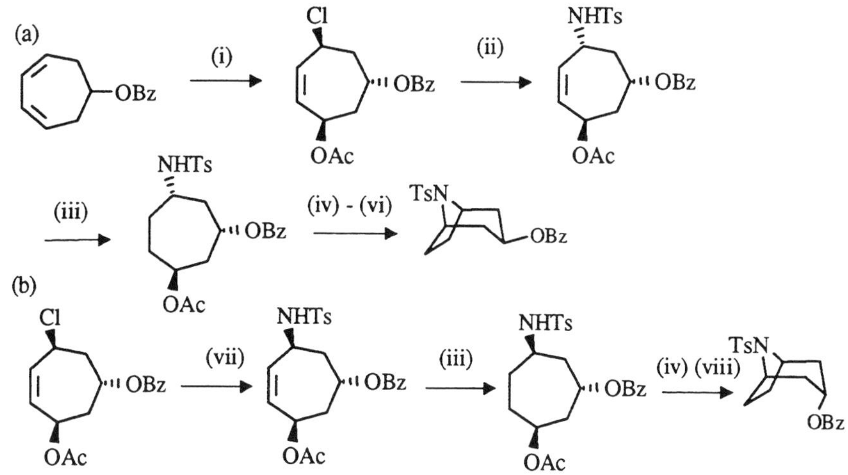

(a)

(i)

(ii)

(iii)

(iv) - (vi)

(b)

(vii)

(iii)

(iv) (viii)

Reagents: (i) Pd(OAc)$_2$/LiCl/LiOAc/1,4-benzoquinone (ii) NaNHTs/HOAc
(iii) H$_2$/RhCl(PPh$_3$)$_2$ (iv) NaOH/MeOH (v) MsCl/Et$_3$N (vi) K$_2$CO$_3$/MeOH
(vii) NaNTs/Pd(PPh$_3$)$_4$ (viii) EtO$_2$CN=NCO$_2$Et/ PBu$_3$/ZnCl$_2$

SCHEME VI

(v) In a related procedure the key intermediate, *trans*-1-tosylamino-4-chlorocycloheptenyl 6-benzyl ether was epoxidized and the product, in turn, cyclized to N-tosylscopine benzyl ether. In two more steps this compound was then hydrogenolized and N-methylated to give scopine. Surprisingly, the epoxide ring endured all these operations (*J. E. Bäckwall et al.*, J. Org. Chem., 1991, **56**, 2769) (SCHEME VII).

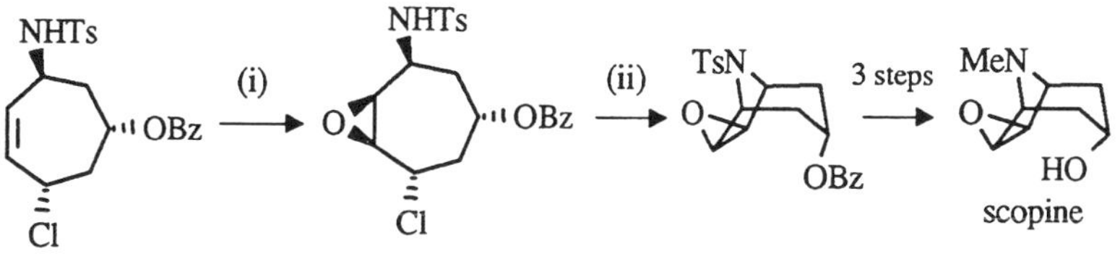

(i)

(ii)

3 steps

scopine

Reagents: (i) MCPBA (ii) K$_2$CO$_3$/MeOH

SCHEME VII

260

(vi) In another approach *Malpass et al.* (Tetrahedron Letters, 1995, **36**, 4689) introduced the nitrogen function into 6-hydroxy-1,3-cycloheptadiene, via the nitroso compound that is generated from N-hydroxybenzyl carbamate and tetramethylammonium periodate. After ring opening of the adducts, two diastereomeric alcohols were obtained that were protected as their TBDMS ethers. These were epoxidised separately with MCPBA to yield the appropriate diastereomeric epoxides and thence through cyclisation and conventional functional group interconversions progressed onto scopine and pseudoscopine (SCHEME VIII).

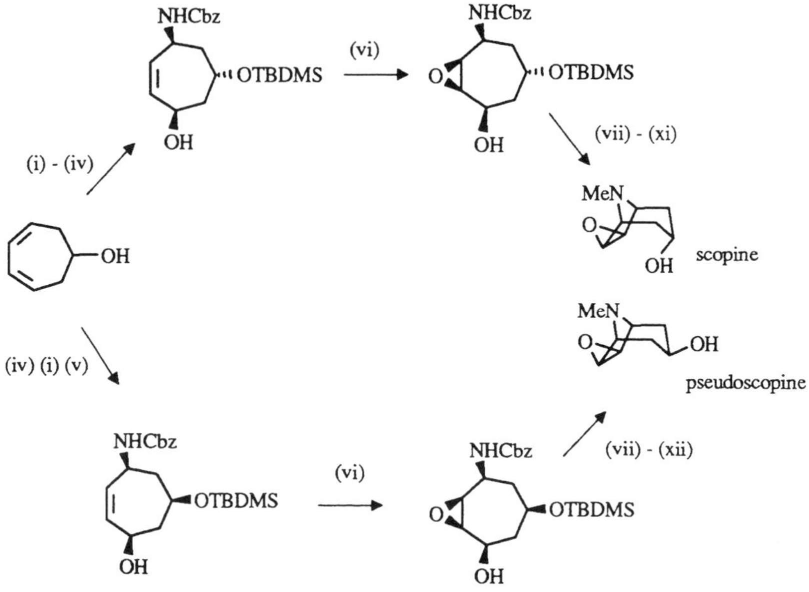

Reagents: (i) BnOCONHOH/Et₄NIO₄ (ii) Jones oxidation (iii) L-Selectride (iv) TBDMSCl/imidazole (v) Na/Hg (vi) MCPBA (vii) BuLi/TosCl (viii) LiCl/DMSO (ix) NaH (x) LAH (xi) TBAF (xii) H₂/Pd

SCHEME VIII

(vii) *Lallemand et al.* have confirmed the structure of different, 1-hydroxytropanes by synthesis. For example, the synthesis of *calystegine A₂* (*J. E. Lallemand et al.*, Synlett, 1992, 357) is reported (SCHEME IX). Starting from the trimethylsilyl enol ether of 4-(N-benzyloxycarbonyl)cyclohexanone, ring expansion with diazomethane led to 5-benzyloxycarbonyl-2-cycloheptene-1-one. Alkaline epoxidation and acid hydrolysis resulted in 2,3-

trans-dihydroxy-5-benzyloxycarbonyl-1-cycloheptanone; hydrogenolysis of the N-protecting group then gave a dihydroxyamino ketone, the tautomeric form of which, 1,2α,3β-trihydroxynortropane is identical with *calystegine A₂*

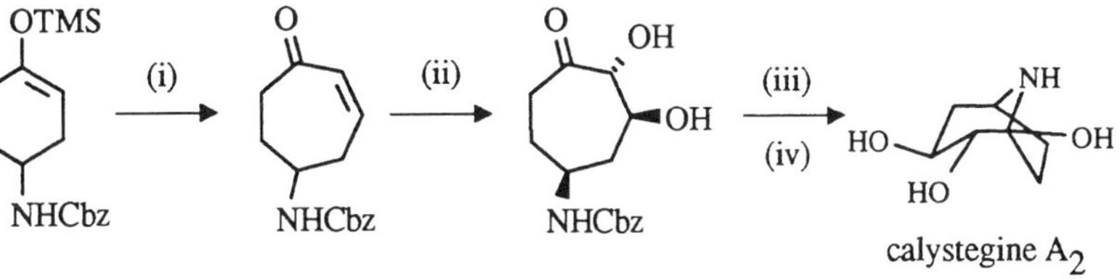

Reagents: (i) CH_2N_2 (ii) $H_2O_2/NaOH$ (iii) H_3O^+ (iv) H_2/Pt

SCHEME IX

II Type Syntheses using the addition of three carbon units to pyrroles, pyrrolines or pyrrolidines are also still of interest, following on from the pioneering work of *Willstätter and Pfannenstiehl, Robinson,* and *Schöpf.* Indeed, the original and elegant synthesis of tropinone (4) by *R. Robinson* (J. Chem. Soc., 1919, **111**, 762), from succindialdehyde methylamine and pentan-3-one 1,5-dicarboxylic acid (SCHEME X) remains to this date the most versatile and practically applicable tropane synthesis. NB The justification for including the R*obinson* synthesis in this group is that it involves the intermediate formation of a pyrroline (*L. A. Paquette and J. W. Heimaster,* J. Amer. Chem. Soc., 1966, **88**, 763).

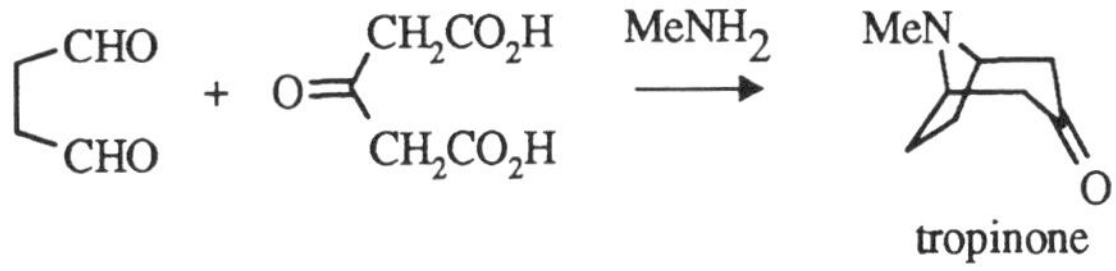

SCHEME X

Other 1,4-dicarbonyl compounds are frequently used and, for example, this has enabled the synthesis of 1-alkenyl and alkynyl tropanes (*G. Fodor and R. Dharanipragada,* J. Chem. Soc. Perkin I, 1986, 545).
Some other illustrations of type II syntheses are included below:
For example, N-methoxycarbonylpyrrole gave N-methoxycarbonyl-6-tropen-3-one in one step when it was reacted with sym-tetrabromoacetone and diiron

nonacarbonyl. Finally , reduction of 6-tropen-3-one with DIBAH gives 6-tropen-3α-ol (SCHEME XI) *(R. Noyori and Y. Hayakawa,* Tetrahedron, 1985, **41**, 5879, *Y. Hayakawa, Y. Baba, S. Maxino and R. Noyori,* J. Amer. Chem. Soc., 1978, **100**, 1786). Unfortunately the elimination of iron compounds, after the reaction caused difficulties, therefore a zinc copper alloy was next used as a catalyst; alternatively EDTA was applied to remove the iron. A further variant of the *Noyori* route has been recommended (*J. Mann and L.-C de A. Barbosa,* J. Chem. Soc. Perkin Trans. I, 1992, 787) using diethylzinc as catalyst for the condensation.

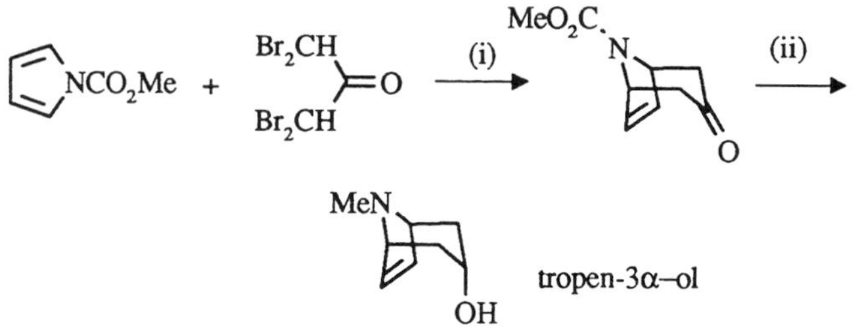

Reagents: (i) $Fe_2(CO)_9$, or Cu/Zn, or $Zn(Et)_2$ (ii) DIBAH

SCHEME XI

N-Methoxycarbonyltropenone may be oxidized by MCPBA to N-methoxylcarbonylnorscopinone. Surprisingly the 0.5 mole excess of the reagent did not cause any *Baeyer-Villiger* product formation. Reduction of the scopinone with DIBAH led to oscine, *i.e.*, 3α, 6α-oxido-7β-hydroxytropane (SCHEME XII).

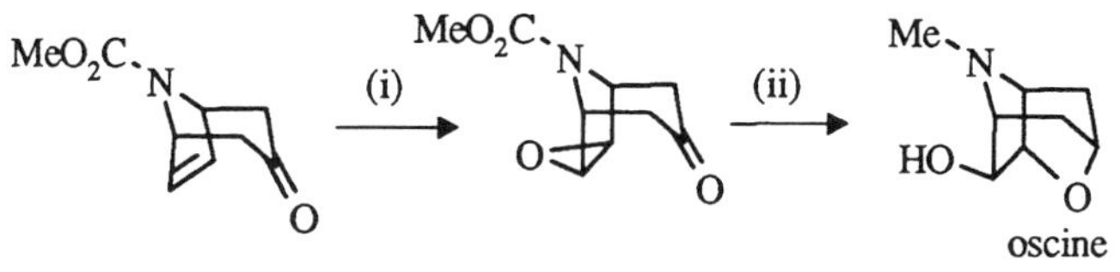

Reagents: (i) MCPBA (ii) DIBAH

SCHEME XII

Vinylcarbenoids, for example methyl-α-diazobutenoate undergo 1,3-dipolar cycloaddition with N-methoxycarbonylpyrrole, or better, N-(2-trimethylsilylethoxy)carbonylpyrrole in the presence of a rhodium catalyst. The 2,6-tropadiene-2-carboxylic esters (X = OMe) obtained as products can be selectively reduced at the more isolated double bond to give the methyl ester of (±)-anhydroecgonine (*Davies et al.*, Tetrahedron Letters, 1989, **30**, 4653; J. Org. Chem., 1991, **56**, 5696). Similarly, 3-diazo-4-penten-2-one (X = Me) with the pyrroles gave rise to 2-acetyl-2-tropene, ferruginine, a lower homologue of anatoxin (SCHEME XIII).

R = CH₂CH₂TMS   X = OMe or Me

anhydroecgonine Me ester (X= OMe)
ferrugine (X = Me)

Reagents: (i) Rh₂(Oalkyl)₄ (ii) H₂/(PPh₃)₂Rh (iii) Bu₄NF (iv) HCHO/Na(CN)BH₃

$$R = CH_2CH_2TMS \quad X = OMe \text{ or } Me$$

Reagents: (i) $Rh_2(Oalkyl)_4$ (ii) $H_2/(PPh_3)_2Rh$ (iii) Bu_4NF (iv) $HCHO/Na(CN)BH_3$

SCHEME XIII

III Type Syntheses of this type were introduced by *A. R. Katritzky*. Thus, 3-hydroxy-N-methylpyridinium betaine when reacted with acrylonitrile gave 6β-cyano-3-tropane-2-one *A. R. Katritzky and Y. Takeuchi*, J. Chem. Soc. C, 1971, 874). This adduct was later used by *M. E. Jung et al.* (J. Org. Chem., 1992, **57**, 3528) to synthesize racemic baogongteng A, by reducing it to 6-cyanotropene-2-one and, after further reduction by converting the cyano group via a Grignard reaction and a Baeyer-Villiger oxidation into 6β-acetoxy-2β-hydroxytropane (SCHEME XIV). The absolute configuration of the alkaloid has been studied by a variety of modern methods, and found to be (2S:6S) (*H. M. He, J. X. Shen, X. Y. Ma and X. P. He*, Jaoxue Xuebao, 1990, **24**, 335; Chem. Abstr., 1990, **112**, 179537).

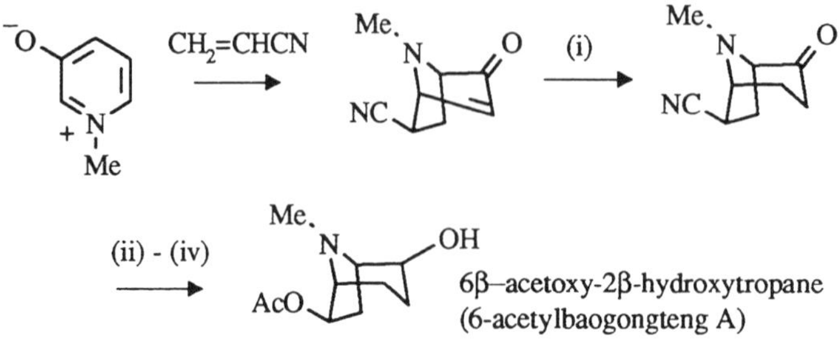

Reagents: (i) H$_2$/Pd (ii) NaBH$_4$ (iii) MeMgBr (iv) MCPBA

SCHEME XIV

An elegant asymmetric synthesis of 2β-hydroxytropane, based on the *Katritzky* approach, uses chiral vinyl tolyl sulfoxide as the dipolarophile in a reaction with 3-hydroxy- N-methyl pyridinium betaine. This gave 6-toluenesulfinyl-2-oxo-3-tropene, which was reduced to the corresponding sulfide and ultimately reduced and desulfurized by Raney-Ni, to optically active 2β-hydroxytropane (*T. Takashashi et al.*, Chem. Lett., 1989, 593) (SCHEME XV).

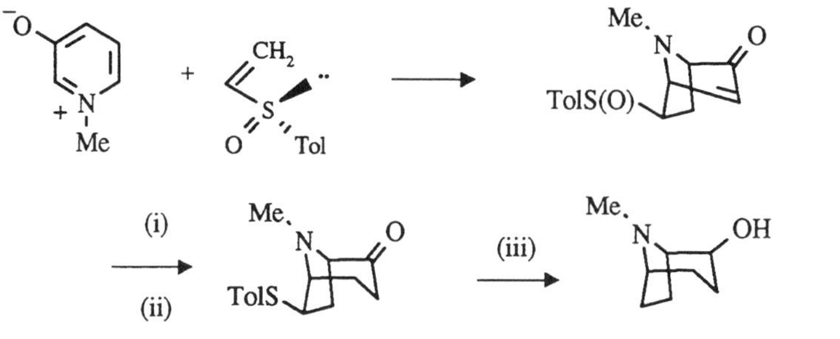

2β-hydroxytropane

Reagents: (i) PBr$_3$ (ii) H$_2$/Pd (iii) Raney Ni

SCHEME XV

Stereochemistry of the Ring Nitrogen in Tropanes

The study of the stereochemistry of tropane alkaloids led to the unexpected recognition that the sequential of introduction of two N-alkyl groups into a nortropane occurs with a remarkable stereoselectivity: the group that enters last, assumes an equatorial position (the A-B rule). These studies, that have involved a large number of tropanium salts, are described in Vol. IVB, C.C.C., 2nd edition; cf. a summary paper by *Fodor et al.* (J. Amer. Chem. Soc., 1971, **93**, 403). With further refinement of NMR techniques and the use of X-ray crystallography these conclusions have received more support, *e.g.*, 6β-hydroxyhyoscyamine base (anisodamine) has the N-methyl group in the equatorial position (*G. Wang et al.*, Chem. Abstr., 1988, **109**, 38009s); and in general, based on the SEL technique most other N-substituted tropane alkaloids show an equatorial preference, ranging from 7:1 to 18:1, depending on the solvent *Glaser et al.* (J. Org. Chem., 1988, **53**, 2172), and *Sarazin et al.* (Magn. Res. Chem., 1991, **27**, 291). NMR spectroscopic data are amply recorded by *M. Loumasmaa and T. Tamminen* (Alkaloids (NY), 1993, **44**, 1).

Established Alkaloids

(-)-6-Hydroxyhyoscyamine has been isolated from *Datura ferox* (*A. Romeike,* Naturwiss., 1962, **49**, 281) and from *Physochlania dubia* (*R. Mirzamatov et al.,* Khim. Prir. Soed., 1972, **8**, 493). Most probably the same compound named "anisodamine" occurs *Przewalskia tangutica* from China and it is also present in at least a dozen plants of the Solanaceae. Two syntheses have been published: a partial synthesis from (-)-hyoscine; followed by a resolution(*G. Fodor, I. Koczor and G. Janzsó,* Arch Pharm., 1962, **295**, 91) and a route from 3,6-tropanediol and acetyltropoyl chloride (*Z. Chen et al.,* Zhongcarya, 1986, **17**, 386). The latter approach received criticism and so a modified synthesis was presented (*C. Zheng and J. Jingxi,* Zhongguo Yijao Gongye Zazhi, 1989, **20**, 990; Chem. Abstr., 1990, **112**, 77642z): starting with rac. 6-methoxymethyl-3-tropanone, reduction to the corresponding α-alcohol, transesterification with methyl hydroxymethylenephenyl acetate, reduction and hydrolysis gave (±)-6-hydroxyatropine (SCHEME XVI). Unfortunately, no steps to resolve this compound were taken.

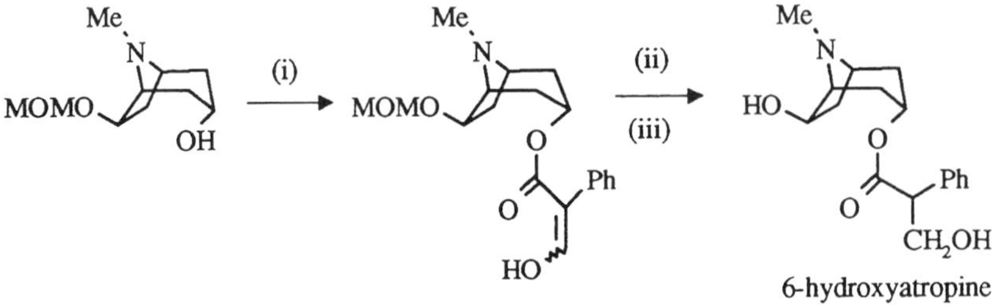

Reagents etc: (i) transesterification with methyl hydroxymethylenephenylacetate
(ii) H_2/Pd (iii) H_3O^+

SCHEME XVI

New Tropane Alkaloids

An almost complete description of all the natural tropanes known up to 1992
was given by *Lounasmaa* (loc. cit.). Some new tropine and pseudotropine
esters, "tropeines" are listed in Table 1, indicating just how many different
types of carboxylic acids form esters with tropanols. A few tropan-3α,6β-diol
esters are also noted in Table 3.

Some 1-hydroxytropanes are listed in Table 2, including (+)-physoperuvine,
the first *seco*-tropane found in *Physalis peruviana* (*A. B. Ray and P. D. Sethi*,
Chem. Ind., London, 1976, 454). This alkaloid was first mis-represented as 3-
methylaminocycloheptanone. However when the hydrochloride was submitted
to X-ray analysis it was found to be the tropane (*A. B. Ray et al.*,
Heterocycles, 1982, **19**, 233).

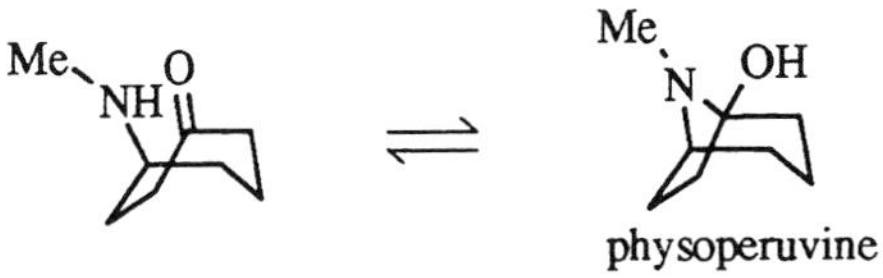

physoperuvine

SCHEME XVII

1-Hydroxytropacocaine is of more recent origin (*J. M. Moore, P. A. Hays,
D. A. Cooper, J. F. Casale and J. Lydon*, Phytochemistry, 1994, **36**, 357) from
the leaves of *Erythroxylum novogratenense*, where it is present to 0.3-0.5%
w/w. It was obtained by ion-pairing and alumina column chromatography. A

mass spectrum taken after esterification of the hydroxyl group by heptafluorobutyric anhydride gave m/z 261 as an apparent molecular ion, 16 amu higher than that of tropacocaine (benzoylpseudotropine); the same increment applied to other ions In addition NMR allowed an unequivocal assignment to the hydroxyl group at C-1. It seems from ^{13}C NMR and IR data that there is no tautomeric ring opening to the corresponding carbinolamine, unlike other 1-hydroxylated tropanes. On these grounds, it is most likely that this compound is a product of oxidative degradation of the long known, mesoid tropane, tropacocaine.

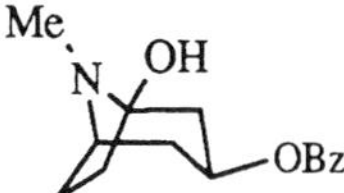

1-hydroxytropacocaine

Calystegines are a group of alkaloids of *Calystegia sepium*; all of which have a hydroxyl group on C-1, and various other hydroxyls on C-2, C-3, C-4 and/or C-6 (36). (*A. Goldmann, J.-Y Lallemand et al.*, Phytochemistry, 1990, **29**, 2125).

Grahamine is a complex alkaloid of *Schizanthus grahamii*. The structure has been determined by an ingenious combination of mass spectrometry and two dimensional NMR (*A. San Martin et al.* Phytochemistry, 1987, **26**, 814; *R. Hartmann, A. San Martin et al.*, Angew. Chem. Int. Ed., 1990, **29**, 366). Fast atom bombardment (FAB) mass spectrometry showed that the molecular mass of grahamine is 871. The ^{1}H-broadband-decoupled ^{13}C NMR spectrum indicated 46 signals, four of which belonged to a monosubstituted benzene ring, therefore 48 carbon atoms were assumed to be present. The DEPT subspectra were consistent with the presence of 7 methyl, 9 methylene, 22 methine, and 10 quaternary carbon atoms. The sum of the mass of these groups was 637, *versus* the total mass of 871. The difference of 234 was accounted for by assuming 3 nitrogen and 12 oxygen atoms. The empirical formula then was deduced to be $C_{48}H_{61}N_3O_{12}$. One and two-dimensional NMR methods revealed that all 61 H atoms were bound to carbons, therefore no hydroxyl or secondary and primary amino groups can be present. The cross signals of an H,H COSY diagram revealed the presence of three 3,6-dihydroxytropane units, also a methylcyclobutane partial structure based on typical CH coupling constants in the coupled ^{13}C NMR spectrum. The connectivity of all six carbonyl carbons was proven in the same way. All these and other data were consistent with two tropane rings linked in the form of a mesaconic diester and that a 2-methyl-4-phenyl-cyclobutane-1,2,3-tricarboxylic acid is at the center of the molecule. The CH COLOC diagram revealed the presence of an angelic ester unit. Combination of all these partial structures gave the structure of grahamine as shown below. That grahamine contains a cyclobutane ring, substituted by 3-carboxyl group is indicative of its formation from a cinnamyltropeine and mesaconic acid by a 2+2 cycloaddition, therefore it is related to the truxillines, but lacks an ecgonine unit.

grahamine

Truxillines Early work on tropane alkaloids, especially from Peruvian *Coca* leaves (*C. Liebermann and W. Drury*, Ber., 1889, **22**, 680) led to the isolation of two isomers of an alkaloid $C_{38}H_{46}N_2O_3$. One was hydrolyzed to 2 moles of (-)-ecgonine, 2 moles of methanol and one mole of "truxillic acid" and it could be re-synthesized from the same fragments. It was named α-truxilline. Another isomer was described as β-truxilline. The truxillic acids were shown to be 1,3-diphenylcyclo-butane-2,4-dicarboxylic acids, i.e. dimers of cinnamic acid. Despite this early work, 100 years elapsed before all 11 possible stereoisomers of truxillines were finally identified by *J. M. Moore* and his associates at the Drug Enforcement Administration in the USA (*J. M. Moore et al.*, J. Chromatography, 1987, **410**, 297). Capillary gas chromatography was used with electron capture as the detection system to identify five stereoisomeric diphenylcyclobutane carboxylic acids and their alkaloidal precursors, α, β, γ, δ, and ε truxillines, in illicit cocaine samples, and in a sample of Bolivian *Coca* leaves. Later the presence of the five remaining possible truxillines, namely of epi, peri, neo, μ, ω, ξ, truxillines was proven both in illicit cocaine samples and in *Coca* leaves (see TABLE IV).

Workers in Britain (*W. C. Evans, M. S. Al Said and R. J. Grout,* J. Chem. Soc. Perkin, 1986, 957) found two new derivatives of truxillines. In the first one the ecgonine moiety was replaced by the 3-tropanyl residue and in the other by an 6β-acetoxy-3-tropanyl group. In the second example the 6β-hydroxy-3-tropanyl residue and the 6β-acetoxytropanyl group replaced the two ecgoninyl groups. Both new alkaloids were found in the leaves of *Erythroxylum hypericifolium.* Unfortunately, no stereochemical information was given to confirm which truxillic acid is present and also the 3,6-dihydroxytropane assignment is arbitrary.

γ-Pyronotropanes. (+)-Bellendine, the first pyronotropane, was isolated from the flowers of an Australian shrub, *Bellendena montana* of the Proteaceae family (*I. R. C. Bick et al.,* Phytochemistry, 1971, **10**, 475). The constitution and the configuration have been elucidated by *Motherwell, Bick et al.* (Chem. Commun., 1971, 134). Racemic bellendine was synthesized first, by *Bick et al.* (Tetrahedron Letters, 1973, **50**, 5059) from tropinone and 3-methoxy-2-methylpropenoyl chloride, a variation is to use acroyl chloride in the same reaction and to oxidise the product by bromination and dehydrobromination (SCHEME XVIII).

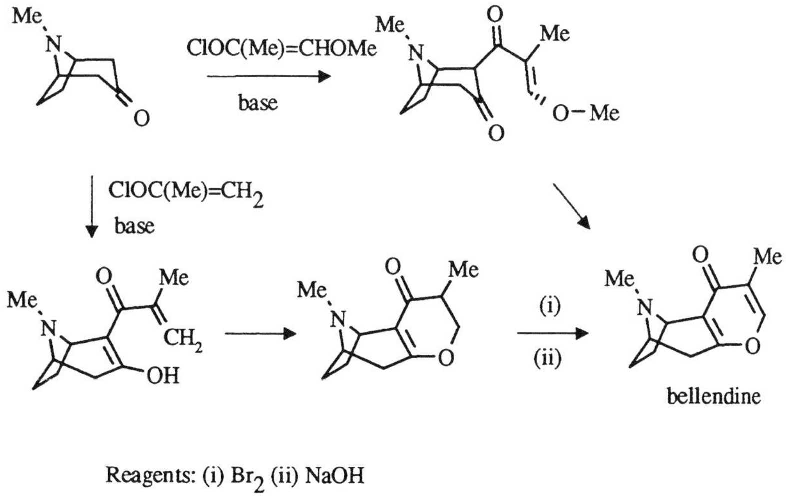

Reagents: (i) Br_2 (ii) NaOH

SCHEME XVIII

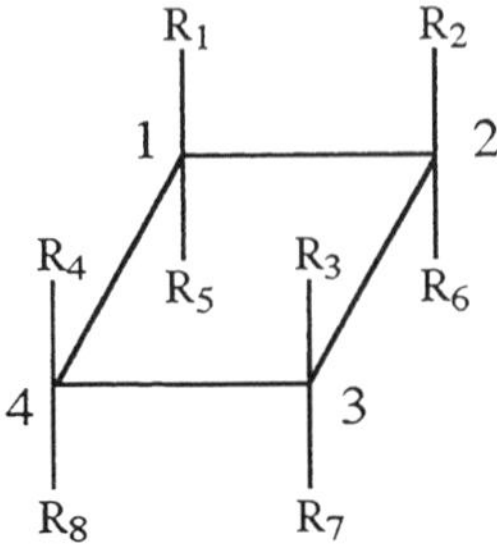

truxillines

1) α (alpha)-, R1 = R7 = Me ecgonine ester, R4 = R6 = phenyl, R2 = R3 =R5 = R8 =H

2) β (beta)-, R5= R6 = Me ecgonine ester, R3 = R4 = phenyl, R1 = R2 =R7 = R8 = H

3) δ (delta)-, R2 =R5 = Me ecgonine ester, R4 = R7 = phenyl, R1 = R3 = R6 = R8 = H

4) ε (epsilon)-, R5 = R7 = Me ecgonine ester R2 = R4 = phenyl, R1 = R3 = R6 = R8 = H

5) μ (mu)- R1 = R6 = ecgonine ester, R4 = R7 = phenyl, R2 = R3 = R5 = R8 = H

6) γ (gamma)-, R1 = R3 = Me ecgonine ester, R4 = R6 = phenyl, R2 = R5 = R7 = R8 = H

7) (neo)-, R2 = R5 = Me ecgonine ester, R3 = R4 = phenyl, R1 = R6 = R7 = R8 = H

8) ζ (zeta)-, R5 = R6 = Me ecgonine ester, R4 = R7 = phenyl, R1 = R2 = R3 = R8 = H

9) (epi)-, R1 = R7 = Me ecgonine ester, R2 = R4 = phenyl, R3 = R5 = H

10) (peri)-, R1 = R3 = Me ecgonine ester, R2 = R4 = phenyl, R5 = R6 = R7 = R8 = H

11) ω (omega)-, R1 = R2 = Me ecgonine ester, R3 = R4 = phenyl, R5 = R6 = R7 = R8 = H

Table IV

An improved general method based on this work has been described
(*M. Lounasmaa, C. Holmberg and T. Langenskiold*, Planta Medica, 1983, **48**,
56) in which the propenoyl chloride is replaced by an acyl cyanide.
Unfortunately, the absolute configuration of bellendine remains unknown.
The same authors describe the synthesis of (±)-isobellendine from the
morpholine enamine of tropinone with diketene (SCHEME XIX).

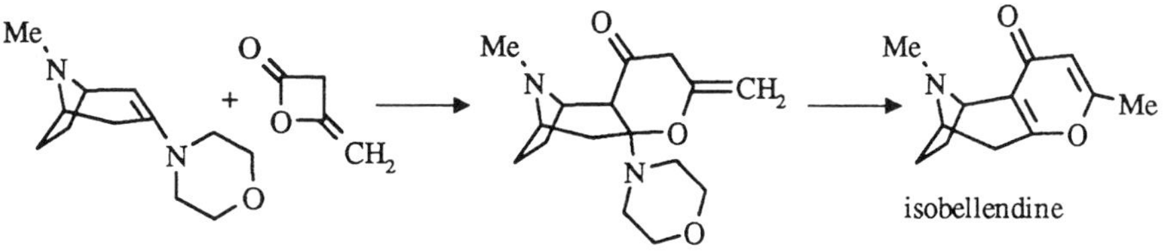

isobellendine

SCHEME XIX

An enantioselective synthesis of γ-pyronotropenes has been outlined
(*M. Majewski and R. Lazny*, Tetrahedron Letters, 1994, **35**, 3653). Here the
deprotonation of tropinone with a chiral N-lithiated amine, followed by
acylation with 2-bromo-3-butenoyl cyanide gave a pyronotropene in one
practical step. Unfortunately the wrong enantiomer formed, but in high optical
purity (SCHEME XX). In similar fashion the non-natural enantiomers of
darlingine, chalkostrobamine and isobellendine were also formed

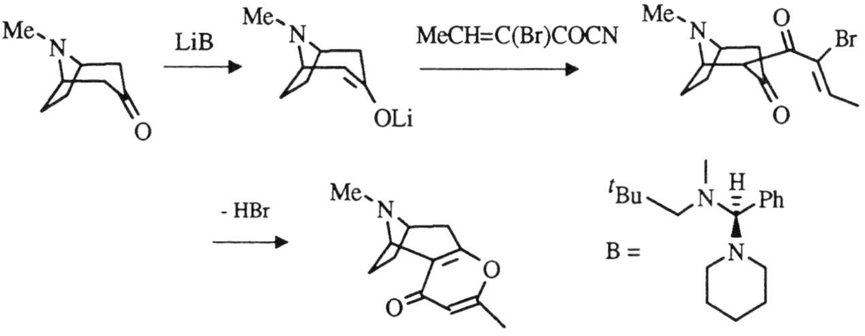

SCHEME XX

New Cocaine Alkaloids

6-Hydroxycocaine has been isolated from Coca leaves by high resolution chromatography and characterized by mass spectrometry (*J. M. Moore et al.*, J. Chromatography, 1987, **410**, 297).

Biosynthesis

The biogeneses of tropinone, the esters of both tropine and pseudotropine and that of cocaine are clear. This knowledge is due, in major part, to the work of (the late) *Edward Leete* (reviewed in Planta Med., 1990, **56**, 339). The common key intermediate is 2-pyrrolinium salt (1), which is formed from enzyme bound putrescine, *via* a N-methylputrescine. The pyrrolinium salt adds acetoacetylcoenzyme A to form the β-ketoester (2), which on the one hand, is decarboxylated to (2*R*)-hygrine (3). This in turn, by a *Michael* type reaction of the corresponding pyrroline (4), cyclizes to tropinone (5). On the other hand, the β-ketoester (2) is also oxidized to the pyrrolinium salt (6), which can undergo ring closure by the same *Michael* pathway to methyl tropinone-2-carboxylate (7). After reduction of (7) to methyl ecgonine, benzoylcoenzyme then acylates the hydroxy group yielding cocaine (8) (SCHEME XXI).

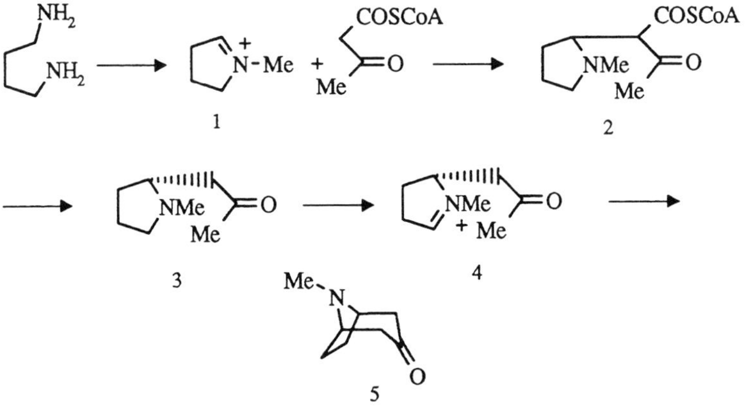

SCHEME XXI continued over page

COSCoA   COSCoA

2   **6**

7   **8**

SCHEME XXI

The biosynthesis of the third, widespread, main-alkaloid, scopolamine and its optically active component, hyoscine occurs through secondary reactions. (The late) *Kurt Mothes*, pioneer in the alkaloid biosynthesis field found that hybrid *Datura ferox* scions on *Cyphomandra betacea* constitute a so-called alkaloid-free plant. Although this hybrid does not produce any tropane alkaloids itself, it is able to interconvert tropane alkaloids which are fed to it. Indeed, if hyoscyamine is fed to this plant, hyoscine can be isolated, which means that the enzyme system is able to replace two H's on the pyrrolidine ring by an epoxide oxygen atom (*A. Romeike,* Flora (Jena), 1956, **143**, 67). On the other hand, by using the[14]C-labeled hyoscyamine (10) in young *Datura stramonium* seedlings, 27% of the radioactivity was recovered in the hyoscine (11) and 17% in the 6β-hydroxyhyoscyamine (12) that the plants produced (*A. Romeike and G. Fodor*, Tetrahedron Letters, 1960, **22**, 1). The reasonable assumption was that hyoscyamine is oxidized first to alkaloid V (6β-hydroxyhyoscyamine). This alkaloid was isolated from *Datura ferox* (*A. Romeike,* Naturwiss., 1962, **49**, 281) and also from *Physochlaina dubia* (*R. Mirzamatov et al.*, Khim. Prirod. Soed., 1972, **8**, 493) and later rediscovered by a Chinese group when it was isolated from *Anisodus tangutica* (*X. Jingsi et al.,* Acta Pharm. Sinica, 1980, **15**, 409) and named "anisodamine". It was supposed that 6-hydroxyhyoscyamine undergoes dehydration to 6-tropenyl-3(*S*)-tropoylate "dehydrohyoscyamine", which is then epoxidized to hyoscine (*G. Fodor*, Alkaloids, 1967, **9**, 298; *E. Leete and D. H. Lucast*, Tetrahedron Letters, 1976, 341). However, an enzyme has been isolated from *Hyoscyamus niger* which carries out the direct oxygenation of the hyoscyamine derivative (15) to the 6β-hydroxyhyoscyamine (16) (*T. Hashimoto and Y. Yamada*, Plant Physiol., 1986, **81**, 619; Eur. J. Biochem., 1987, **164**, 277). The existence of this enzyme disproves the above assumption. Furthermore, when ^{18}O-6-hydroxyhyoscyamine (13) was fed to a shoot culture of *Duboisia myoporoides*, the whole of the label was recovered in the hyoscine (14) which was isolated (*T. Hashimoto, J. Kohno and Y. Yamada*, Plant Physiol., 1987, **84**, 144; Phytochemistry, 1989, **28**, 1077) (SCHEME XXII). Therefore insertion of the 6β-hydroxyl in the "alkaloid free plant" only leads to the levorotatory form of atropine, *i.e.*, (-)-hyoscyamine is converted to the epoxide (*Romeike*, loc. cit.).

SCHEME XXII

Biological Activity of Some Tropane Alkaloids

A full discussion of the biological effects of these alkaloids is inappropriate here, however, since cocaine is a drug of abuse a basic knowledge of its effects are of interest to the chemist. Similarly the properties of `harmless' alkaloids atropine (hyoscyamine) and scopolamine (hyoscine) should not be ignored. Atropine that has a mydriatic effect, is still being used in ophthalmology (*J. L. Kohrick et al.*, Aviat. Space Environ. Med., 1990, **61**, 622). In addition, its importance in cardiology has been reviewed (*A. Hollman,* Br. Heart J., 1991, **66**, 367).

Atropine. Atropine is useful in counteracting the toxic effects of organophosphorus agents because it helps to reactivate cholinesterase levels (*N. N. Finer,* Crit. Care Med., 1990, **18**, 1307). Atropine also delays the onset of organophosphorus-induced delayed neurotoxicity (*A. Mortensen and O. Lodefoged,* Neurotoxicology, 1992, **13**, 347). Atropine has a central action on cardiovascular reflexes in humans (*S. J. Wall et al.,* J. Pharmacol. Exp. Ther., 1992, **262**, 584).

Cocaine. Cocaine is a strong central nervous system stimulant that is either injected as its hydrochloride or sniffed as a base, crack-cocaine. The toxicity of the alkaloid has been recently reviewed (*D. W. Yu et al.,* J. Med. Chem., 1992, **35**, 2178). Cocaine self administration can be controlled by treating drug dependence with concurrent administration of an agonist with an antagonist. Cocaine affects lung, heart, kidney and particularly, the brain. In view of all these, and many other ill-effects, a momentous struggle is going on to combat cocaine abuse (see *R. L. Clarke,* Alkaloids (NY), 1977, **16**, 147-153; *J. M. Moore, J. F. Casale, G. Fodor and A. B. Jones,* Forensic Science Review, 1995, **7**, 77-102; Cocaine, an Annotated Bibliography, Vols. 1-2 (*Carlton Turner,* Ed., University Press of Mississippi, 1988)).

Scopolamine and Hyoscine. While scopolamine was used for some time to prevent motion sickness, its major application these days is as an amnesia model in experimental animals. A general paper was published on scopolamine, cognition and dementia (*A. Parrott,* J. Psychopharmacol., 1992, **6**, 541). Transdermal scopolamine patches stop nausea and vomiting, caused by alfentanil anesthesia (*V. Bosek and J. B. Downs,* Anesthesiol. Rev., 1992, **19**, 19).

Anisodamine (Alkaloid V; 6β-hydroxyhyoscyamine) was studied recently in China and a property of dilating arterioles and improving microcirculation was reported (*X. Jing-xi et al.,* Chemistry of Natural Products, Proceedings American-Sino Symposium, 1980, pp. 131-134). Fulminant epidemic meningitis, toxic bacillary dysentery and septic shock were successfully treated.

Quaternary salts of tropanes do not occur in Nature, except for N-oxides, however, several hundreds of these derivatives have been synthesized. Thus, quaternary tropanium salts have been subject to many pharmacological studies (Arzneimittelforschung, 1976, **26**, 959; *R. C. Clarke,* Alkaloids (NY), 1977, **16**, 151). Among the large number of synthetic tropanium salts a new bronchospasmolytic agent, N-methyl-N-2-fluoroethylnortropanium bromide, has been developed. Both N-epimers of this compound have been described. They show different bioactivities (*R. Banholzer et al.,* Arzneim.-Forsch., 1986, **36**, 1161).

The biological activities and pharmacological applications of tropane alkaloids, including those of non-natural tropanes, have been summarized regularly in Natural Products Reports (*G. Fodor and R. Dharanipragada*, 1988, **5**, 67; 1990, **7**, 539; 1991, **8**, 603; 1993, **10**, 199; 1994, **11**, 443).

Second Supplements to the 2nd Edition of Rodd's Chemistry
of Carbon Compounds, Vol.IV B, edited by M. Sainsbury
© 1997 Elsevier Science B.V. All rights reserved.

Chapter 12

PYRROLE PIGMENTS

KEVIN M. SMITH

1. Scope

For the purposes of this review, pyrrole pigments are classified as porphyrins, chlorins, phlorins, corroles, sapphyrins, pentaphyrins, porphycenes, corrins, and open-chain biliverdins. The review is based upon a Chemical Abstracts search performed on January 3, 1996. The search period was January 1, 1985, through December 31, 1995; the simple search terms "pigment" with "synthesis" or "reaction" gave several thousand retrievals. Of necessity, therefore, the review deals with carefully selected highlights, since 1985, of the pyrrole pigment literature; it is unfortunate that it will not be possible to mention all of the significant contributions which have been made, in the past ten years, to the field of pyrrole pigments.

2. Nomenclature

The IUPAC system for numbering of the chromophores in pyrrole pigments will be used throughout. Typical examples (**1-4**) are given below. In the case of the chlorin macrocycle **2**, Fischer's tradition of presenting the reduced pyrrole subunit as ring D will be maintained.

3. Porphyrins

(a) Meso-Tetra-arylporphyrins and Other Meso-arylporphyrins

A very mild method for syntheses of 5,10,15,20-tetra-arylporphyrins (e.g. tetraphenylporphyrin, **5**), which gives improved yields, has been reported by Lindsey and coworkers (J.S. Lindsey et al., Tetrahedron Lett., 1986, **27**, 4969; J. Org. Chem., 1987, **52**, 82; Tetrahedron, 1994, **50**, 8941); the approach involves a two step procedure in which pyrrole and arylaldehyde are condensed at high dilution with trace acid catalysis (e.g. $BF_3 \cdot Et_2O$) to give a porphyrinogen, followed by oxidation to porphyrin, usually using a high-potential quinone. A number of other investigators have described similar two-step approaches (A.M.d'A.R.

Gonsalves et al., J. Heterocycl. Chem., 1985, **22**, 931; 1991, **28**, 635; A.W. Van der Made et al., Recl. Trav. Chim. Pays-Bas, 1988, **107**, 15). Lindsey et al. (J. Org. Chem., 1994, **59**, 579) also showed that the troublesome high dilution used in their approach can be offset, to a certain extent, by increase in the concentration of the acid catalyst, but 5,10,15,20-tetra-arylporphyrins can also be made at higher working concentrations by using high-valent transition metal salts (A. Gradillas et al., J. Chem. Soc., Perkin Trans. 1, 1995, 2611). Using $BF_3 \cdot EtOH$ co-catalysis, ortho-substituted tetra-arylporphyrins, such as 5,10,15,20-tetra-mesitylporphyrin can be synthesized (J.S. Lindsey and R.W. Wagner, J. Org. Chem., 1989, **54**, 828; R.W. Wagner et al., Tetrahedron Lett., 1991, **32**, 1703). A new route to an number of meso-aryl and meso-alkyl porphyrins has also been reported (Y. Kuroda et al., Tetrahedron Lett., 1989, **30**, 2411).

Porphyrin
1

Chlorin
2

Pheophorbide
3

Biliverdin
4

Using a modification of the well-known tetraphenylporphyrin synthesis, two groups have isolated and identified so-called "N-confused" porphyrins **6**, in which one of the monopyrrole subunits is inverted, as a by-product along with tetraphenylporphyrin **5** itself (H. Furuta et al., J. Am. Chem. Soc., 1994, **116**, 767; P.J. Chmielewski et al., Angew. Chem. Int. Ed. Engl., 1994, **33**, 779).

5   6

Making use of its mesoporosity, a number of investigators have used Montmorillonite K10 clay as an acid catalyst in the synthesis of 5,10,15,20-tetra-arylporphyrins (P. Laszlo and J. Luchetti, Chem. Lett., 1993, 449; M. Onaka et al., Tetrahedron Lett., 1993, **34,** 2625; Stud. Surf. Sci. Catal., 1994, **90**, 85; T. Shinoda et al., Chem. Lett., 1995, 493).

8

[R]

7

The task of synthesizing 5,10,15,20-tetra-arylporphyrins (e.g. **7**) in which all four meso-aryl groups are different and are also arranged in a unique pre-defined array, is daunting; however, making use of the basic aroylpyrrole building blocks **8**, such a synthetic approach has been accomplished in good overall yield (D.M. Wallace and K.M. Smith, Tetrahedron Lett. 1990, **31**, 7265; D.M. Wallace et al., J. Org. Chem., 1993, **58**, 7245). Some years later, C.-H. Lee et al. (Tetrahedron, 1995, **51**, 11645) also reported accomplishing the same objective using very similar methodology.

It is possible to obtain a 92% abundance of Collman's now famous $\alpha,\alpha,\alpha,\alpha$-5,10,15,20-tetra-aryl- ("picket-fence") porphyrin by heating zinc(II) 5,10,15,20-tetra(o-nitrophenyl)porphyrins in naphthalene for four days, followed by reduction of the nitro groups to amines (E. Rose et al., J. Org. Chem., 1995, **60**, 3919). $\alpha,\beta,\alpha,\beta$-Picket fence and "basket-handle" porphyrins have been synthesized, and the relative rigidity in the systems has been investigated (J.P. Renaud et al., New J. Chem., 1987, **11**, 279). Basket handle porphyrins were first prepared by Momenteau and his collaborators in Orsay, France; a number of new syntheses of heavily facially encumbered basket handle porphyrins have been reported (M. Momenteau et al., J. Chem. Soc., Perkin Trans. 1, 1988, 283; P. Maillard et al., J. Chem. Soc., Perkin Trans. 1, 1988, 3285), including "inter-locked" basket handle examples (M. Momenteau et al., Tetrahedron Lett., 1994, **35**, 3289) and the ligation characteristics of various metal derivatives have been studied. Using prefunctionalized benzaldehydes, tetra-arylporphyrins (e.g. **9**) with facially encumbering groups have been synthesized (R.W. Wagner et al., Tetrahedron, 1994, **50**, 11097); so-called "bis-pocket" porphyrins (e.g. **10**) bearing carboxylic acid functions on both faces of the macrocycle have also been obtained via hydrolysis of the corresponding arylnitriles (Z. Gross and I. Toledano, J. Org. Chem., 1994, **59**, 8312).

9

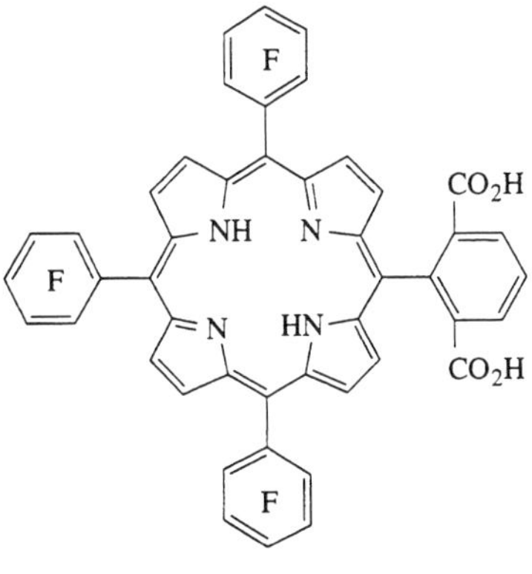

10

"Capping" the face of a porphyrin is another way of inducing selective facial encumbrance, and along these lines a number of capped porphyrin systems bearing aryl rings have been synthesized and studied as potential models for active sites of heme proteins (W. Ma et al., J. Org. Chem., 1993, **58**, 6349; 1995, **60**, 8081). Other examples of encumbered tetra-arylporphyrins reported during the review period include tetraanthracenylporphyrins (H. Volz and H. Schaeffer, Chem.-Ztg. 1985, **109**, 308), aryl-benzyloxy analogues (N. Datta-Gupta et al., J. Heterocycl. Chem., 1987, **24**, 629), tetraphenylporphyrins mono-substituted with glycerol ester units (J. Sliwiok and P. Kus, Monatsh. Chem., 1992, **123**, 1149), tetrasubstituted with sugar units (P. Maillard et al., J. Am. Chem. Soc., 1989, **111**, 9125) (which turned out to be unstable) (P. Maillard, Tetrahedron Lett., 1992, **33**, 8081), tetrabiotinylated-tetrapyridinium porphyrins (H. Fukushima et al., Langmuir, 1995, **11**, 3523), a 5,10,15,20-tetra-[(4-diphenylphos-phino)phenyl]porphyrin and a caged double decker porphyrin (G. Maerkl et al., Angew. Chem. Int. Ed. Engl., 1995, **34**, 2230), and tetraphenylporphyrin peptides (S.E. Matthews et al., J. Chem. Soc., Chem. Commun., 1995, 1809). A route to D4-symmetric chiral tetra-arylporphyrins **11** derived from R-(+)-nopinone has been described (J. Barry and T. Kodadek, Tetrahedron Lett., 1994, **35**, 2465). The interestingly named "octopus" porphyrin has also been synthesized in which four alkyl phosphocholine groups have been placed on each side of an amphiphilic tetraphenylporphyrin (T. Komatsu et al., J. Chem. Soc., Chem. Commun., 1993, 728). Syntheses and properties of a number of cationic (water soluble) tetra-arylporphyrin derivatives have also been described (D.K. Lavalee et al., J. Org. Chem., 1993, **58**, 6000; T. La et al., Inorg. Chem., 1994, **33**, 3159; S.V. Vodinskii et al., Zh. Org. Khim., 1989, **25**, 1529). As if to emphasize, adopt and adapt to the

282

current frantic interest in "buckyball" chemistry, a tetraphenylporphyrin has been conjugated with fullerene C60 (T. Drovetskaya et al., Tetrahedron Lett., 1995, **36**, 7971); through space interaction between the two chromophores was established using electronic spectroscopy.

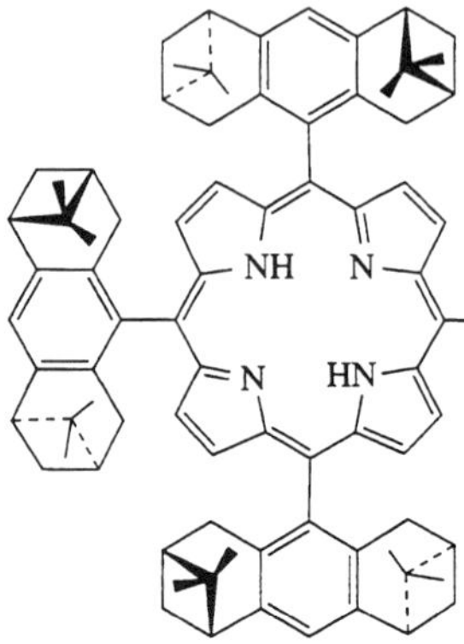

11

Reaction of N-aryl or N-styryl-5,10,15,20-tetraphenylporphyrins with amine cation radicals (or under anodic oxidation conditions) affords N,N'-1,2-phenylene or N,N'-1,2-vinylidene bridged porphyrins (H.J. Callot et al., J. Chem. Soc., Chem. Commun., 1986, 767; J. Am. Chem. Soc., 1987, **109**, 2946).

A number of other syntheses of meso-aryl "capped" or strapped porphyrins have been reported, either rigidly capped (J.A. Wytko et al., J. Org. Chem., 1992, **57**, 1015), calixarene capped (T. Nagasaki et al., Chem. Lett., 1994, 989), bis-calixarene strapped (D. Rudkevich et al., Tetrahedron Lett., 1994, **35**, 7131), steroid capped (R.P. Bonar-Law and J.K.M. Sanders, J. Chem. Soc., Chem. Commun. 1991, 574), or hydrocarbon-encapsulated (W.F.K. Schatter et al., Tetrahedron, 1991, **47**, 8687). A quinone capped porphyrin has been obtained by way of a self-photosensitized one-pot reaction (A. Osuka et al., Chem. Lett., 1986, 479). Other examples of strapped and capped porphyrins include a doubly strapped porphyrin which was, as expected, reluctant to undergo metalation (A. Osuka et al., Chem. Lett., 1991, 1687; T. Nagata, Bull. Chem. Soc. Jpn., 1992, **65**, 385), a chiral bistetralin-strapped twin coronet porphyrin (Y. Naruta et al., Chem. Lett., 1991, 1933), a bicyclo[2,2,2]octane-capped tetra-arylporphyrin (H. Zhang et al., J. Am. Chem. Soc., 1992, **114**, 6621), and extremely short chain basket-handle porphyrins (U. Simonis et al., J. Am. Chem. Soc., 1987, **109,** 2659).

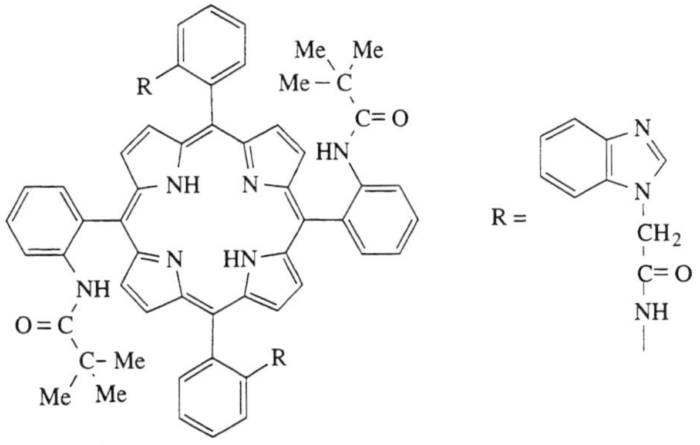

"α,α,α,α-atropisomer"

12

A tetra-arylporphyrin **12** bearing two appended benzimidazole arms, termed a "pincer-porphyrin" has been synthesized, and the conformational preferences of the superstructure have been investigated (S.J. Rodgers et al., Inorg. Chem., 1987, **26**, 3647). In a similar vane, a porphyrin-based "molecular tweezer" **13** bearing two benzo-crown ethers has been prepared and shown to bind bipyridinium guest molecules (M.J. Gunther and M.R. Johnston, Tetrahedron Lett., 1992, **33**, 1771).

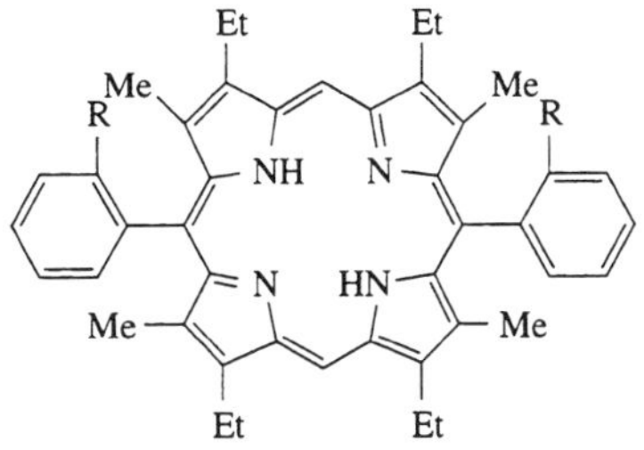

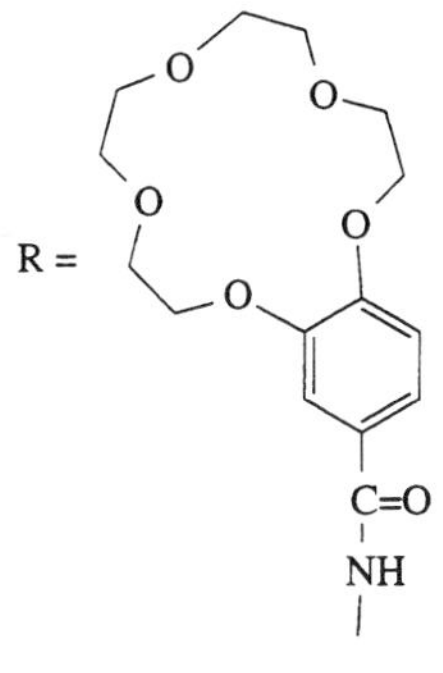

"α,α-atropisomer"

13

Cofacial-porphyrin systems are of great interest in connection with π-interactions between the two stacked chromophores. In the tetra-arylporphyrin series, a "shopping basket" bis-porphyrin was synthesized (R. Karaman et al., J. Org. Chem., 1992, **57**, 2169), and Collman's group has reported facile syntheses of meso-tetra-aryl cofacial ("Pac-man"-

284

type) diporphyrins (J.P. Collman et al., J. Org. Chem., 1995, **60**, 1926); the method proceeds in a rational step-wise fashion, and provides access to a number of cofacial hetero- and homo-diporphyrin ligands, the metal salts of which appear to display interesting catalytic chemistry. Sterically hindered strapped porphyrins have been prepared (C.H. Lee and C.K. Lee, Bull. Korean Chem. Soc., 1992, **13**, 352). Synthetic routes to linear bridged porphyrin oligomers have also been described (A. Osuka et al., Chem. Lett., 1993, 1505), while Crossley's group has reported the synthesis of rigid laterally-bridged bis-porphyrins, e.g. **14**, (M.J. Crossley and P.L. Burn, J. Chem. Soc., Chem. Commun., 1987, 39) and a very ingenious bisporphyrin possessing a large chiral cavity with a bimetallic binding site (M.J. Crossley et al., J. Chem. Soc., Chem. Commun., 1995, 1077). The possibility of synthesizing porphyrins in which appended (non-meso) substituents can be directed towards the porphyrin center (and an active metal catalytic center) has been investigated and accomplished by way of a 5,10,15,20-tetra-arylporphyrin bearing a 2-(o-hydroxyphenyl group) **15** (M.J. Crossley et al., Tetrahedron Lett., 1988, **29**, 1597); the 2-substituent was attached by nucleophilic attack upon the corresponding 2-nitroporphyrin **16**, a substituted porphyrin which has been of great interest in Crossley's group in recent times.

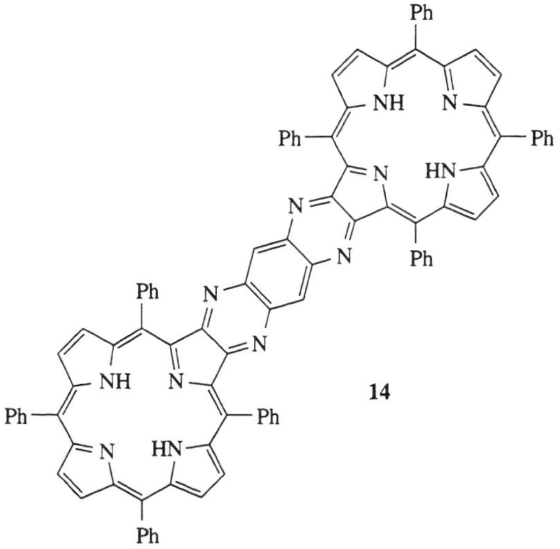

Tetra-arylporphyrins have also been featured in several papers in which large numbers of boron atoms have been attached; the eventual use of such highly boronated porphyrins is associated with boron neutron capture therapy of tumors, and this field has been reviewed (D. Gabel et al., Neutron Capture Ther., Proc. Int. Symp., 2nd 1986, pp 37-45, H. Hatanaka, ed., Nishimura, Niigata, Japan; S.B. Kahl and P. Mica, ibid., pp 61-67). Porphyrins with phenylboronic acid groups have been synthesized (H. Toi et al., Chem. Lett., 1993, 1043), while certain tetra-aryl and other carboranylporphyrins (e.g. **17,18**) (M. Miura et al., Tetrahedron Lett., 1990, **31**, 2247) and **19** (A.S. Phadke and A.R. Morgan, Tetrahedron Lett., 1993, **34**, 1725) have been prepared and tested; see also S.B. Kahl et al. (Adv. Neutron Capture Ther., Ed. A. Soloway et al., Plenum, New York, 1993, 301).

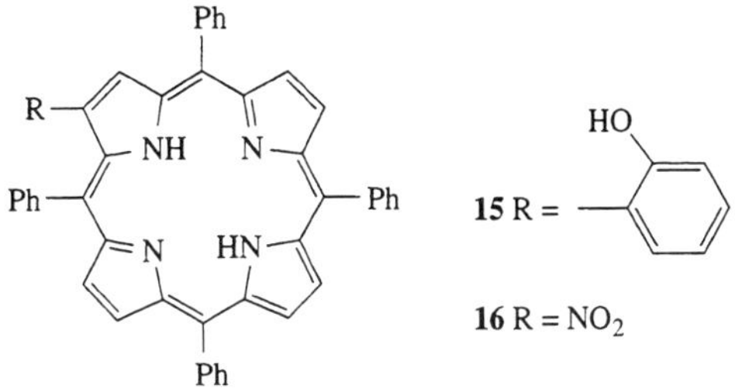

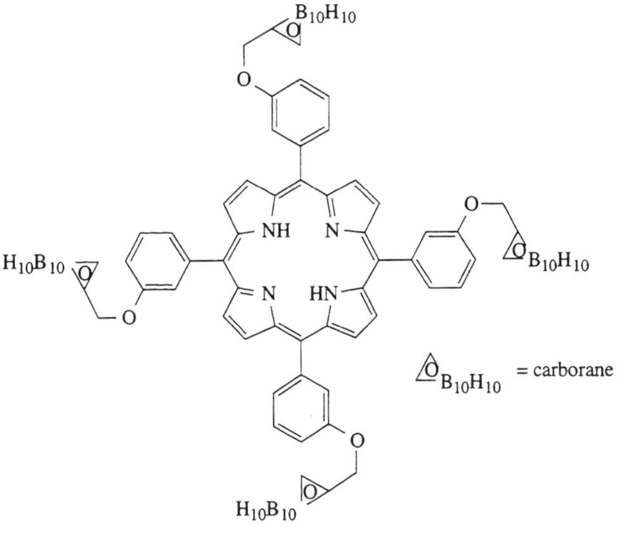

15 R =

16 R = NO$_2$

= carborane

17

An efficient synthesis of 5-substituted dipyrromethanes, and their use in the preparation of a number of 5,15-disubstituted porphyrins have been reported (C.-H. Lee and J.S. Lindsey, Tetrahedron, 1994, **50**, 11427; D.

A. Lee et al., Heterocycles 1995, **40**, 131). A "direct" synthesis of 5-aryldipyrromethanes has also been described (S.J. Vigmond et al., Tetrahedron Lett., 1994, **35**, 2455). 5-Aryldipyrromethanes have been used for general syntheses of a number of interesting 5-monoaryl, 5,15-

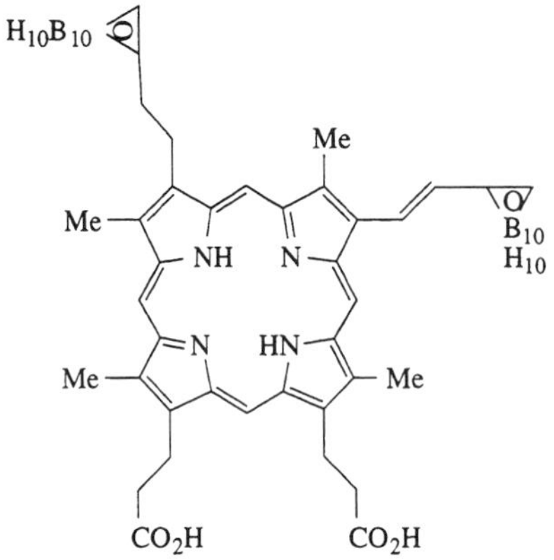

18

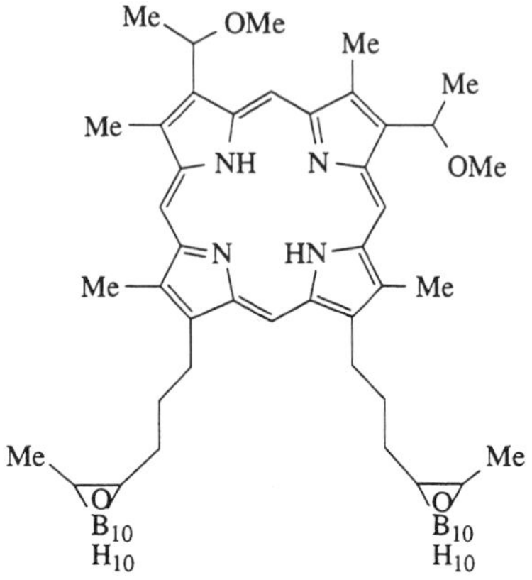

19

disubstituted and 5,10,15-trisubstituted porphyrins using the MacDonald-type "2+2" methodology, or variations thereof; examples include **20** (H.K. Hombrecher et al., Tetrahedron, 1992, **48**, 9451), **21** (J. Zindel and D.A. Lightner, J. Heterocycl. Chem., 1995, **32**, 1219) and **22** (T. Ema et al., Tetrahedron Lett., 1991, **32**, 4529).

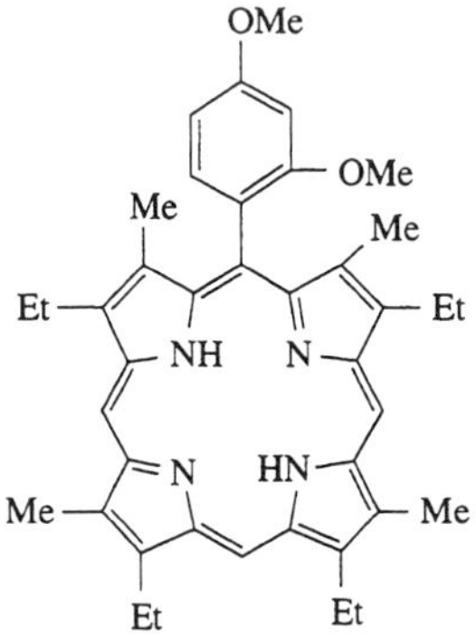

20

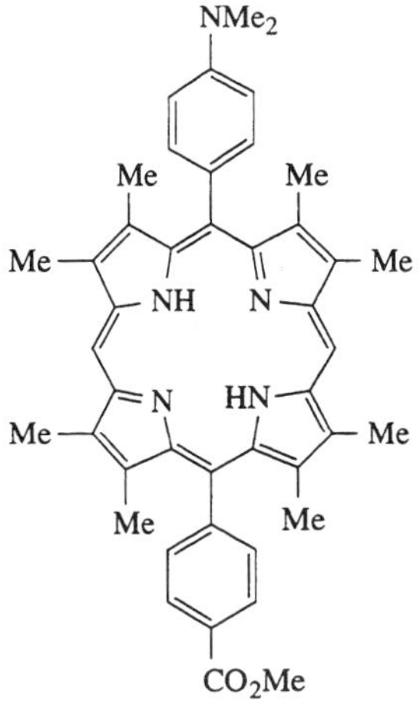

21

Iodination and Heck alkynation of 5,15-diphenylporphyrins has been shown to afford 10-alkynyl-5,15-diphenylporphyrins (e.g. **23**) (R.W. Boyle et al., J. Chem. Soc., Chem. Commun., 1995, 527). 1,4-Bis-dipyrromethanyl-benzenes have also been used for the syntheses of both symmetrical and also unsymmetrically substituted 5,15-diarylporphyrin

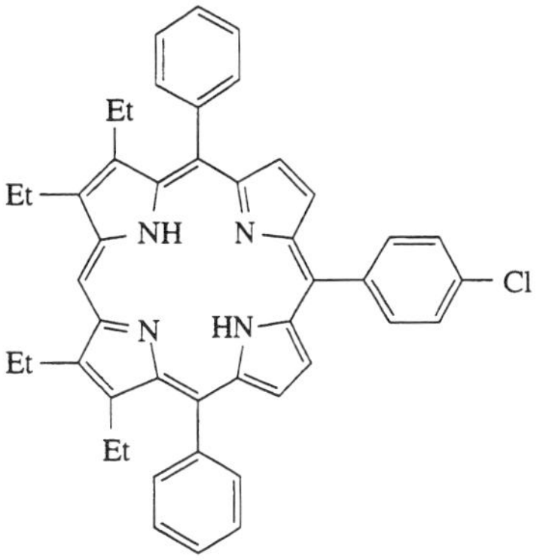

22

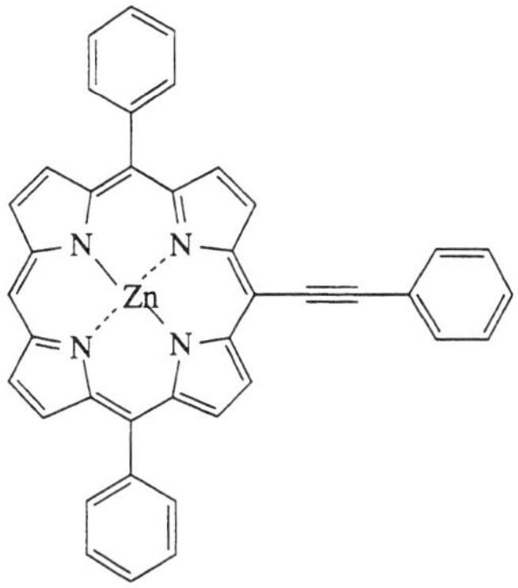

23

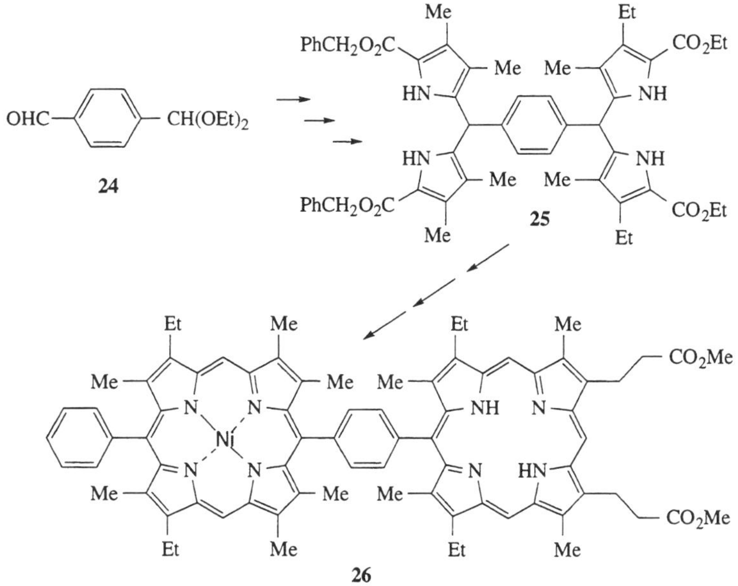

24

25

26

pseudo-dimers (e.g. J.L. Sessler et al., J. Am. Chem. Soc., 1990, **112**, 9310; R. K. Pandey et al., Tetrahedron Lett., 1992, **33**, 5315; R. Paolesse et al., J. Am. Chem. Soc., 1996, **118**, 3869); for an unsymmetrical example see **24** → **26** (R.K. Pandey et al., Tetrahedron Lett., 1992, **33**, 5315).

5,15-Bis-(8-quinolinyl)- and 5,15-bis-(2-pyridyl)-porphyrins have been prepared, the former giving stable and isolable atropisomers (Y. Aoyama et al., Tetrahedron Lett., 1987, **28**, 2143); chiral octaethylporphyrins bearing 5,15-(2-hydroxynaphthyl)- or 2-hydroxyphenyl-groups have been synthesized and resolved into their individual enantiomers using HLPC (H. Ogoshi et al., Tetrahedron Lett., 1986, **27**, 6365; T. Mizutani et al., J. Am. Chem. Soc., 1994, **116**, 4240). Finally, reaction of the 5,15-diarylporphyrin **27** with 1,4-(ClOC)$_2$-benzene afforded a very rigid so-called "gyroscope" porphyrin due to linkages of an opposite pair (or pairs) of amino groups (B. Boitrel et al., J. Chem. Soc., Chem. Commun., 1985, 1820).

27

Montmorillonite K10 clay has been shown to be useful and mild in the synthesis of meso-tetra-*alkyl*porphyrin systems (M. Onaka et al., Chem. Lett., 1993, 117) as well as certain bilanes and porphyrinogens of biological interest (C. Pichon and A.I. Scott, Tetrahedron Lett., 1994, **35**, 4497).

There has been an explosion in interest with regard to porphyrins which possess large numbers of sterically crowding substituents, mostly for the reason that steric strain is relieved in these systems by the adoption, by the macrocyclic core, of highly non-planar conformations. Typical examples of non-planar dodecasubstituted porphyrins are derivatives of octaethyltetraphenylporphyrin **28** and dodecaphenylporphyrin **29**.

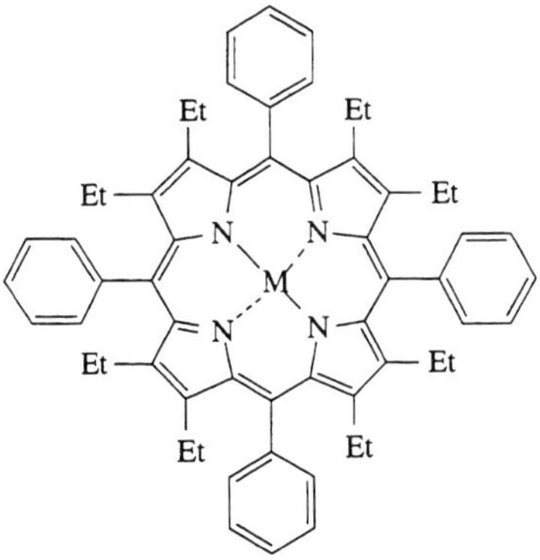

28 M = 2H
30 M = Cu
33 M = Zn

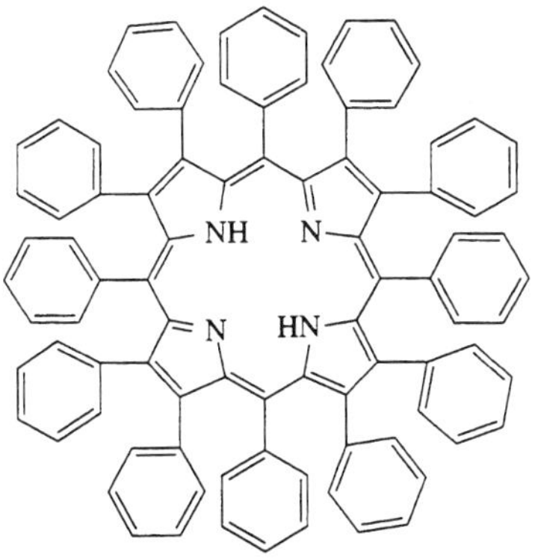

29

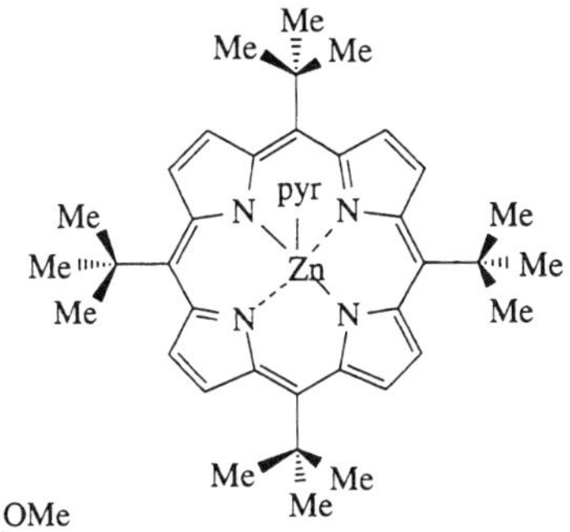

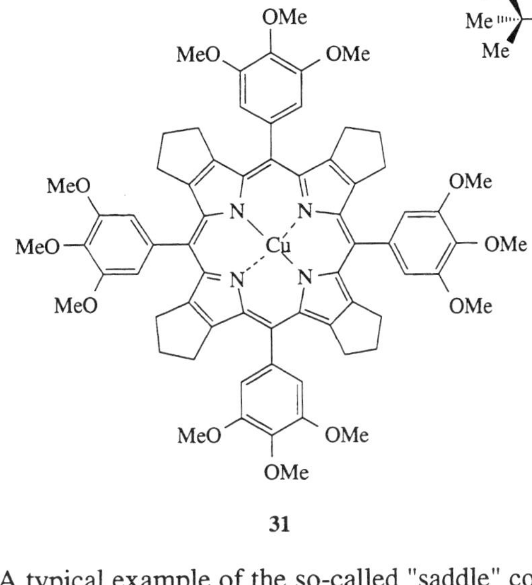

31

32

A typical example of the so-called "saddle" conformation is presented in Figure 1 [for copper(II) octaethyltetraphenylporphyrin **30**]. Not all dodeca-substituted porphyrins show non-planar behavior; the

290

macrocyclic core of the copper(II) tetrapropanoporphyrin **31** (Figure 2) is almost perfectly planar because the angles in the cyclopentane rings minimize steric interaction between the α-methylenes and the abutting meso-phenyl substituents (M. O. Senge et al., Inorg. Chem., 1993, **32**, 1716); the methoxyl groups in **31** were added to the aryl rings merely to serve the purpose of increasing solubility in organic solvents.

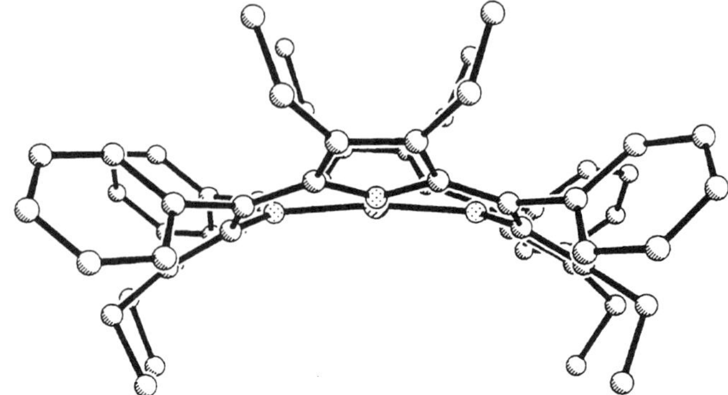

Figure 1: X-ray structure of severely "saddled" copper(II) octaethyltetraphenylporphyrin **30**.

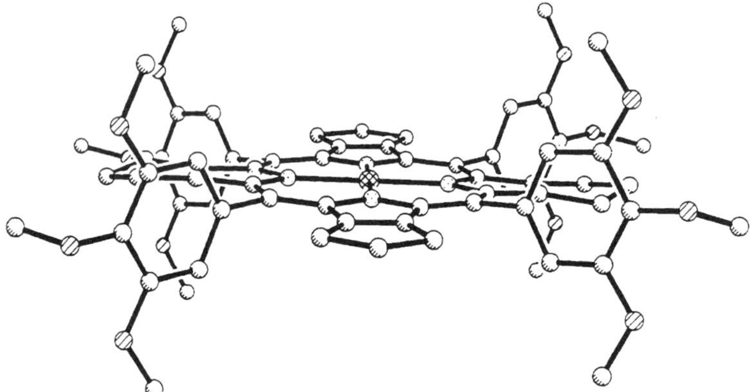

Figure 2: X-ray structure of "flat" tetra-aryl-tetrapropanoporphyrin **31** (M. O. Senge et al., Inorg. Chem., 1993, **32**, 1716).

The so-called "ruffled" porphyrin core conformation is exemplified in Figure 3 with the structure of the zinc(II) 5,10,15,20-tetra-(t-butyl)porphyrin pyridine adduct **32**.

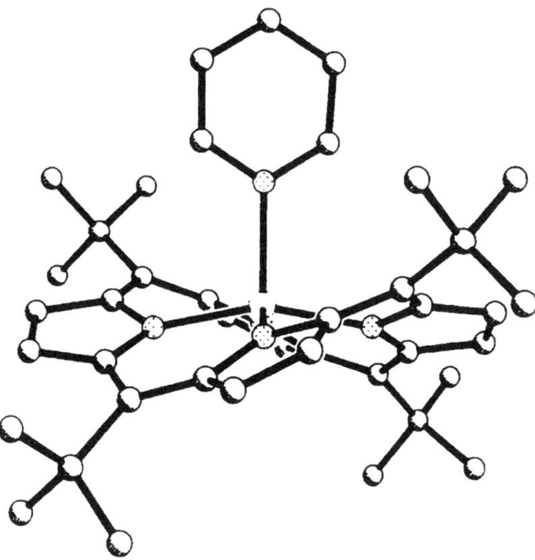

Figure 3: X-ray structure of severely "ruffled" zinc(II) pyridine 5,10,15,20-tetra(t-butyl)-porphyrin **32** (M.O. Senge et al., J. Chem. Soc., Chem. Commun., 1995, 733).

The first X-ray structure of a dodecasubstituted porphyrin, namely zinc(II) octaethyltetraphenylporphyrin **33**, was reported in 1990 (K. M. Barkigia et al., J. Am. Chem. Soc., 1990, **112**, 8851). Subsequently, syntheses of dodecaphenylporphyrin **29** (C.J. Medforth and K.M. Smith, Tetrahedron Lett., 1990, **31**, 5583; J. Takeda et al., Chem. Phys. Lett., 1991, **183**, 384), other dodeca-arylporphyrins (J. Takeda and M. Sato, Chem. Pharm. Bull., 1994, **42**, 1005), deca- and undecaphenylporphyrins (J. Takeda and M. Sato, Chem. Lett., 1994, 2233), hexa-, octa-, and deca-phenylporphyrins (J. Takeda and M. Sato, Tetrahedron Lett., 1994, **35**, 3565), and 5,10,15,20-tetra-(t-butyl)porphyrin (e.g. zinc complex **32**) (T. Ema et al., Angew. Chem. Int. Ed. Engl., 1994, **33**, 1879; M.O. Senge et al., J. Chem. Soc., Chem. Commun., 1995, 733) were reported. 5,10,15,20-Tetra(perfluoroalkyl)porphyrins have also been synthesized (S.G. DiMagno et al., J. Org. Chem., 1994, **59**, 6943). meso-Tetra-vinylporphyrins have been prepared (L. Hevesi et al., J. Chem. Soc., Chem. Commun., 1986, 1725). Porphyrins with six, eight and ten phenyl groups were obtained via chromatographic separation of the mixture obtained using mixed condensation with pyrrole, 3,4-diphenylpyrrole and benzaldehyde, whereas the deca- and undeca-phenylporphyrins were obtained by mixed condensation with 3,4-diphenylpyrrole with benzaldehyde and formaldehyde. Octabromotetraphenylporphyrin **34**

292

[obtained by octabromination of copper(II) 5,10,15,20-tetraphenyl-porphyrin] has also been synthesized and predicted to have a very nonplanar macrocyclic core (P. Bhyrappa and V. Krishnan, Inorg. Chem., 1991, **30**, 239), as was subsequently confirmed in an X-ray analysis of the dication salt (M. O. Senge et al., Angew. Chem. Int. Ed. Engl. 1994, **33**, 2485); brominated basket-handle porphyrins have also been synthesized (D. Reddy et al., J. Chem. Soc., Dalton Trans., 1993, 3575). β-Aryl-substituted porphyrins can be efficiently prepared by way of palladium catalyzed Suzuki cross-coupling reactions of β-bromo-porphyrins with arylboronic acids (K.S. Chan et al., Tetrahedron, 1995, **51**, 3129). Finally, the octa-acetic-tetra-phenylporphyrin **35** has also been synthesized and spectroscopically characterized (M. Miura et al., Inorg. Chem., 1994, **33**, 6078).

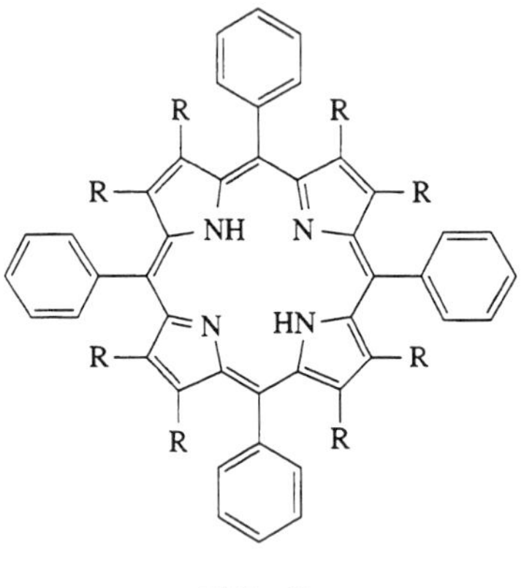

34 R = Br
35 R = CH$_2$CO$_2$H

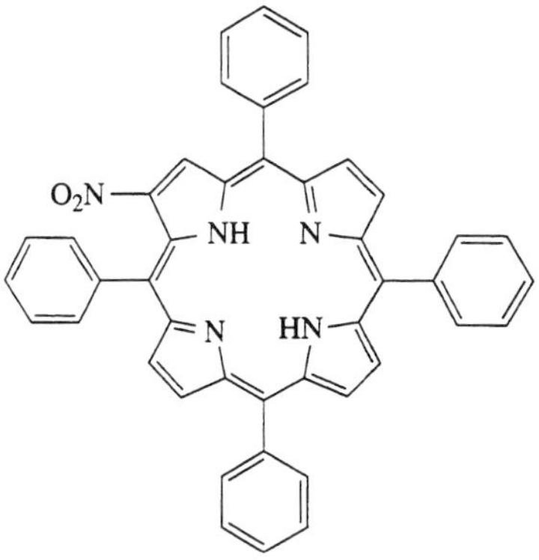

36

5,10,15,20-Tetraphenylporphyrin undergoes reaction with osmium tetroxide, as do regular octaalkylporphyrins, to give the corresponding cis-2,3-diol (C. Brueckner and D. Dolphin, Tetrahedron Lett., 1995, **36**, 3295). Two approaches to the deuteration of tetra-arylporphyrins have been described. In the first, EtOD is used to deuterate pyrrole positions, while the phenyl rings can be deuterated using the porphyrin and deuteriosulfuric acid (Z. Gross and L. Kaustov, Tetrahedron Lett., 1995, **36**, 3735). In the second approach, free-base and metalated 2-nitro-5,10,15,20-tetraphenylporphyrins (e.g. **36**) are first reduced with borohydride to give the 2,3-dihydro-2-nitroporphyrin (chlorin); treatment with tributyltin hydride gives the denitriated chlorin, or elimination of nitrous acid gives the porphyrin. Choice of appropriate deuterated reductants allows synthesis of regioselectively deuterated analogues (M.J. Crossley and L.G. King, J. Org. Chem., 1993, **58**, 4370).

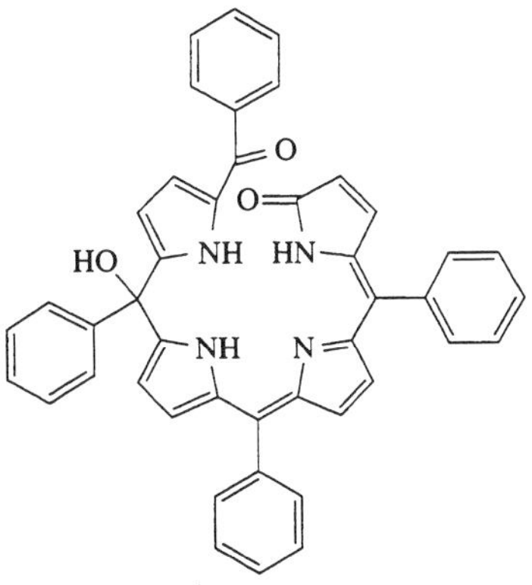

37

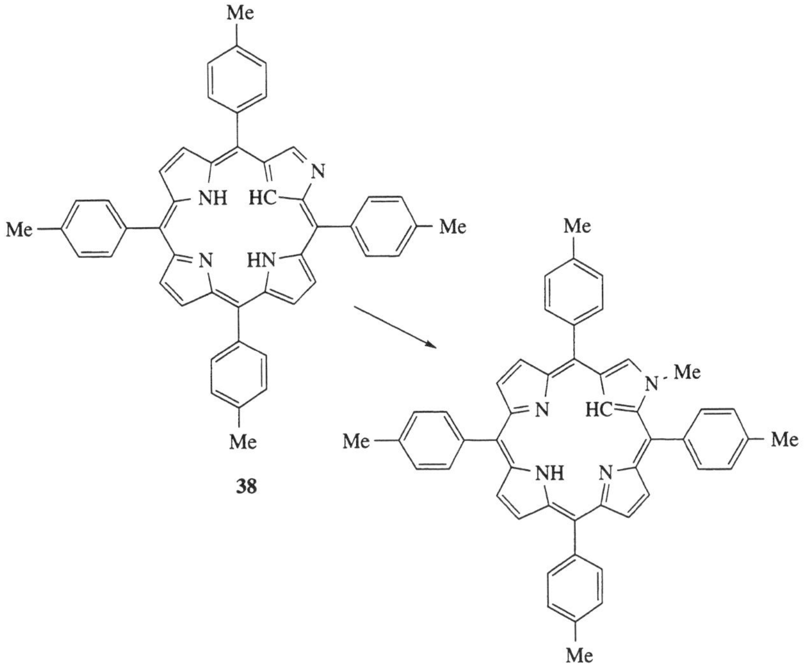

38

39

Zinc(II) β-pyridinium tetraphenylporphyrin can be obtained by electrolysis of the parent porphyrin at platinum electrodes in the presence of pyridine (L. El Kahef et al., J. Chem. Soc., Chem. Commun., 1986,

294

621). Aerial oxidation of certain 4-hydroxyphenyl-porphyrins has been studied, and the redox chemistry of these systems has been investigated (L.R. Milgrom et al., J. Chem. Soc., Perkin Trans. 2, 1989, 301). Photo-oxygenation of 5,10,15,20-tetraphenylporphyrin affords a ring-opened product **37** possessing a violinoid structure (J.A.S. Cavaleiro et al., J. Chem. Soc. Chem. Commun., 1986, 142). Mild methylation of 5,10,15,20-tetraphenyl-21-carbaporphyrin **38** affords a novel isomer **39** of N-methyl-5,10,15,20-tetraphenylporphyrin, and the chemistry of this new system has been investigated (P.J. Chmielewski and L. Latos-Grazynski, J. Chem. Soc., Perkin Trans. 2, 1995, 503). A new type of "naphthochlorin" **40** has been synthesized by way of acid catalyzed intramolecular cyclization of nickel(II) 2-vinyl-5,10,15,20-tetraphenylporphyrin **41** (M.A. Faustino et al., Tetrahedron Lett., 1995, **36**, 5977). β-Alkoxytetra-arylporphyrins **42** (alkoxy = OMe, OPr, OiPr, Ot-Bu) can be synthesized by treatment of the β-N=NOH porphyrin **43** with alcohols in sulfuric acid/chloroform (H.K. Hombrecher et al., J. Prakt. Chem., 1994, **336**, 542).

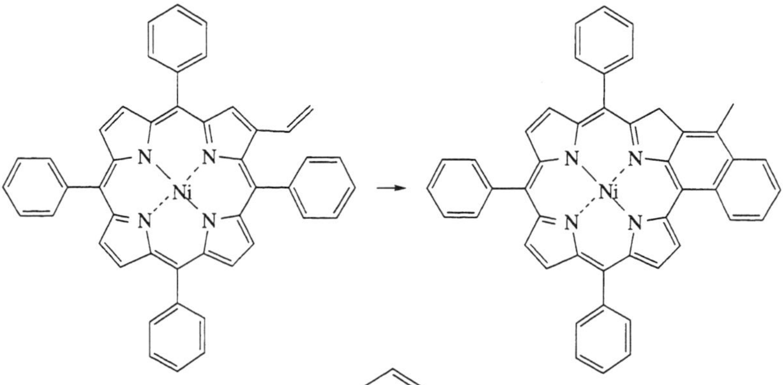

41   **40**

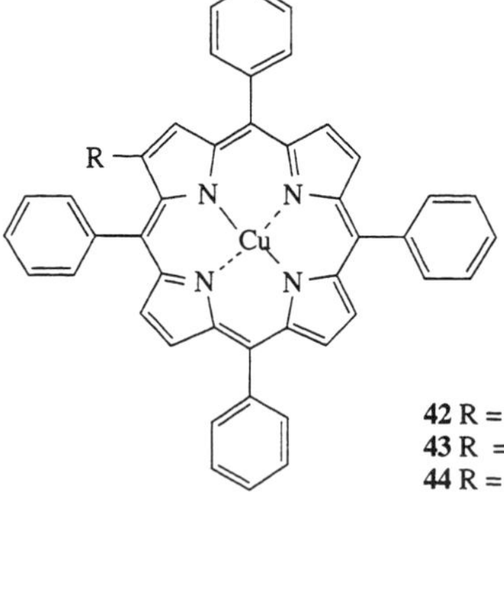

42 R = OR
43 R = N=N-OH
44 R = OH

Tautomerism in 2-hydroxy-5,10,15,20-tetraphenylporphyrin and its zinc and copper complexes (e.g. **44**) has been investigated (M.J. Crossley et al., J. Org. Chem., 1988, **53**, 1132; Tetrahedron, 1987, **43**, 4569); proton NMR spectroscopy indicates multiple tautomeric forms in solution, and that the phenomenon is solvent dependent.

(b) Octa-substituted and Related Porphyrins

Methodology, from Clezy's group in Australia, for synthesis of β-substituted porphyrins has been reviewed (P.S. Clezy, Aust. J. Chem., 1991, **44**, 1163), as has the synthesis and structure of biomimetic porphyrins by Dolphin's group in Vancouver (B. Morgan and D. Dolphin, Struct. Bonding (Berlin), 1987, **64**, 115). Substituent manipulations which can be carried out on protoporphyrin IX have also been reviewed (J.A.S. Cavaleiro and K.M. Smith, Heterocycles, 1987, **26**, 1947). In connection with the elucidation of structure/function relationships in reconstituted heme proteins, synthetic procedures for obtaining mono-, di- and tri-propionic porphyrins (R.K. Pandey et al., J. Chem. Res., Synop. 1987, 262), for protoporphyrins IX bearing 2- and 7-methyl deuterated groups (K.M. Smith et al., J. Chem. Res., Synop. 1986, 324), carbon-13 enriched vinyl groups (K.M. Smith and E.M. Fujinari, J. Labelled Compd. Radiopharm. 1986, **23**, 73), and carbon-13 enriched propionates (K.M. Smith et al., J. Chem. Res., Synop. 1986, 402) have been described. Jackson's group also published their results on the synthesis of hepta-, hexa-, and penta-carboxylic porphyrins related to uroporphyrin-I (A.H. Jackson and D. Supphayen, J. Chem. Soc., Perkin Trans. 1, 1987, 277). A number of studies designed to investigate the mechanism of oxidative cyclization 1,19-disubstituted b-bilenes (P.S. Clezy et al., Aust. J. Chem., 1986, **39**, 399) and a,c-biladienes (K.M. Smith and R.K. Pandey, Synth. Commun. 1986, **16**, 929; J. Chem. Soc., Perkin Trans. 1, 1987, 1229; K.M. Smith and O.M. Minnetian, J. Chem. Soc., Perkin Trans. 1, 1986, 277; P.A. Liddell et al., J. Org. Chem., 1993, **58**, 6681) to give porphyrins have been published. Of particular interest is the observation that a,c-biladienes with bulky 1,19-substituents (e.g. **45**) afford fully characterized novel compounds with interrupted conjugation **46** as well as chlorins **47** and **48**, and meso-disubstituted porphyrins **49**, by way of remarkable stepwise rearrangement processes (P.A. Liddell et al., Tetrahedron Lett., 1990, **31**, 2685; J. Org. Chem., 1993, **58**, 6681).

1-Bromo-19-methyl-a,c-biladienes have been shown to be useful open-chain tetrapyrroles for cyclizations to porphyrins (T.P. Wijesekera and D. Dolphin, Synlett., 1990, 235; J.B. Paine III, et al., J. Org. Chem., 1988, **53**, 2796), and the corresponding 1,19-dibromo-a,c-biladienes e.g. **50** can

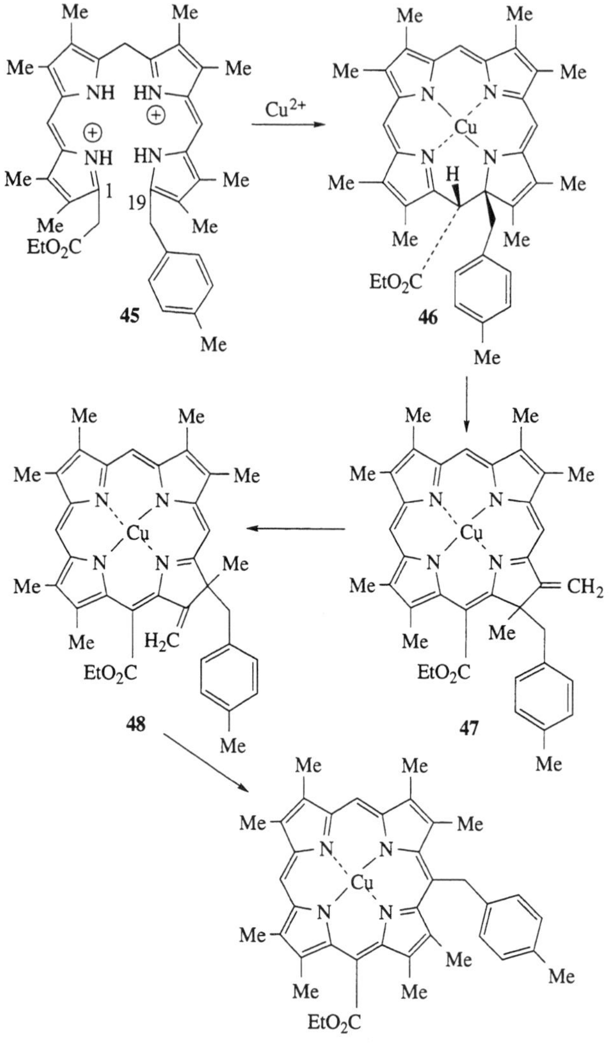

45 46 47 48 49

be hydrolyzed to provide a very efficient route to biliverdins **51**, or else cyclized, depending upon the reaction conditions, to give corroles **52** or azaporphyrins **53** (R.K. Pandey et al., J. Chem. Soc., Chem. Commun., 1992, 183; J. Chem. Soc., Perkin Trans. 1, 1994, 971). Azaspirographis porphyrin has been synthesized and converted into an azachlorin (B.

Gerlach and F.P. Montforts, Tetrahedron Lett., 1993, **34**, 6369). An improved synthesis of mono-azaporphyrins from biladienes has also been reported (J.P. Singh et al., Tetrahedron Lett., 1995, **36**, 1567).

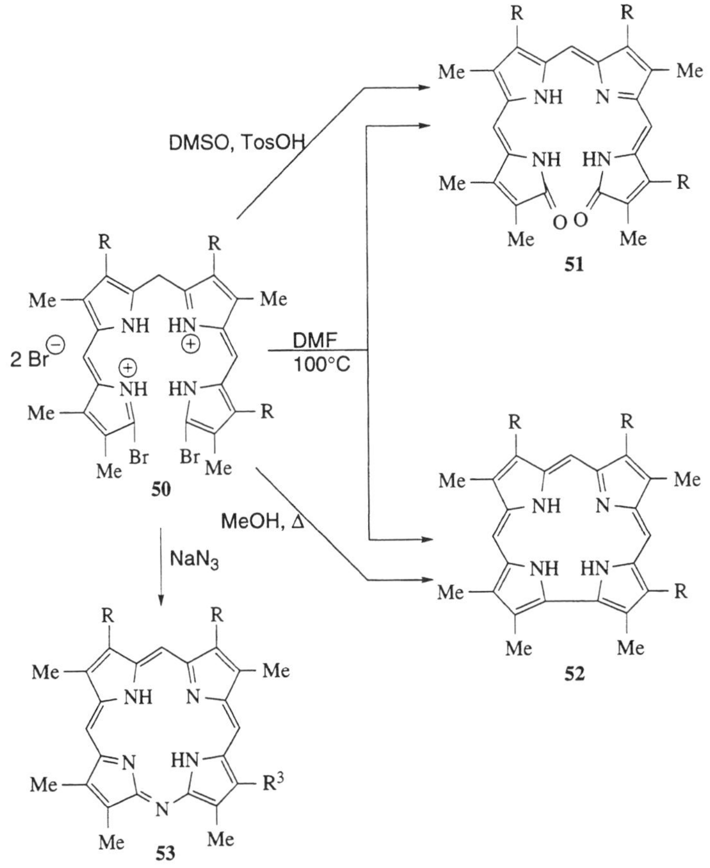

An optimized procedure for synthesis of 2,3,7,8,12,13,17,18-octaethylporphyrin and its monopyrrole precursor, 3,4-diethylpyrrole has appeared (J.L. Sessler et al., Org. Syn., 1992, **70**, 68). The so-called "symmetrical" route to octaalkylporphyrins, and particularly to octa-ethylporphyrin, has also been optimized by the Strasbourg synthesis group, and then applied to the synthesis of meso-carbon-13 and N-15 analogues (A. Rohrer et al., Synthesis, 1994, 923). Certain other octa-alkyl- and octa-aryl-porphyrins can also be prepared from monopyrroles, which are in turn obtained following the Barton-Zard procedure (D. H.

R. Barton et al, Tetrahedron, 1990, **46**, 7587) from nitroalkenes (N. Ono and K. Maruyama, Chem. Lett., 1988, 1511; N. Ono et al., Tetrahedron, 1990, **46**, 7483). Syntheses of 2-(substituted methyl)-3,4-disubstituted pyrroles and their conversion into porphyrins have also been described (H. Kinoshita et al., Bull. Chem. Soc. Jpn., 1992, **65**, 2660). Jeandon and Callot (Bull. Soc. Chim. Fr., 1993, **130**, 625) have also published a revealing paper on the tetramerization of Knorr-type pyrroles to give porphyrins, and have shown that all is not as it seems, both with regard to the identity of the monopyrrole and the proportions of the various porphyrin type isomers produced due to pyrrole redistribution reactions under acidic conditions. Pure type-I porphyrins can be obtained from monopyrrole tetramerizations by ingenious use of steric encumbrance; in this way, type-I tetramethyltetraphenyl-, tetramethyl-tetra(2,4,6-iso-propyl)- and tetramethyltetraterphenyl- porphyrins were obtained (C.K. Chang and N. Bag, J. Org. Chem., 1995, **60**, 7030; N. Bag et al., Tetrahedron Lett., 1995, **36**, 6409). Substantially pure type-isomers of symmetric porphyrins can also apparently be obtained using an improved monopyrrole tetramerization reaction in which pyrrole-carboxylates are treated with KOH and then acetic acid and oxygen (L. Cheng and J. Ma, Org. Prep. Proced. Int., 1995, **27**, 224). Pyrrole redistribution reactions can be eliminated by use of 2,5-di(dimethylaminomethyl)-pyrroles **54**, and a new one pot procedure has been described which affords porphyrins (e.g. **55**) with like-rings opposite (L.T. Nguyen et al., Tetrahedron Lett., 1994, **35**, 7581). The methodology can be expanded to a "3+1" approach by use of a tripyrrane and 2,5-di(dimethyl-aminomethyl)pyrroles (L.T. Nguyen et al., J. Org. Chem., 1996, **61**, 998). The same "3+1" methodology to porphyrins (e.g. **56**) has been developed independently using a tripyrrane **57** and a 2,5-diformylpyrrole **58** as the monopyrrole unit (A. Boudif and M. Momenteau, J. Chem. Soc., Chem. Commun., 1994, 2069; Y. Lin and T.D. Lash, Tetrahedron Lett., 1995, **35**, 2493; T.D. Lash, Angew. Chem. Int. Ed. Engl., 1995, **34**, 2533).

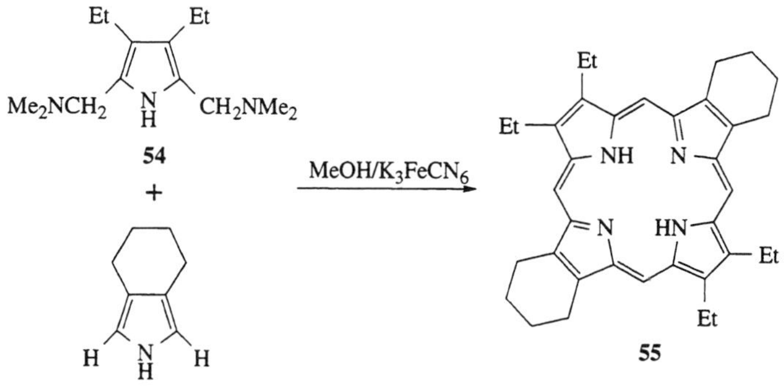

Novel methodology for functionalization of the ethyl groups in octaethyl-porphyrin has been reported. Chang's group has primarily used the OsO_4 reaction of octaethylporphyrin **57** to give the 2,3-diol **58** (C.K. Chang and C. Sotiriou, J. Org. Chem., 1987, **52**, 926), while the UC Davis group modified an ethyl group in octaethylporphyrin to give porphyrins **59** bearing H, CHO, CH=CH$_2$, CH=CHBr, and CH(OR)Me groups (M.G.H. Vicente and K.M. Smith, Synlett, 1990, 579; Tetrahedron, 1991, **47**, 6887).

57 R = Et
59 R = H, CHO, CH=CH$_2$,
CH=CHBr, CH(OR')Me

58

A number of research groups have put considerable effort into the synthesis of fluorinated porphyrins. Examples include fluorinated analogues of hematoporphyrin-II (M. Omote et al., Heterocycles, 1994, **39**, 381), of protoporphyrin-IX (e.g. **60**) (K.M. Snow and K.M. Smith, J. Org. Chem., 1989, **54**, 3270), and of 2-demethyl-mesoporphyrin-IX **61** (A. Suzuki et al., Heterocycles, 1992, **33**, 87); in the first two cases the fluorine was added to an intact porphyrin, while the third porphyrin **61** was prepared by total synthesis using ethyl 3-ethyl 4-fluoro-5-methylpyrrole-2-carboxylate **62** as the ring A precursor. Other work has

300

involved the synthesis of more symmetrical fluorinated porphyrins from the correspondingly fluorinated monopyrroles (H. Onda, Tetrahedron Lett., 1985, **26**, 4221; N. Ono et al., Bull. Chem. Soc. Jpn., 1989, **62**, 3386; W. Huang and W. Ma, Chin. J. Chem., 1990, 474).

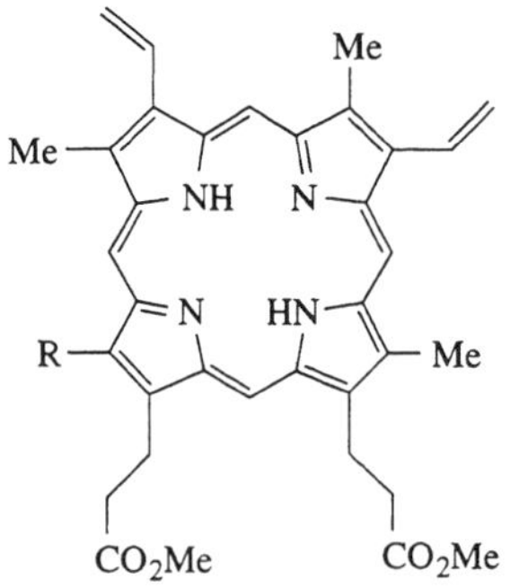

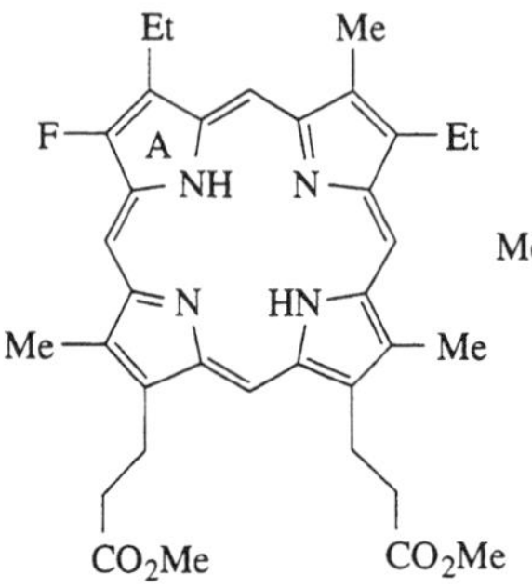

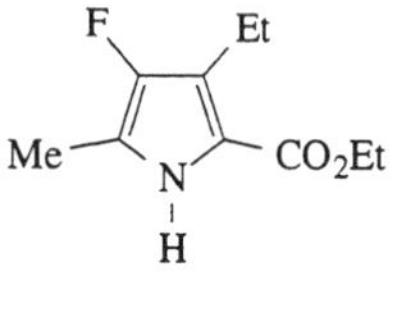

60 R = CH_2F, CHF_2
72 R = Me

61

62

A "2+2" approach to *opp*-dibenzoporphyrins **63**, by way of dipyrromethenes **64**, has been reported; the method for dipyrromethene synthesis utilizes 1-formyl-3-haloisoindoles and α-free pyrroles (R. Bonnett and K.A. McManus, J. Chem. Soc., Chem. Commun., 1994, 1129).

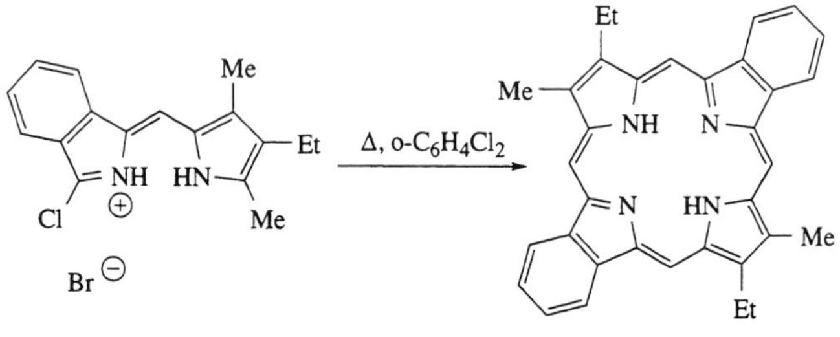

64

63

As part of T.D. Lash's series of papers on porphyrins with exocyclic rings (see also *Petroporphyrins*) naphthoporphyrins **65** (T.D. Lash and T.J. Roper, Tetrahedron Lett., 1994, **35**, 7715; T.D. Lash and C.P. Denny, Tetrahedron, 1995, **51**, 59), phenanthroporphyrins **66** (T.D. Lash and B.H. Novak, Angew. Chem., Int. Ed. Engl., 1995, **34**, 683; Tetrahedron Lett., 1995, **36**, 4381) and phenanthrolinoporphyrins **67** (via the "3+1" approach, vide supra) (Y. Lin and T.D. Lash, Tetrahedron

Lett., 1995, **36**, 9441) have been synthesized. An attempt to prepare a pyrroloporphyrin **68** by cyclization of a porphyrin amino ester **69** has failed (R.F.C. Brown et al., Aust. J. Chem., 1995, **48**, 1447), but in the tetraphenyl series, fused pyrroloporphyrins (e.g. **70**) can be obtained by application of the Barton-Zard pyrrole synthesis to a 5,10,15,20-tetraphenyl-2-nitroporphyrin **71** (L. Jaquinod et al., J. Chem. Soc., Chem. Commun., 1996, in press).

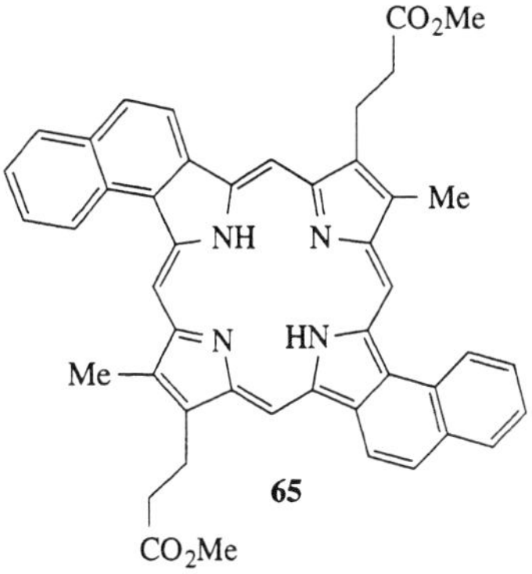

65

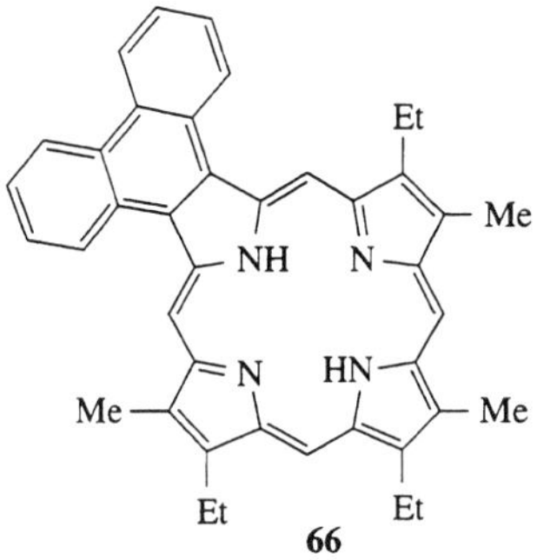

66

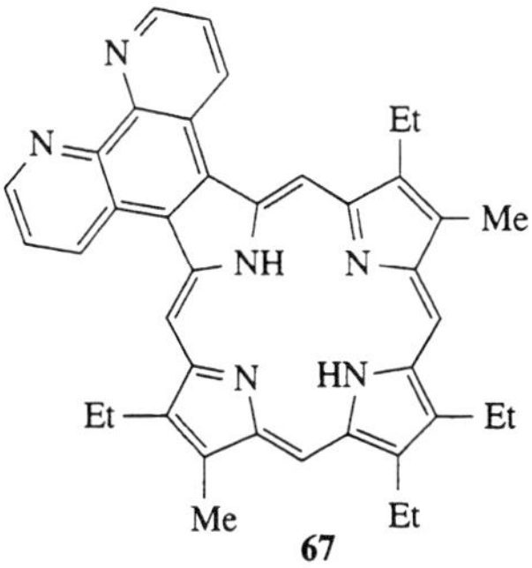

67

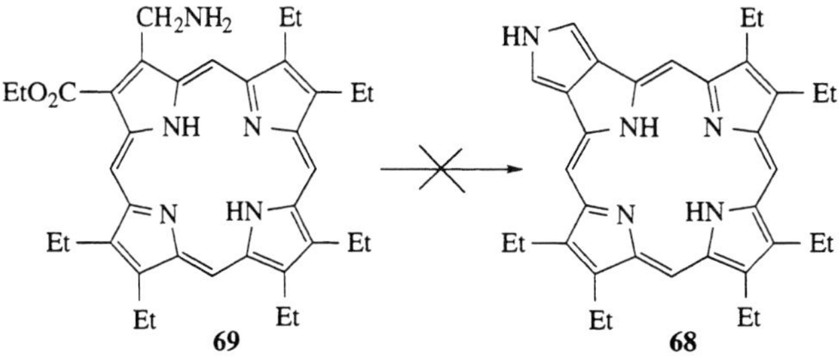

69 **68**

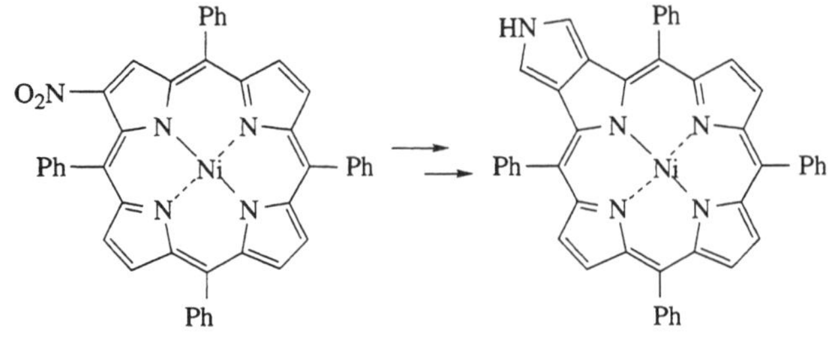

71 → **70**

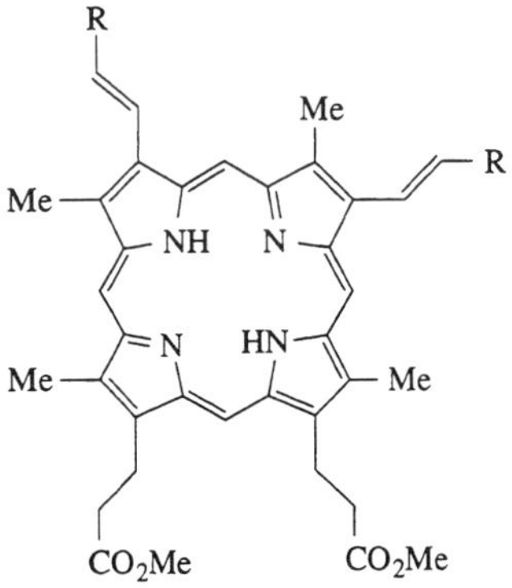

72 R = H
73 R = CH$_2$NMe$_2$
74 R = CH$_2$NMe$_3$ I$^{\ominus}$

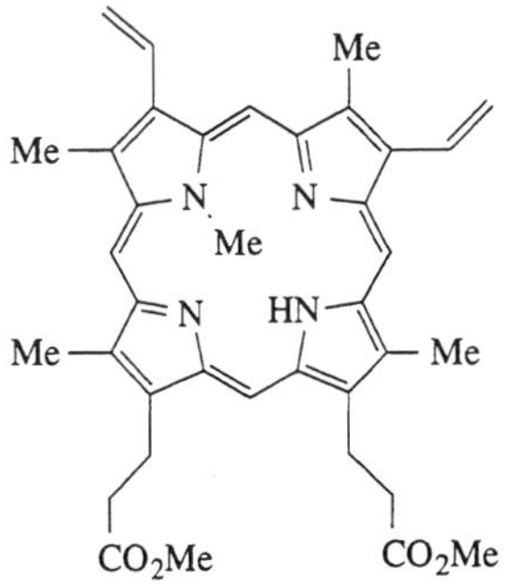

75

Virtually any vinyl-substituted porphyrin (e.g. protoporphyrin-IX dimethyl ester **72**) or vinyl-chlorin can be rendered water soluble by treatment with Eschenmoser's reagent (N,N-dimethylmethylene-ammonium iodide (to give **73**) followed by quaternization with alkyl iodides to give **74** (R.K. Pandey et al., J. Chem. Soc., Chem. Commun., 1991, 1637; Tetrahedron, 1992, **48**, 7591). N,N'-Etheno-bridged porphyrins in the octaethylporphyrin series have been prepared from the reaction of diaquocobalt(III) porphyrin perchlorates with alkynes (J. Setsune et al., Organometallics, 1991, **10**, 1099), and these N,N'-bridged systems (etheno-, ethano-, or vinylideno-) undergo regio- and stereo-selective reduction to give stable 5H-phlorins in which the reduction has taken place at the meso-carbon between the pyrrole subunits bearing the N,N'-bridge (J. Setsune et al., J. Am. Chem. Soc., 1988, **110**, 6572). Total syntheses, from open-chain tetrapyrrolic N-methylated precursors, of all four isomeric N-methylprotoporphyrins (e.g. **75**) have been reported (R.K. Pandey et al., J. Chem. Soc., Perkin Trans. 1, 1991, 1211); these compounds were shown in earlier work to have differential activity as ferrochelatase inhibitors. N-(Methoxycarbonylmethyl)-porphyrins, upon treatment with nickel(II) salts affords the nickel(II) meso-CH_2CO_2Me porphyrin due to intramolecular migration of the N-substituent as well as a novel nickel chlorin (K.M. Smith et al., J. Chem. Soc., Chem. Commun., 1986, 1498).

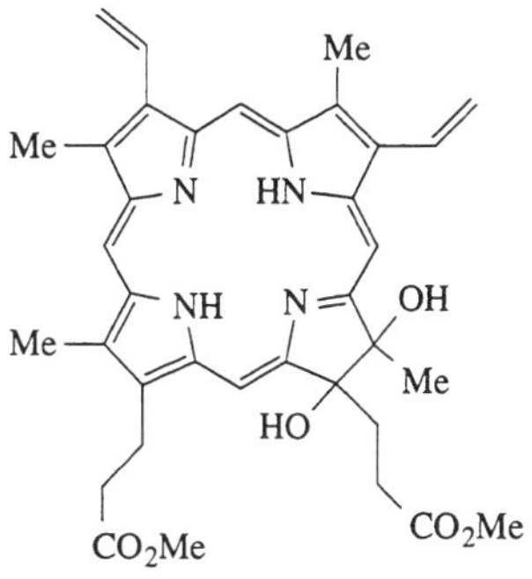

76

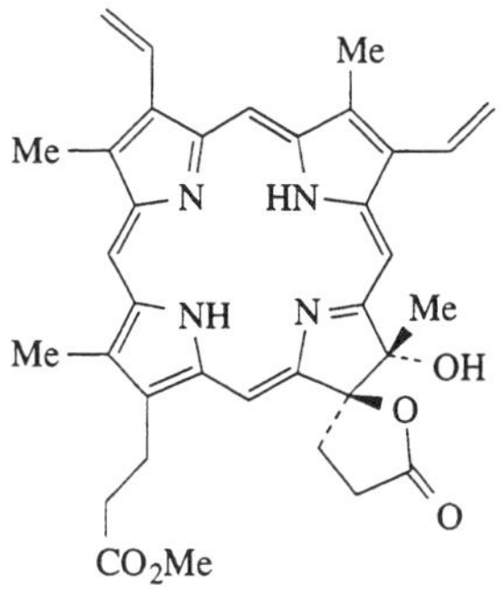

77

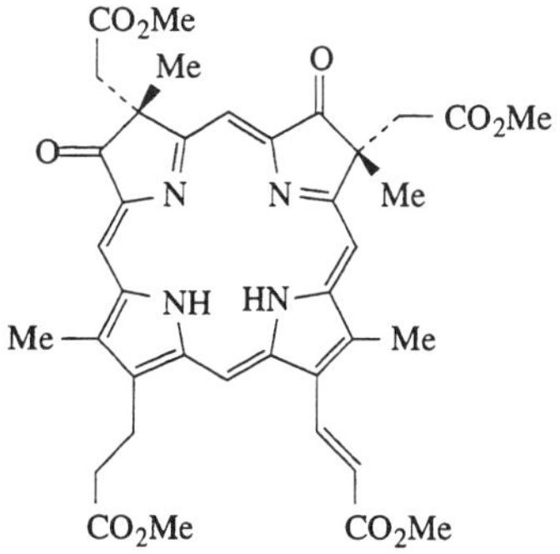

78

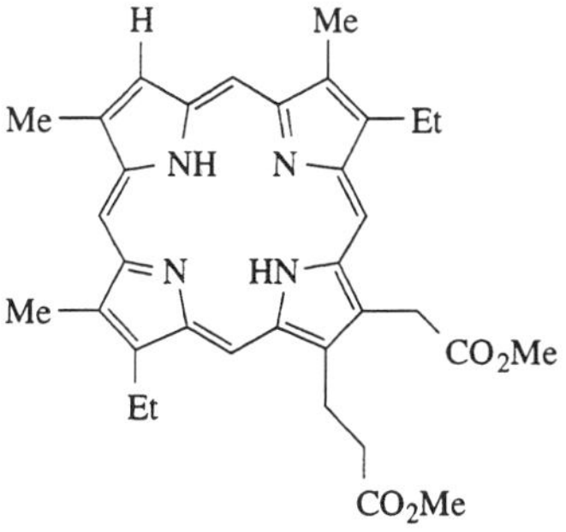

79

Chang's group has published model methodology studies and total syntheses of the methyl ester of the metal-free heme-d prosthetic group of bacterial terminal oxidase, which appears to be a diol **76** (M.R. Vavra et al., Arch. Biochem. Biophys., 1986, **250**, 461) and readily lactonizes to give e.g. **77** (C.K. Chang and C. Sotiriou, J. Org. Chem., 1985, **50**, 4989; C. Sotiriou and C.K. Chang, J. Am. Chem. Soc., 1988, **110**, 2264; L.A. Andersson et al., J. Am. Chem. Soc., 1987, **109**, 258); the synthesis proceeds by way of OsO_4 oxidation of a protoporphyrin-IX derivative.

The so-called "dioneheme", the green heme prosthetic group in cytochrome cd_1 dissimilatory nitrite reductase has also been synthesized (as its metal-free tetramethyl ester **78**), again by way of OsO_4 oxidation of appropriate synthetic precursor porphyrins (C.K. Chang and W. Wu, J. Biol. Chem., 1986, **261**, 8593; W. Wu and C.K. Chang, J. Am. Chem. Soc., 1987, **109**, 3149). Very similar chiral dioxobacteriochlorins have also been prepared from hematoporphyrin dimethyl ester by way of the Claisen-rearrangement methodology (F.P. Montforts et al., Angew. Chem., 1989, **101**, 471). Migratory aptitudes in the acid catalyzed pinacol-pinacolone rearrangement of the 2,3-diols to oxochlorins have been established; they follow the order alkyl, propionate > H > Me > acetic side chains (C.K. Chang and C. Sotiriou (J. Heterocycl. Chem., 1985, **22**, 1739). The structure and redox chemistry of model isobacteriochlorin diones has also been investigated (C.K. Chang et al., J. Am. Chem. Soc., 1986, **108**, 1352).

New porphyrins have been isolated from the ethanol extract of a coral sea demosponge, and have been named corralistin-A, -B, -C, -D, and -E. Corralistin-A, as its dimethyl ester, **79** has been synthesized by P. Yon-Hin and A.I. Scott (Tetrahedron Lett., 1991, **32**, 4231), and corralistins -B, -C and -E have been synthesized by R.K. Pandey et al. (Tetrahedron Lett., 1994, **35**, 8093) as the methyl ester derivatives; several syntheses of corralistin-D (3-ethyldeuteroporphyrin-IX) are already in the literature. In a the major accomplishment, the combined Battersby and Clezy groups have reported the total synthesis of the dimethyl ester of porphyrin-a, the iron-free prosthetic group of cytochrome-c oxidase (A.R. Battersby et al., J. Chem. Soc., Perkin Trans. 1, 1985, 135). An improved synthesis of porphyrin-c from hematoporphyrin-IX has been published (P.A. Scourides et al., J. Chem. Soc., Chem. Commun., 1986, 1817).

A general synthesis of 3,4-dialkoxypyrroles has been reported, and upon reaction with aldehydes, these pyrroles afford octaalkyloxyporphyrins in moderate yield (A. Merz et al., Synthesis, 1995, 795). As mentioned above, β-alkoxytetra-arylporphyrins have also been synthesized (H.K. Hombrecher et al., J. Prakt. Chem., 1994, **336**, 542).

Catanane and rotaxane porphyrins have been synthesized (P.R. Ashton et al., J. Chem. Soc., Chem. Commun., 1992, 1128; J.C. Chambron et al., J. Chem. Soc., Chem. Commun., 1992, 1131; M.J. Gunther et al., J. Am. Chem. Soc., 1994, **116**, 4810).

In connection with biological studies of heme oxygenase activity, the four (5-, 10-, 15-, 20-) meso-methylmesoporphyrins (e.g. **80**) have been synthesized, separated, and identified (J.W. Torpey and P.R. Ortiz de Montellano, J. Org. Chem., 1995, **60**, 2195).

Et Me Me
Me
NH N
Et
N HN
Me Me
CO_2H CO_2H

80

Strapped and capped porphyrins continue to occupy the interests of several synthetic groups; examples include highly deformed tightly strapped porphyrins (T.P. Wijesekera et al., J. Org. Chem., 1988, **53**, 1345), derivatized capped porphyrins (H. Tang et al., Can. J. Chem., 1992, **70**, 1366), hydrocarbon-strapped porphyrins containing quinone and phenolic groups (B. Morgan and D. Dolphin, J. Org. Chem., 1987, **52**, 5364), thioether strapped porphyrins (B. Morgan et al., J. Org. Chem., **1987**, *52*, 4628), durene-capped porphyrins (T.K. Wijesekera et al., Can. J. Chem., 1988, **66**, 2063), and a picnic-basket porphyrin with a substituent in the interior of the pocket (T. Michida et al., Chem. Pharm. Bull., 1992, **40**, 3157).

Metal-mediated (e.g. Heck- or Stille-type) cross coupling reactions in porphyrin systems (O.M. Minnetian et al., J. Org. Chem., 1989, **54**, 5567; I.K. Morris et al., J. Org. Chem., 1990, **55**, 1231) have been a topic of renewed interest (S.G. DiMagno et al., J. Org. Chem., 1993, **58**, 5893). β-Substituted porphyrins can also be prepared using palladium-catalyzed reactions (H. Ali and J.E. van Lier, Tetrahedron, 1994, **50**, 11933), as can alkynylphenylporphyrins (K.S. Chan and C.S. Chan, Synth. Commun., 1993, **23**, 1489).

Quite considerable effort has focused on synthetic methodology for conversion of porphyrin into chlorin systems. This has often been associated with attempts to red-shift optical spectra of "natural" type porphyrins in connection with the cancer modality known as photodynamic therapy; in this modality, the aim of investigators looking into syntheses of so-called "second generation photosensitizers" is to obtain compounds with long-wavelength absorbance (λ_{max} 650-800 nm) because low energy light travels further through tissues than does short wavelength light, and so it should be possible to treat larger tumors using photodynamic therapy. Cost is also an issue, since a number of inexpensive light sources are available in the long-wavelength region, compared with argon-pumped dye lasers which need to be used at 630 nm or below. Literally thousands of papers on porphyrin photodynamic therapy have been published in the past ten years, and only a very brief series of comments can be made here, though more information can be found in the chlorin and bacteriochlorin paragraphs of this review. Thus, reactions of porphyrinones with organolithiums give red-shifted optical spectra (A.S. Phadke et al., Tetrahedron Lett., 1993, **34**, 6359) and the Diels-Alder reaction of protoporphyrin IX which affords the commercially successful "benzoporphyrin derivative monoacid" (e.g. isomer **81**) (A.M. Richter et al., J. Natl. Cancer Inst., 1987, **79**, 1327; D. Dolphin, Can. J. Chem., 1994, **72**, 1005; I. Meunier et al., Bioorg. Med. Chem. Lett., 1992, **2**, 1575) has been further investigated using triazolinediones (A.R. Morgan and D.H. Kohli, Tetrahedron Lett., 1995, **36**, 7603) or using rhodoporphyrins as the substrate (R.K. Pandey et al., Bioorg. Med. Chem. Lett., 1993, **3**, 2615); rhodoporphyrins, which possess only one vinyl group and one propionic side-chain proved to be the ideal Diels-Alder substrate (giving, for example **82**, with dimethyl acetylene dicarboxylate) because no mixtures are produced (unlike when protoporphyrin-IX is used in synthesis of benzoporphyrin derivative monoacid). Using X-ray crystallography, both kinetic (**83**) and thermo-dynamic (**84**) products have been characterized stereochemically in the base catalyzed transformation of the initial Diels-Alder adduct **85** into **84**; this established the previously unrecognized ring A stereochemistry in **81** (I Meunier et al., Bioorg. Med. Chem. Lett., 1992, **2**, 1575).

81   **82**

A new porphyrin derivative **86** bearing a tetra-methine bridge and which can be efficiently used in Diels-Alder reactions (to give, for example, **87**) has been synthesized (P.A. Liddell et al., Tetrahedron Lett., 1994, **35**, 995).

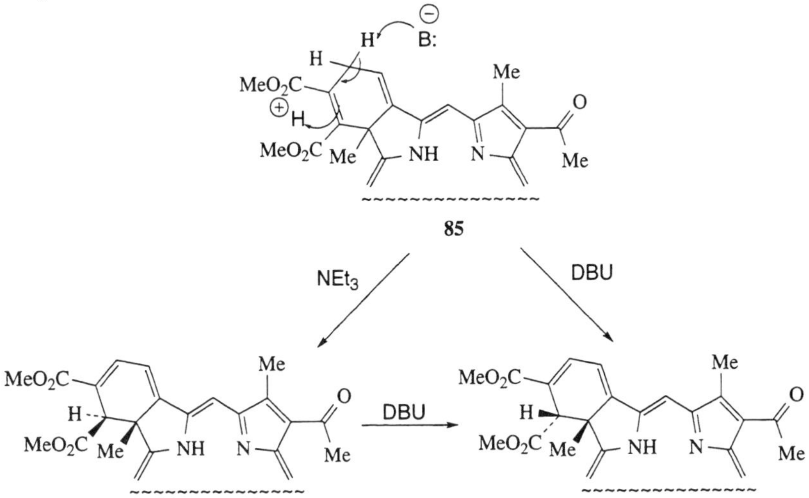

85

NEt$_3$

DBU

83

DBU

84

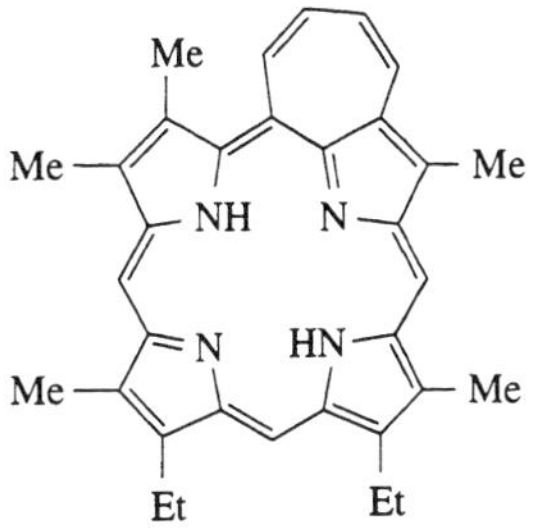

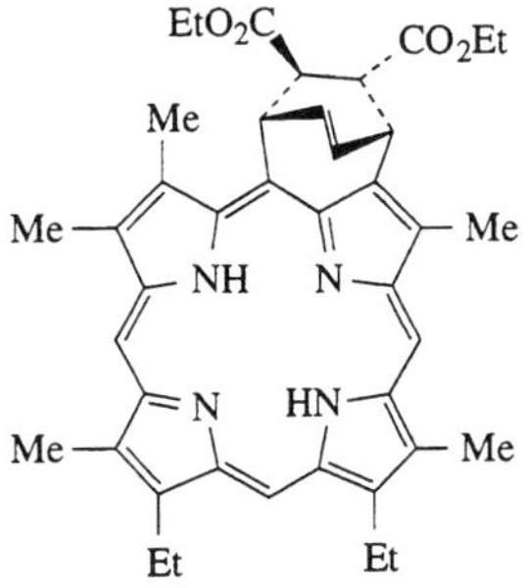

86

87

Reduction of iron porphyrins with sodium in isopentyl alcohol affords a mixture of porphyrins (depending upon which subunit is reduced), and these have been identified in a number of cases (D.H. Burns et al., J. Chem. Soc., Perkin Trans. 1, 1988, 3119). Certain iron complexes of these chlorins have been reconstituted into proteins, and studied using magnetic circular dichroism (A.M. Bracete et al., Inorg. Chem., 1994, **33**, 5042). On the other hand, photoreduction of zinc porphyrins can be specifically directed to a particular pyrrole subunit; for example, ascorbic acid/base photoreduction of zinc(II) porphyrin **88** gives the cis-

308

reduced vinylchlorin **89** (K.M. Smith and D.J. Simpson, J. Chem. Soc., Chem. Commun., 1987, 613; P. Iakovides et al., Photochem. Photobiol., 1991, **54**, 335; K.M. Smith et al., J. Am. Chem. Soc., 1986, **108**, 6834). The inverse reaction (oxidation of chlorins to porphyrins) has been shown to be very effectively accomplished by treatment of various chlorophyll derivatives with benzoyl chloride in DMF (L. Ma and D. Dolphin, Tetrahedron Lett., 1995, **36**, 7791).

(c) Petroporphyrins

The review period has featured a great deal of new work associated with petroporphyrins, pigments derived from chlorophylls which are found in petroleum and other oil deposits. Perhaps the most famous petroporphyrin is deoxophylloerythroetioporphyrin (**90**, DPEP), and an improved chemical preparation of this from chlorophyll-a has been described (C. Jeandon et al., Tetrahedron Lett., 1993, **34**, 1791). Other synthetic work on cyclopentanoporphyrins and chlorophyll-c fossils of sedimentary origin has also been reported by the same group (C. Bauder et al., Tetrahedron Lett., 1991, **32**, 2537; Tetrahedron, 1992, **48**, 5135). The biogeochemistry of chlorophyll, with emphasis on DPEP, has also been discussed (J.W. Louda and E.W. Baker, ACS Symp. Ser., 1986, **305**, 107). P.S. Clezy et al. (Aust. J. Chem., 1993, **46**, 1193) have reported partial syntheses of 13^1-methyl-DPEP **91** and the corresponding 13^2-isomer **92**; both **91** and **92** (as nickel complexes) were shown to be present as fossil petroporphyrins in Julia Creek oil shales. The same group has also presented further evidence (based on total syntheses via the a,c-biladiene route) that the etiopetroporphyrins in Julia Creek shales are derived from chlorophyll (P.S. Clezy et al., Aust. J. Chem., 1989, **42**, 775). An abundant chlorophyll-a transformation product bearing fused five and seven membered rings has been identified, and linked to other suggested chlorophyll degradation pathways involving such fused systems (P.G. Harris et al., Org. Geochem., 1995, **23**, 183). A sedimentary metallochlorin in German Messel oil shale has been shown to be the nickel complex **93** of mesopyropheophorbide-a (W.G. Prowse et al., Org. Geochem., 1990, **16**, 1059). A naturally occurring benzo-petroporphyrin **94** has been prepared via the b-bilene ring synthesis route (P.S. Clezy and C.W.F. Leung, Aust. J. Chem., 1993, **46**, 1705).

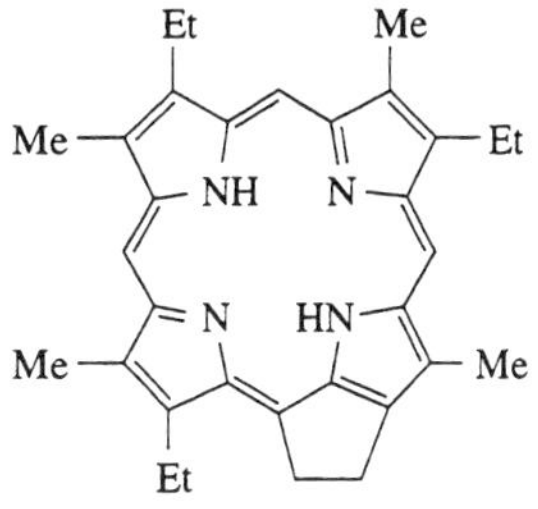

90

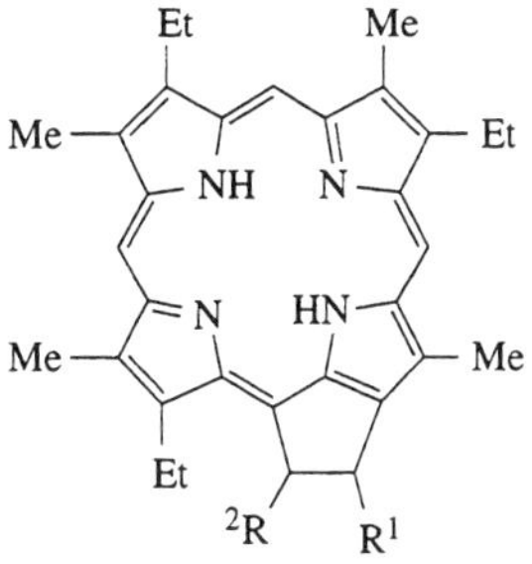

91 R^1 = H; R^2 = Me
92 R^1 = Me; R^2 = H

The four de-ethyl analogues of etioporphyrin-III **95** have been synthesized (P.J. Clewlow et al., J. Chem. Soc., Chem. Commun., 1985, 724; P.J. Clewlow and A.H. Jackson, J. Chem. Soc., Perkin Trans. 1, 1990, 1925), and HPLC analyses showed the so-called C30-de-ethyletioporphyrin from Gilsonite and Serpiano oil shales to be the ring C de-ethyl isomer. On the other hand, a demethyl analogue of DPEP found in Julia Creek oil shale (as its nickel complex) was shown to lack the methyl in ring A of **90** (P.S. Clezy et al., Aust. J. Chem., 1992, **45**, 703).

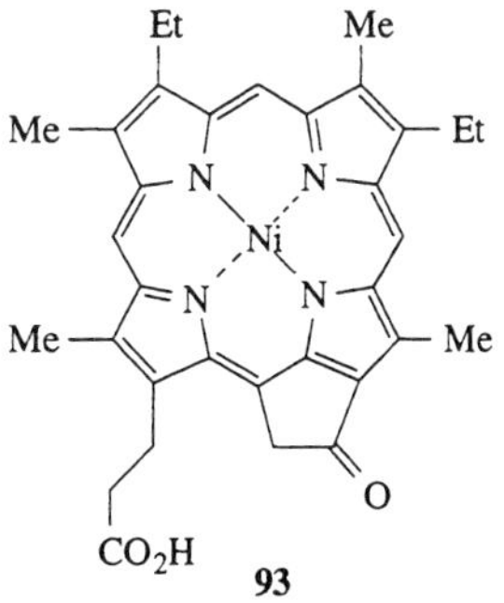

93

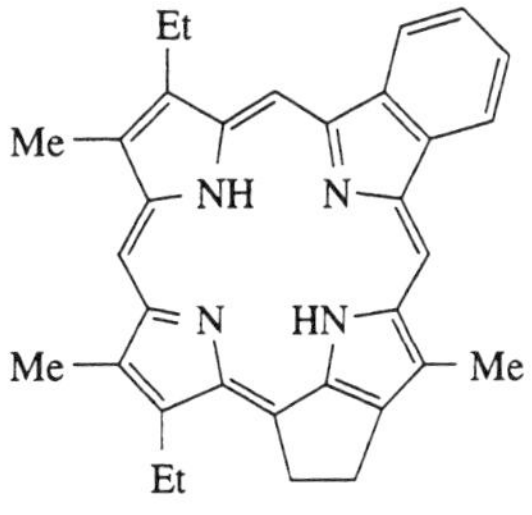

94

310

Though DPEP **90** contains a five-membered exocyclic ring, there is abundant evidence to suggest that exocyclic rings of different sizes are present in petroleum residues; as a result, much effort has been expended on the development of synthetic methodology to prepared porphyrins bearing exocyclic rings of various sizes. For example, P.S. Clezy et al. (J. Chem. Soc., Chem. Commun., 1988, 83; Aust. J. Chem., 1990, **43**, 839) accomplished partial syntheses of petroporphyrins bearing six and seven membered rings, while, in a series of papers, Lash's group described preparations of porphyrins (e.g. **96**) with as many as four exocyclic rings (T.D. Lash et al., Tetrahedron Lett., **1987**, *28*, 1135; J. Org. Chem., **1992**, *57*, 4809), of so-called type-II DPEPs (T.D. Lash and J.J. Catarello, Tetrahedron, 1993, **49**, 4159), and of propano- and methyl-propanoporphyrins of sedimentary origin (T.D. Lash and T.J. Perun, Jr., Tetrahedron Lett., 1987, **28**, 6265; 1990, **31**, 7545; T.D. Lash et al., Energy Fuels, 1993, **7**, 172).

95

96

97

The bacteriochlorophylls -c, -d, and -e are unusual chlorophylls produced by green or brown sulfur bacteria (K.M. Smith, Photosyn. Res., 1994, **41**, 23). The unusual 8-propyl and 8-isobutyl substituents present in petroporphyrins from Messel oil shale have been suggested to be derived from bacteriochlorophyll-d precursors (R. Ocampo et al., J. Chem. Soc., Chem. Commun., 1985, 200), and partial syntheses of these sedimentary pigments (e.g. **97**) have been accomplished from samples of the natural bacteriochlorophylls-d (N.W. Smith and K.M. Smith, J. Chem. Soc., Perkin Trans. 1, 1989, 188; Energy Fuels, 1990, **4**, 675); analogous bacteriopetroporphyrins from the bacteriochlorophyll-c series, though not yet isolated from oil shales, were also synthesized.

(d) Porphyrin Oligomers

Usually based upon the argument that the only way to effectively mimic the photosynthetic reaction center or antenna systems of photosynthetic systems is to study porphyrin oligomeric systems, there have been published literally hundreds of ingenious papers detailing synthetic methodology, spectroscopic properties, and photophysical character-ization of oligomeric porphyrin systems. Space does not allow any discussion of synthetic tactics or rationale for carrying the syntheses forward, but a brief listing is provided below as a means of getting a flavor for what has been going on. Thus, tetrameric and hexameric porphyrin arrays (G.M. Dubowchik and A.D. Hamilton, J. Chem. Soc., Chem. Commun., 1986, 1391; 1987, 293), square multiporphyrin arrays (C.M. Drain and J.-M. Lehn, J. Chem. Soc., Chem. Commun., 1994, 2313), an orthogonal porphyrin triad (K. Susumu et al., Chem. Lett., 1995, 929), porphyrins linked by alkyne spacers (H. Imahori et al., Bull. Chem. Soc. Jpn., 1994, **67**, 2500; D.P. Arnold and L.J. Nitschinsk, Tetrahedron, 1992, **48**, 8781; V.S.Y. Lin et al., Science, 1994, **264**, 1105; H.L. Anderson et al., J. Chem. Soc., Perkin Trans. 1, 1995, 2275; S. Anderson et al., J. Chem. Soc., Perkin Trans. 1, 1995, 2255; D.W. McCallien et al., J. Am. Chem. Soc., 1995, **117**, 6611), azobenzene linked dimers (H.K. Hombrecher and K. Luedtke, Tetrahedron, 1993, **49**, 9489), phenylene bridged diporphyrins (A. Osuka et al., J. Am. Chem. Soc., 1993, **115**, 4577; Chem. Lett., 1993, 161), bis-porphyrins with phenanthroline attachments (S. Noblat-Chardon and J.P. Sauvage, Tetrahedron, 1991, **47**, 5123; M.J. Crossley et al., J. Chem. Soc., Chem. Commun., 1995, 1921), hydrocarbon-linked systems (E.B. Fleischer and A.M. Shachter, J. Heterocycl. Chem., 1991, **28**, 1693), self-associating base-paired conjugates (A. Harriman et al., J. Phys. Chem., 1991, **95**, 1530; see also J. Am. Chem. Soc., 1992, **114**, 388), ruthenium templated systems (A. Harriman et al., J. Am. Chem. Soc., 1995, **117**, 9461), strapped, dimeric and trimeric porphyrins synthesized via intramolecular macrocyclization reactions (A. Osuka et al., Bull. Soc. Chem. Jpn., 1991, **64**, 1213), quinone-linked oligomers (J.L. Sessler and S. Piering, Tetrahedron Lett., **1987**, *28*, 6569; A. Osuka et al., Tetrahedron, 1989, **45**, 4815; Chem. Lett., 1991, 481; P. Leighton and J.K.M. Sanders, J.

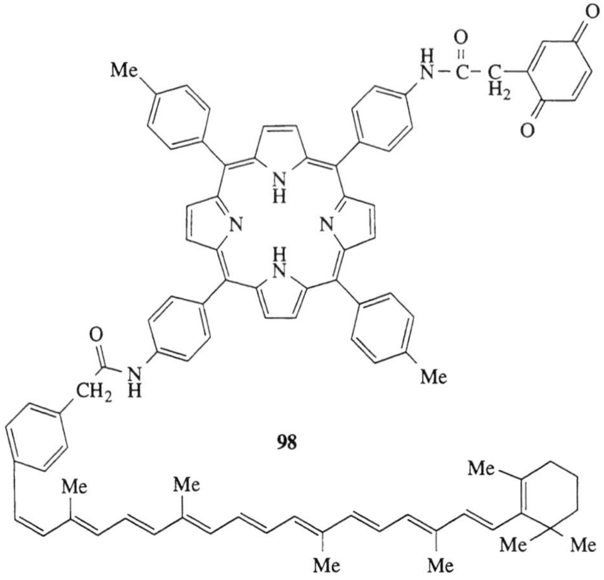
Me
H
N
C
O
C
H2
O
O
N
H
N
N
H
N
O
N
H
CH2
Me
Me
Me
Me
Me
Me
Me
Me

98

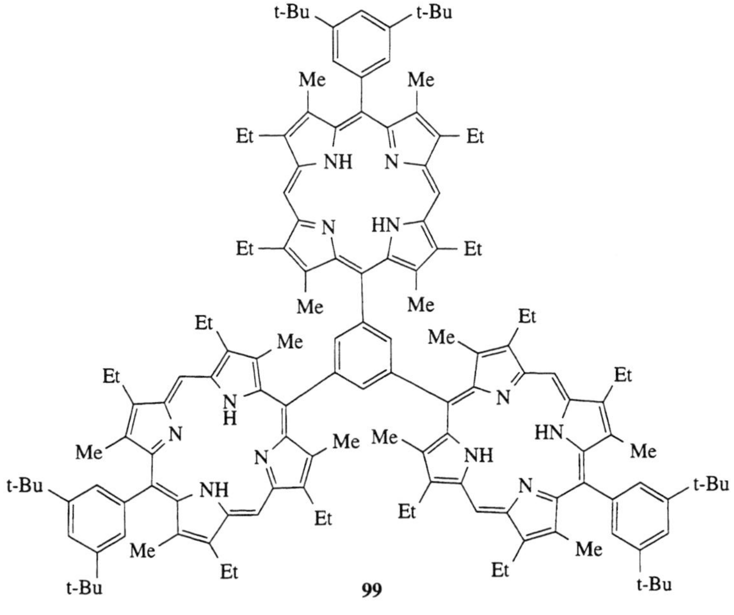
t-Bu
t-Bu
Me
Me
Et
Et
NH
N
N
HN
Et
Et
Me
Me
Et
Me
Me
Et
Et
Et
N
H
N
Me
N
Me
Me
HN
NH
Me
t-Bu
NH
N
t-Bu
Et
Et
Me
Et
Me
Et
t-Bu
t-Bu

99

Chem. Soc., Perkin Trans. 1, 1987, 2385; L. Sun et al., Tetrahedron, 1995, **51**, 3535), carotenoid-linked porphyrins (A. Osuka et al., Chem. Lett., **1990**, 1905), carotenoid-porphyrin-quinone systems (e.g. **98**) (D. Gust et al., Tetrahedron, 1989, **45**, 4867; J. Am. Chem. Soc., 1987, **109**, 846; Methods Enzymol., 1992, **213**, 87), a porphyrin-oxochlorin-pyromellitidimide triad (A. Osuka et al., Bull. Chem. Soc. Jpn., 1995, **68**, 262), pyromellitimide-bridged porphyrins (J.A. Cowan and J.K.M. Sanders, J. Chem. Soc., Chem. Commun., 1985, 2435), conformationally constrained oligomers (A. Osuka et al., Chem. Lett., 1989, 741; T. Nagata et al., J. Am. Chem. Soc., 1990, **112**, 3054; E.J. Atkinson et al., Tetrahedron Lett., 1993, **34**, 6147), a rigid/terpyridine system (J.P. Collin et al., Tetrahedron Lett., 1991, **32**, 5977), a 1,3,5-triporphyrinylbenzene **99** (A. Osuka et al., J. Org. Chem., 1993, **58**, 3582), and symmetrical and/or unsymmetrical porphyrin dimers (K. Maruyama et al., Bull. Chem. Soc. Jpn., 1989, **62**, 3167; J.L. Sessler et al., J. Am. Chem. Soc., 1990, **112**, 9310; R.K. Pandey et al., Tetrahedron Lett., 1992, **33**, 5315) and oligomers (A. Vidal-Ferran et al., J. Chem. Soc., Chem. Commun., 1994, 2657) have been described. Diphenylethane linked porphyrins have been made for the first time, using an organometallic coupling reaction of porphyrinyl meso-substituted benzyl halides (D.A. Lee et al., Heterocycles, 1995, **40**, 131). Oligomeric porphyrin systems can also be efficiently prepared via stepwise nucleophilic substitution of aminoporphyrins to cyanuric chloride (K. Ichihara and Y. Naruta, Chem. Lett., 1995, 631).

In literally dozens of papers, investigators have accomplished the coupling of various biologically important residues (including for example aminoacids, nucleosides, sugars, etc.) to photoactive porphyrins. Examples include porphyrinyl nucleosides (L. Czuchajowski et al., Tetrahedron Lett., 1991, **32**, 7511; J. Heterocycl. Chem., 1992, **29**, 479; Tetrahedron Lett., 1993, **34**, 5409; M. Hisatome et al., Chem. Lett., 1993, 1357; Heterocycles, 1993, **36**, 441; M. Cornia et al., J. Org. Chem., 1995, **60**, 4964; X. Jiang et al., Tetrahedron Lett., 1995, **36**, 365) and sugar conjugates (K. Kohata et al., Chem. Lett., 1992, 477; N. Ono et al., Tetrahedron Lett., 1992, **33**, 1629; G. Casiraghi et al., Nat. Prod. Lett., 1992, **1**, 45; J. Org. Chem., 1994, **59**, 1801; P. Maillard et al., J. Org. Chem., 1993, **58**, 2774; A. Bourhim et al., Synlett., 1993, 563); a cyclodextrin-sandwiched porphyrin has also been reported (Y. Kuroda et al., Chem. Lett., 1994, 2361). Other porphyrin conjugates involve C60 (P.A. Liddell et al., Photochem. Photobiol., 1994, **60**, 537; H. Imahori et al., Chem. Lett., 1995, 265), flavins (J. Takeda et al., J. Am. Chem. Soc., 1987, **109**, 7677; cyanine dyes (J.S. Lindsey et al., Tetrahedron, 1989, **45**, 4845), salens (K. Maruyama et al., Bull. Chem. Soc. Jpn., 1991, **64**, 29) and barbituric acid (B.C. Robinson and A.R. Morgan, Tetrahedron Lett., 1993, **43**, 3711).

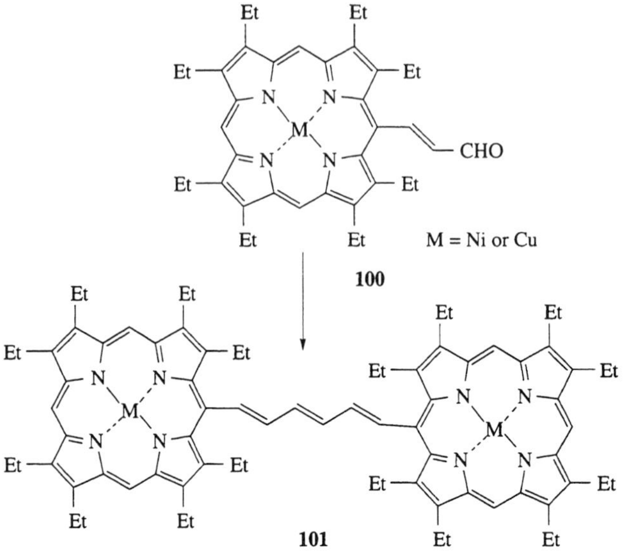

100

101

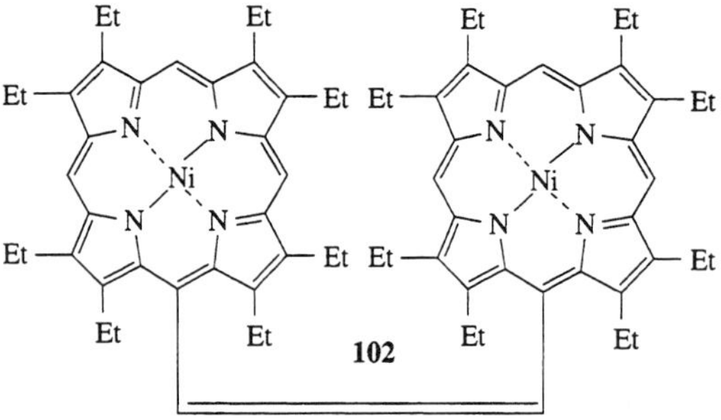

102

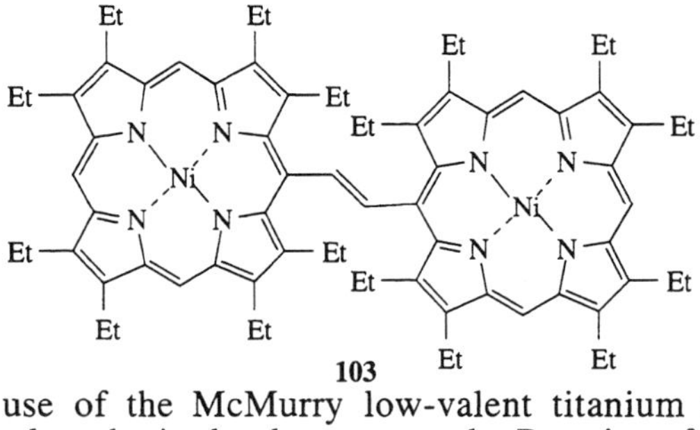

103

Widespread use of the McMurry low-valent titanium reaction for coupling formylporphyrins has been reported. Reaction of copper(II) or nickel(II) porphyrins with dimethylaminoacrolein/POCl$_3$ gives the corresponding acrolein porphyrins **100** which can be coupled under standard McMurry conditions to give the bis-porphyrin **101** (M.G.H. Vicente and K. M. Smith, J. Org. Chem., 1991, **56**, 4407); when meso-formyl-porphyrins are used, the corresponding cis (**102**) or trans (**103**) 1,2-diporphyrinyl-ethenes are obtained (M.G.H. Vicente and K. M. Smith, J. Org. Chem., 1991, **56**, 4407; M. O. Senge et al., Angew. Chem. Int. Ed. Engl., 1993, **32**, 750; Inorg. Chem., 1994, **33**, 5625). The same type of compounds can be prepared from the corresponding bis-porphyrinyl-ethanes (D.V. Yashunsky et al., Tetrahedron Lett., 1995, **36**, 8485), and ethene bridged trimers have also been described (H. Higuchi et al., Tetrahedron Lett., 1995, **36**, 5359).

Cofacial bis-porphyrin systems have generated a great deal of interest, mostly with regard to their use in catalysis, ligand binding, and as mimics of various components of the photosynthetic apparatus in plants and bacteria. Interesting work has centered on cofacial biphenylene or anthracene pillared bis-porphyrins (originally know as "Pac-man" porphyrins), and particularly on heterobimetallic examples. Thus, the novel cobalt/aluminum heterobimetallic biphenylene pillared cofacial diporphyrin has been prepared and characterized (R. Guilard et al., J. Am. Chem. Soc., 1992, **114**, 9877), as has the copper/manganese analogue (R. Guilard et al., J. Am. Chem. Soc., 1994, **116**, 10202); the latter compound, in the biphenylene series, reversibly binds dioxygen at room temperature. A series of other metalated examples of Pac-man porphyrins have been made and examined as electrocatalysts for proton reduction (J.P. Collman et al., J. Am. Chem. Soc., 1993, **115**, 9080). EPR spectroscopy of bis-copper anthracene pillared Pac-man porphyrins has been used to estimate the Cu-Cu distance, and these data have been compared with X-ray parameters (S.S. Eaton et al., J. Am. Chem. Soc., 1985, **107**, 3177). A triple-deckered triporphyrin has also been prepared (I. Abdalmuhdi and C.K. Chang, J. Org. Chem., 1985, **50**, 411) in a one-step condensation between a 1,9-diunsubstituted dipyrrylmethane and 8-porphyrinyl-1-anthracenecarboxaldehyde. Other cofacial bisporphyrins have featured in electrochemical studies (J.A. Cowan and J.K.M. Sanders, J. Chem. Soc., Perkin Trans. 1, 1987, 2395) and in excited state electron transfer investigations (D. Heiler et al., J. Am. Chem. Soc., 1987, **109**, 604). Conformations of water soluble cofacial porphyrin dimers have been studied using proton NMR, spectrophotometry, and emission spectroscopy (R. Karaman and T.C. Bruice, J. Org. Chem. 1991, **56**, 3470). The first example of heterometallic mixed triads formed using multiple bonds between metals in metalloporphyrins has been reported (J.P. Collman et al., J. Am. Chem. Soc., 1994, **116**, 9761), as have some quite remarkable 1 to 1 sandwich type complexes between

316

phthalocyanines and porphyrins (M. Lachkar et al., New J. Chem. 1988, **12**, 729).

Porphyrinyl 2,3-diones **104** and 2,3,7,8- (**105**) or 2,3,12,13-tetra-ones have been prepared, and have been shown to be excellent precursors for the simple construction of quite remarkable structural arrays (e.g. **106**) (M.J. Crossley and P.L. Burn, J. Chem. Soc., Chem. Commun., 1987, 39; M.J. Crossley et al., J. Chem. Soc., Chem. Commun., 1991, 1567; 1995, 2379).

104   **105**

106

4.   Chlorins and Bacteriochlorins

(a) Chlorophylls and Derivatives

Without doubt, a major occasion in the field of tetrapyrrole synthesis is the publication of the full details of R.B. Woodward's synthesis of chlorin-e$_6$ trimethyl ester **107**, and by relay, of chlorophyll-a **108** (R.B. Woodward et al., Tetrahedron, 1990, **46**, 7599); the article is not only a masterpiece of chemistry, but is also an historical document which should be read from front to back by all porphyrinologists.

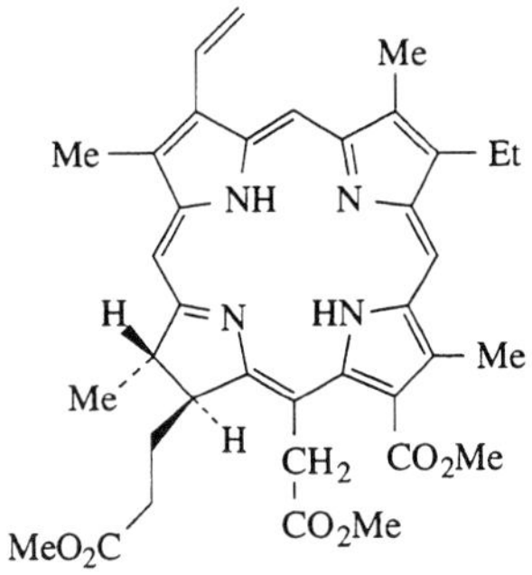

107

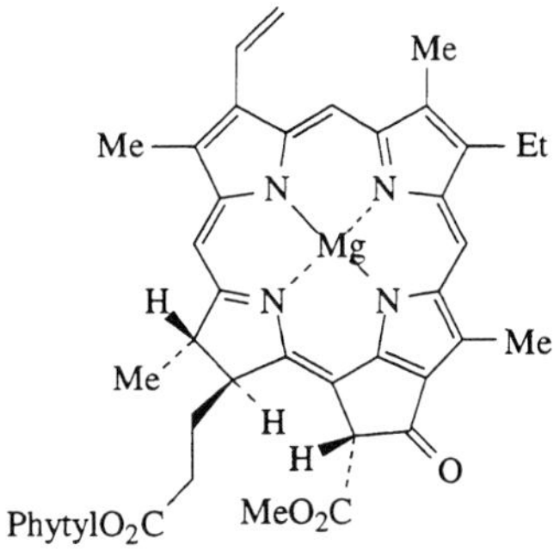

108

Major monographs on chlorophylls ("Chlorophylls", H. Scheer, Ed., CRC Press, Boca Raton, 1991, pp 1-1257) and on bacterial chlorophylls ("Anoxygenic Photosynthetic Bacteria", R.E. Blankenship et al., Eds., Kluwer, Dordrecht, 1995, pp 1-1331) have appeared.

Tunichlorin **109**, a nickel chlorin isolated from the Caribbean tunicate *Trididemnum solidum* has been isolated and identified (K.L. Rinehart et al., Proc. Natl. Acad. Sci U.S.A., 1988, **85**, 4582; J. Nat. Prod., 1988, **51**, 1).

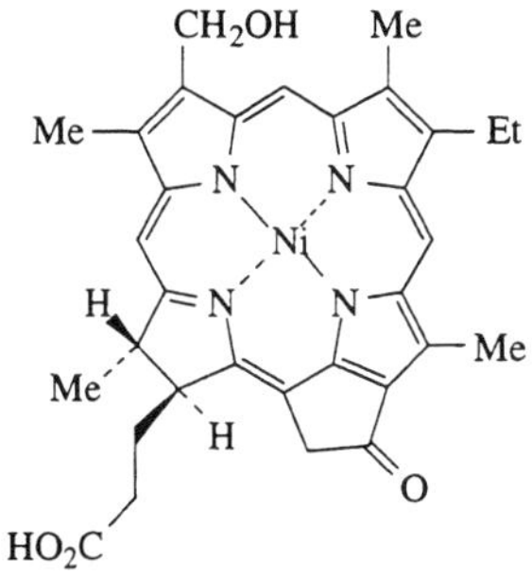

109

The bacteriochlorophylls -c **110**, -d **111**, and -e **112** (Tables 1-3) occur in certain green and brown bacteria; the striking difference between these bacteriochlorophylls and chlorophyll-a appears in certain of the

318

peripheral substituents, which are further alkylated (above the chlorophyll-a **108** level) by methionine (F. W. Bobe et al., Biochemistry, 1990, **29**, 4340; M.S. Huster and K.M. Smith, Biochemistry, 1990, **29**, 4348). A systematic nomenclature for the various bacteriochlorophyll -c, -d, and -e homologues has been proposed and formally adopted (K.M. Smith, Photosyn. Res., 1994, **41**, 23).

Table 1: Structural Assignments for the Bacteriochlorophylls *c* **110**. (K.M. Smith et al., J. Chromatogr., 1983, **281**, 209).

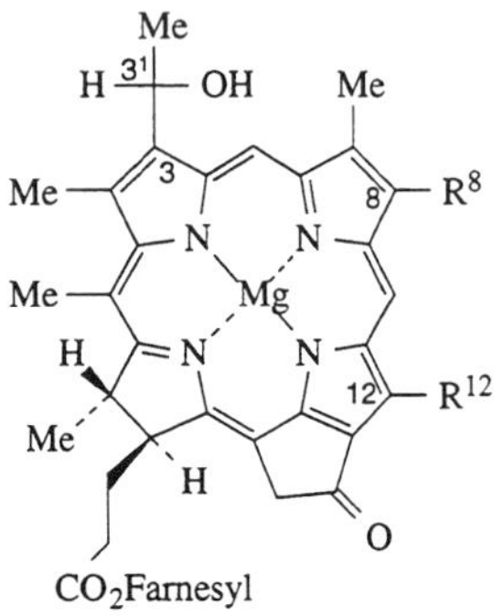

110

Homologue	Holt's Band #[a]	R^8	R^{12}	Configuration at C-3^1
110a	6	Et	Me	R
110b	5	Et	Et	R
110c	4	Pr	Et	R
110d	3	Pr	Et	S
110e	2	iBu	Et	R
110f	1	iBu	Et	S

$Pr = CH_2CH_2CH_3$; $iBu = CH_2CH(CH_3)_2$

[a] (A.S. Holt et al., Can. J. Chem., 1966, **44**, 88).

Table 2: Structural Assignments for the Bacteriochlorophylls d **111**. (K.M. Smith and D.A. Goff, J. Chem. Soc., Perkin Trans. 1, 1985, 1099).

111

Homologue	Holt's Band #[a]	R^8	R^{12}	Configuration at C-3[1]
111a	6	Et	Me	R
111b	4	Et	Et	R
111c	5	Pr	Me	R
111d	2	Pr	Et	R
111e	3	iBu	M	S
111f	1	iBu	Et	S
111g	-	NeoPn	Me	S
111h	-	NeoPn	Et	S

$NeoPn = CH_2C(CH_3)_3$

[a] (A.S. Holt and D.W. Hughes, J. Am. Chem. Soc., 1961, **83**, 499; D.W. Hughes and A.S. Holt, Can. J. Chem., 1962, **40**, 171).

A direct route for transformation of the bacteriopheophorbides-d **113** into the bacteriopheophorbides-c **114** has been described, along with several other reactions for accomplishment of 20-substitution in **113** (K.M. Smith et al., 1985, **107**, 4946).

Table 3: Structural Assignments for the Bacteriochlorophylls e̲ **112**
(D.J. Simpson and K.M. Smith, J. Am. Chem. Soc., 1988, **110**, 1753).

112

Homologue	R^8	Configuration at C-3^1	Percent of R or S for each R^8 homologue
112a	Et	R	95
112b	Et	S	5
112c	Pr	R	40
112d	Pr	S	60
112e	iBu	R	2
112f	iBu	S	98

113

114

115

Methyl bacteriochlorophyllide-d **115** in chloroform forms a tight dimer which is long-lived on the NMR time scale (K.M. Smith et al., J. Am. Chem. Soc., 1986, **108**, 1111); the interplane separation in the dimer was calculated, using a double dipole ring current model, to be approximately 3.5Å, and the dimer is formed by use of bidirectional pairs of interactions between the chiral 3^1-hydroxyl and the magnesium in an adjacent component of the dimer. The tight dimer can be broken into monomer units by use of competitive ligands such as methanol and pyridine. Functional groups responsible for the interactions which are involved in solution aggregation phenomena of various chlorophyll derivatives (e.g. pyrochlorophylls) have been investigated using partial synthesis from chlorophyll degradation products, and proton NMR spectroscopy (R.J. Abraham et al., J. Chem. Soc., Perkin Trans. 2, 1993, 1047).

(b)   Other Chlorins

Sulfhemoglobin and sulfmyoglobin are green heme proteins in which the prosthetic group has been altered by introduction of a sulfur atom into the porphyrin macrocycle. The various sulf-pigments possess the optical spectra of chlorins. Structures for three forms of so-called "sulfhemins" have been proposed (M.J. Chatfield et al., J. Am. Chem. Soc., 1986, **108**, 7108), and metal-free oxygen analogues **116** and **117** of sulfhemins A and C, respectively, have been synthesized using Mitsunobu-type chemistry (P. Iakovides and K.M. Smith, Tetrahedron Lett., 1990, **31**, 3853; Tetrahedron, 1996, **52**, 1123). Compound **116** represents the first example of a porphyrin epoxide, because C.K. Chang et al. (Angew. Chem. Int. Ed. Engl., 1992, **31**, 70) have shown that a compound previously reported by Johnson et al. in 1971 to be a porphyrin epoxide is actually a novel furan-containing macrocycle **118**. A dimer containing an epoxide has been reported (G.V. Ponomarev and A.M. Shul'ga, Khim. Geterotsikl. Soedin., 1992, 126).

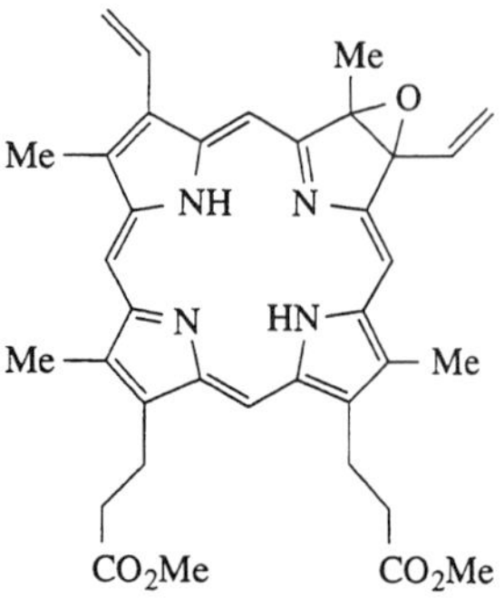

116

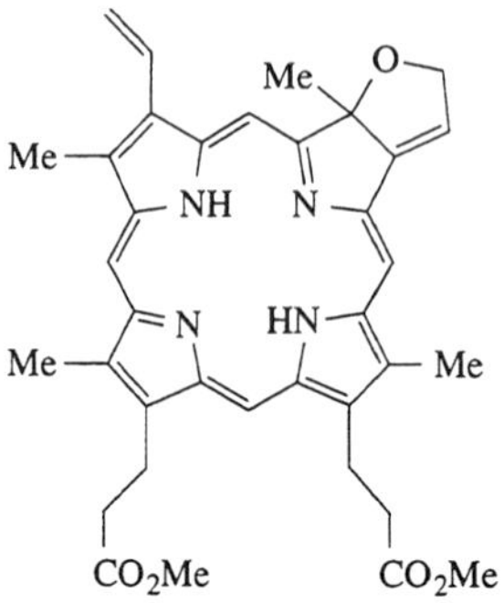

117

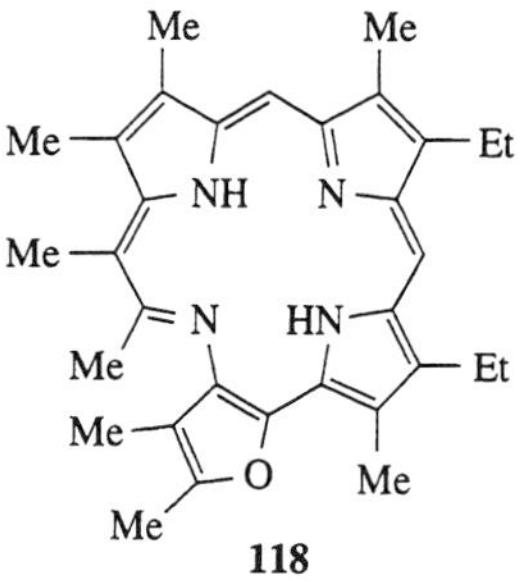

118

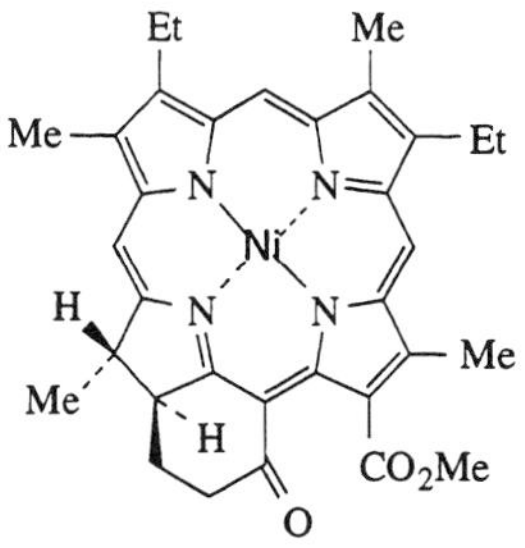

119

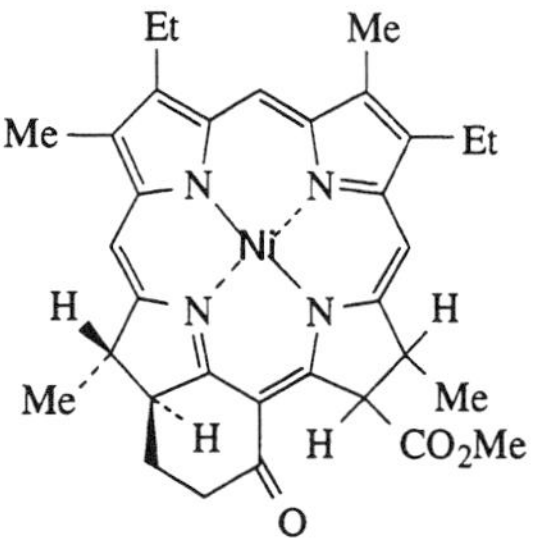

120

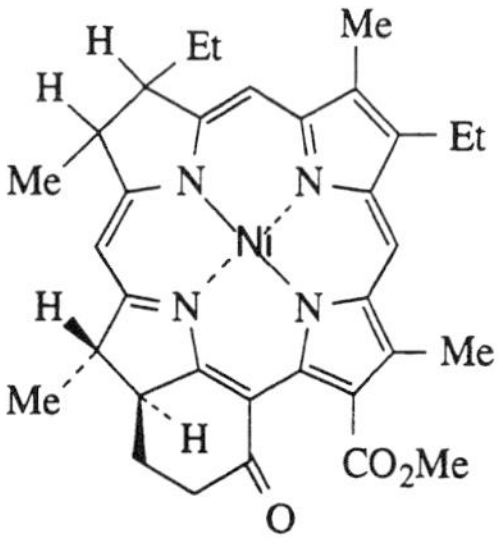

121

Raney nickel reductions of the nickel "anhydro"-chlorin **119** affords a number of interesting new pigments in the isobacteriochlorin (**120** and **121**), hexahydroporphyrin (**122** and **123**) octahydroporphyrin **124** series (D.J. Simpson and K.M. Smith, J. Am. Chem. Soc., 1988, **110**, 2854).

The series of pigments **119-124** proved to be good substrates for investigation of ring-current diminutions of progressively more reduced chlorophyll-type macrocycles (R.J. Abraham et al., J. Chem. Soc., Perkin Trans. 2, 1988, 1365) and as models for Factor 430 from methanogenic bacteria (M.W. Renner et al., J. Am. Chem. Soc., 1991, **113**, 6891).

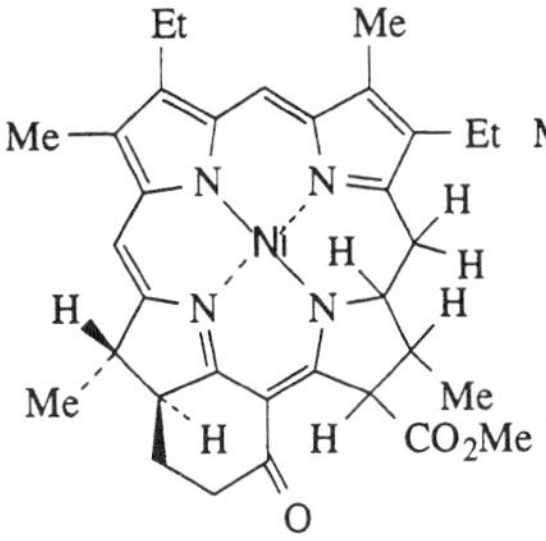

122

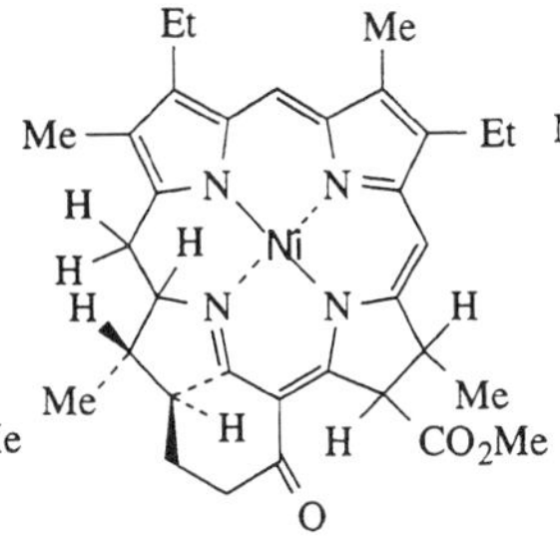

123

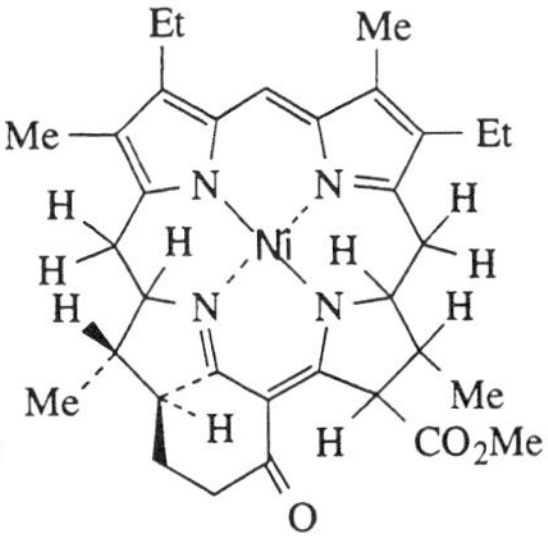

124

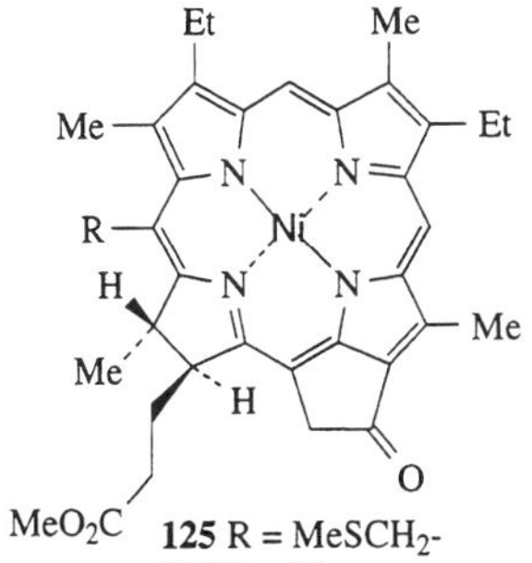

125 R = $MeSCH_2$-
126 R = Me

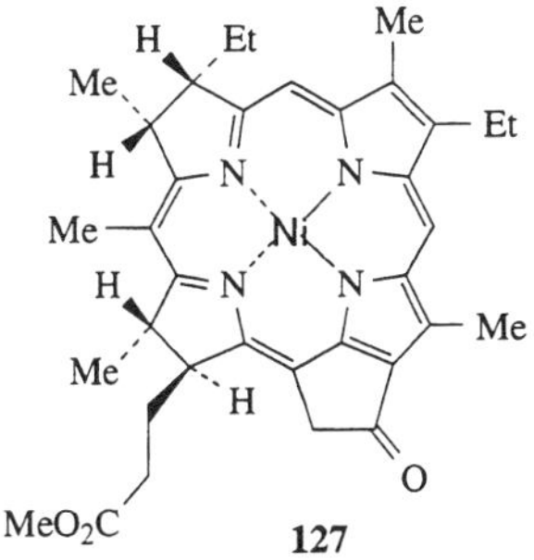

127

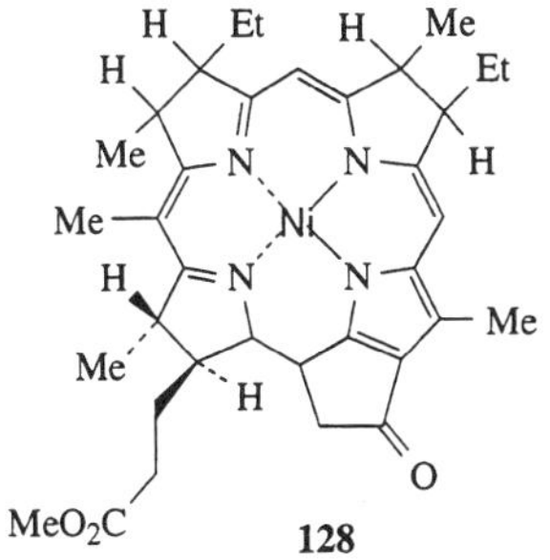

128

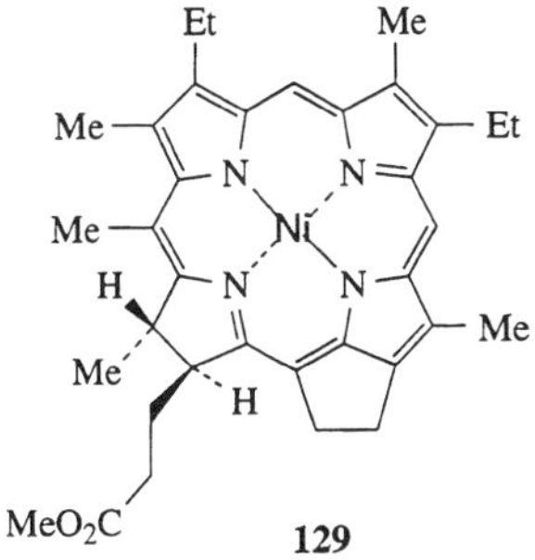

129

Similar treatment of nickel(II) 20-[(methylthio)methyl]mesopyropheophorbide a **125** with Raney nickel gives the expected 20-methyl

324

compound **126**, along with a nickel isobacteriochlorin **127**, a nickel pyrrolocorphin **128** and the nickel 20-methyl-13^1-deoxopyropheophorbide **129** (K.M. Smith and D.A. Goff, J. Am. Chem. Soc., **1985**, *107*, 4954).

Ascorbic acid photoreduction of zinc(II) vinylporphyrins **130** affords the cis-chlorin **131** in which the pyrrole subunit containing the vinyl group has been reduced; subsequent rearrangement takes place (depending upon solvent) to afford vinylidene chlorins **132** and eventually ethylporphyrins **133** (K.M. Smith and D.J. Simpson, J. Chem. Soc., Chem Commun , 1987, 613; P. Iakovides et al., Photochem. Photobiol., 1988, **54**, 335); the sequence of reactions **130-133** was proposed as a model for 8-vinyl reduction in the biosynthesis of chlorophyll-a **108** from protoporphyrin-IX.

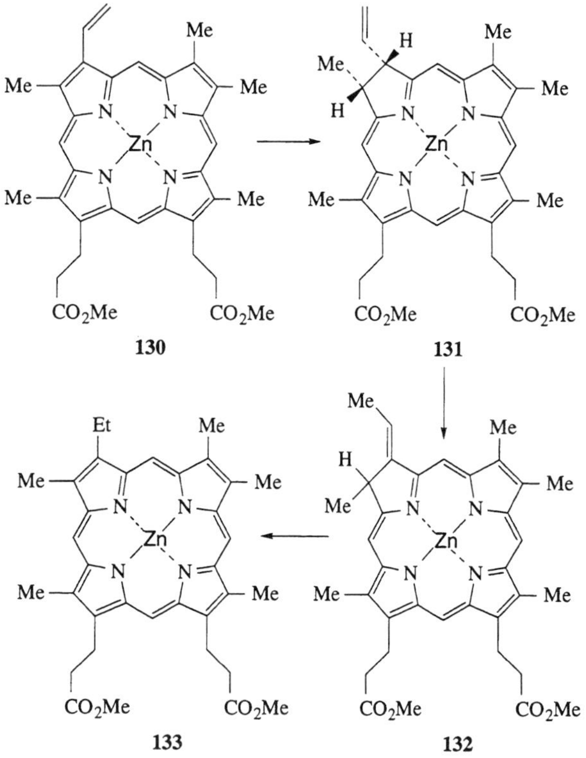

The same ascorbic acid photoreduction reaction, when performed upon zinc(II) chlorins, affords regiochemically pure zinc(II) isobacteriochlorins which can be demetalated to provide a very efficient access to metal-free isobacteriochlorins (K.M. Smith et al., J. Am. Chem. Soc., 1986, **108**, 6834; D.J. Simpson and K.M. Smith., J. Am. Chem. Soc., 1988, **110**, 2854).

The OsO_4 reaction to give vic-diols has been shown, not surprisingly, to work on tetraphenylporphyrin (C. Brueckner and D. Dolphin, Tetrahedron Lett., 1995, **36**, 3295). OsO_4 oxidation of zinc chlorins gives a very simple access to isobacteriochlorins, whereas OsO_4 oxidation of metal free chlorins gives uniquely bacteriochlorins (C.K. Chang et al., J. Chem. Soc., Chem. Commun., 1986, 1213).

Chlorins related to chlorophyll-a can be dehydrogenated to afford the corresponding porphyrins by treatment with benzoyl chloride in DMF (L. Ma and D. Dolphin, Tetrahedron Lett., 1995, **36**, 7791). Related chlorins can be conjugated with nucleoside units, using Heck-type chemistry (X. Jiang et al., Tetrahedron Lett., 1995, **36**, 365) or rendered water-soluble by treatment of the 3-vinyl group with Eschenmoser's reagent, followed by quaternization with methyl iodide (R.K. Pandey et al., J. Chem. Soc., Chem. Commun., 1991, 1637).

134 **135**

As a means of probing the mechanisms and pathways of chlorophyll degradation in senescent leaves, photo-oxidation of regiochemically pure oxochlorins **134** (obtained from the corresponding 20-trifluoroacetoxy-chlorins) affords ring-opened tetrapyrroles **135** by way of a mechanism involving two dioxygen molecules (M.S. Huster and K.M. Smith, Tetrahedron Lett., 1988, **29**, 5707). 20-Trifluoroacetoxy- and 20-chloro-chlorophyll derivatives form oxoniachlorins (e.g. **136**) when photo-oxidized (J. Iturraspe and A. Gossauer, Photochem. Photobiol., 1991, **54**, 43). Natural chlorophyll degradation has been shown, in a series of definitive investigations outside the scope of this review (e.g. B. Kraeutler et al., Angew. Chem. Int. Ed. Engl., 1991, **30**, 1315; J.

326

Iturraspe et al., Photochem. Photobiol., 1993, **58**, 116) to involve ring opening of the macrocycle at the 5-position. Two examples of oxidative ring cleavage of chlorophyll derivatives at the 5-position, which mimic the natural process, have recently been described; the first involves oxidative cleavage of a cadmium(II) chlorophyll derivative (J. Iturraspe and A. Gossauer, Tetrahedron, 1992, **48**, 6807; Helv. Chim. Acta, 1991, **74**, 1713), and the second involves a less biomimetic oxidative ring cleavage of a zinc(II) 10-trifluoroacetoxy-2-oxochlorin (M. Isaac et al., J. Chem. Soc., Perkin Trans. 1, 1995, 705).

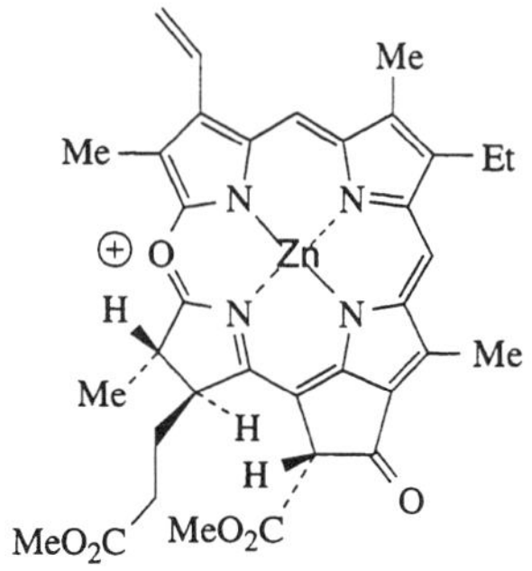

136 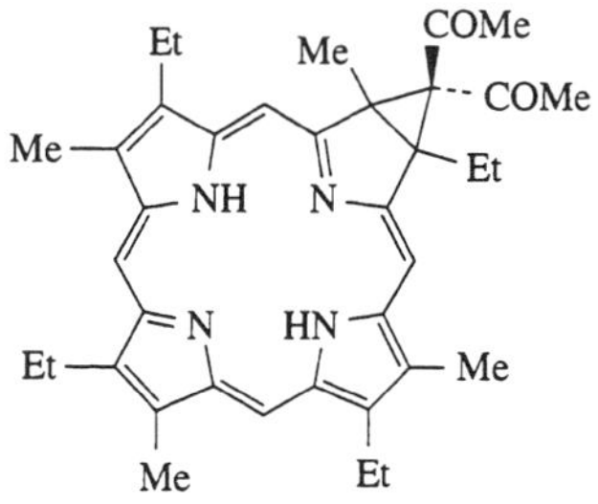   **137**

Reduction of unsymmetrically substituted hemins using sodium in isopentyl alcohol gives a mixture of all four possible chlorins (D.H. Burns et al., J. Chem. Soc., Perkin Trans. 1, 1988, 3119). Treatment of mesoporphyrin-IX dimethyl ester with 6% hydrogen peroxide in sulfuric acid gives four oxochlorins, four dioxoisobacteriochlorins, and one dioxobacteriochlorin (C.K. Chang and W. Wu, J. Org. Chem., 1986, **51**, 2134). When etioporphyrin-I 2,3-diol is heated with zinc acetate in acetylacetone, chlorin dimers and a diacetylcyclopropylchlorin **137** are obtained (G.V. Ponomarev and A.M. Shul'ga, Khim. Geterotsikl. Soedin., 1992, 126). A rational ring-synthesis of a chlorin (17,18-di*hydro*-porphyrin) has been described (D.H. Burns et al., Tetrahedron Lett., 1993, **34**, 2883); chlorins are usually obtained by dihydrogenation of porphyrin systems under a variety of different conditions.

Simple syntheses of a number of chlorins have been worked out and reported, either involving allylboration of formylporphyrins (F. Hoeper and F.P. Montforts, Liebigs Ann. Chem., 1995, 1033), or through reactions of thiones (F.P. Montforts and U.M. Schwartz, Liebigs Ann. Chem., 1985, 1228). Geminal dialkyl chlorins **138** have been synthesized in good yield by treatment of mono-(1-hydroxyethyl)-porphyrins (e.g. **139**) with N,N-dimethylacetamide dimethylketal,

Claisen rearrangement (to give **140**), and hydrogenation (F.P. Montforts and G. Zimmermann, Angew. Chem. Int. Ed. Engl., 1986, **25**, 458); dioxobacteriochlorins **141** have also been prepared by the same group, starting with hematoporphyrin dimethyl ester **142** (F.P. Montforts et al., Angew. Chem., 1989, **101**, 471). Optically active chlorins and iso-bacteriochlorins with geminal dialkyl substituents can be obtained by chirogenic, enantioselective reduction, followed by Claisen rearrangement with transfer of chirality (K. Kusch et al., Angew. Chem. Int. Ed. Engl., 1995, **34**, 784).

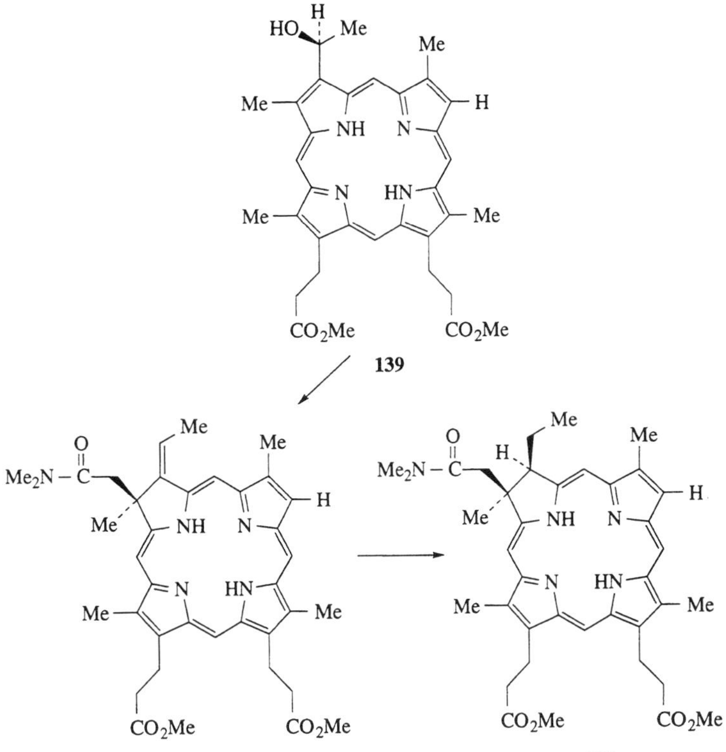

139

140

138

Meso-aminomethylporphyrins **143** in ethyl iodide afford 1,2-bis-porphyrinylethanes **144** in high yield, and this same reaction can be applied to the synthesis of 1,2-bis-chlorinylethanes (G.V. Ponomarev, Khim. Geterotsikl. Soedin., 1993, 1692). Vilsmeier formylation of copper(II) trans-octaethylchlorin, followed by treatment of the intermed-

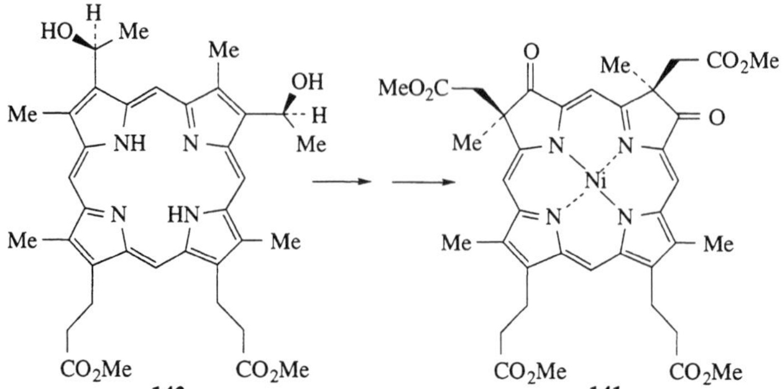

328

142

141

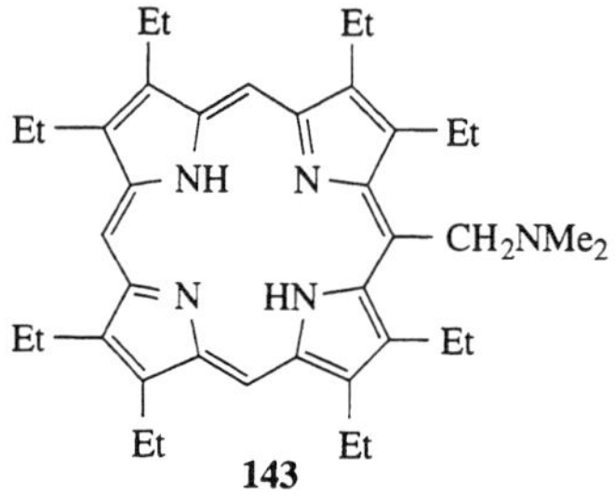

143

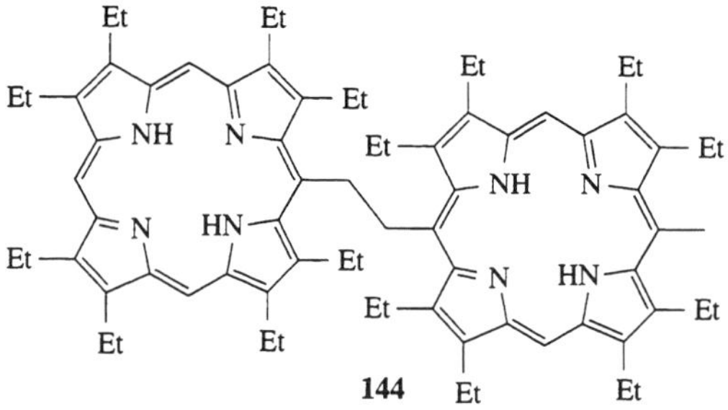

144

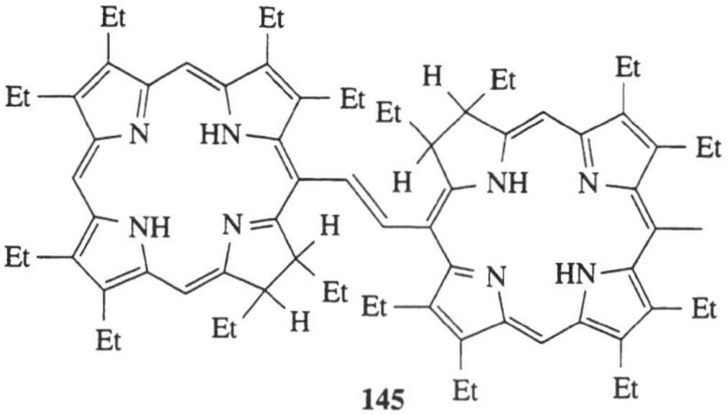

145

iate imine with methyl iodide and reduction gives the dimethyl-aminomethylchlorin; dimerization in methyl iodide afforded the 1,2-bis-chlorinylethane which could be demetalated and oxidized to give the 1,2-trans-bis(trans-octaethylchlorin)ethene **145** (G.V. Ponomarev, Khim. Geterotsikl. Soedin., 1993, 1693). The same molecule **145** can be prepared by low-valent titanium (McMurry-type) self-condensation of 20-formyl-trans-octaethylchlorin metal complexes, followed by demetalation; the crystal structure of bis-chlorin **145**, prepared in that manner, has been reported (M.O. Senge et al., J. Chem. Soc., Perkin Trans. 2, 1993, 11).

N-(Methoxycarbonylmethyl)porphyrins **146**, upon treatment with nickel(II) salts, afford a nickel(II) meso-CH_2CO_2Me porphyrin due to intramolecular migration of the N-substituent, as well as a novel nickel chlorin **147** (K.M. Smith et al., J. Chem. Soc., Chem. Commun., 1986, 1498). Similar chlorins **148** and **149** are also obtained when, for example, 1,19-di(p-tolylmethyl)-a,c-biladienes are cyclized to give species such as **150**, and then heated in dimethylformamide (P.A. Liddell et al., Tetrahedron Lett., 1990, **31**, 2685; J. Org. Chem., 1993, **58**, 6681).

146 R = CH_2CO_2Me **147**

(b) Chlorins and Bacteriochlorins in Photodynamic Therapy

Because of their long-wavelength absorbance and concomitant ability to sensitize singlet oxygen formation at increased depths in biological tissues, chlorins and bacteriochlorins have featured in a host of publications detailing new second and third generation PDT sensitizers. By analogy with the conclusion that the active component in Photofrin®, the only sensitizer currently approved by the US Federal Drug Administration for use in PDT, consists of a mixture of porphyrin oligomers joined by ether bonds, investigators have attempted to link chlorin and porphyrin macrocycles in the same way; (E.G. Levinson and

A.F. Mironov, Bioorg. Khim., 1995, **21**, 230; E.G. Levinson et al., Mendeleev Commun., 1995, 44; Tetrahedron, 1992, **48**, 6485; R. K. Pandey et al., Proc. SPIE, 1991, **1426**, 356).

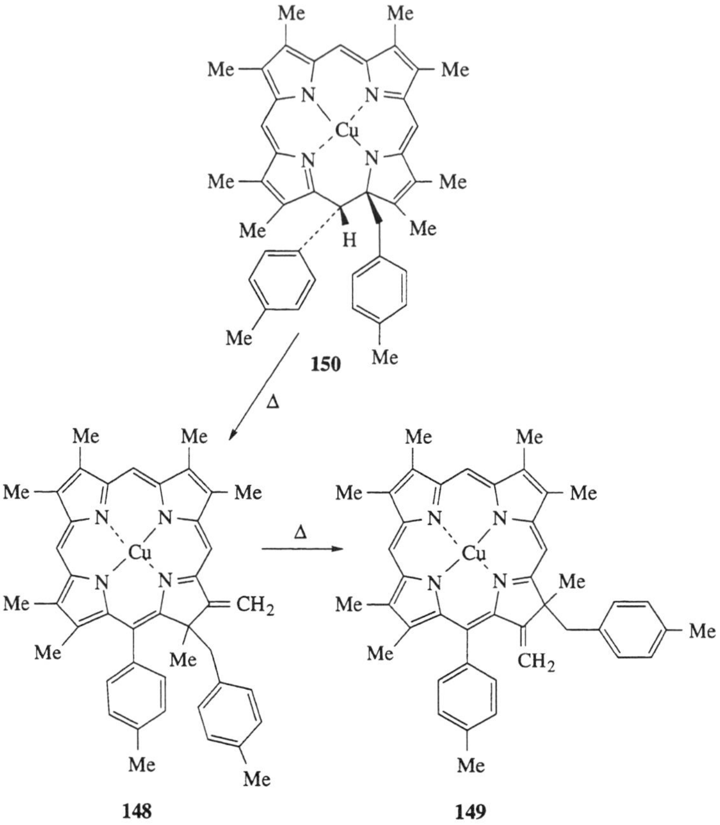

150

148

149

However, the clinically most advanced second generation photosensitizer is without doubt "benzoporphyrin derivative mono-acid" **81**, which is obtained from protoporphyrin-IX dimethyl ester **72** by Diels-Alder cycloaddition with dimethyl acetylenedicarboxylate, followed by chromatographic separations (V.S. Pangka et al., J. Org. Chem., 1986, **51**, 1094). If rhodoporphyrin-XV dimethyl ester is used as the Diels-Alder substrate, biologically active ring A mono-adducts which can be preferentially hydrolyzed in ring D to give mono-acids **82** can be prepared without any troublesome separation procedure (R.K. Pandey et al., Bioorg. Med. Chem. Lett., 1993, **3**, 2615; Tetrahedron, 1996, **52**,

5349). Use of divinylporphyrins **151** in which the vinyl groups are on opposite pyrrole subunits, as Diels-Alder dienes results in the formation of BPD-type molecules **152** possessing the bacteriochlorin chromophore (F.-Y. Shiau et al., Proc. SPIE, 1991, **1426**, 330; P. Yon-Hin et al., Tetrahedron Lett., 1991, **32**, 2875; R.K. Pandey et al., J. Chem. Soc., Perkin Trans. 1, 1992, 1377).

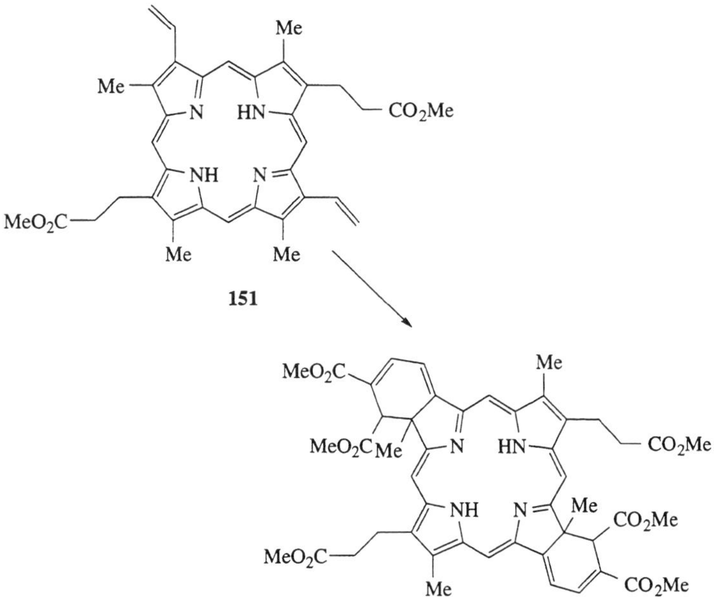

151

152

Efficient syntheses of benzochlorins (e.g. **153**) which absorb at long wavelength have been described. Thus, treatment of nickel or copper porphyrins **154** with dimethylaminoacrolein/POCl$_3$ gives the acrolein derivatives **155**, which can be cyclized in acid to give benzochlorins **153**; obvious expansion of the procedure leads to dibenzobacteriochlorins **156** and dibenzoisobacteriochlorins **157** (M.G.H. Vicente et al., Tetrahedron Lett., 1990, **31**, 1365; M.G.H. Vicente and K. M. Smith, J. Org. Chem., 1991, **56**, 4407). Use of the dimethylaminoacrolein methodology on 5,15-diarylporphyrins yields similar benzochlorins (M.J. Gunter et al., Tetrahedron, 1991, **47**, 7853), while copper(II) 5,15-diphenylporphyrin is converted into 5,15-diaryl-7-oxobenzochlorin **158** using the dimethylamino-acrolein/POCl$_3$ reaction, cyclization, and demetalation (R.W. Boyle and D. Dolphin, J. Chem. Soc., Chem. Commun., 1994, 2463). Benzochlorin monomers, dimers, and porphyrin-benzochlorin heterodimers have similarly been prepared, using the vinylogous Vilsmeier formylation, from 5-aryl- and 5,15-diaryl-octaethylporphyrins (A. Osuka et al., Bull. Chem. Soc. Jpn., 1992, **65**, 3322), or by the MacDonald approach from 5,15-disubstituted octa-alkylporphyrins (D.A. Lee et al., Heterocycles, 1995, **40**, 131), while benzochlorins and chlorins have also been obtained from appropriately meso-substituted porphyrins (D.H. Kohli and A.R. Morgan, Bioorg. Med. Chem. Lett., 1995, **5**, 2175).

153

154 R = H
155 R = CH=CH-CHO

M = Ni, Cu

156

157

Biologically active benzochlorins have been prepared from the chlorophyll-a derivative, methyl 9-deoxopyropheophorbide-a (R.K. Pandey et al., Bioorg. Med. Chem. Lett., 1995, **5**, 857). A number of cationic benzochlorin iminium salts, which appear to be very active in PDT, have been synthesized (D. Skalkos et al., Int. Congr. Ser. - Excerpta Med., 1992, **101**, 860; Photochem. Photobiol., 1994, **59**, 175; D. Skalkos and J.A. Hampton, Med. Chem. Res., 1992, **2**, 276; A.R. Morgan and S. Gupta, Tetrahedron Lett., 1994, **35**, 5347).

Purpurins, such as compound **159** (A.R. Morgan and N.C. Tertel, J. Org. Chem., 1986, **51**, 1347), are currently well-advanced in clinical studies as PDT sensitizers .

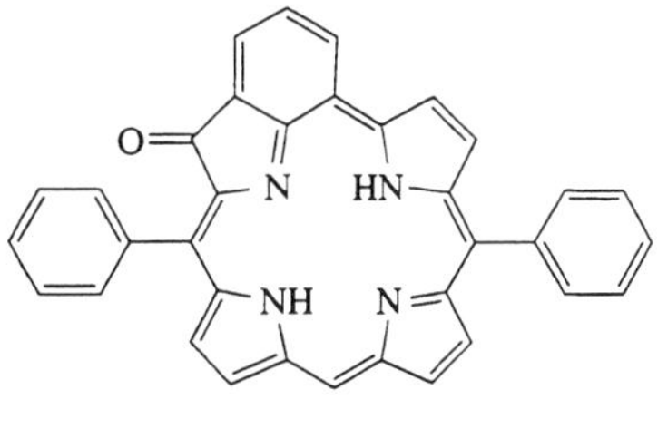

158

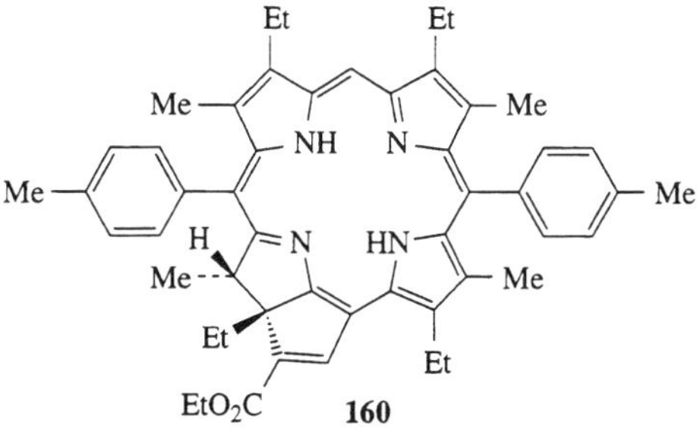

160

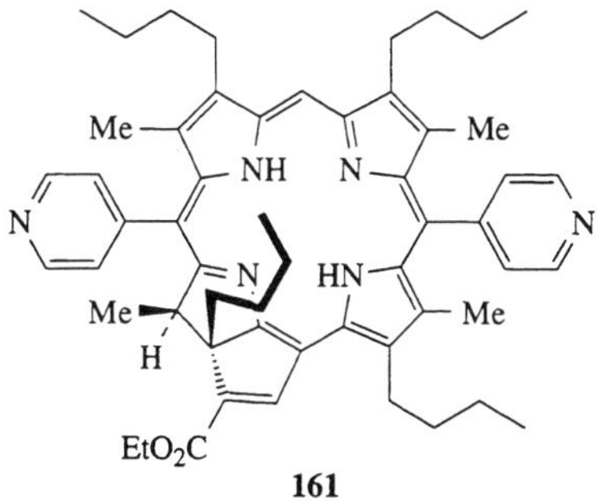

161

5,15-Diarylpurpurins **160** have been synthesized by way of formylation of the 5,15-diarylporphyrin, followed by Vilsmeier formylation, Wittig reaction, and cyclization (M.J. Gunter and B.C. Robinson, Aust. J. Chem., 1990, **43**, 1839). The related bis-pyridyl-purpurin **161** was also obtained in a similar manner, and was quaternized to produce water soluble analogues; the same paper reports the first X-ray structure of a purpurin (T.P. Forsyth et al., Tetrahedron Lett. 1995, **36**, 9093). By appropriate modification of reaction conditions it is possible to cyclize meso-substituents on porphyrins to afford benzopurpurins at the chlorin, bacteriochlorin or isobacteriochlorin level (A.R. Morgan and S. Gupta, Tetrahedron Lett., 1994, **35**, 4291).

162   **163**

Because of the ease of isolation of chlorophyll-a derivatives from *Spirulina* alga (K.M. Smith et al., J. Am. Chem. Soc., 1985, **107**, 4946), naturally occurring chlorins have been used as starting materials for a number of new, biologically active PDT sensitizers; the alga contains no chlorophyll-b derivatives. Thus, for example, lysyl chlorin-p_6 **162** can be isolated readily from ring-opening of the anhydride ring in purpurin-18 methyl ester **163** by lysine amino-acid derivatives (K.M. Smith et al., J. Chem. Soc., Perkin Trans. 1, 1993, 2369; Int. Congr. Ser. - Excerpta Med., 1992, **1011**, 769). Amino-acids can also be attached to the propionic side-chain of purpurin-18 and other chlorophyll-a derivatives (A.F. Mironov, Proc. SPIE, 1994, **2078**, 186; Proc. SPIE, 1993, **1922**, 204). Bacteriochlorophyll analogues of purpurin-18 and other chlorins, obtained using OsO_4, have also been reported and studied as photosensitizers (R.K. Pandey et al., Bioorg. Med. Chem. Lett., 1994, **4**, 1263; K.M. Smith, Proc. SPIE, 1992, **1645**, 274). Chlorin-e_6 dimers and trimers have been synthesized (T. Ando et al., Int. Congr. Ser. - Excerpta Med., 1992, **1011**, 764), and chlorin-e_6 can be rendered water soluble after reaction at the 3-vinyl with Eschenmoser's reagent, followed by quaternization (R.K. Pandey et al., Tetrahedron, 1992, **48**, 7591). Bonnett's group, in London, has synthesized a number of amphiphilic hydroxychlorins, (e.g. **164, 165**), from derivatives of chlorophyll-a (R. Bonnett et al., Int. Congr. Ser. - Excerpta Med., 1992, **1011**, 866; Proc. SPIE, 1994, **2078**, 171); hydroxylated octaethylchlorins and octaethylbacteriochlorins were likewise shown to be active in PDT (R. Bonnett et al., J. Photochem. Photobiol., B, 1990, **6**, 29; K.R. Adams et al., J. Chem. Soc., Perkin Trans. 1, 1992, 1465), and one of the more successful second generation sensitizers appears to be 5,10,15,20-tetra-(m-hydroxyphenyl)-2,3-chlorin **166** (R. Bonnett, Chem. Soc. Rev., 1995, **24**, 19; R. Bonnet et al., Biochem. J., 1989, **261**, 277).

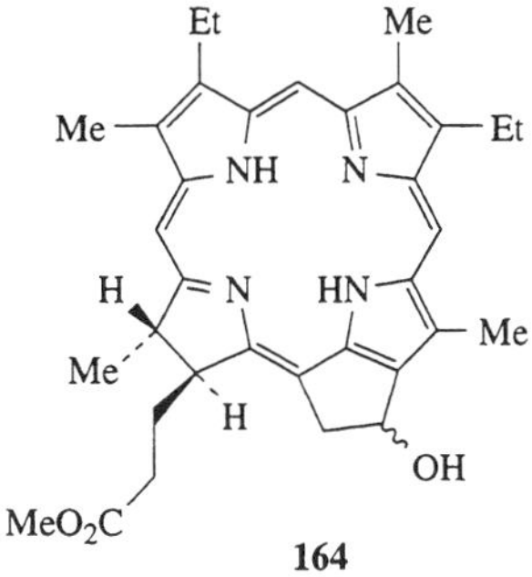

164

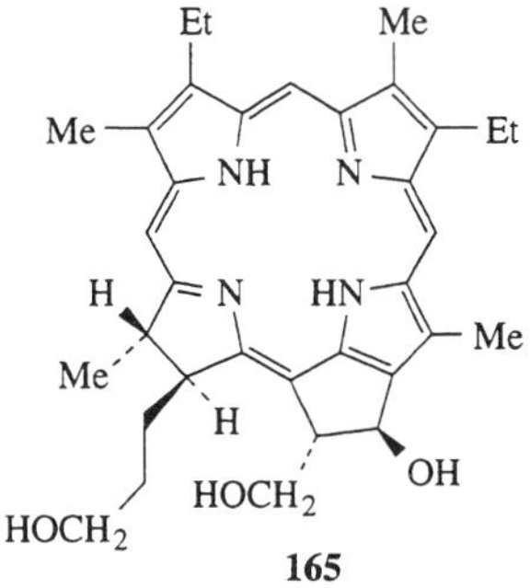

165

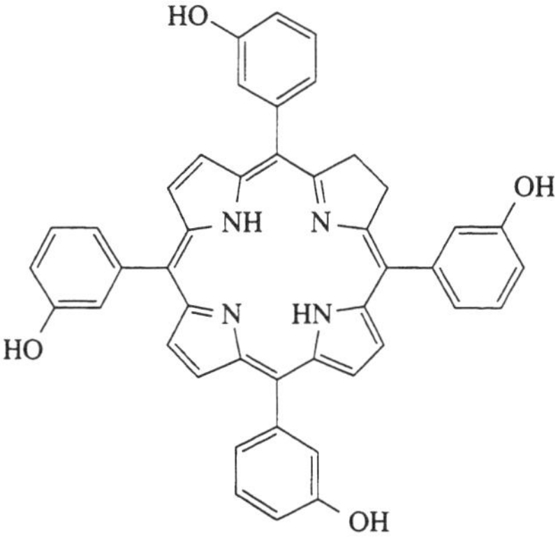

166

In a particularly important development, Chang et al. have shown that OsO_4 oxidation of zinc(II) chlorins gives isobacteriochlorins, whereas OsO_4 oxidation of metal free chlorins gives bacteriochlorins (C.K. Chang et al., J. Chem. Soc., Chem. Commun., 1986, 1213). Substituent effects on the regiospecificity with which OsO_4 reacts with certain porphyrins and chlorins has been investigated; in the presence of electron-withdrawing groups, the oxidation takes place in a pyrrole subunit in the quadrant opposite to the electron-withdrawing group (R.K. Pandey et al., Tetrahedron Lett., 1992, **33**, 7815). The OsO_4 reaction, followed by acid catalyzed pinacol-pinacolone rearrangement has also been used in the 5,15-diarylporphyrin series (A. Osuka et al., Bull. Soc.

Chem. Jpn., 1993, **66**, 3837). Morgan's group has synthesized a number of bacteriochlorins from porphyrindiones and has reported upon their biological activity (A.R. Morgan et al., J. Photochem. Photobiol., B, 1990, **6**, 133). Starting from azaporphyrins, the OsO_4 reaction, or the Diels-Alder reaction with electron-deficient dieneophiles, have been used to synthesize a number of new chlorins and bacteriochlorins with very long wavelength absorbance (K.R. Gerzevske et al., Heterocycles, 1994, **39**, 439). Azachlorins have also been prepared from the corresponding azaporphyrins via bilirubins (F.P. Montforts and B. Gerlach, Tetrahedron Lett., 1992, **33**, 1985; B. Gerlach and F.P. Montforts, Tetrahedron Lett., 1993, **34**, 6369; Liebigs Ann. Chem., 1995, 1509). Porphyrin and chlorin derivatives, upon functionalization with barbituric acid (e.g. **167**, **168**), have been shown to have astonishingly red-shifted long-wavelength absorptions (in the region 700-870 nm) (B.C. Robinson and A.R. Morgan, Tetrahedron Lett., 1993, **34**, 3711).

167

168

In a series of elegant papers, Mortforts' group in Bremen has demonstrated that a number of chlorins (e.g. **138**, **140**) can be tailored for PDT, as well as for studies of model systems for photosynthesis (F.P. Montforts et al., Tetrahedron Lett., 1991, **32**, 3481; Angew. Chem. Int. Ed. Engl., 1992, **31**, 1592; Biol. Eff. Light 1993, Proc. Symp., 3rd., 1994,

322; G. Haake et al., Liebigs Ann. Chem., 1992, 325; D. Kusch et al., Liebigs Ann. Chem., 1995, 1027; J.W. Bats et al., Liebigs Ann. Chem., 1995, 1617).

5.   Phlorins and Oxophlorins

(a) Phlorins

Phlorins **169** are dihydroporphyrins bearing the additional hydrogen atoms on a meso carbon and on a nitrogen. Gold(III) tetraphenyl-porphyrin **170** undergoes direct nucleophilic addition with hydroxide to give the corresponding phlorin **171**, which is nominally equivalent to a Meisenheimer complex (H. Segawa et al., J. Am. Chem. Soc., 1992, **114**, 7564).

169

170

171

N,N'-Bridged porphyrins with a diarylvinylidene, ethylene, vinylene, or trimethylene bridge have been synthesized, and like most N-alkylated porphyrins, shown to possess very basic nitrogens; hydride reduction stereoselectively and regioselectively affords stable 5-H phlorins in which the reduction has taken place at the meso-carbon between the pyrrole subunits bearing the N,N'-bridge. The phlorins can be re-oxidized (T. Kitao and J. Setsune, Kenkyu Hokoku - Asahi Garasu Zaidan, 1990,

56, 95; Chem. Abs., **115**, 8393; J. Setsune et al., Chem. Lett., 1990, 1351; J. Am. Chem. Soc., 1988, **110**, 6572). N,N'-Etheno-bridged tetraphenyl-porphyrins (J. Setsune et al., Chem. Lett., 1990, 1351) and N,N'-etheno-bridged octaethylporphyrins (J. Setsune et al., Organometallics, 1991, **10**, 1099) have been prepared; when the N,N'-(1,2-diphenyletheno)-octaethyl-5H-phlorin is air-oxidized to the porphyrin, the 5-exo-H was lost. If the corresponding 5-exo-alkyl-5H-phlorin was oxidized, with Cu(BF$_4$)$_2$, the product is the 5-alkylporphyrin which can be deprotonated in the alkyl group to give a methylene phlorin (J. Setsune et al., Tetrahedron Lett., 1990, **31**, 5057).

The isophlorin **172** was obtained by two-electron reduction of N,N',N'',N'''-tetramethyloctaethylporphyrin diperchlorate; protonation gave the phlorin **173**, which upon treatment with NaCN gave **174** (M. Pohl et al., Angew. Chem. Int. Ed. Engl., 1991, **30**, 1693).

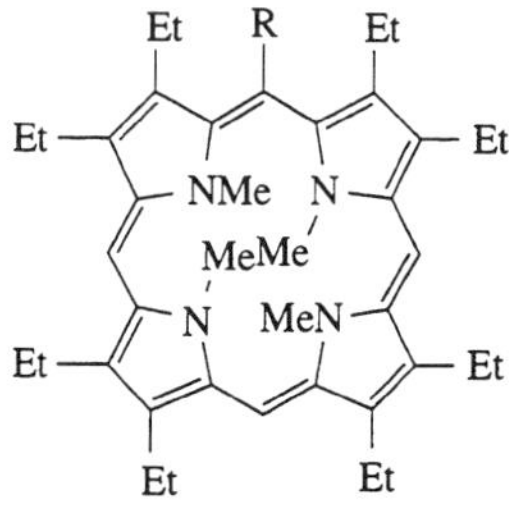

172 R = H
174 R = CN

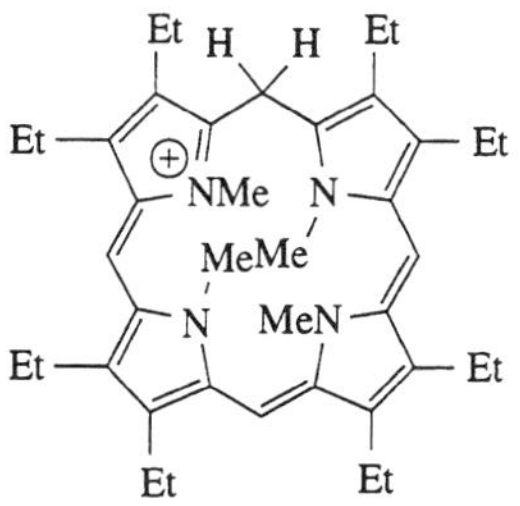

173

(b) Oxophlorins

Oxophlorins **175** can be prepared in a number of ways, for example by direct ring synthesis from open-chain tetrapyrroles or from MacDonald dipyrromethane/dipyrroketone condensations, or by oxidation of intact porphyrins or metalloporphyrins.

A new method for synthesis of oxophlorins **176** and β-hydroxy-porphyrins **177** involves nucleophilic ipso displacement of meso-nitro groups in the corresponding 5-nitro-octaethyl- or 2-nitro-tetraphenyl-porphyrin copper or nickel porphyrins by the sodium salt of E-benzaldoxime, followed by demetalation (M.J. Crossley et al., Tetrahedron, 1987, **43**, 4569). A number of studies on iron oxophlorins and so-called verdohemes have been carried out (A.L. Balch et al., J. Am. Chem. Soc., 1993, **115**, 11846; Inorg. Chem., 1995, **34**, 1395), indicating the importance of such compounds in the catabolism of heme to give biliverdin. Nickel (A.L. Balch et al., J. Am. Chem. Soc., 1993, **115**, 2531; Inorg. Chem., 1993, **32**, 4730) and cobalt complexes (A.L. Balch et al., Inorg. Chem., 1993, **32**, 4737) have also been reinvestigated using modern spectroscopic and X-ray diffraction techniques.

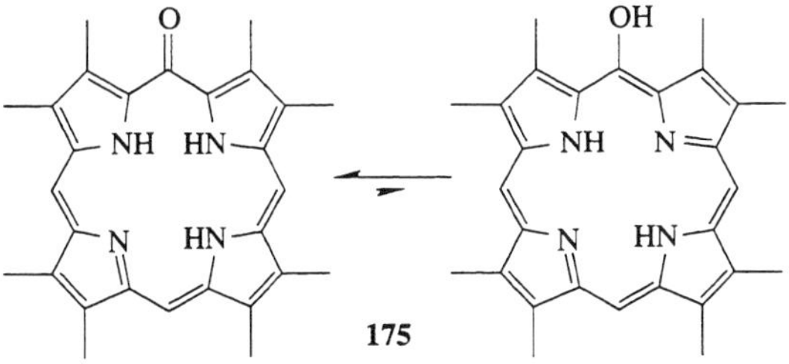

175

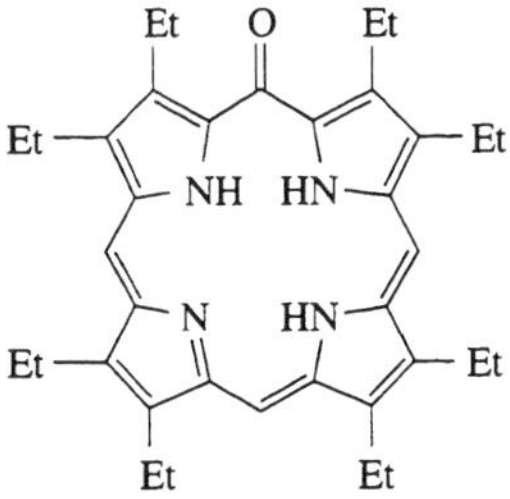

176

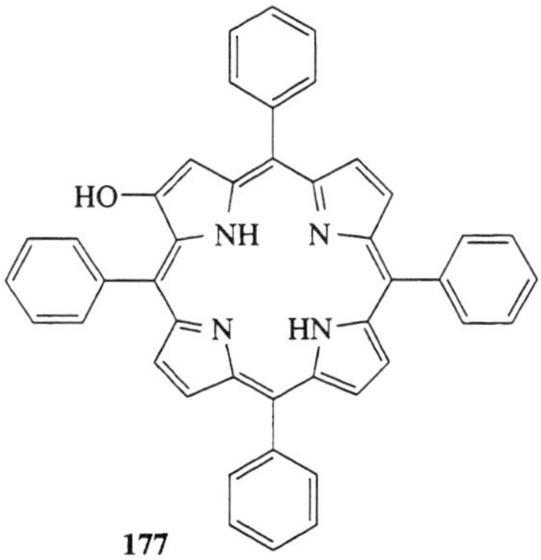

177

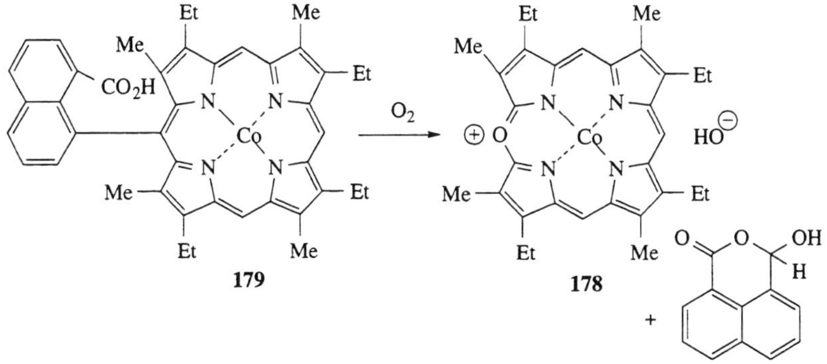

179 **178**

A verdoheme-like oxaporphyrin **178** has been obtained by oxygenation of a cobalt(II) porphyrinyl naphthoic acid **179** (C.K. Chang et al., J. Am. Chem. Soc., 1994, **116**, 12127); the mechanism was proposed to mimic that proposed in heme degradation to give biliverdins by way of verdo-hemes.

6. Corroles

Corrole has the carbon skeleton shown in **180**, and like phlorins and oxophlorins also possesses three NHs. Corrole's importance derives from the fact that the macrocycle features the direct link between carbons 1- and 19-, as does vitamin B_{12}. Because of this direct link between rings A and D, the central core "hole" is smaller than normally found in porphyrins. Metal complexes of corroles and other corrinoids have been reviewed (S. Licoccia and R. Paolesse, Struct. Bonding (Berlin), 1995, **84**, 71) - in this particular article, the synthesis and chemistry of corroles and metallocorroles are reviewed in detail.

MeOH, Δ

181   **180**

The synthesis of corroles **180** from 1,19-dibromo-a,c-biladiene salts **181** has been described (R.K. Pandey et al., J. Chem. Soc., Chem. Commun., 1992, 183; J. Chem. Soc., Perkin Trans. 1, 1994, 971); biliverdins are a by-product in this reaction. Reaction of 2-hydroxyalkylpyrroles with cobalt salts produces both corroles and porphyrins, the identity of the major product depending upon the steric situation within the starting monopyrrole; for example, treatment of pyrrole **182** with cobalt ions gives the cobalt corrolate **183** (S. Licoccia et al., Inorg. Chim. Acta, 1995, **235**, 15; R. Paolesse et al., Inorg. Chem., 1994, **33**, 1171; Inorg. Chim. Acta, 1993, **203**, 107), and this is the first example of such a monopyrrole self-condensation to give corrole.

Making use of the small core size in the corrole macrocycle, treatment of octaethylcorrole with $Fe_2(CO)_9$, followed by oxidation gave the μ-oxo complex, which then afforded **184**; both the μ-oxo complex and **184** are examples of a metallocorrole with a formally tetravalent iron atom (E. Vogel et al., Angew. Chem. Int. Ed. Engl., 1994, **33**, 731). Iron corroles containing nitrosyl axial ligands (M. Autret et al., J. Am. Chem. Soc.,

1994, **116**, 9141) and cobalt corroles with triphenylphosphine ligands (V.A. Adamian et al., Inorg. Chem., 1995, **34**, 532) have been synthesized, and their electrochemical behavior has been studied.

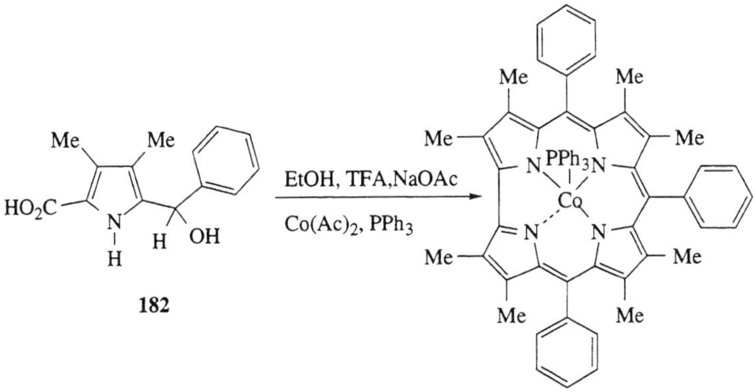

182

EtOH, TFA, NaOAc

Co(Ac)$_2$, PPh$_3$

183

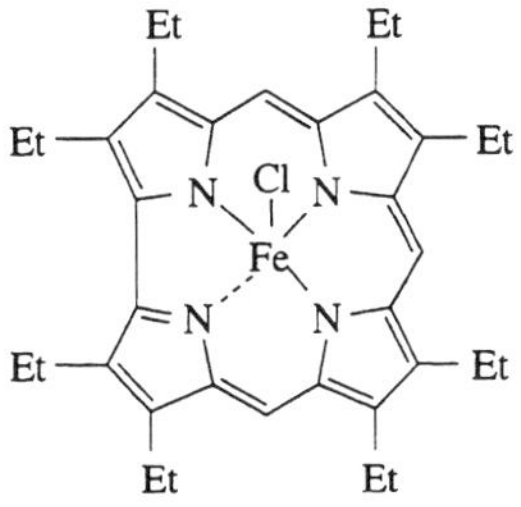

184

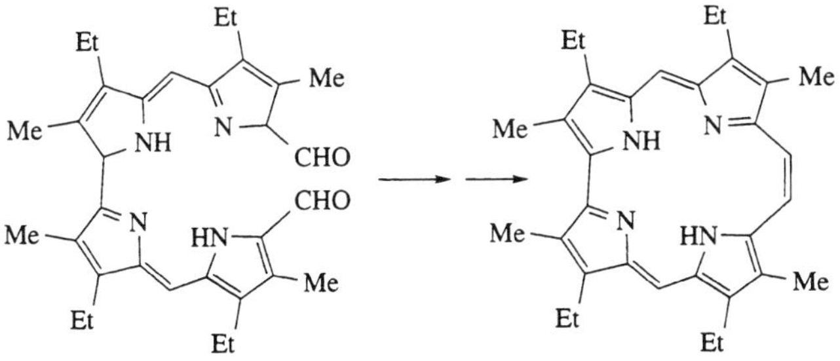

186

185

A number of other metal derivatives (e.g. M = Sn, Ge, In, Zn, Rh) of octamethylcorrole have been synthesized by way of cyclization of 2,3,7,8,12,13,17,18-octamethyl-a,c-biladiene salts in the presence of the

342

appropriate metal ions, and their spectroscopic properties (IR, NMR, and optical) have been studied (R. Paolesse et al., Inorg. Chim. Acta, 1990, **178**, 9; T. Boschi et al., Inorg. Chim. Acta, 1988, **141**, 169; J. Chem. Soc., Dalton Trans., 1990, 463).

The expanded corrole, corrphycene **185**, has been synthesized (J.L. Sessler et al., Angew. Chem. Int. Ed. Engl., 1994, **33**, 2308; see also H. Falk and Q.-Q. Chen, Monatsh. Chem., 1996, **127**, 69). The same compound **185**, under the name etioporphycerin, has been synthesized by the Dijon group (M.A. Aukauloo and R. Guilard, New J. Chem., 1994, **18**, 1205); both the Cologne and Dijon routes employed reductive coupling of the terminal aldehyde groups in **186**. Another expanded corrole, hemiporphycene, a [2,1,1,0]-system, has also been obtained (vide infra) by ring contraction of a nickel homoporphyrin (H.J. Callot et al., New J. Chem., 1995, **19**, 155).

Finally, "isocorrole" **187** has been synthesized from porphycene **188** by treatment with bromine (to give **189**) followed by rearrangement in the presence of base (S. Will et al., Angew. Chem. Int. Ed. Engl., 1990, **29**, 1390).

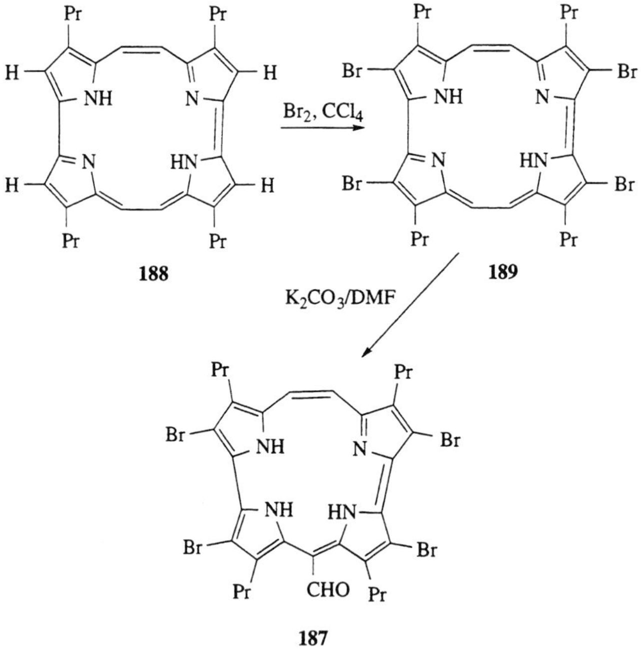

7. Expanded Systems: Sapphyrins and Pentaphyrins

Expanded porphyrin systems have been reviewed (J.L. Sessler and A.K. Burrell, Top. Curr. Chem., 1991, 177). It cannot escape notice that many of the recent discoveries involving porphyrin isomers and expanded porphyrin systems have appeared first in Angewandte Chemie. Discovery and studies of new expanded porphyrin systems have intensified in the past decade; typical examples include sapphyrins and pentaphyrins (featured in this Section) as well as, for example, texaphyrins (J.L. Sessler et al., J. Am. Chem. Soc., 1988, **110**, 5586; Acc. Chem. Res., 1994, **27**, 43), hexaphyrins (R. Charriere et al., Heterocycles, 1993, **36**, 1561), platyrins (E. LeGoff and O.G. Weaver, J. Org. Chem., 1987, **52**, 710), rosarins (J.L. Sessler et al., J. Am. Chem. Soc., 1992, **114**, 8306), rubyrins (J.L. Sessler et al., Angew. Chem. Int. Ed. Engl., 1991, **30**, 977), turcasarin (J.L. Sessler et al., Angew. Chem. Int. Ed. Engl., 1994, **33**, 1509), orangarin (J.L. Sessler et al., Chem. Eur. J., 1995, **1**, 56), invertopentaphyrinogen (K.H. Schumacher and B. Franck, Angew. Chem., 1989, **101**, 1292), figure-of-eight octapyrroles (E. Vogel et al., Angew. Chem. Int. Ed. Engl., 1995, **34**, 2511), extended porphycenes {i.e. [22]porphyrin(2,2,2,2)} (E. Vogel et al., Angew. Chem. Int. Ed. Engl., 1990, **29**, 1387), and others (e.g. B. Franck and A. Nonn , Angew. Chem. Int. Ed. Engl., 1995, **34**, 1795; B. Franck et al., Liebigs Ann. Chem., 1994, 503). Unfortunately, most of these interesting new systems, for reasons of space, must be outside the bounds of this review, but much of the current work in this area has been reviewed (J.L. Sessler et al., NATO ASI., Ser. C, 1994, **448**, 391).

(a) Sapphyrins

Since their discovery during Woodward's pyrrole pigment heyday, the sapphyrin macrocycle **190** has enjoyed a comeback (J.L. Sessler et al., Synlett, 1991, 127). A new synthesis of meso-diarylsapphyrins **191**, using Lindsey-type conditions, has been published (J.L. Sessler et al., J. Org. Chem., 1995, **60**, 5975). The Rothemund-type reaction between pyrrole and benzaldehyde yields tetraphenylporphyrin, a "N-confused porphyrin" (vide supra), and 5,10,15,20-tetraphenylsapphyrin **192** (P.J. Chielewski et al., Chem. Eur. J., 1995, **1**, 68), showing that pentapyrrolic macrocycles are accessible by way of Rothemund-type procedures. An improved synthesis of sapphyrins, e.g. **193**, from tripyrranes **194** and diformylbipyrroles **195** has been reported (J.L. Sessler et al., J. Am. Chem. Soc., 1990, **112**, 2810; Tetrahedron, 1992, **48**, 9661); new procedures for dioxa- and dithia-sapphyrin syntheses were also reported. 3,8,12,13,17,22-Hexaethyl-2,7,18,23-tetramethylsapphyrin, as its mixed fluoride-hexafluorophospate salt is capable of encapsulating fluoride within its penta-aza macrocyclic core (J.L. Sessler et al., J. Am. Chem. Soc., 1990, **112**, 2810; 1992, **114**, 5714; Pure Appl. Chem., 1993, **65**, 393).

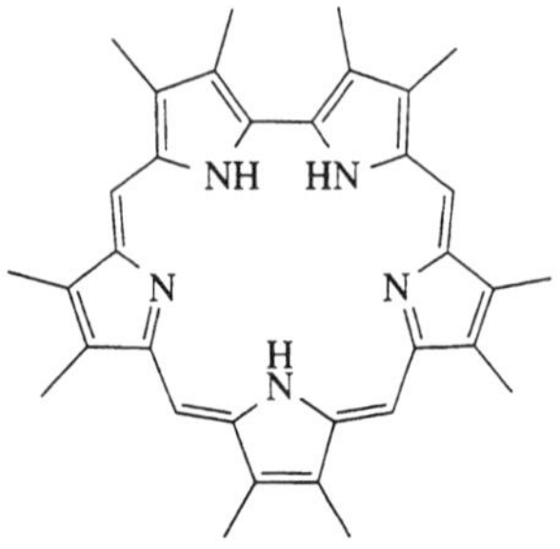

190

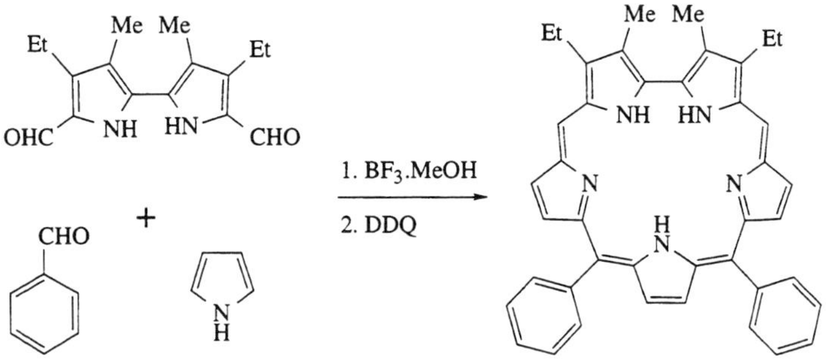
Et
Me Me
Et
OHC
NH
HN
CHO
CHO
+
N
H
1. BF3.MeOH
2. DDQ
Et
Me Me
Et
NH
HN
N
N
H
N

191

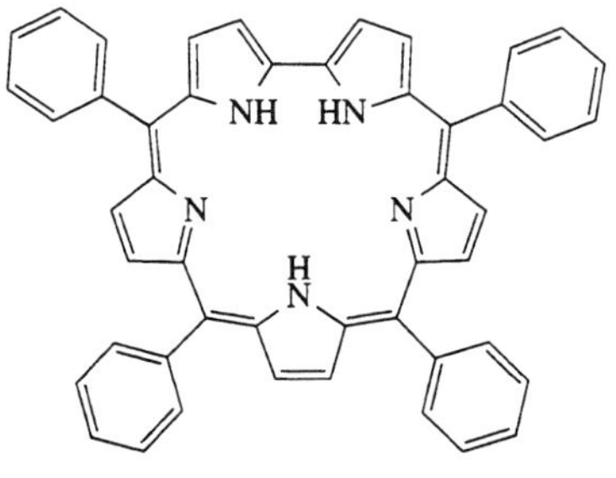
NH
HN
N
N
H
N

192

Synthesis of the first example of a β-to-meso linked porphyrin-sapphyrin cofacial pseudo dimer has been reported, and the molecule, when excited at wavelengths at which the porphyrin absorbs, undergoes efficient

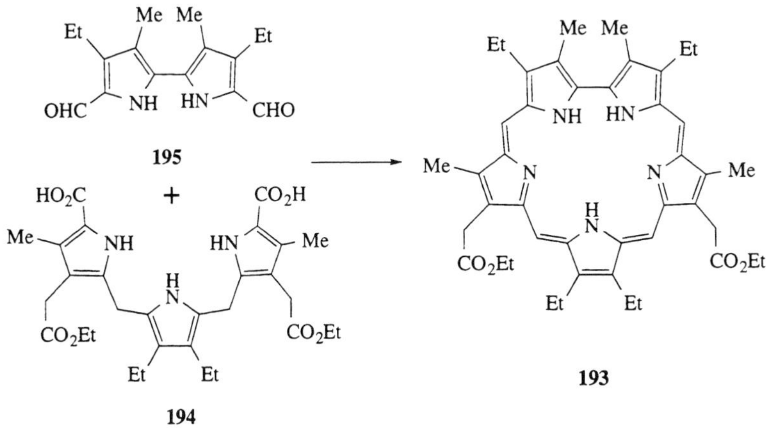

195

194

193

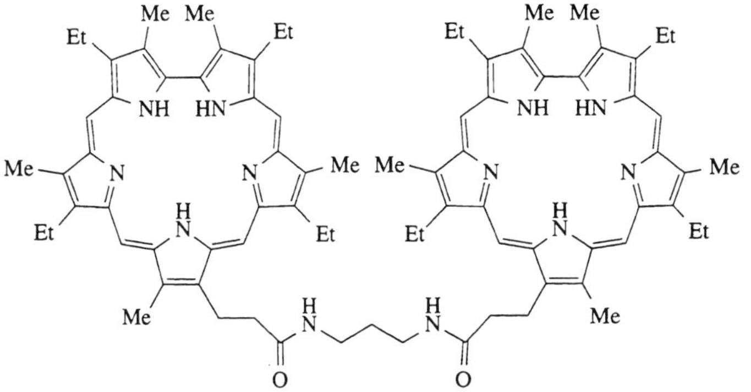

196

intramolecular singlet energy transfer from the porphyrin to the sapphyrin component (J.L. Sessler et al., Supramol. Chem., 1994, **4**, 35). The synthesis of a cyclophane containing a sapphyrin and a crown ether has been reported; the molecule acts as a receptor for the concurrent complexation of both cationic and anionic substrates (J.L. Sessler and E.A. Brucker, Tetrahedron Lett., 1995, **36**, 1175). A covalently-linked sapphyrin dimer **196**, when protonated, acts as an effective receptor for certain dicarboxylate ions (V. Kral et al., J. Am. Chem. Soc., 1995, **117**, 2953). A sapphyrin-EDTA conjugate has also been synthesized by the Texas group, using methodology involving modification of an intact sapphyrin; the iron complex accomplished concentration dependent cleavage of a supercoiled plasmid (B.L. Iverson et al., J. Org. Chem., 1995, **60**, 6616). A new sapphyrin macrocycle **197** in which one of the N atoms of the inner core has been replaced with a selenium atom has

346

been synthesized; its chemistry resembles that of normal (penta-aza) sapphyrins (J. Lisowski et al., Inorg. Chem., 1995, **34**, 3567). Finally, sapphyrins react with UO_2^{2+} to give a stable complex which is no longer aromatic due to nucleophilic addition of methanol to a meso-position (A.K. Burrell et al., J. Chem. Soc., Chem. Commun., 1991, 1710).

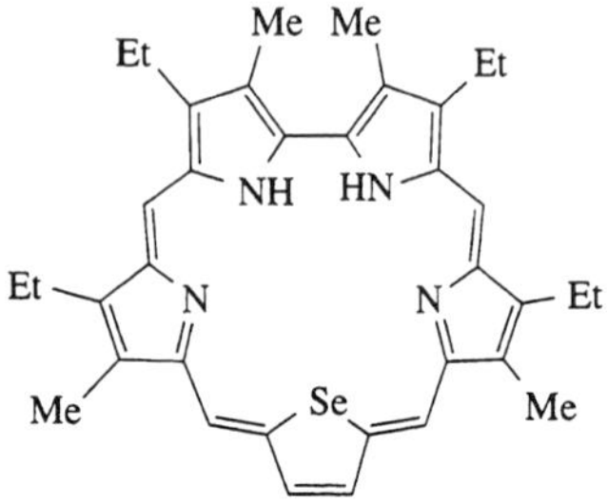

197

(b) Pentaphyrins

Pentaphyrins **198** have turned out to be the poor cousins of sapphyrins, and so the relative amount of literature on these pentapyrroles is small.

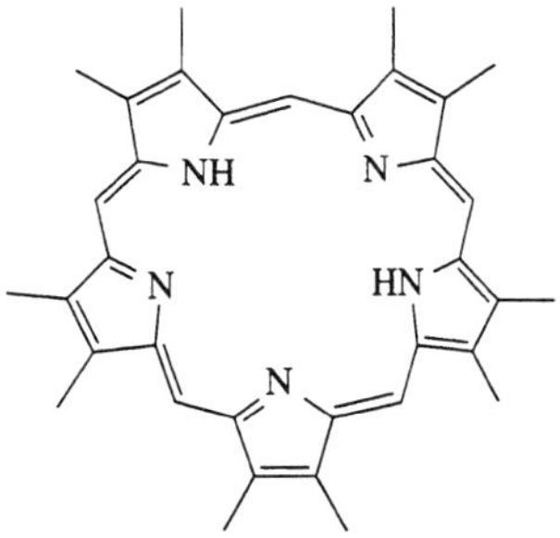

198

Two novel sapphyrin-like pentaphyrins have been synthesized and characterized spectroscopically and using X-ray diffraction; the compounds are [22]dehydropentaphyrin-(2.1.0.0.1) **199** and [22]pentaphyrin-(2.1.0.0.1) **200** (S.J. Weghorn et al., Tetrahedron Lett., 1995, **36**, 4713). A uranylpentaphyrin has been synthesized, and the molecular structure has been determined using X-rays (A.K. Burrell et al., J. Am. Chem. Soc., 1991, **113**, 4690). Finally, a water-soluble tetrahydroxypentaphyrin **201** which exhibits cytotoxicity has been synthesized (V. Kral et al., Bioorg. Med. Chem., 1995, **3**, 573).

199

200

201

8. Porphycenes

Porphycenes **202** are the (2,0,2,0)-meso isomers of porphyrins (1,1,1,1) **203**, and they have received great recent attention on account of their interesting spectroscopic, chemical, and structural properties. Porphyrin isomers have been reviewed (J.L. Sessler, Angew. Chem. Int. Ed. Engl., 1994, **33**, 1348). Vogel has also reviewed his novel annulene chemist's outlook on porphyrin chemistry which led to the discovery and synthesis of porphycenes and their analogues (E. Vogel, Pure Appl. Chem., 1990, **62**, 557), and has also reported the expansion of his work to afford figure-of-eight octapyrroles (E. Vogel et al., Angew. Chem. Int. Ed. Engl., 1995, **34**, 2511). The history of the porphycene field is contained completely within the current review period, and began in 1986 when E.

348

Vogel et al. (Angew. Chem. Int. Ed. Engl., 1986, **25**, 257) published their short paper detailing a 2-3% yield of **202** obtained by McMurry type titanium coupling of 5,5'-diformyl-2,2'-bipyrrole **204**. In 1987, three new porphycenes, including 2,7,12,17-tetrapropylporphycene **205** were reported, and yields were raised to 10% (E. Vogel et al. Angew. Chem. Int. Ed. Engl., 1987, **26**, 928); a number of other porphycenes followed subsequently (E. Vogel et al., Angew. Chem. Int. Ed. Engl., 1993, **32**, 1600). The rhenium complex of 2,7,12,17-tetrapropylporphycene has also been synthesized and studied by X-ray diffraction (C.-M. Che et al., Inorg. Chem., 1995, **34**, 984). Porphycenes have been characterized electrochemically, and have been compared with their porphyrin isomers (C. Bernard et al., Inorg. Chem., 1994, **33**, 2393).

202

203

204

205

Hydrogenation of porphycene to give the analogue of "chlorin" in the porphycene series, 2,3-dihydroporphycene **206** (presumably via **207**), the analogue of "chlorin", has also been reported (E. Vogel et al., Angew. Chem. Int. Ed. Engl., 1987, **26**, 931). Hydrogenation of a tetrapropyl-porphycene **208** actually affords a pyrrolophanediene **209**, similar to **207**, which is oxygen sensitive, but which undergoes tautomerism in acid to give **210** (E. Vogel et al., Angew. Chem. Int. Ed. Engl., 1989, **28**, 1655). Use of 5,5'-diformyl-2,2'-difuran **211** (instead of the bipyrrole analogue) in the McMurry type reaction affords tetraoxaporphycene **212**, which has been studied structurally (E. Vogel et al., Angew. Chem. Int. Ed. Engl.,

1988, **27**, 411) as well as spectroscopically and theoretically (R. Bachmann et al., J. Am. Chem. Soc., 1993, **115**, 10286); with 5,5'-diformyl-2,2'-bithiophene **213** the corresponding macrocycle **214** is obtained, which cannot be oxidized to give the tetrathiaporphycene (F. Ellinger et al., Monatsh. Chem., 1993, **124**, 931). A dithiaporphycene **215** has also been synthesized (G. De Munno et al., Tetrahedron, 1993, **49**, 6863).

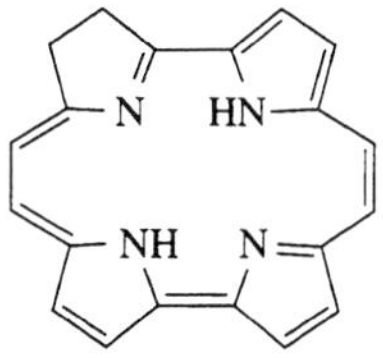

206

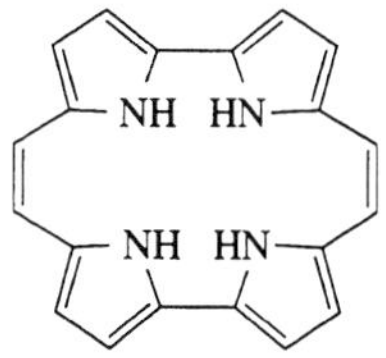

207

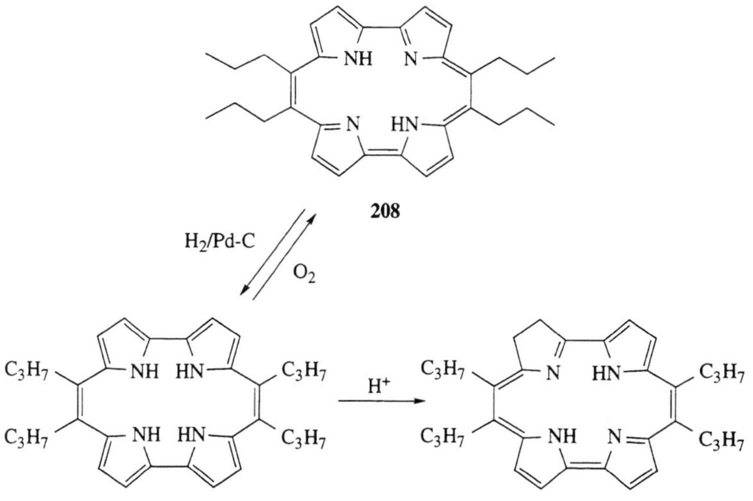

208

H$_2$/Pd-C

O$_2$

C_3H_7 C_3H_7
C_3H_7 C_3H_7

209

H$^+$

C_3H_7 C_3H_7
C_3H_7 C_3H_7

210

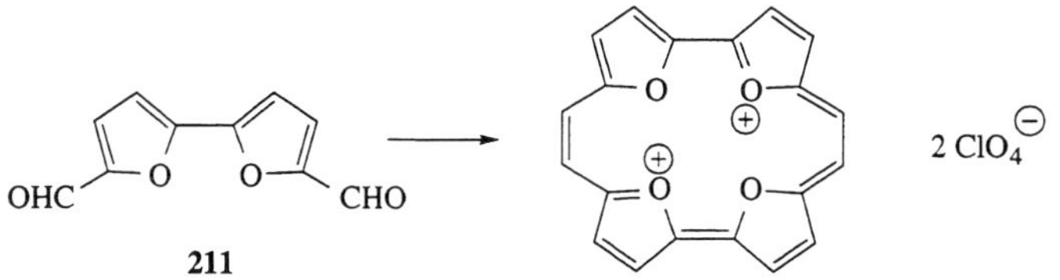

OHC — CHO

211

$2\,ClO_4^{\ominus}$

212

Tetraphenylporphycene **216** has been synthesized (S. Nonell et al., Tetrahedron Lett., 1995, **36**, 3405) using a modification of the standard procedure. It has been shown that tetrapropylporphycene **205** reacts with various Vilsmeier reagents to give N,N'-bridged analogues, e.g. **217**, and this compound was characterized using X-ray diffraction (C.K. Chang et al., J. Chem. Soc., Chem. Commun., 1995, 1173).

213

214

215

216

217

219

218

Isomers of the porphycene/porphyrin systems have also been synthesized. For example, corrphycene **185** has been synthesized (J.L. Sessler et al., Angew. Chem. Int. Ed. Engl., 1994, **33**, 2308; H. Falk and Q.-Q. Chen, Monatsh. Chem., 1996, **127**, 69); the same isomer of porphycene and porphyrin, named porphycerin **185**, has also been synthesized by the Dijon group (M.A. Aukauloo and R. Guilard, New J. Chem., 1994, **18**, 1205). The so-called hemiporphycene **218**, a (2,1,1,0)-system, has also been obtained by ring contraction of a nickel homoporphyrin **219** (H.J. Callot et al., New J. Chem., 1995, **19**, 155).

Bromination of the porphycene **188** followed by treatment with base affords the so-called isocorrole **187** (S. Will et al., Angew. Chem. Int. Ed. Engl., 1990, **29**, 1390). Finally, an acetylene-cumulene-porphyrinoid **220** has been synthesized in 18% yield by McMurry coupling of the dialdehyde **221** (N. Jux et al., Angew. Chem. Int. Ed. Engl., 1990, **29**, 1385).

9. Corrins

The dust is beginning to settle on what has perhaps been the major enterprise on the biological side of "pyrrole pigment" studies in the past thirty years, namely the elucidation of the biosynthesis of vitamin B_{12}; a few loose ends remain to be cleared up, but there seems a definite sense that a job well-done is making its way into history. Somewhat sadly for organic chemistry, though a tremendous amount of world class organic chemistry in the early years heralded the beginning of the assault on the biosynthetic problem, it was genetics which eventually allowed closure to this task, a problem which was beset with riddles even until the last moment. Space does permit a detailed account of the discoveries uncovered in vitamin B_{12} biosynthetic studies; indeed the contracted 25-50 page space allocation for this article, even after expansion, has permitted virtually no discussion of porphyrin, heme, chlorophyll, or factor 430 biosynthesis at all! But some reviews and final developments can be noted. Reviews of the colossal accomplishments in vitamin B_{12} biosynthesis have been published by the principals (F. Blanche et al., Angew. Chem. Int. Ed. Engl. 1995, **34**, 383; A.R. Battersby, Science, 1994, **264**, 1551; A.I. Scott, Tetrahedron, 1994, **50**, 13315; Synlett, 1994,

871). The biosynthesis of factor 430, the tetrapyrrole-derived nickel complex (from methanogenic bacteria) which is involved in one of the catalytic steps in the reduction of carbon dioxide to methane, has also been reviewed (R.K. Thauer and L.G. Bonaker, CBI Found. Symp., 1994, **180**, 210). Eschenmoser has published a review detailing with some of his work following the "non-enzymatic" synthesis of vitamin B_{12} (A. Eschenmoser, Ciba Found. Symp., 1994, **180**, 309).

The mechanistics of vitamin B_{12} action have been reviewed in the period covered by this report (B. Kraeutler, NATO ASI Ser., Ser. C, 1989, **257**, 435).

222

223

Gossauer's group in Fribourg (Switzerland) has published a series of papers describing the reactions and chemistry of xanthocorrinoids; the main product from the reaction of heptamethyl dicyanocobyrinate with oxygen in the presence of ascorbic acid is a stable yellow corrinoid **222** (B. Gruening et al., Helv. Chim. Acta, **1985**, *68*, 1754). In the first example of a pinacol-type rearrangement in the corrin series, the c-acetic side-chain of hexamethyl dicyanocobyrinate can be converted into an ethyl group by reduction, mesylation, iodination and hydrogenolysis; upon reaction with oxygen in presence of ascorbic acid, migration of the C-5 methyl to the vicinal position at C-6 takes place concomitantly with introduction of a carbonyl group at C-5 (B. Gruening et al., Helv. Chim. Acta, 1985, **68**, 1771). C-15 hydroxylation of the corrin chromophore to give **223** takes place when cyanocob(III)alamin is treated with Udenfriend's reagent (G. Holze et al., Helv. Chim. Acta, 1986, **69**, 1567), and the 5,6-trans-diol derivatives are produced when cis-5,6-dihydrocobyrinate c,8-lactam is treated with sulfuric acid in methanol (G. Holze et al., Helv. Chim. Acta, 1991, **74**, 1287).

10. Biliverdins

Biliverdins **224** are fully conjugated open-chain tetrapyrroles produced biologically by catabolism of hemes. The most common methods for synthesis of biliverdins involve ring-opening of porphyrins or modification of open-chain tetrapyrroles.

R = CO_2t-Bu

The urobiliverdin isomers have been synthesized by oxidation of urohemes -I and -III, and natural bactobilin was shown to be urobiliverdin-I **225** (A. Valasinas et al., J. Org. Chem., 1985, **50**, 2398). The most useful method for simple biliverdin (e.g. **224**) synthesis is treatment of b-bilene di-tert-butyl esters **226** with bromine in TFA; a number of copro- and uro-biliverdin isomers (including bactobilin) have been synthesized in this way (A. Valasinas et al., J. Org. Chem., 1986, **51**, 3001), as have a series of similar biliverdins (J. Awruch and B. Frydman, Tetrahedron, 1986, **42**, 4137), and observations on the mechanism of the procedure have been reported (J. Awruch,

Tetrahedron, 1990, **46**, 1171). Using the same basic approach, biliverdins with stable extended conformations have been synthesized and their spectroscopic properties have been correlated with structure; the conformational rigidity was introduced by way of cyclization reactions involving modification of 2-chloroethyl substituents (J. Iturraspe et al., J. Am. Chem. Soc., 1989, **111**, 1525; Tetrahedron, 1995, **51**, 2243; S.E. Bari et al., Tetrahedron, 1995, **51**, 2255).

A new route for the synthesis of biliverdins **227** from open-chain tetrapyrroles involves treatment of symmetrical or unsymmetrical 1,19-dibromo-a,c-biladiene salts **228** with dimethyl sulfoxide in presence of acid (R.K. Pandey et al., J. Chem. Soc., Chem. Commun., 1992, 183; J. Chem. Soc., Perkin Trans. 1, 1994, 971).

228   DMSO/TsOH   **227**

Using a 1,19-dibromo-a,c-biladiene with an inverted pyrrole subunit, the biliverdin **229** with an inverted pyrrole has been synthesized (R.K. Pandey et al., Tetrahedron Lett., 1994, **35**, 8995).

229

Total synthesis of racemic cis-2,3,18^1,18^2-tetrahydroprotobiliverdin-IXα dimethyl ester **230** has been described; the compound isomerizes under neutral conditions to give racemic Z-phycocyanobilin dimethyl ester **231**

(A. Gossauer et al., Helv. Chim. Acta, 1989, **72**, 518), indicating that if tetrahydroprotobiliverdin-IXα (dicarboxylic acid of **230**) is in fact a biosynthetic precursor of phycocyanobilin, it would spontaneously isomerize into the natural product. The syntheses and conformational properties of some oxa- and thia-deazabiliverdins have been described (P. Nesvadba et al., Helv. Chim. Acta, 1994, **77**, 1837).

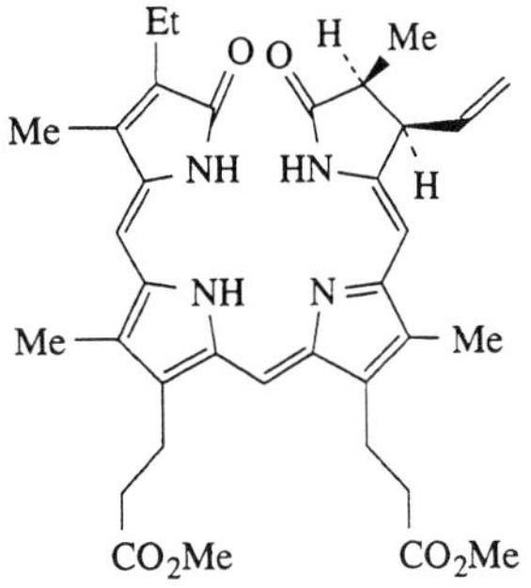

230

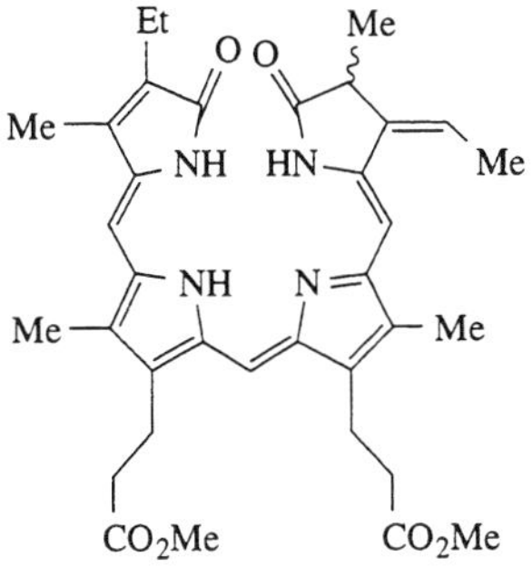

231

Biliverdin-IIIα dimethyl ester **232**, when treated with methanolic acid, undergoes ring closure of the exocyclic vinyl groups to give the 2,18-bridged helical biliverdin **233** (D. Krois and H. Lehner, J. Chem. Soc., Perkin Trans. 1, **1989**, 2179); chiral diastereomers of **233** have been separated (D. Krois and H. Lehner, J. Chem. Soc., Perkin Trans. 2, **1989**, 2085).

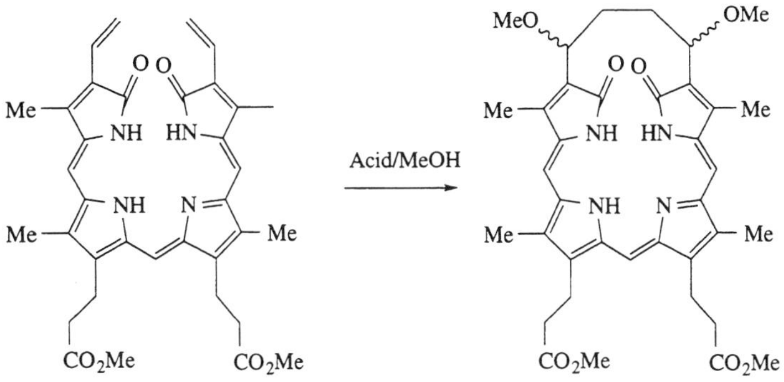

Acid/MeOH

232

233

Yet another 2,18-bridged biliverdin **234** has been obtained under basic conditions during dehydrohalogenation of a 3,18-bis(2-chloroethyl)-

biliverdin dimethyl ester **235** (E.D. Sturrock et al., J. Chem. Soc., Chem. Commun., 1993, 872). Biliverdin and bilirubin cyclic esters have also been synthesized (S.E. Boiadjiev et al., Tetrahedron: Asymmetry, 1995, **6**, 901; J.M. Ribo et al., Tetrahedron, 1994, **50**, 3967), as have bilirubin analogues with N,N'-methylene bridges (K.-O. Hwang et al., Tetrahedron, 1994, **50**, 9919).

235   **234**

Biliverdin-IXα dimethyl ester undergoes various acid-promoted dimerizations at the vinyl groups to give a cyclobutane, diastereomeric cyclohexene linked dimers, and a butane and butene linked dimer (D. Krois et al., J. Chem. Soc., Perkin Trans. 1, 1992, 1043). Biliverdins and bilirubins have been known, for some time, to undergo a number of reactions under oxidative conditions, producing compounds with an array of colors; these reactions have been reinvestigated using nitric acid oxidation and thallium(III) and lead(IV) salt oxidations, and structures have been assigned to the various products (C. Acero et al., Monatsh. Chem., 1993, **124**, 401). Photo-oxidation of 20-trifluoroacetoxy- and 20-chloro-chlorophyll derivatives gives oxoniachlorins **136**, which can be readily cleaved in the presence of nucleophiles; the relevance of this reaction to chlorophyll catabolism has been discussed (J. Iturraspe and A. Gossauer, Photochem. Photobiol., 1991, **54**, 43). Oxochlorin intermediates have been isolated from 20-trifluoroacetoxy-chlorins, and using enriched oxygen-18 studies, these were shown to be cleaved to give dihydrobiliverdins by way of a two oxygen molecule mechanism (M.S. Huster and K.M. Smith, Tetrahedron Lett., 1988, **29**, 5707).

Over the years, H. Falk's group has published literally dozens of papers dealing with synthetic, synthetic and spectroscopic studies of verdins, their analogues, and precursors; part XCVII in his series on the Chemistry of Pyrrole Pigments has just been published (H. Falk et al., Monatsh. Chem., 1996, **127**, 77), and other leading papers dealing with dipyrrinonylethenes (Q.-Q. Chen et al., Monatsh. Chem., 1995, **126**, 983), dipyrrinonylidene-ethanes (Q.-Q. Chen and H. Falk, Monatsh. Chem., 1995, **126**, 1097), dehydrobiladienes-a,b (J.S. Ma et al., Monatsh. Chem., 1995, **126**, 201) and, for example, reconstitution of apomyoglobin with bile pigments (H. Falk et al., Monatsh. Chem., 1990, **121**, 893) have recently appeared.

11. Acknowledgments

Though, because of space limitations, only a small proportion of the papers from my own research group over the past ten years have been mentioned in this review, I would like to take this opportunity to thank all of my graduate students, post-doctorals, and senior collaborators for their freely given enthusiasm, dedication, and talent. I would also like to thank both the National Institutes of Health (HL 22252) and the National Science Foundation (CHE-93-05577) for uninterrupted support of my own pyrrole pigment research over the past twenty years.

Second Supplements to the 2nd Edition of Rodd's Chemistry of Carbon Compounds, Vol.IV B, edited by M. Sainsbury

Chapter 13

FIVE-MEMBERED MONOHETEROCYCLIC COMPOUNDS (CONTINUED): AZAPORPHYRINS; BENZOPORPHYRINS; BENZOAZOPORPHYRINS; PHTHALOCYANINES AND RELATED STRUCTURES

This chapter has not been supplemented since the subjects that would have been dealt with here are largely covered in Chapters 12 and 15.

Chapter 14

FIVE-MEMBERED MONOHETEROCYCLIC COMPOUNDS: THE INDIGO GROUP

M.SAINSBURY

1. *Introduction*

Although the production of indigo is still a major commercial interest (see M.R.Fox and J.H.Pierce, *Textile Chemist and Colourist*, 1990, **22**, 13; A.S.Travis, *ibid.*, 1990, **22**, 18; R.J.H.Clark *et al.*, *Endeavour*, 1993, **17**, 191), since the 2nd supplement of this volume was published (1985) only a limited amount of new chemistry dealing with indigo in the ground state has appeared. However, there is considerable activity in the photochemical behaviour of indigoid dyestuffs - a fact that probably has its origins in the fashionable photo-fading of indigo-dyed textiles (J.Fabian and H.Hartmann, *Light Absorption of Organic Colorants*, Springer, Berlin, 1980, p32, p115, p162).

Indigo

It has been demonstrated that the electronic absorption characteristics of many dye molecules containing an unsaturated five-membered ring system can be explained in terms of their anti-aromaticity (R.Brelow, *Acc. Chem. Res.*, 1973, **6**, 393). Since these criteria were enunciated, factors which define anti-aromaticity in more complex examples, including the indigoid dyes, have been specified. These guidelines are rather less restrictive than those laid down in the original Beslow definition (F.Dietz, N.Tyutyulkov and M.Rabinovitz, *J. Chem. Soc., Perkin Trans. 2*, 1993, 157).

362

2. Synthesis

A one pot synthesis of indigo involves the oxidation of indole with an organic hydroperoxide in the presence of a molybdenum complex (Y.Inoue *et al., Stud. Surface Sci. Cat.*, 1994, **82**, 615). See also section 7 below.

An adventitious synthesis of a pyrrole-based indigoid compound arose when the adduct (2) of 2-(ethylthio)-1-azetidinone (1) and diphenylcyclopropene was thermolysed in boiling 1,2-dichlorobenzene. It seems probable that the first step in the reaction is the thermal fragmentation of the four membered ring affording the ethylthiopyrrolone (3), which eliminates ethene to give the thione (4) and thence, through desulphuration, the indigo analogue (5) (K.Hemming *et al., Tetrahedron Lett.*, 1992, **33**, 2231).

(1) (2)

(3) (4) (5)

3. Metal Complexation

The dianion of octahydroindigo (1) reacts with copper(II) and zinc(II) salts to afford insoluble polymeric-bridged complexes of the type (2; M = Cu, or Zn), while chloro-bridged complexes [(μ-Cl)M(Cl)(5-C$_5$Me$_5$)$_2$] give dimeric compounds of the type (3; M = Pd, or Pt). The latter are much more soluble than the former and their electronic spectra show large bathochromic shifts of the absorption maxima, relative to those of indigo itself.

Corresponding compounds of each class [*i.e.* (5) and (6] are formed by similar reactions upon the dianion of 'so-called' 4,4′-dipropyl-5,5′-dimethylpyrrole indigo (4) (C.Schmidt, H.-U.Wagner and A.Beck, *Chem. Ber.*, 1992, **125**, 2347).

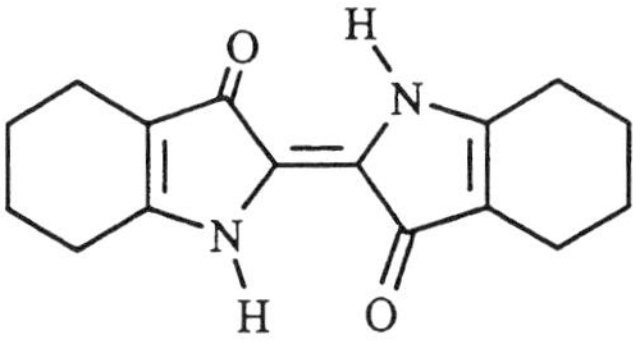

(1)

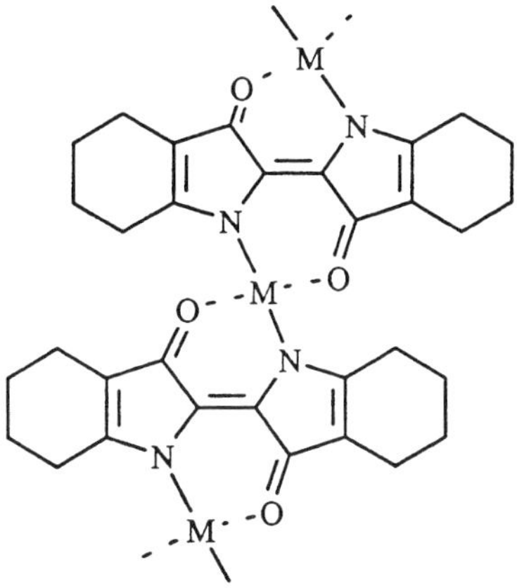

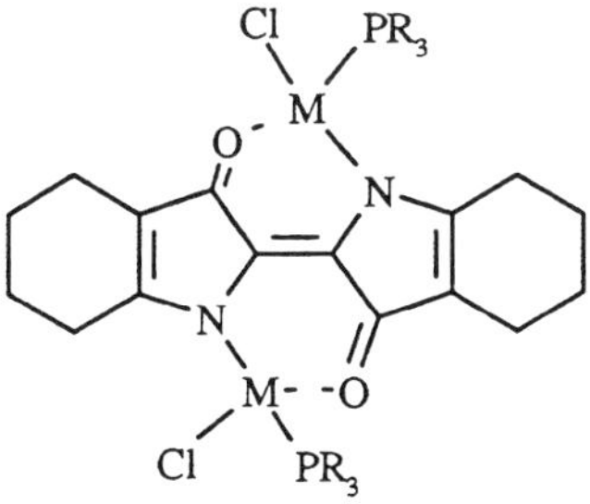

(3)

(2)

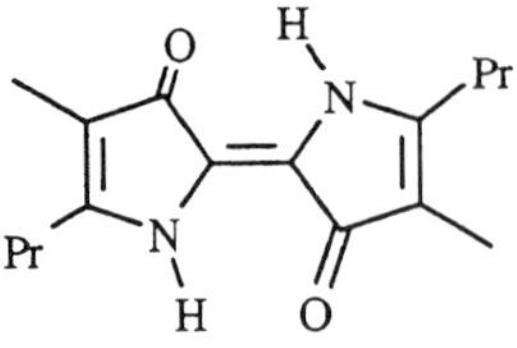

(4)

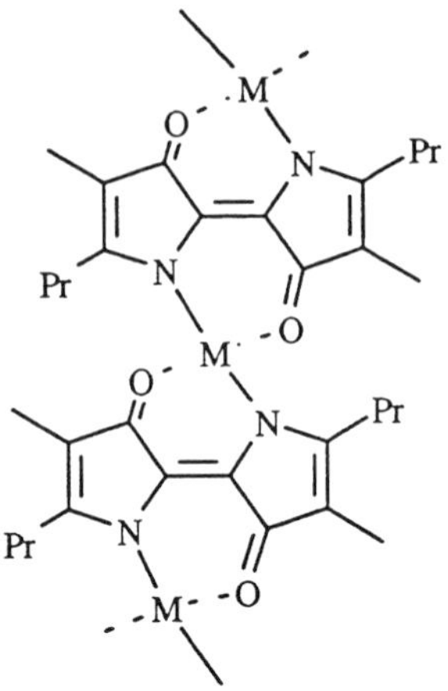

(5)

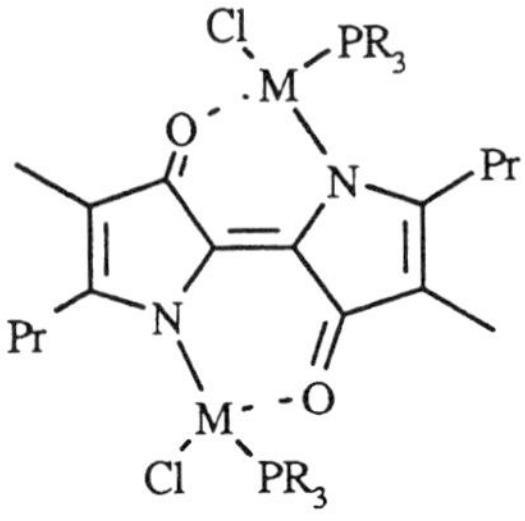

(6)

A dimeric platinum triethylphosphine complex (8) has also been of obtained from the dianion of epindolidione [dibenzo[*b,g*][1,5]naphthyridin-6,12(5*H*,11*H*)-dione] (7), a structural isomer of indigo (K.Polborn *et al.*, *Zeitschrift für Naturforschung, Sect. C*, 1992, **47**, 1393).

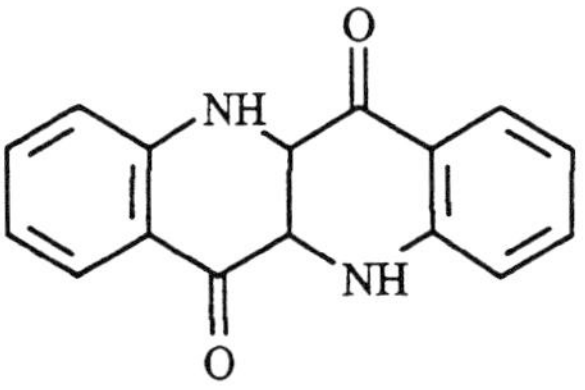

(7)

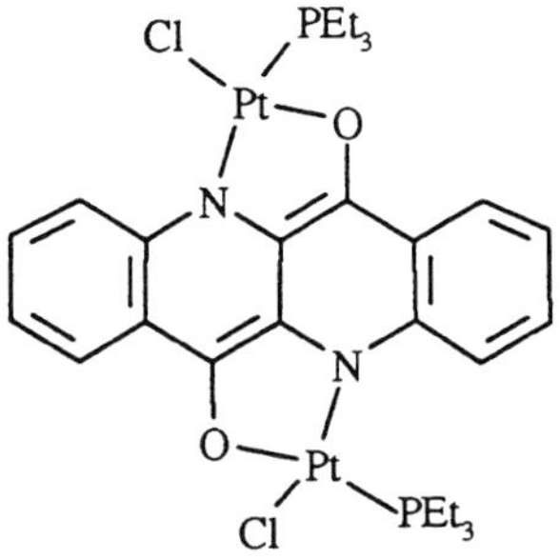

(8)

4. *Photoisomerism and photochemistry of indigo, its derivatives and related compounds*

Indigo is stable to photoinduced *trans* → *cis* [(1, R = R′ = H) → (2, R = R′ = H)] isomerisation, but when the hydrogen atoms of the NH groups are replaced by methyl, or acetyl, functions phototropic mobility is induced. This is normally explained by loss of hydrogen-bonding to the oxygen atoms of the carbonyl units (W.R.Brode, E.G.Pearson and G.M.Wyman, *J. Amer. Chem. Soc.*, 1954, **76**, 1034; J.Weinstein and Wyman, *ibid.*, 1956, 78, 4007), but a proton-transfer between NH and CO groups in the excited singlet state has also been proposed as a pathway for rapid radiationless decay (Wyman and B.M.Zarnegar, *J. Phys. Chem.*, 1973, **77**, 1204).

(1)   (2)

Furthermore, inspection of models suggests that *N*-substitution should lead to a deformation of the planar chromophore, since strong steric interactions are expected to occur between the *N*-substituents and/or between the substituents and the carbonyl units (G.M.Wyman and B.M.Zarnegar, *J. Phys. Chem.*, 1973, **77**, 831).

X-Ray diffraction studies of *N,N′*-diacetylindigo (1, R = R′ = Ac) confirm the distortion from planarity, showing that the molecule is twisted around the central bond and exhibits twisting and pyramidalisation angles of 26° and 9.5°, respectively (G.Miehe *et al.*, *Angew. Chem. Int. Ed. Engl.*, 1991, **30**, 964). The extent of torsional twisting has also been examined by Smith and his colleagues (B.D.Smith, M.-F.Paugam and K.J.Haller, *J. Chem. Soc., Perkin Trans.* 2, 1993, 165), they find a slightly smaller twist angle of 20° for *trans-N,N′-*

bis(chloroacetyl)indigo (1, R = R′ = ClCH$_2$CO). The structure of *N*-chloroacetylindigo (1, R = H; R′ = ClCH$_2$CO) is more planar still, with a twist angle of only 8°. Despite the difference in twist angles, both of these compounds have the same central bond length.

NMR studies confirm that the *N*-acyl groups in *trans-N,N′*-bis(chloroacetyl)-indigo and *N*-chloroacetylindigo suffer restricted flexibility and hindered rotation. In further work of this type X-ray studies upon *trans-N,N′*-diacetylindigo, *N,N′*-bis(chloroacetyl) indigo and *N*-acetyl-*N′*-chloroacetylindigo show that in the solid state there is hydrogen-bonding between the acyl substituents and the carbonyl groups of the indigo unit (Smith, Haller and M.Stang, *J. Org. Chem.*, 1993, **58**, 6905).

Semi-empirical MNDO calculations have been carried out by Takahashi (J.Abe, Y.Nagasawa and H.Takahashi, *J. Chem. Phys.*, 1989, **91**, 3431) upon *N,N′*-diacetylindigo, and indigo itself has been studied by Hückel molecular orbital and PPP methods (see W.Lüttke and M.Klessinger, *Tetrahedron*, 1963, **19**, 315). Recently AM1-SCF and AM1-SDCI calculations have been performed upon indigo and also upon three conformers of *N,N′*-diacetylindigo (in their *cis* and *trans* configurations). Potential energy curves for the *cis-trans* isomerisation reaction in several electronic states were obtained, and for *N,N′*-diacetylindigo the ground state curve shows a maximum at a torsion angle of the central C-C bond of 90°; whereas the S_1 and T_1 state curves show minima at the corresponding geometry. For indigo there are significant barriers to photoisomerisation in the S_1 and T_1 excited states, because of increased hydrogen bonding. The calculated enthalpies of activation for the isomerisation processes *cis*→ *trans* in the ground state agree well with values deduced by experimentation (G.Grimme *et al.*, *Chem. Ber.*, 1993, **126**, 1015).

The *cis-trans* photoisomerisation of *N,N′*-diacetylindigo is considered to be a classical example of this type of interconversion and the compound and its derivatives have been considered for possible use in solar energy conversion and storage systems. The *trans*-isomer, which is pink, absorbs at λ 562 nm, whereas the yellow *cis*-form absorbs at λ 430 nm. In aprotic solvents the photoisomerisation is clean and reversible over many cycles, but in the presence of electron donors, such as tertiary amines, and in the absence of oxygen single-electron transfer photochemistry predominates, leading to dihydro derivatives (for a discussion of this type of reaction see under thioindigo, below) (X.Ci and D.G.Witten, in *Photoinduced Electron Transfer*, eds., M.A.Fox and F.M.Chanon,

Elsevier, 1988, Chapter 4.9; J.Setsune *et al.*, *Chem. Letters*, 1986, 1393; Setsune, T.Fujiwara and T.Kitao, *Chem. Express*, 1986, **1**, 299).

In addition to the established photoisomerisation of *N,N'*-diacylindigos, for certain derivatives (3) a new reaction, an [1,3] alkyl migration, has been discovered. This is considered to involve initial photoexcitation to a singlet state, and thence to a pentacyclic biradical intermediate (4) of unknown multiplicity. If the alkyl group R in such an intermediate has a sufficiently high migratory aptitude (*i.e.* there is a sufficiently low CO-R bond dissociation energy: *e.g.* R = CH$_2$OMe; CH$_2$SMe, CH$_2$Ar, or CH$_2$OCH$_2$OMe) the biradical may be quenched through a 1,3-sigmatrophic shift of R giving the corresponding lactone (5).
The overal photochemical process may then be rationalised as follows: *trans* to *cis* isomerism occurs when the the excited *trans* singlet state, obtained upon direct irradiation at λ 530 nm, twists to form a p* state. This may relax *via* a twisted p-configuration to the *cis*-ground state. Although the *cis* configuration is unaffected by irradiation at λ 530 nm, it is unstable and thermally isomerises back to the *trans* form. Competing with the isomerisation reaction is the photochemical 1,3-migration *via* the biradical. Should the alkyl group not exhibit sufficient migratory aptitude, as is the case when R = Me; Ph, or CH$_2$Cl, the energy of the migration step is too high and only reversion to the *trans*-ground state is observed (B.D.Smith *et al.*, *J. Org. Chem.*, 1993, **58**, 6493).

Indigo and *N,N'*-dimethylindigo undergo oxidative photodegradation to give isatin and *N*-methylisatin, respectively. For these reactions a dye-sensitised *singlet oxygen* mechanism has been proposed leading to the appropriate oxetane (6). This intermediate then undergoes ring fission giving isatin (7, R = H), or *N*-methylisatin (7, R = H) (N.Kuramoto and T.Kitao, *Dyers Colour*, 1979, **95**, 257; G.Miehe *et al., Angew. Chem. Int. Ed. Engl.*, 1991, **30**, 964).

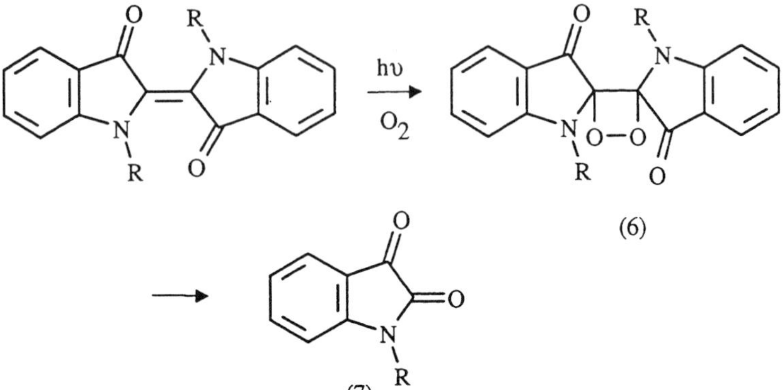

Interestingly, *N,N'*-diacetylindigo, when irradiated with light in the presence of oxygen undergoes oxygen insertion and *N to O* acyl transfer, by a *radical* mechanism. Evidence in support of this conclusion is: (a) when the reaction is carried out in toluene benzyl hydroperoxide is produced, (b) the reaction occurs in the dark if an initiator is added, and (c) attempts to obtain the photooxidised products, which are the *O*-acylated compound (8) and the epoxide (9), using singlet chemistry fail. The reactions leading to these two products are quantitative and, although slow at room-temperature, at 100° C are complete in a few hours. A possible mechanism for the formation of (8) and (9) is shown below (B.D.Smith *et al., J. Org. Chem.*, 1994, **59**, 8011).

$(3, R = Ac) \longrightarrow$

(9)

(8)

Thioindigo (TI), like the *N,N'*-diacylindigos, has been shown to undergo photo-reduction to yield a dihydro derivative in the presence of electron donors, such as triethylamine (TEA) and *N*-benzyl-1,4-dihydronicotinamide. Thus, when it is irradiated at λ 546 nm with triethylamine (TEA) the dye is converted into an excited singlet, which abstracts an electron from the donor to form an excited complex, consisting of the dye, as a radical anion, and the radical cation of TEA [TI•⁻,TEA•⁺][1,3]. This complex can decay back to TI and TEA, or undergo H^+ transfer to afford an excited pair of the radicals of monohydrothioingio and monodehydro TEA [TIH•,$CH_3CH•NEt_2$][1,3]. From this pair diffusion may lead to the free radicals TIH• and $CH_3CH•NEt_2$. The former may give dihydrothioindigo TIH_2, plus the parent TI by a *dark* disproportionation reaction, while the $CH_3CH•$ NEt_2 radical may undergo simultaneous electron-transfer and deprotonation to afford the iminium salt $CH_3CH=N^+Et_2$.

In an alternate pathway the radical pair [TIH•,$CH_3CH•NEt_2$][1,3] may participate in a second electron-transfer reaction giving the anion (TI⁻) and the iminium salt. The anion can then add a proton to form TIH_2.

Thioindigo
(TI)

Dihydrothioindigo
(TIH_2)

In the case of oxalylindigo (10) a similar photo-reduction gives rise a product, which exhibits a 18 line ESR spectrum; this is consistent with that expected for the ESR spectrum of the semi-reduced radical (11) (K.S.Schanze *et al.*, *J.Amer. Chem. Soc.*, 1986, **108**, 2646).

When 7,7′-bis(chloroacetyl)thioindigo (12) is reacted with hydroxyalkyl thiobenzoates the corresponding thioindigo thioesters (13; n = 1, 2, or 3) are formed. Hydrolysis of the thioesters by treatment with hydrazine monohydrate, in the absence of air, affords the thiols (14; n = 1, 2, or 3). However, in the case of the thioester (14; n = 3), if air is admitted, the *trans*-annular disulphide (15) is obtained. Irradiation of the thioesters in argon-bubbled 1,2-dichloroethane causes bleaching of the absoption bands at λ 534 and 316 nm in the electronic spectra, with the simultaneous development of new bands at λ 390 and 268 nm. If the irradiated solutions are left in the dark, in the absence of air, the original spectra are regenerated. The irradiation products are considered to be the disuphide leuco-compounds (16; n = 1, 2, or 3), which form through hydrogen (proton) abstraction from the thiol groups of the starting compounds, followed by intramolecular S-S coupling. Supporting evidence comes from the fact that if the leuco-compound (16, n = 3) is oxidised with chloroanil the bridged disulphide (15) is obtained in 50% yield. In this case the side chains are sufficiently long, apparently, to allow the disulphide bridge to form across the molecule rather than on the same face (S.M.F.-Raham and K.Fununishi, *Chem. Soc., Chem. Commun.*, 1992, 1740).

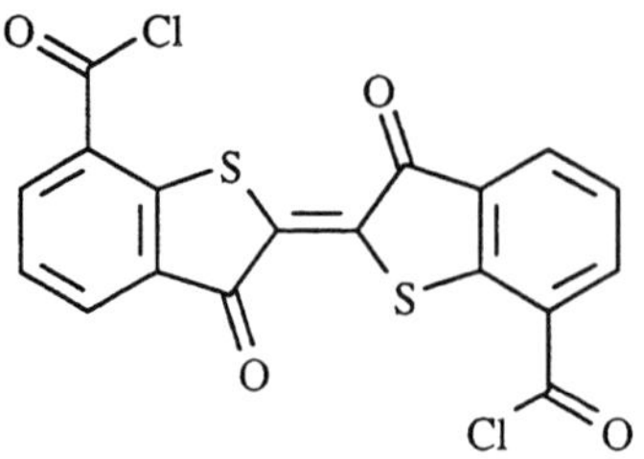

O, Cl

(12)

O, (OCH$_2$CH$_2$)nSCOPh

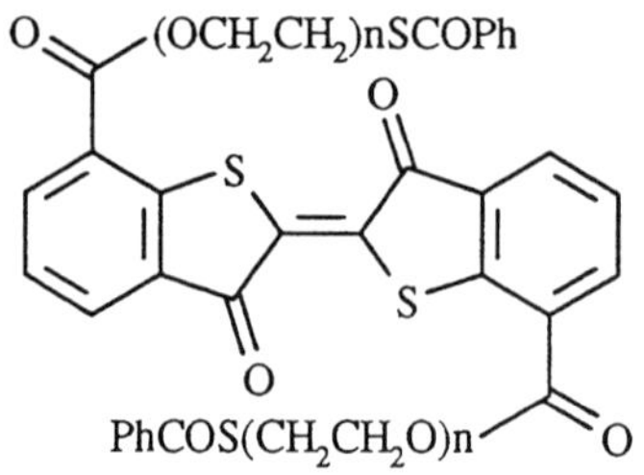

PhCOS(CH$_2$CH$_2$O)n

(13)

O, (OCH$_2$CH$_2$)nSH

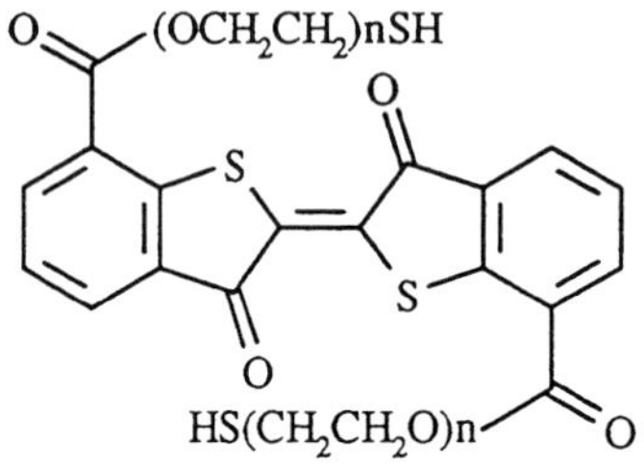

HS(CH$_2$CH$_2$O)n

(14)

O, (OCH$_2$CH$_2$)$_3$S

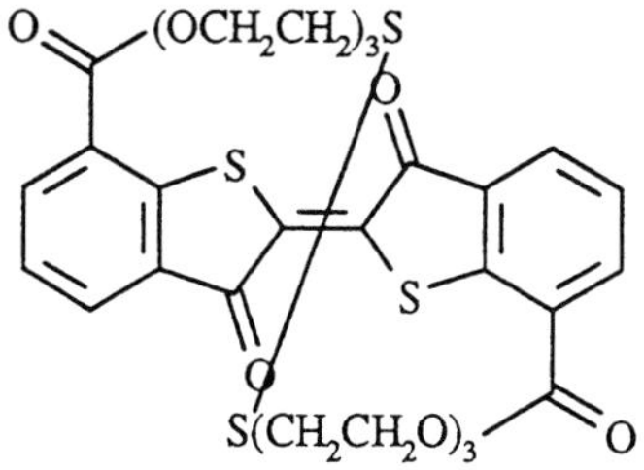

S(CH$_2$CH$_2$O)$_3$

(15)

O(CH$_2$)nSS(CH$_2$)nO

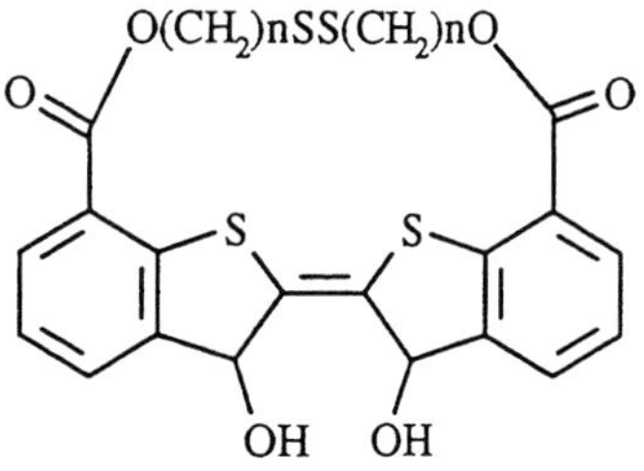

OH OH

(16)

5. *Thermal isomerisation*

Unlike thioindigo, indigo fails to undergo thermal or photochemical *trans-cis* isomerism and dry distillation decomposes the compound into aniline. However, vapour phase vacuum pyrolysis at 450 °C gives epindolidione (see section 3). This isomerisation probably involves the scission of the two C-CO bonds of indigo possibly *via* a dyotope rearrangement (intramolecular symmetrical transfer of two sigma bonds). However, since this concerted reaction is symmetry forbidden with very high activational energy (610 kJ mol^{-1}) required for the development of the transition state a step wise process is not excluded (G.Haucke and G.Graness, *Angew. Chem. Int. Ed. Engl.*, 1995, **34**, 67). Analogous isomerisations occur for 5,5'-dibromo-, 4,5,4',5'-tetrabromo- and 5,7,5',7'-tetrabromoindigo, but not for thioindigo nor thinaphthenindigo. Traces of epindolindione were detected when a solution of indigo in DMSO were heated.

transition state

Epindolididione

6. *Other indigo analogues*

The bis(cyanimino) derivatives (3) of indigo are available through the oxidative dimerisation of the dicyanomethylene compounds (2), formed *in situ* through the reaction of indoles (1) with cyanogen azide and the tetracyano (5) equivalent is obtained by the oxidative dimerisation of 1-acetyl-3-dicyanomethyleneindoline (4)

374

in the presence of silver acetate. However, a reaction between 3-(dicyanomethyl)thioindole (6) and its 2-[*N*-(phenyl)imino] derivative (7) in the presence of acetic anhydride and acetic acid yields the dark blue compound (8), rather than the expected product (9) (R.Gompper *et al., Angew. Chem. Int. Ed. Int.*, 1995, **34**, 464). When treated with butyl lithium and then with iodine the thioindole (6) affords the unstable bis(thioindole) (10), which fails to undergo further oxidation to the bis(dicyanomethylene)thioindigo (9), but gives instead the heptatriene (11). Treatment of (6) with bromine and then with trietylamine affords the bi(benzothiophene) (12).

All carbon compounds which mimic the chromophoric system of the indigos are, of course, well known. Some of these compounds *e.g* (13) and (14) are highly crowded and the pi-systems they contain are severely twisted. In the case of these two compounds the dihedral angles between the planes of the two units on either side of the central double bond are 49.7 ° and 53 °, respectively (A.Beck *et al., Chimia*, 1994, **48**, 493).

(1; R = H, Br or OMe)

(2)

(3)

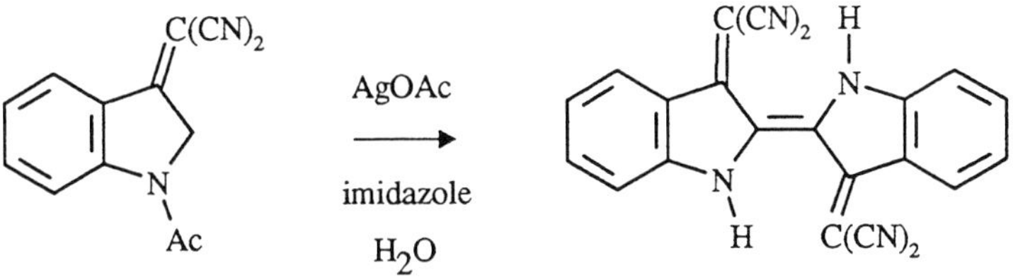

AgOAc

imidazole

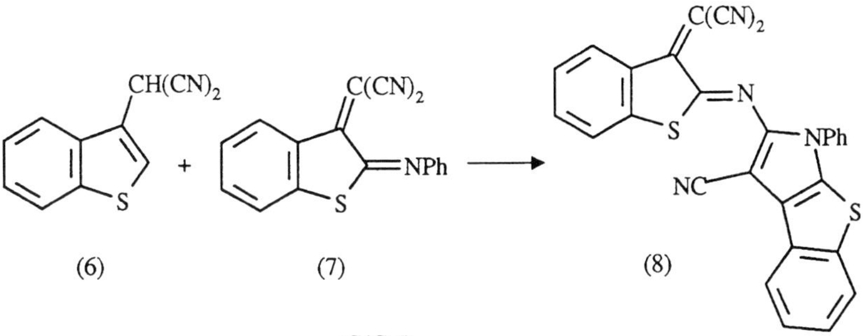

H_2O

(4)

(5)

(6)

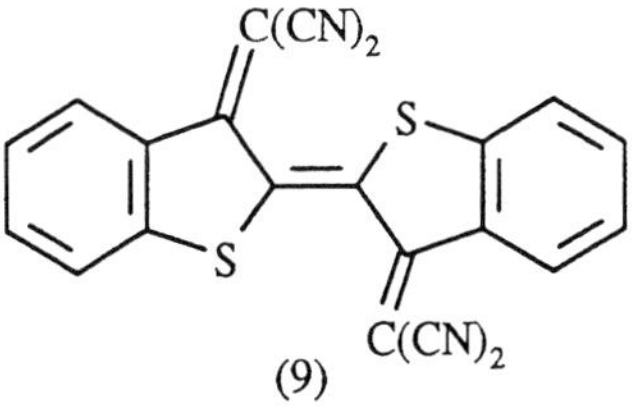

+

(7)

(8)

(9)

376

(i) BuLi

(6) $\longrightarrow$

(ii) I$_2$

(10)

(i) Br$_2$

(ii) Et$_3$N

(12)

(11)

(13)

(14)

It has been announced recently (W.Schroth *et al.*, *Chem. Ber.*, 1994, **127**, 401) that 3-mercaptobenzo[*b*]thiophene (15) in contact with an amine (*e.g.* morpholine, acting as a base) is transformed into bis(benzo[4,5]thieno)[3,2-*c*:2$^{'}$,3$^{'}$-*e*]-1,2-dithiin (18), rather than its valence isomer *trans*-dithioxoindigo as previously thought. The reaction by which this compound is formed *may* require initial oxidative coupling to generate the disulphide (16). This then undergoes a sigmatrophic rearrangement to dihydrodithioxoindigo (17), which after thioenolisation, deprotonation and further oxidation affords the dithiin. A more direct route is not excluded, but it is signifivant that the same deep red compound is also formed when thioindigo is reacted with Lawesson's reagent. Reduction of this product with sodium borohydride yields the dithiol (19; R = H), which undergoes alkylation and acylation with appropriate reagents at the thiol sulphur atoms (giving, *e.g.*, 19, R = Me or Et; or 19, R = Ac). Contrary to the behaviour of most 1,2-dithiins, sulphur is not extruded when the compound (18) is heated under mild conditions; however, this does take place at high temperatures when the product formed is bis(benzo[*b*]thieno)thiophene (20). A reaction with hydrazine gives the pyridazine (21), treatment with triethylphosphite affords (22).

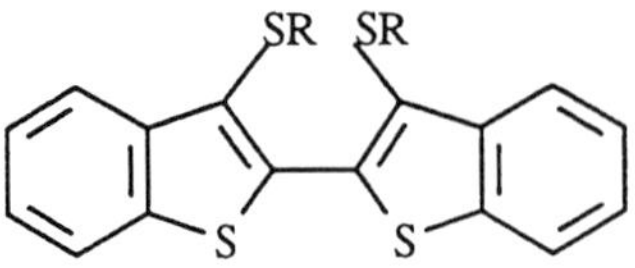

(19; R = Me, Et or Ac)

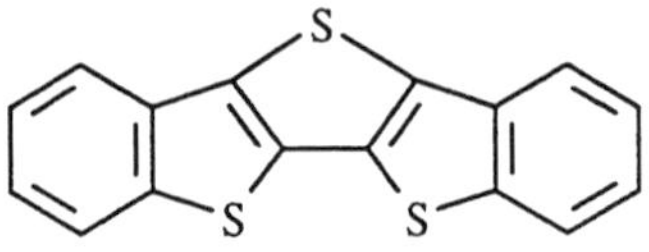

(20)

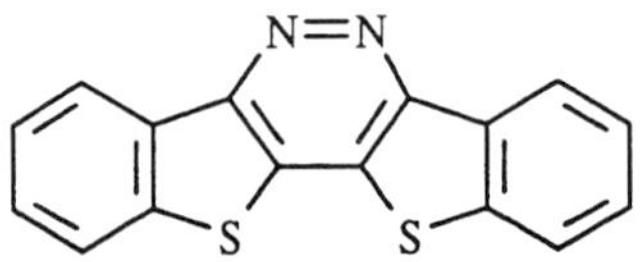

(21)

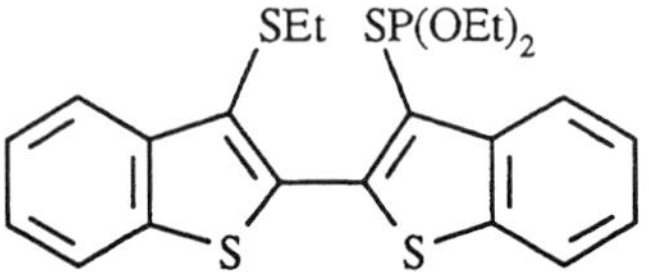

(22)

7. *Oxidation and chemical reduction of indigo derivatives*

The disodium salt of indigo carmine (1) is oxidatively degraded to give sodium isatin-4-sulphonate (4) by treatment with *N*-haloamines. In the case of *N*-bromamines the first reaction is to form the bromohydrin (2), which hydrolyses to the diol (3). This product undergoes further oxidation to a dihydroxyindoxyl, the immediate precursor of the isatin sulphonate (D.S.M.Puttaswamy and R.S.Rangappa, *Bull. Soc. Soc. Jpn.*, 1989, **62**, 3343).

The electrochemical oxidation of indigo carmine at different pH values has also received attention. In acid or neutral media (pH 1-7) the product is dehydroindigo, in basic solution (pH > 9) at the first oxidation potential (+ 0.4 V) 2,2′-dihydroxyindigo carmine is formed and at the second (+ 0.8 V) isatin is produced (G.Beggiato *et al.*, *Ann. Chim.*, 1993, **83**, 355).

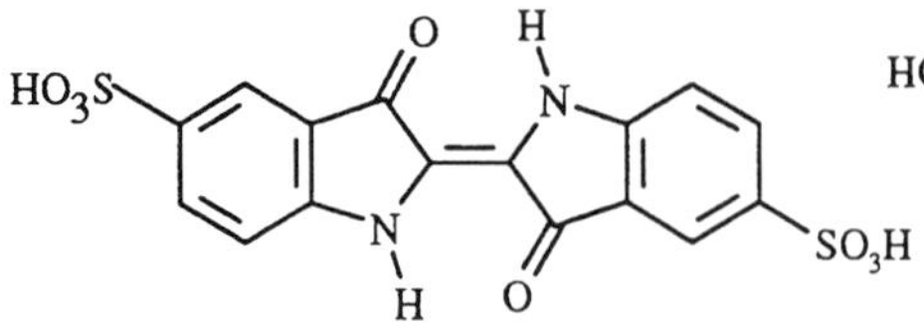

Indigo carmine

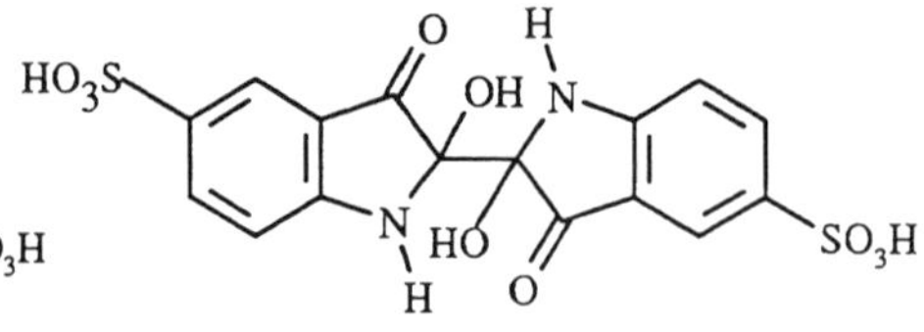

2,2'-Dihydroxyindigo carmine

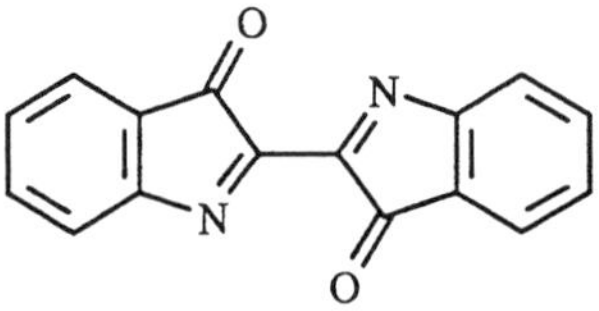

Dehydroindole

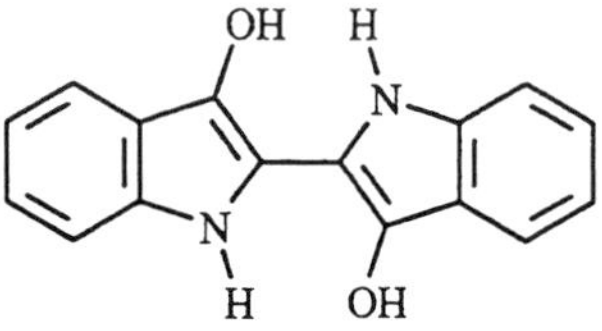

Leucoindigo

The reduction of water insoluble indigo to the soluble sodium salt of leucoindigo, is crucial to the application of indigo as a dye for textiles. Many reductants are used, a common one being sodium dithionate. Once employed, such reductants cannot be recovered and this has prompted the application of an indirect electrochemical method. This utilises triethanolamine and iron(II) salts as mediators in alkaline aqueous solution, thus indigo is introduced to this system as a finely divided suspension and then undergoes reduction by the iron-ethanolamine complex. The latter is oxidised into its Fe(III) form, and this is then recycled back to the original complex at the surface of a copper cathode (T.Bechtold *et al.*, *Angew. Chem. Int. Ed. Engl.*, 1992, 1068).

The X-ray structures of *threo*-(1,1,3,3-tetramethyl)leucoisoindigo [bis(1,3-dimethyloxindol-3-yl)] (5) and its rotational isomer have been determined (T.Suyama *et al.*, *Heterocycles*, 1994, **37**, 1069). The results complement earlier conclusions, about the conformational preferences of these compounds, which were based upon NMR evidence (T.Kato *et al.*, *Chem. Pharm. Bull.*, 1985, **33**, 5270).

Reduction of indigo with zinc in acetic acid-acetic anhydride gives mixtures of 2,2'-bisindole (6), its 1,1'-diacetyl and 3-acetoxy derivatives. The relative amounts of these products depends on the length of time the reaction is carried out and also upon the amounts of zinc used (M.Somei *et al., Heterocycles*, 1995, **41**, 2161).

(5)   (6)

8. *Biosynthesis and the isolation of indigoids from natural sources*

There are contradictory accounts in the literature concerning the role of L-tryptophan in the biosynthesis of the indigo precursors indican (1) and isatan B (2). These arguments are now resolved and it has been shown that in higher plants indole, and not tryptophan, is the biosynthetic precursor of the indoxyl derivatives (Z.-Q.Xia and M.H.Zenk, *Phytochem.*, 1992, **31**, 2695). The biosynthesis of indigo from glucose by genetically manipulated microorganisms can be sufficiently rapid to make this route one of potential commercial interest (B.D.Ensley, *Chimia,* 1994, **48**, 491).

Indigo, 6-bromoindigo and 6,6′-dibromoindigo and indirubin have been detected in the hypobranchial glandular secretions of *Murex trunculus* and *M. brandaris* (J.Wouters and J.Verhecken, *J. Soc. Dyers Colourists*, 1991, **107**, 266; R.H.Michel, J.Lazar and P.E.McGovern, *ibid.*, 1992, **108**, 145). Indigoid pigments are also present in the urine of humans (A.H.Jackson *et al., Clinica Chim. Acta,* 1988, **172**, 245) and the indigo pigment candidine is found in the urine and in the heamofiltrate of uremic patients (H.Laatch and H.Ludwigkohn, *Ann.*, 1986, 1847).

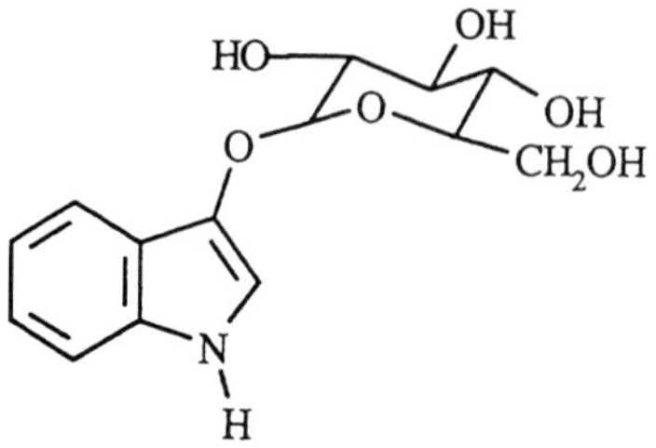

(1)

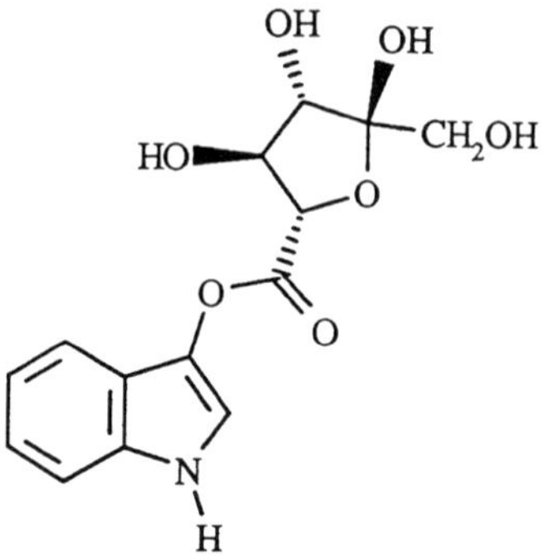

(2)

Second Supplements to the 2nd Edition of Rodd's Chemistry
of Carbon Compounds, Vol.IV B, edited by M. Sainsbury
© 1997 Elsevier Science B.V. All rights reserved.

Chapter 15

CYANINE DYES AND RELATED COMPOUNDS

G. BACH AND S. DAEHNE

Introduction

This contribution emphasizes advances in the chemistry of cyanine dyes achieved since the topic was reviewed by D.J. Fry in *Chemistry of Carbon Compounds* (2nd Ed., 1st Suppl., Vol. IV^B, *Heterocyclic Compounds*, pp. 267-296, Elsevier, Amsterdam 1985). Whereas in the past the term "cyanines" had been restricted to polymethine dyes terminated by nitrogen atoms, according to formula (1), it is now more or less generally used also for polymethine dyes terminated by other heteroatoms X,Y having a higher electronegativity than the methine atoms. Consequently, in the following the expressions "cyanines" and "polymethine dyes" are used synonymously.

$$X - (CR)_{N-2} - Y \qquad (N \pm 1)\pi \tag{1}$$

As in principle any deeply colored organic molecule contains one or even more structural units (1) further restrictions of the following review are necessary. In the polymethine dyes the index N of formula (1) is an odd, whole number ensuring that the dyes exist in non-radicaloid, diamagnetic form. However, N may be also even-numbered yielding polymethine radicals like the violenes and semiquinones which will be not considered in this Chapter.

Essential for the realization of energetically stabilized polymethine structural units (1) is the occupation of the N-numbered atomic chain with (N+1) or (N-1) π-electrons. The surplus of one π-electron is usually provided by linkage of a donor and a acceptor substituent via a conjugated hydrocarbon chain. The precondition for energetic stabilization of (1) in case of the $(N+1)\pi$ polymethines is that the electronegativity of the terminal X,Y atoms is higher than that of the methine atoms. In case of polymethines having terminal atoms of equal or lower electronegativity than carbon, e.g. silicon, boron, etc., chain-shaped structures (1) are energetically stabilized only when they are occupied with $(N-1)$ π-electrons. The allyl cation and its vinylogous derivatives are typical examples of $(N-1)\pi$ polymethines, but such dyes will be not reviewed here.

For that reason the following Chapter will be mainly occupied with the cyanines (1; $X,Y = N^{1/2+}$), the merocyanines (1; $X = N^{1/2+}$; $Y = O^{1/2-}$) and the oxonols (1; $X,Y = O^{1/2-}$) plus their derivatives which possess an odd-numbered polymethine chain having $(N+1)$ π-electrons. Each methine atom can be arbitrarily modified by any substituent R or even replaced by heteroatoms such as N, P, As, O, S, etc.

Great progress has been accomplished in molecular physics and in important applications of cyanine dyes together with new synthetic methodology. For instance, in the last decade the number of publications cited in *Chemical Abstracts* on the topic "Cyanine Dyes" has enlarged from about 100 to 200 a year. Recently a comprehensive review about the synthesis of methine dyes and their application especially in the textile industry has been written by R. Raue (*Ullmann's Encyclopedia of Industrial Chemistry*, 5th Ed., Vol. A 16, pp. 487-534, VCH, Weinheim 1990). This review mainly covers the patent literature. Furthermore general reviews were published by A.D. Kachkovski [*Stroenie i Tsvet Polimetinovykh Krasitelei (Structure and Color of Polymethine Dyes)*, Naukova Dumka, Kiev, 1989; *Chem.Abstr.* 1990, **112**, 200590k], H. Zollinger (*Color Chemistry*, 2nd ed., VCH, Weinheim, 1991), N. Tyutyulkov, J. Fabian, A. Mehlhorn, F. Dietz, A. Tadjer (*Polymethine Dyes - Structure and Properties*, St. Kliment Ohridski University Press, Sofia, 1991), H. Kampfer (*Ullmann's Encyclopedia of Industrial Chemistry*, 5th Ed., Vol. A20, VCH, Weinheim, 1992, pp. 10-17), and A.A. Ishchenko [*Stroenie i Spektral'no-luminestsentnye Svoistva Polimetinovykh Krasitelei (Structural, Spectral, and Luminescent Properties of Polymethine Dyes)*, Naukova Dumka, Kiev, 1994].

Emphasis on the *Analytical Chemistry of Synthetic Colorants* is laid in *Adv. Color Chem. Ser.*, Eds. A.T. Peters, H.S. Freeman, Chapman and Hall,

London etc., 1995, Vol. 2. Properties and applications of dyes absorbing in the near infrared (NIR) region are described in *Infrared Absorbing Dyes*, *Top. Appl. Chem.*, Ed. M. Matsuoka, Plenum Press, New York, London, 1990. The position and absorptivity of thousands of dyes are compiled in the handbook from M. Okawara, T. Kitao, T. Hirashima, M. Matsuoka, *Organic Colorants*, Phys. Sci. Data, Kodansky, Tokyo, and Elsevier, Amsterdam, Oxford, New York, Tokyo, 1988, Vol. 35, and in M. Matsuoka, *Absorption Spectra of Dyes for Diode Lasers*, Bunshin Publ., Tokyo, 1990. Dyes recently used in modern high technology fields are considered in *Chemistry of Functional Dyes*, Eds. Z. Yoshida, T. Kitao, Mita Press, Tokyo, 1989, and in *The Chemistry and Application of Dyes*, Eds. R.D. Waring, G. Hallas, Plenum Press, New York, London, 1990.

The classical field of spectral sensitization of photographic silver halides by cyanine dyes has been widely extended to other semiconductors as well as to photovoltaic processes in order to simulate photosynthesis in bacteria and plants and to realize artificial solar energy conversion and storage processes like water splitting. Also interesting improvements have been realized in the application of cyanine dyes for coloration of textiles, especially for polyacrylonitrile fibers, as well as in histological staining, as therapeutical drugs, and as indicators in analytical chemistry.

However, the most important progress in the application of cyanine dyes has been achieved e.g. in laser physics, nonlinear optics, optical data storage, liquid crystal displays (LCD's), dye sensor techniques, initiating photopolymerization, photochromism, electrochromism, electrophotography, phototherapy, as well as in fluorescence labeling in analytical chemistry, DNA sequencing, immunoassay techniques, and for the determination of membrane potentials. Also the understanding of aggregation effects and energy propagation phenomena in cyanine dyes has been widely improved. These developments have even justified the creation of the term "Functional Dyes" for modern applications (J. Griffiths, *Chimia (Aarau)*, 1991, **45**, 304; *Chem. Unserer Zeit*, 1993, **27**, 21).

In basic research a more general concept of color and constitution has been developed by S. Daehne [*Science(Washington) 1978*, **199**, 1163; *Progr. Phys. Org. Chem.* 1985, **15**, 1; *ibid* 1990, **18**, 1;. *Chimia (Aarau)* 1991, **45**, 288]. An up-to-date historical account on color and constitution was recently given by the same author (*Die Allianz von Wissenschaft und Industrie: A.W. Hofmann (1818-1892)*, Eds. C. Meinel, H. Scholz, VCH, Weinheim, 1992, pp. 257-286).

Therefore, this review will cover the following topics:
1. Theory of color and constitution
2. Syntheses of cyanine dyes
3. Properties of cyanine dyes
4. Applications of cyanine dyes

1. Theory of color and constitution
(a) The ideal polymethine state
The concept of the ideal polymethine state now well accepted in the literature has already been reviewed by D.J. Fry (*loc. cit.*). It describes the characteristic features of cyanines such as their bond order equalization [which is identical with bond lengths equalization in respect of C-C bonds and which is the reverse parameter of bond order alternation (BOA)], strong electron density alternation (identical with charge alternation in most cases) and charge reversal on light excitation, energetic stabilization, and their deep color. The energetic stabilization of polymethine-like compounds has recently been explained by the so-called charge stabilization rule by B.M. Grimarc (*J. Am. Chem. Soc.* 1983, **105**, 1979) and Y. Cohen et al. (*J. Am. Chem. Soc.*, 1988, **110**, 4634). Although so far no theoretical deduction has been given, the rule shows that charge alternation and stabilization must be related to each other.

Additional support of the polymethine concept came from theoretical results obtained by G.G. Dyadyusha et al. who considered the influence of carbo- and heterocyclic terminal groups on the color and energy levels of polymethine chains (*Ukr. Khim. Zh.*, 1975, **41**, 1176; *J. Mol. Struct.*, 1990, **217**, 195; *Dyes Pigm.*, 1991, **15**, 191.). Through definition of topological indices Φ_0 and the effective length L the dependence of the light absorption on the molecular structure and electron donor ability (i.e. the basicity) of the end groups at polymethine chains can be quantified. As the numerical values of Φ_0 and L are not only available by quantum chemical calculations but also by simple graph theoretical inspections (G.G. Dyadyusha, A.D. Kachkovski, *Teor. Eksp. Khim.*, 1979, **15**, 152; *J. Inform. Record. Mater.*

1985, **13**, 95) the model considerations are easy to handle and confirm the existence of the ideal polymethine state in case of $\Phi_0 = 45^0$ (A.D. Kachkovski, M.L. Dekhtyar, *Dyes Pigm.*, 1993, **22**, 83; M.L. Dekhtyar, *Dyes Pigm.*, 1995, **28**, 261)

(b) Triad theory

The concept of the ideal polymethine state has been further extended to an approach designated "triad theory" which embraces all conjugated organic compounds [COC's, S. Daehne, *Science (Washington)* 1978, **199**, 1163; *Z. Chem., 1981*, **21**, 58; *Progr. Phys. Org. Chem.* 1985, **15**, 1; *Chimia (Aarau)* 1991, **45**, 288; C. Reichardt, *J. Phys. Org. Chem.*, 1995, **8**, 761]. This theory comprises both the results of modern quantum chemistry, the immense amount of experimental facts, and the manifold empirical rules formulated through the last century describing special structure-property relationships in COC's. In this way, a qualitatively new understanding of the physical and chemical properties is attained also of the cyanine dyes which enables molecules now to be designed with intended features. Hundreds of examples are given by S. Daehne (*Progr. Phys. Org. Chem., loc. cit.*).

According to triad theory, the structure of conjugated organic compounds, and hence also of cyanines, can be derived from three fundamental principles. These are:

> 1. *Triad principle*:
> The diversity of COC's is interpreted in terms of intermediates between three ideal states: the aromatic state, the polymethine state, and the polyene state.
> 2. *Energy principle*:
> The structure of COC's derives from their tendency to form as many as unbranched, energetically stabilized, highly symmetric, aromatic and/or polymethine structural units as possible.
> 3. *Bonding principles*:
> 3a. The sum of bond strengths of an atom is approximately constant (Gebhard-Pauling rule),
> 3b. The bond lengths between equicharged, neighboring atoms are additionally enlarged and those between oppositely charged atoms are additionally shortened (Kulpe's rules).

Some important properties in the ground and first excited singlet state of the three basic states are summarized in Table 1.1 (S. Daehne, *Progress and Trends in Applied Optical Spectroscopy*, Eds. D. Fassler, K.H. Feller, B. Wilhelmi, *Teubner Texte zur Physik*, Vol. 13, Leipzig, 1987, pp. 178-196). A more detailed analysis of excited state properties was given by M. Klessinger (*J. Mol. Struct.*, 1992, **266**, 53).

TABLE 1.1.
CHARACTERISTIC FEATURES OF THE THREE IDEAL STATES OF CONJUGATED COMPOUNDS

Properties of the ground state S_0

Property	aromatic state	polymethine state	polyene state
π-bond orders	equal	equal	alternating
π-electron densities	equal	alternating	equal
resonance energy	high	medium	low
polarizability	low	high	medium
chemical reactivity	low	substitution	addition
$S_0 \longrightarrow S_1$ transition energy	high	low	medium

Properties of the first excited singlet state S_1

π-bond orders	alternating	equal	equal
π-electron densities	equal	alternating	equal
polarizability	medium	high	high

TABLE 1.1. continued

Changes of properties on $S_0 \longrightarrow S_1$ excitation

Change	aromatic state	polymethine state	polyene state
π-bond orders	medium	minor	strong
π-electron densities	no	strong	no
polarizability	medium	small	strong

By analogy to the well-known alternant and nonalternant hydrocarbons, two basically different types of polymethines have been defined [S. Daehne, S, Kulpe, *J. Prakt. Chem.*, 1978, **320**, 395; S. Daehne, F. Moldenhauer, *Progr. Phys. Org. Chem.*, 1985, **15**, 1; S. Daehne, *Chimia (Aarau)* 1991, **45**, 288]. On structural modification of cyanine dyes either the characteristic π-electron density alternation through the polymethine chain is maintained yielding *alternating polymethines*, or it is disturbed giving *nonalternating polymethines*. Some examples of both types are given in Fig. 1.1.

The (non)alternant hydrocarbons and the (non)alternating polymethines may be identical. However, as per definition the charge alternation in polymethines by all means starts with negative sign at each heteroatom having a higher electronegativity than carbon, and *vice versa*; there may occur also differences between both types.

Based on the definition of the alternating and nonalternating polymethines, the manifold color rules described in the literature can be reduced to five general rules which are valid for all strongly conjugated, planar molecules. These rules read:
   1. *Polymethines* are deeper in color in comparison to *aromatics* and *polyenes* having the same number of π-electrons.
   2. *Alternating polymethines* absorb at slightly shorter wavelengths than the longest constituent ideal polymethine unit in the molecule.

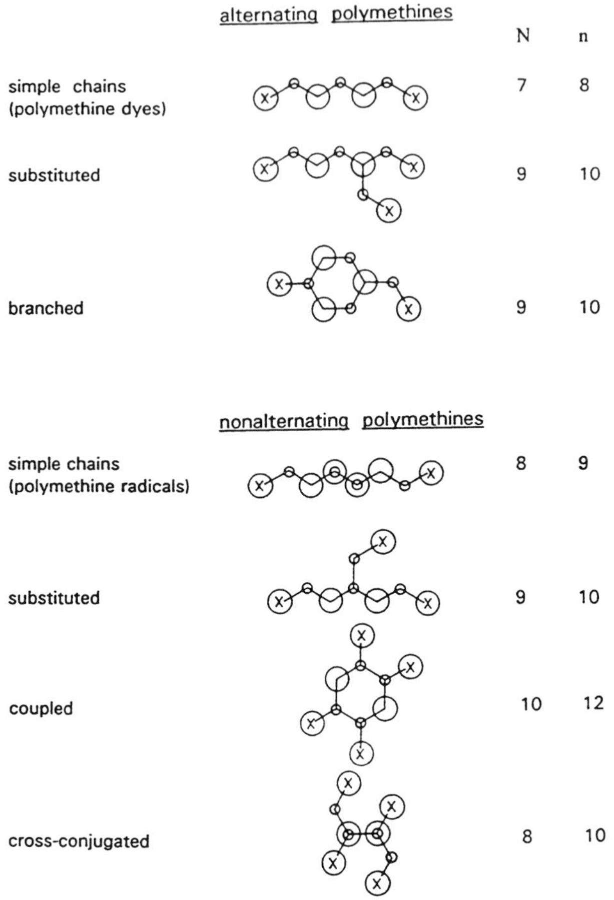

390

FIG. 1.1. CLASSIFICATION OF ALTERNATING AND NON-ALTERNATING POLYMETHINES.

The π-electron density alternation is indicated by large and small circles, respectively. X stands for any donor substituent, such as NR_2, SR, OR, O^-, etc., whereas C=X stands for any acceptor substituent, such as NO_2, $C=NR_2^+$, CHO, CN, COO^-, $C=OR^+$. N is the number of constituent atoms and n the number of π-electrons in the conjugated molecule. (Taken from S. Daehne, F. Moldenhauer, *Progr. Phys. Org. Chem.*, 1985, **15**, 1. Copyright 1985 by John Wiley and Sons, Chichester. Reprinted by permission of the copyright owner).

3. *Nonalternating polymethines* absorb at substantially longer wavelengths than the longest constituent ideal polymethine unit in the molecule.
4. If the polymethine structure in alternating or nonalternating polymethines is weakened in favour of competing *aromatic structures* the molecules absorb at substantially shorter wavelengths than the longest constituent ideal polymethine unit in the molecule, in agreement with color rule 1.
5. If structural effects or terminal groups of unlike electronegativity remove the symmetrical π-electron distribution along the polymethine chain, molecules of *polyenic structure* are formed which absorb at slightly shorter wavelengths than the longest constituent ideal polymethine unit in the molecule, in agreement with color rule 1.

If *exceptions* from these general color rules happen, they indicate the existence of *second order effects* such as sterical distortion, *cis-trans*-isomerization, formation of complexes, molecular aggregation, etc. which influence color likewise.

A special case are cyanine dyes which cannot be written in classical Kekulé structures. An example are the 3,3'-pyridocyanine dyes (1.1) (S. Daehne, F. Moldenhauer, *Progr. Phys. Org. Chem.*, 1985, **15**, 1). Quantum chemical considerations showed that the non-Kekuléan dyes are nevertheless closely related to the isoconjugated classical polymethine dyes. However, they should absorb at longer wavelengths than the classical one and should have a lower transition probability. Also a certain biradicaloid character is to be expected with non-Kekuléan dye structures (J. Fabian, *J. Prakt. Chem.*, 1989, **331**, 637).

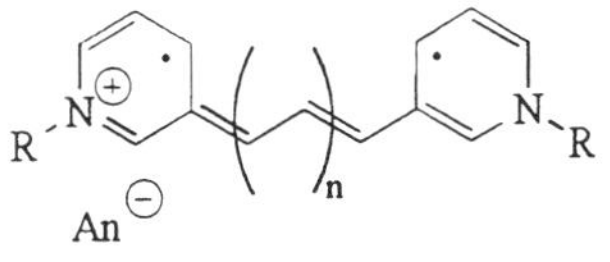

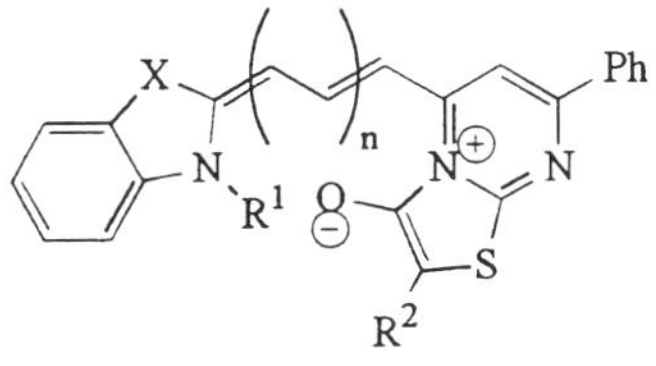

Another exception takes place with cyanine dyes which are terminated by mesoionic groups, e.g. the thiazolo[3,2-a]pyrimidinium 3-oxide (1.2). Due to their antiaromatic character such groups eventuelly absorb at longer wavelengths than the constituent polymethine system, thus governing the light absorption (K.V. Fedotov et al., *Zh. Org. Khim.*, 1988, **24**, 1310).

Striking results demonstrating the superiority of triad theory in theoretical organic chemistry are, e.g., the explanation of the physical background of deviations and exaltations observed in case of strongly conjugated substituents in linear free-energy relationships which are caused by formation of polymethine-like structures (S. Daehne, K. Hoffmann, *Progr. Phys. Org. Chem.* 1990, **18**, 1), the possibility of controlling electrophilic and nucleophilic substitution reactions (S. Daehne, K. Hoffmann, *loc. cit.*), the design of organic compounds with promising nonlinear optical properties as will be shown in Sect. 4(*d*), or molecular modelling of NIR dyes.

(c) Significance of NIR dyes
One of the most important topics in current chemistry of cyanine dyes is the development of dyes which absorb in the near infrared region (NIR dyes). This had been initiated by the availability of small, easily to handle, and low-priced laser diodes which are useful monochromatic and coherent light sources of high intensity. They render possible miniaturization of uv/vis spectroscopy as well as in situ investigations using light conducting glass fibers. In one, the spectroscopic detection limit of dyes can be extended over several orders of magnitude, which is one of the conditions in the development of new optical sensors (*Fiber Optic Chemical Sensors and Biosensors*, Ed. O.S. Wolfbeis, CRC Press, Boca Raton, Florida, 1991, Vols. 1 and 2; O.S. Wolfbeis, G. Boisdé, *Sensors*, Eds. W. Goepel, J. Hesse, J.N. Zemel, VCH, Weinheim, New York, Basel, Cambridge, 1992; *Probe Design and Chemical Sensing*, Ed. J.R. Lakowicz, Plenum Press, New York, London, *Top. Fluor. Spectrosc.*,1994, Vol. 4). The only restriction in the application of laser diodes is, up to now, that their emission lines are positioned at wavelengths longer than 630 nm necessiating dyes, which absorb in the NIR region.

In order to design new NIR dyes, both triad theory (see preceding Section) and the concept of antiaromaticity (J. Fabian, R. Zahradnik, *Angew. Chem.* 1989, **101**, 693; *Angew. Chem. Int. Ed. Engl.* 1989, **28**, 677) proved to be useful. Apart from the books mentioned in the Introductionary Section the present state of the art on NIR dyes has been comprehensively reviewed by

J. Fabian (*J. Prakt. Chem.* 1991, **333**, 197; J. Fabian et al., *Chem. Rev.*, 1992, 92, 1197). Fig. 1.2, taken from the first cited review, gives some examples of vinylogously lengthened cyanine dyes (1.3, N = odd-numbered) and polymethine radicals usually designated violene radicals (1.4, N = even-numbered), whose longer derivatives absorb in the NIR region.

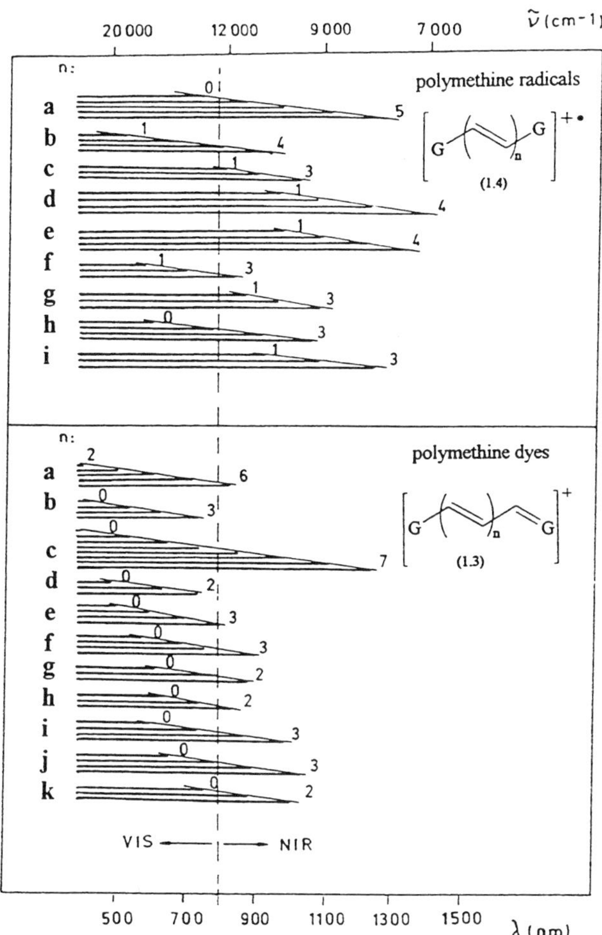

FIG. 1.2. MAXIMA OF THE LONGEST WAVELENGTH ABSORPTION BAND OF VINYLOGOUS POLYMETHINE (OR CYANINE) DYES (1.3) AND POLYMETHINE (OR VIOLENE) RADICALS (1.4) IN DEPENDENCE ON THE CHAIN LENGTH.

FIG. 1.2. continued
The top number indicates the longest vinylogous entity of each series. The inclined lines at each column demonstrate the constancy of the vinlyene-shift which amounts to about 100 nm. Dyes absorbing at wavelengths longer than 600 nm are already useful for applications in high technology fields. (Taken from J. Fabian, *J. Prakt. Chem.*, 1991, **333**, 197, by courtesy of Huethig-Fachverlag, Heidelberg)

The letters represent dyes with the following terminal groups G:

polymethine or cyanine dyes   (1.3)

a: G =   ==NMe$_2$

b: G =

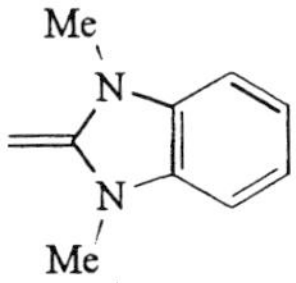

c: G =

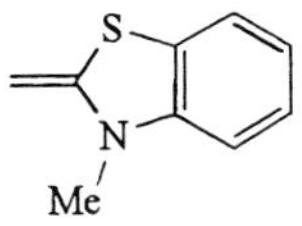

d: G =

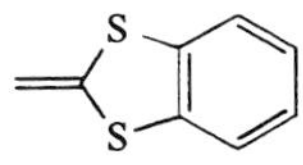

e: G =

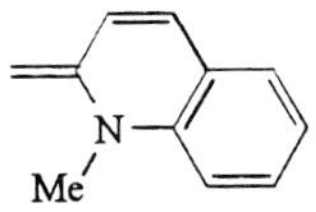

f: G =

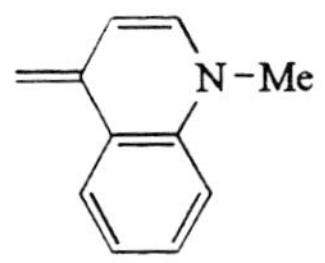

g: G =

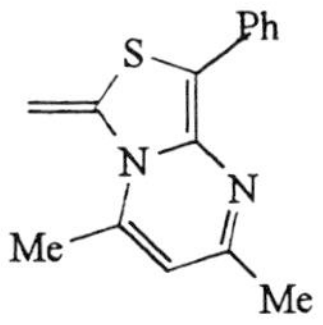

h: G =

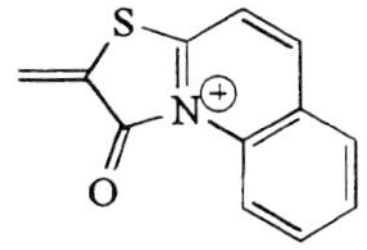

i: G =

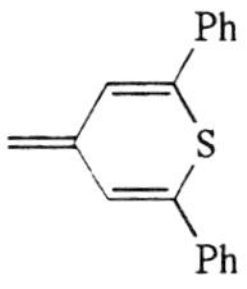

j: G =

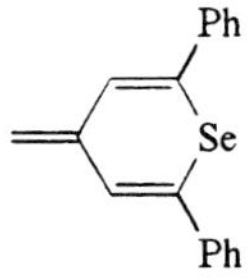

k: G =

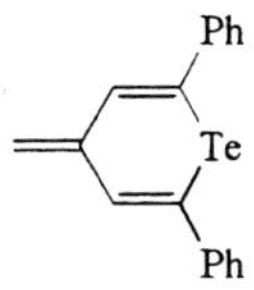

FIG. 1.2. continued

polymethine or violene radicals (1.4)

a: G = 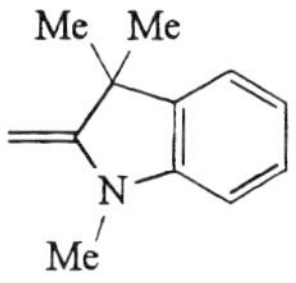

e: G = 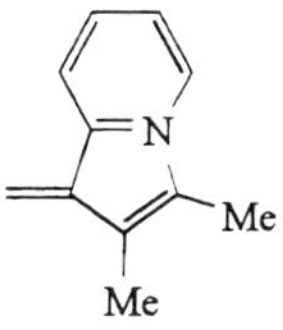

b: G = 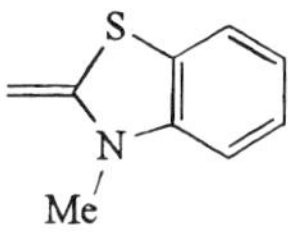

f: G = 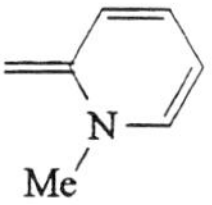

g: G = 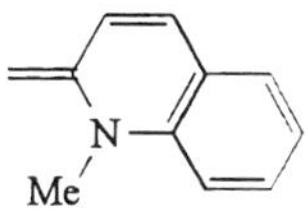

c: G = 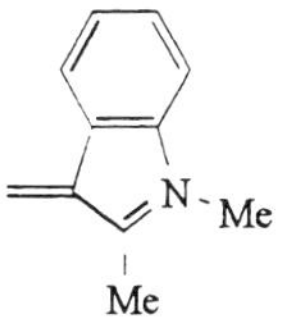

h: G = 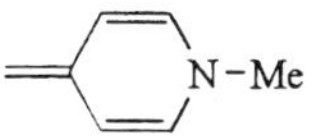

d: G = 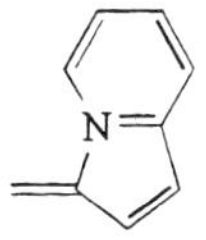

i: G = 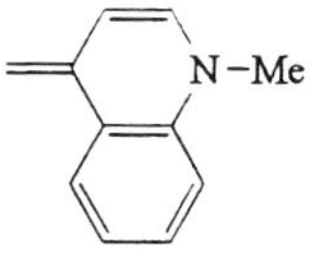

In agreement with color rule 3, especially strong red-shift of light absorption is realized through multifold substitution of polymethine chains in a nonalternating manner. As shown in Table 1.2, in comparison to the non-substituted cyanine dye (1.5), doubly nonalternating substitution by one hydroxylate group and one keto group giving the squarylium dyes (1.6) causes a red-shift of about 10 nm (K.-Y. Law, F.C. Bailey, *J. Org. Chem.*, 1992, **57**, 3278), whereas trifold nonalternating substitution with one hydroxylate and two keto groups giving the croconate dyes (1.7) yields shifts up to 126 nm (S. Yasui et al., *Dyes Pigm.*, 1988, **10**, 13).

TABLE 1.2.
LONGEST WAVELENGTH ABSORPTION MAXIMUM λ_{MAX} [nm] OF
DYES (1.5) TILL (1.9)

X	(1.5)	(1.6)	(1.7)	(1.6)-(1.5)	(1.7)-(1.5)
CH=CH	706	724	832	18	126
S	651	663	771	12	120
$CMe_2^{+)}$	638	629	764	-9	126
(1.8)	860	-	(1.9) 1014	-	$154^{++)}$

$^{+)}$ N-Me substituted for N-Et; $^{++)}$ difference (1.9) - (1.8).

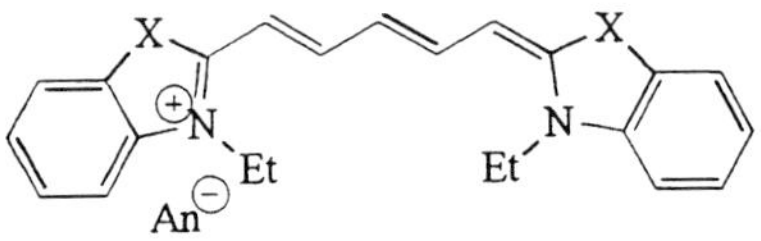

(1.5)

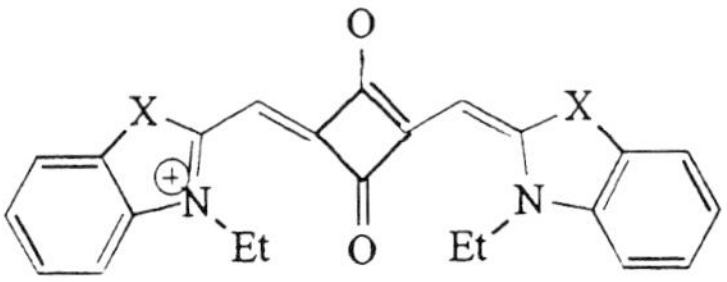

(1.6)

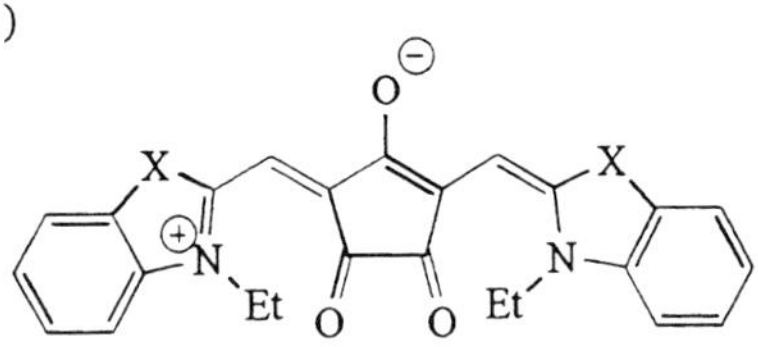

(1.7)

With dyes (1.8) and (1.9) even a bathochromic shift of 154 nm is realized
[J. Griffiths, *Chimia (Aarau)*, 1991, **45**, 304].

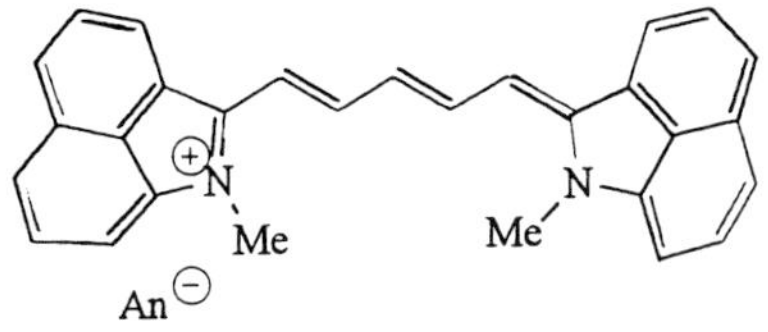

(1.8)

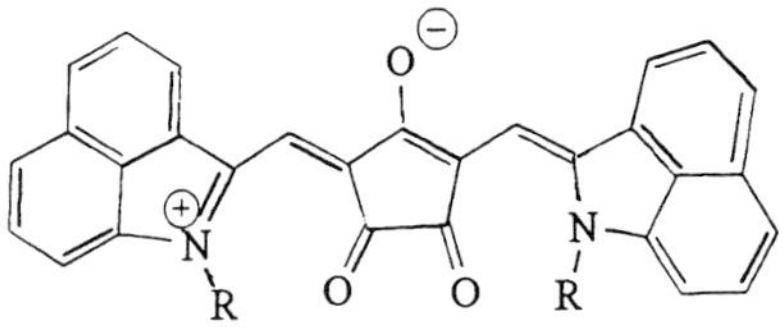

(1.9) $R = C_{10}H_{21}$

In view of applications in laser physics [cf. Sect. 4(c)] and nonlinear optics [cf. Sect. 4(d)], asymmetrical cyanine dyes such as (1.10) and (1.11) were developed (N.A. Derevyanko et al., *Mendeleev Commun.*, **1991**, 91), the absorption and fluorescence bands of which are filed in Table 1.3 in comparison to the symmetrical dyes.

(1.10)

(1.11)

TABLE 1.3.
LONGEST WAVELENGTH ABSORPTION AND FLUORESCENCE MAXIMA (λ_{max}) AND (λ_{fl}), RESPECTIVELY, OF PYRIDOPYRYLIUM POLYCARBOCYANINE DYES (1.10) and (1.11) IN DICHLORO-METHANE

dye	X	Y	n	λ_{max} [nm]	λ_{fl} [nm]
(1.10)	NMe	NMe	0	945	1012
	NMe	O	0	730	970
	NMe	O	1	760	--
(1.11)	NMe	NMe	-	1015	1030

Although according to color rule 5, the asymmetrical dyes absorb at appreciably shorter wavelengths than the symmetrical ones, they exhibit fluorescence at rather long wavelengths. Similar observations were made with asymmmetrical thia- and imidacarbo-, dicarbo, and tricarbocyanine dyes (A.A. Ishchenko et al., *Dyes Pigm.*, 1992, **19**, 169).

398

2. Progress in syntheses of cyanine dyes

(a) Basic synthetic routes

Nearly all synthetic routes of polymethine dyes containing the structural unit (2.1) are based on condensation reactions independent of the nature of their terminal groups,

$$X \diagdown\!\!\!\diagup \left(\diagup\!\!\!\diagdown \right)_n \diagdown Y \quad \longleftrightarrow \quad X \diagup \left(\diagdown\!\!\!\diagup \right)_n \diagup\!\!\!\diagdown Y \qquad (2.1)$$

like cyanines ($X,Y = N^{1/2+}$-cycles); oxonols ($X,Y = O^{1/2-}$-atoms): merocyanines ($X = N^{1/2+}$-cycles or noncyclic substituents; $Y = O^{1/2-}$-atoms); streptocyanines ($X,Y = N^{1/2+}$-noncyclic substituents); hemicyanines ($X = N^{1/2+}$-cycles; $Y = N^{1/2+}$-noncyclic substituents); styryl dyes ($X = N^{1/2+}$-cycles, $Y = N^{1/2+}$-phenylene analogues); and (thia)pyrylium dyes ($X,Y = O(S)^{1/2+}$ in six-membered rings). The synthetic routes are also independent of the length of the polymethine chain including its bridging or branching, and of both the polymethine chains' substituents and the substitution of heteroatoms for the methine atoms, like N (azacyanines), P (phospha-cyanines), or As (arsacyanines). Thus, all feasible syntheses and substitutions are founded on reactions between electrophilic and nucleophilic partners, taking into account the principles of vinylene or phenylene analogous compounds as well as possible polarizations of the π-electron system through the polymethine chain. However, this conclusion does not make allowance for sterical effects which have to be considered separately.

A recently described example of synthesis of carbocyanines (2.4a) including their protonated forms (2.4b) is shown in Scheme 2.1, where for better understanding partly carbenium and partly ammonium limiting structures are depicted. The weakly acidic methylcycloammonium salts (2.2) are condensed with dichloromethoxymethane (2.3) in acetonitrile and/or dichloromethane using solid sodium hydroxide as auxiliary base (D. Schelz et al., *Dyes Pigm.*, 1986, 7, 187). Unlike common syntheses of carbocyanines the formation of the intermediate product (2.5) does not play any part. However, at higher temperatures or with the deficiency of (2.3) the trinuclear tetramethinecyanines (2.6a/b) are formed. With imidazolium salts only the latter, or even the four-nuclear dyes (2.7a/b) are produced through addition of (2.2). Table 2.1 displays some results taken from D. Schelz et al. (*loc.cit.*).

SCHEME 2.1

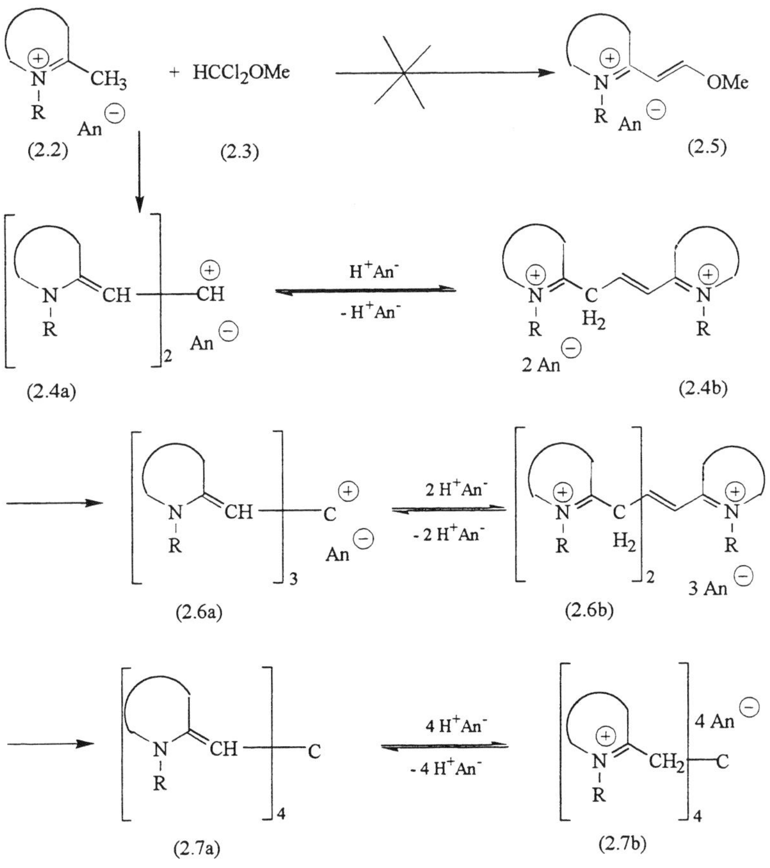

The most favorable reaction medium for synthesizing symmetrical and asymmetrical cyanine dyes is pyridine which acts simultaneously as solvent and auxiliary base. Instead of pyridine triethylamine or sodium hydroxide are often used in combination with solvent mixtures of, e.g., acetic anhydride alone or in combination with nitrobenzene, phenols, alcohols, acetonitrile, dimethylsulfoxide, or dimethylacetamide.

400

TABLE 2.1.

REACTION YIELDS OF DYES (2.4), (2.6), AND (2.7), ACCORDING TO SCHEME 2.1.

Heterocycle (2.2)		yield in percent				
No.		(2.4a)	(2.4b)	(2.6a)	(2.6b)	(2.7b)
a	1,3-dimethylbenzimidazole-2	50	69	-	-	-
b	1-octadecyl-3-phenylbenz-imidazole-2	7	9	-	-	-
c	1,2-dimethylindazole-3	22	-	-	21	-
d	1-methyl-2-octadecylindazole-3	45	19	-	-	-
e	1-methylpyridine-2	-	49	-	29	-
f	1-octadecylpyridine-2	17	-	30	21	-
g	1-octadecylpyridine-4	-	59	-	-	-
h	1,3-dimethyl-4,5-diphenyl-imidazole-2	-	-	-	40	-
i	1,3-dimethylimidazole-2	-	-	-	-	77

The dominant influence of acetanhydride on the reactivity of hemicyanines has been proven with the benzimidazolo derivatives (2.8a,b,c) by determination of the reaction rate constant k_0 of a typical dye condensation reaction, shown in Scheme 2.2 using nmr and uv-vis spectroscopy (Table 2.2, E. Kleinpeter et al., *J. Prakt. Chem.*, 1974, **316**, 761)

TABLE 2.2.

REACTION RATE CONSTANT k_0 FOR THE FORMATION OF MEROCYANINE DYES (2.8d) ACCORDING TO SCHEME 2.2.

R	An	k_0 10^{-3} [mol/l min]
$COCH_3$	CH_3CO_2	immediate reaction
$COCH_3$	Cl	306.5
H	CH_3CO_2	35.9
(2.8b)	--	15.0
H	Cl	7.1
H	I	6.1

SCHEME 2.2

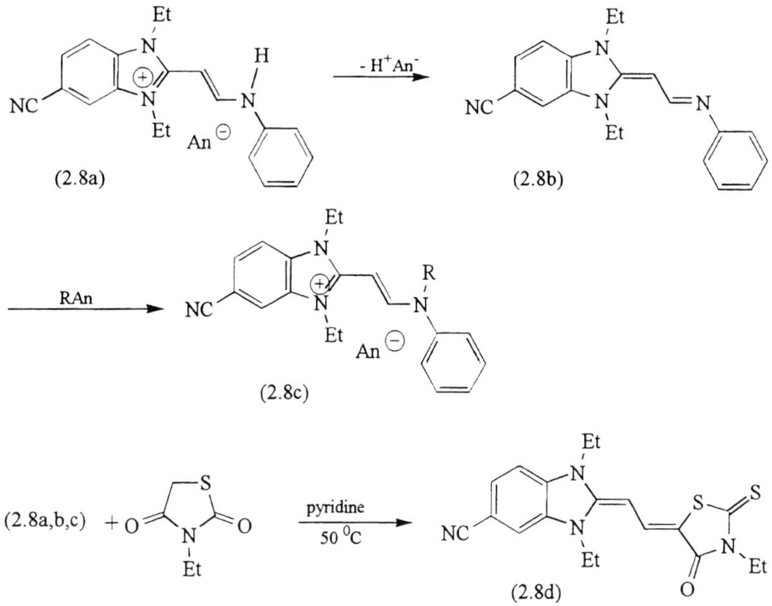

The heteroatoms X of the classical 2-methylcyclammonium salts (2.9) has been now extended even to tellurium (W.H.H. Gunther et al., *Eur. Pat. Appl. EP* 136,847/1985; *US. Pat.* 4,576,905/1983; *Chem. Abstr.*, 1985, **103**, 72624z).

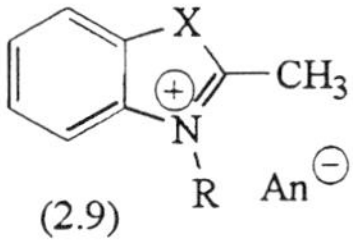

R = alkyl or substituted alkyl
X = CMe$_2$, CH=CH, NR, O, S, Se, Te

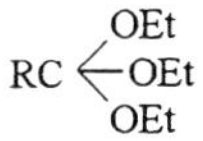

(2.9)

R = H, Me, Et

(2.10)

Variations of the reaction media, especially the conditions of deprotonation and activation, enables further syntheses:

- asymmetric imidacarbocyanines by reaction of equimolar amounts of (2.9; X = NR) with both (2.8b) and acetic anhydride in acetonitrile without adding any auxiliary base. According to Scheme 2.2 the intermediate product (2.8c; R = COCH$_3$, An = CH$_3$CO$_2$) is formed (G. Bach, J. Eckert, unpublished results);

- symmetrical and asymmetrical monomethine cyanines according to Scheme 2.3 via the intermediate compound (2.11) (R.E. Bernard, *Res. Discl.* 1976, **152,** 48).

SCHEME 2.3

(2.9; X = O, S)
+
(F)Cl—⟨benzene ring⟩—NO$_2$; O$_2$N

$\xrightarrow{\text{NEt(Me}_2\text{CH)}_2}$

(2.11)

(2.9) + (2.11) $\longrightarrow$

An$^{\ominus}$

- symmetrical thia-, oxa-, and imidacarbocyanines using (2.9; X = S, O, NR; R = Et, sulfoalkyl) in combination with (2.10) and sodium acetate or triethylammonium acetate in nitrobenzene [G. Bach et al., *Ger.(East) DE* 87,109/1970; *Chem. Abstr.*, 1973, **78**, 73668u];

-symmetrical thiadicarbocyanines from (2.9; X = S) and 1,1,3,3,tetra-ethoxypropane or its 2-Me(Et)-derivatives in pyridine in the presence of phase transfer catalysts, e.g. dibenzo-18-crown ether, pyridine-dichloroacetate, tributyl- or benzyltrimethylammonium bromide (N. Monich et al., *U.S.S.R. SU* 1,002,329/1981; *Chem. Abstr.*, 1983, **98**, 199 852f);

- symmetrical imidatricarbocyanines from (2.9; X = NR) and the salts of bis-di-N-methylaniline-glutaconaldehyde with sodium alcoholate in dimethyl-sulfoxide. Such dyes absorb in the NIR region between 685 nm and 740 nm and have strong fluorescence (V.M. Zubarovski et al., *Khim. Geterocycl. Soedin.* **1976**, 1562);

- bridged thiadicarbocyanines from (2.9; X = S) and 3-ethoxy-5-phenyl-2-cyclohexenone with magnesium sulfate at 160 °C (T. Kato, *Jpn.Kokai Tokkyo Koho JP 05,311,083 [*93,311,083/1992]; *Chem. Abstr.*, 1994, **120**, 220 392r);

- symmetrical pentamethinoxonols from 1-(2,4-disulfophenyl)-3-carbethoxypyrazolin-5-one and salts of glutaconaldehydedianil with 2-dimethylaminoethanol in dimethylformamide. The yield and purity is higher than 95% (S. Hoshimiaya et al., *Jpn.Kokai Tokkyo Koho JP* 04,114,063 [92,114,063/1990]; *Chem. Abstr.*, 1992, **117**, 173 383q);

- asymmetrical oxacarbocyanines from (2.9; X = O, R = sulfoalkyl) and (2.12) in dimethylacetamide using molecular sieves as catalyst (A. Ikegawa, *Jpn. Kokai Tokkyo Koho JP 62,192,465 [*87,192,465/1986]; *Chem. Abstr.*, 1988, **108**, 39 661q);

(2.12)

- chiral, chain-substituted indacarbocyanines from 1,3,3-trimethyl-2-ethylindoleninium salts. Activation with boron trifluoride gives with the proper methylene base and orthoformic acid triethylester the expected α,α'-dimethyl-substituted achiral dye (2.13), whereas in the presence of pyridine a Wagner-Meerwein rearrangement takes place resulting in the chiral α-monomethyl derivative (2.14) which contains in the 3-position besides the 3-methyl group an 3-ethyl substituent (Scheme 2.4, C. Reichardt et al., *Chem. Ber.* 1990, **123**, 565).

Chiral carbo- and dicarbocyanines as well as chiral trinuclear cyanines have been also synthesized from the (+)-(3R,15S)-3-*sec*-butyl-1,2,3-trimethylindoleninium salt. This was made from 2-methyl-3-*sec*-butylindole by a zeolite-catalyzed Fischer synthesis, with further alkylation and subsequent separation of the diastereomeric salts (C. Reichardt et al., *Liebigs Ann. Chem.* **1995**, 329; C. Reichardt, *J. Phys. Org. Chem.*, 1995, **8**, 761). Also simple chain-shaped pentamethinium salts having terminal

404

ephedrine groups as chirality center are described (C. Payrastre et al., *Tetrahedron Lett.*, 1994, **35**, 3059).

SCHEME 2.4

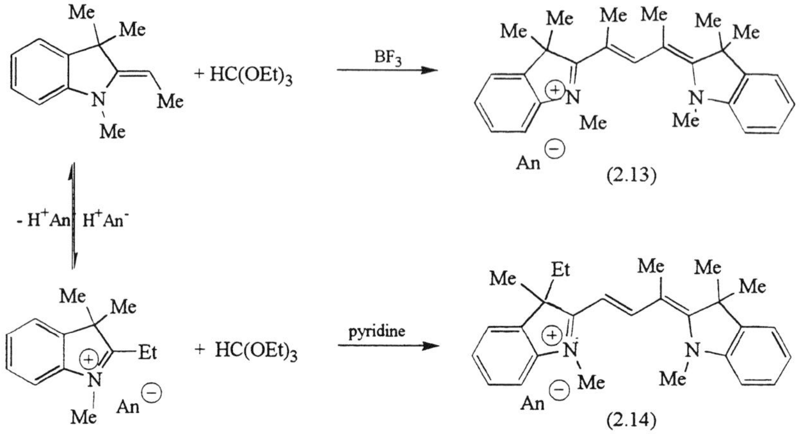

+ HC(OEt)$_3$ $\xrightarrow{\text{BF}_3}$ (2.13)

$- \text{H}^+\text{An}$ | H^+An^-

+ HC(OEt)$_3$ $\xrightarrow{\text{pyridine}}$ (2.14)

Special problems arise with the reactions hitherto considered if substituents R larger than methyl are introduced at the nitrogen atom of (2.9). Because of steric hindrance in the neighboring positions, very reactive, and hence often toxic reagents RAn for alkylation have to be employed. A completely new alkylation method is the reaction of protonated heterocycles (2.15; X = O, S) with the well-known esters of *ortho*-carbonic acids (2.16a), diethoxy-methylacetates (2.16b), 2-ethoxy-1,3-dioxolanes (2.16c), or with the acetyl derivatives of ethylene and propylene glycols (2.16d) to give the 2-methylcyclammonium salts (2.9a) shown in Scheme 2.5. Using more rigorous reaction conditions in case of (2.16a,b), the 2,2-ethoxyalkenyl-3-alkyl derivatives (2.17) are obtained (G. Bach et al., *J. Signal Aufz. Mater.* 1980, **8**, 137).

SCHEME 2.5

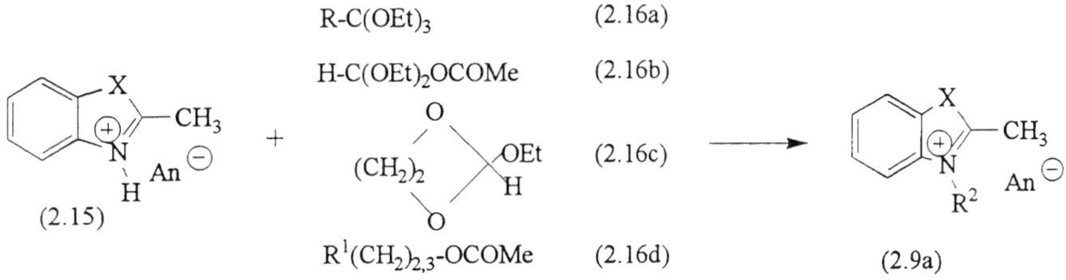

R-C(OEt)$_3$ (2.16a)

H-C(OEt)$_2$OCOMe   (2.16b)

(2.16c)

R^1(CH$_2$)$_{2,3}$-OCOMe   (2.16d)

(2.15) + (2.16...) $\longrightarrow$ (2.9a)

R = H, alkyl
R^1 = OH; OR; Hal; OCOMe
R^2 = Et, (CH$_2$)$_{2,3}$OCOMe
An = ClO$_4$; BF$_4$

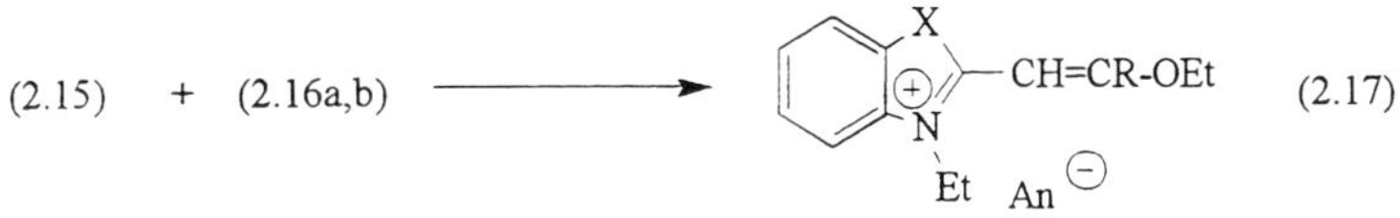

(2.15) + (2.16a,b) $\longrightarrow$ (2.17)

Steric hindrance can be substantially avoided if the alkyl groups are already introduced in the primary amine, like the alkylation of ortho-aminophenol (2.18a) according to Scheme 2.6. In this case the temperature of alkylation can be lowered (G. Bach et al., *J. Inf. Record. Mater.*, 1986, **14**, 373).

SCHEME 2.6

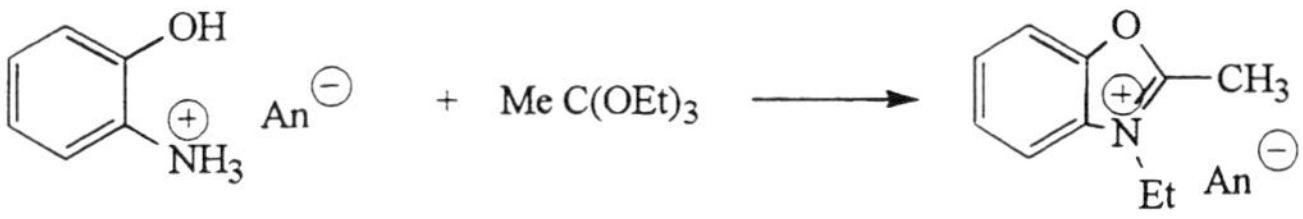

(2.18a)   An = ClO$_4$, BF$_4$

Furthermore, this procedure can be modified by usage of sulfoalkylated ortho-aminophenols (2.18b) reacting with trans-glutaconic acids (2.19) or glutaconic acid anhydrides (2.20) as shown in Scheme 2.7. On addition of trimethylsilylpolyphosphate (PPSE) as activator, in dichloromethane at 40 °C, symmetrical and asymmetrical N-sulfoalkyloxacarbocyanine dyes are obtained (F. Hoppe et al., *Ger. (East) DD* 245,190/1983; *Chem. Abstr.*, 1988, **108**, 39660p; F. Hoppe *Thesis*, Technische Hochschule Leuna-Merseburg, 1984).

SCHEME 2.7

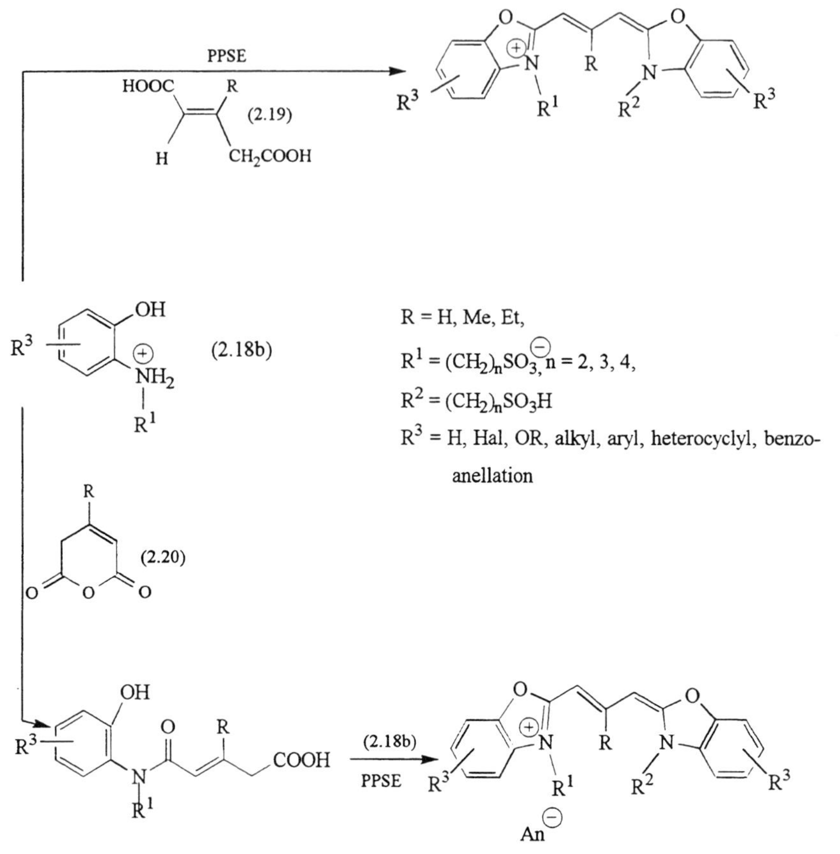

PPSE
HOOC
R
(2.19)
H
CH2COOH
R3
R1
R2
R3
OH
NH2
(2.18b)
R = H, Me, Et,
R1 = (CH2)nSO3, n = 2, 3, 4,
R2 = (CH2)nSO3H
R3 = H, Hal, OR, alkyl, aryl, heterocyclyl, benzo-anellation
(2.20)
O
COOH
(2.18b)
PPSE
An

(b) Cyanines having halogeno and cyano substituents at the polymethine chain

Carbocyanines with *meso*-fluoroalkyl substituents have been photochemically or electrochemically synthesized as shown in Schemes 2.8 and 2.9, respectively.

SCHEME 2.8

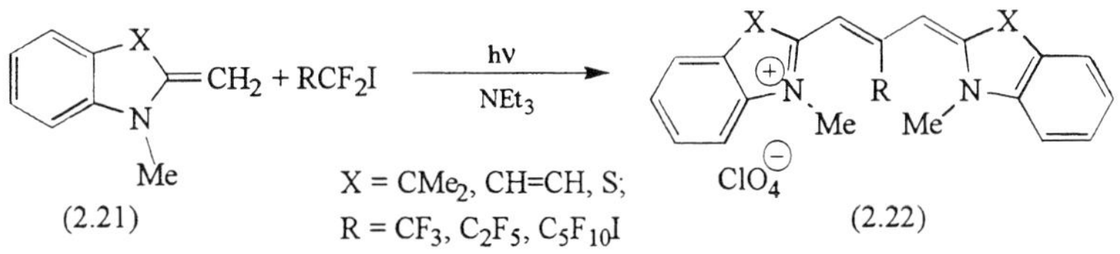

X = CMe₂, CH=CH, S;
R = CF₃, C₂F₅, C₅F₁₀I

(2.21)   (2.22)

(L.M. Yagupolski et al., *Zh. Org. Khim.* 1983, **19**, 2223):

SCHEME 2.9

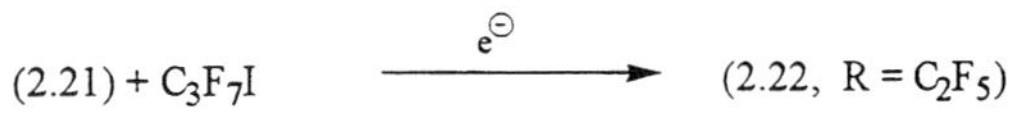

$$(2.21) + C_3F_7I \xrightarrow{e^\ominus} (2.22, \ R = C_2F_5)$$

(N.V. Ignat'ev et al. *Zh. Org. Khim.*, 1990, **26**, 1740.
Chain-substituted cyanines (2.23) having two different substituents such as
R^1 = F, Cl, R^2 = F, Cl, CN, CF₃, and CCl₃, are synthesized according to
Scheme 2.10, whereby acetylenic cyanines (2.24) can be obtained likewise
(E.A. Chaika et al., *Zh. Org. Khim.* 1985, **21**, 846).

SCHEME 2.10

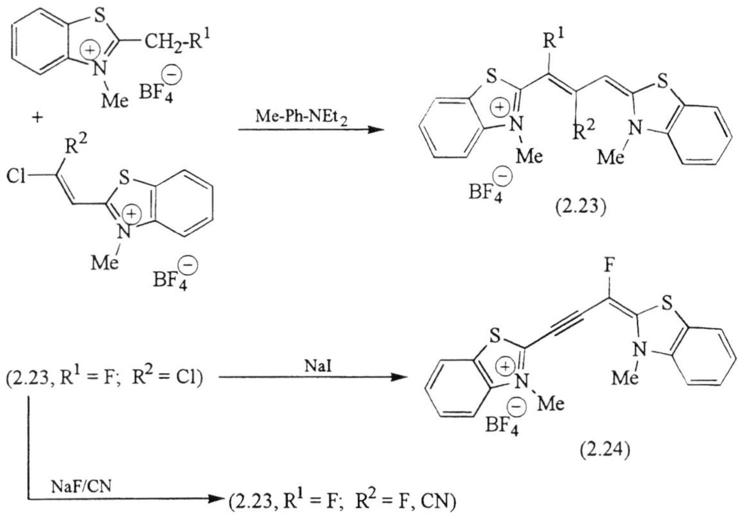

β-Cyano-substituted merocyanines (2.26) are prepared from α-ethoxyethylidene compounds (2.25) through adding substances which generate hydrocyanic acid as well as electrophilic compounds (E), according to Scheme 2.11 (G. Bach et al., *J. Signal Aufz. Mater.* 1981, **9**, 221).

SCHEME 2.11

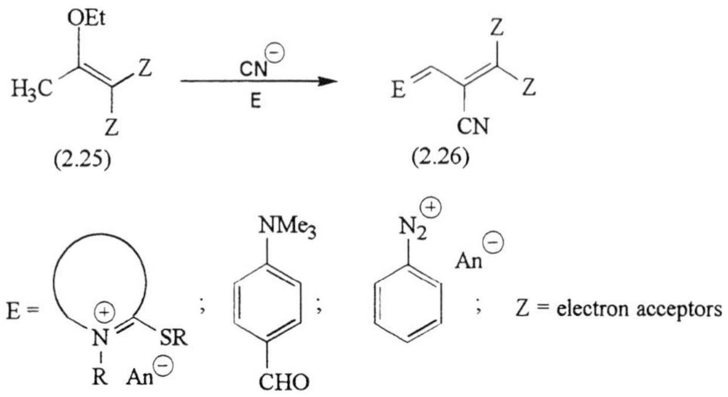

Cyano-substituted benzylidene dyes (2.28) are obtained through reaction of dimethylaminobenzaldehyde with cyanide and a ketomethylene compound such as 1-phenyl-3-methylpyrazolin-5-one, giving the intermediate product (2.27), which is subsequently dehydrogenated as shown in Scheme 2.12 (R. Stolle et al., *J. Signal Aufz. Mater.* 1981, **9**, 31):

SCHEME 2.12

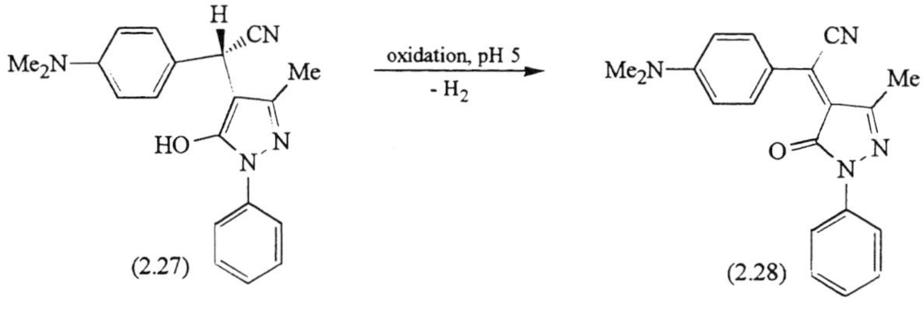

As the π-electron alternation is nonalternatingly perturbed by the cyano group, the absorption maximum of the cyano-substituted merocyanines and benzylidene dyes is bathochromically shifted by about 100 nm.

(c) Chain-bridged cyanines

An important field of research is the synthesis of cyanine dyes having a bridged polymethine chain, because such dyes possess improved stability against water, oxygen, light, and heat, as well as better optical and spectral properties favouring them for applications in the NIR region [cf. Sect. 1(*c*)].

Besides improvements of the well-known synthetic routes taking cyclic ketones and their derivatives for making the polymethines rigid in the polymethine chain (c.f. D.J. Fry, in *Chemistry of Carbon Compounds*, 2nd Ed., 1st Suppl., Vol. IVB, p. 270), bridged vinylogous carbonic acids such as squaric acid (2.29) and croconic acids (2.30), as well as analogous derivatives of benzene, naphthalene, and anthracene, proved to be useful.

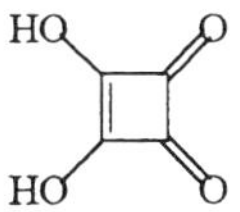

(2.29)

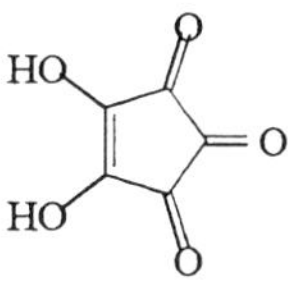

(2.30)

A synthetic route of fundamental importance has been described by Tolmachev et al. (*Zh. Org. Khim.* 1975, **11**, 392; *ibid.*, 1983, **19**, 2389). Scheme 2.13 shows the synthesis of a fully bridged tetracarbocyanine according to W. Lüttke et al. (*Chem. Ber.* 1986, **119**, 3102). Birch-reduction of 2,7-dimethoxyanthracene (2.31) gives 1,4,5,8,9,10-hexahydro-2,7-dimethoxy-anthracene (2.32). This is saponified and isomerized to the vinylogous carbonic acid (2.33), which can be immediately converted with 2-methyl-cycloammonium salts (2.9) to the intermediate product (2.34). From this and further (2.9), the symmetrical or asymmetrical tetracarbocyanines (2.35) are obtained.

SCHEME 2.13

(2.31)

Li_{met} in liq. NH_3

(2.32)

OH^-

(2.33)

(2.9)

(2.32) or (2.33)

(2.34)

(2.9; X ≠ Y) + (2.34)

(2.35)

Examples of the rather high yields of this reaction as well as the absorption wavelength λ_{max}, absorptivity ε_{max}, and fluorescence maxima λ_{fl} are given in Table 2.3 (W. Lüttke et al., *loc. cit.*, and *Chem. Ber.* 1988, **121**, 407).

TABLE 2.3.
REACTION YIELD AND UV-VIS SPECTROSCOPIC DATA OF SOME FULLY BRIDGED TETRACARBOCYANINES (2.35).

Compound (2.35) X,Y	yield [%]	λ_{max} [nm]	ε_{max} [10^5 mmol/cm^2]	λ_{fl} [nm]
O,O	39	793	1.36	814
S,S	73	855	1.55	887
Se,Se	75	872	1.21	894
S,O	57	840	0.99	884
S,Se	88	863	0.61	892
S,CMe$_2$	28	864	1.13	890

An one-step synthesis of the chain bridge simultaneously with the formation of the cyanine dyes is an exceptional case, realized by reaction of the bis-pyridiniumpropane derivatives, like (2.36), with α-diketones or α-ketoesters giving the 4,4'-pyridinocyanines (2.37a,b) (A.R. Katritzky et al., *J. Heterocyclic Chem.* 1988, **25**, 1311, Scheme 2.14).

SCHEME 2.14

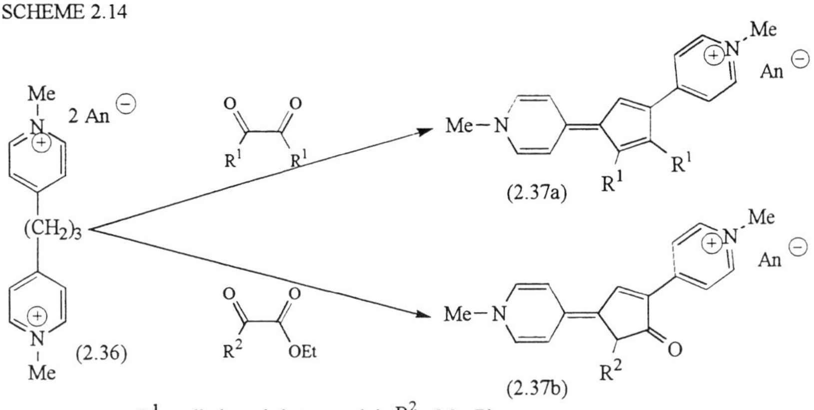

R^1 = alkyl, aryl, heterocyclyl; R^2= Me, Ph

Whereas according to Scheme 2.15 the analogous benzthiazolium derivative (2.38) reacts with α-diketones to give only the non-bridged carbocyanine (2.39), with 1,2-cyclohexanedione the bridged carbocyanine dye (2.40) is obtained. Analogous ring closures are realized with chloranil, 2,3-dichloro-quinoxaline, and 3,4-dichloromaleimide yielding dye (2.41) as shown in Scheme 2.15 likewise (A.R. Katritzky et al., *J. Heterocyclic Chem.* 1988, **25**, 1315).

SCHEME 2.15

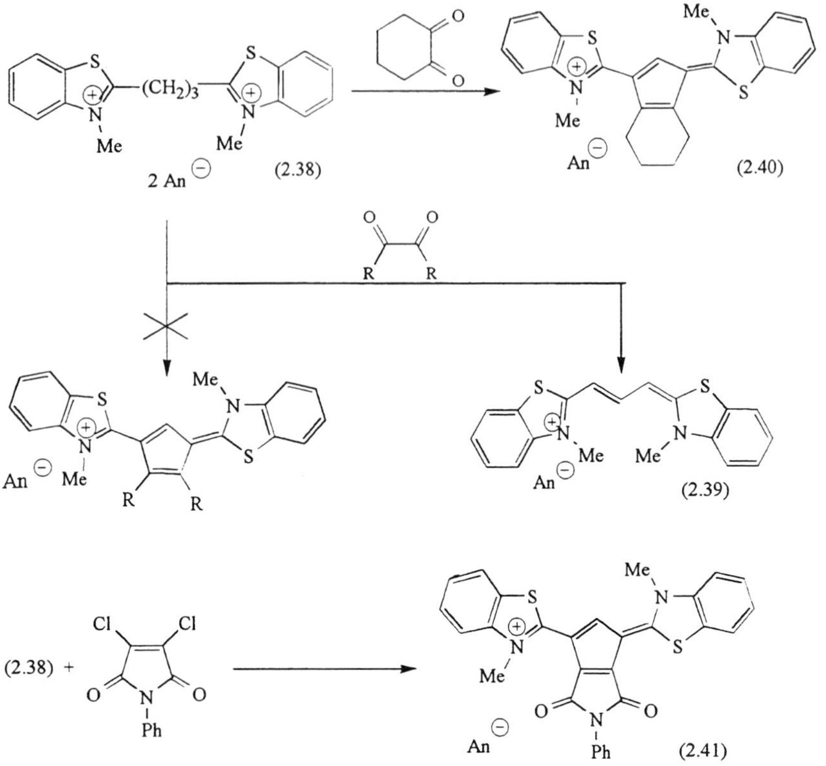

Symmetrical and asymmetrical squaraine dyes (2.42) are prepared according to Scheme 2.16 (K.-Y. Law et al., *J. Org. Chem.* 1992, **57**, 3278).

Vinylene-bridged anionic polymethine dyes (2.44) were obtained by oxidation of dye (2.43) with chloranile according to Scheme 2.17 (Yu.L. Slominsky et al., *Dyes Pigm.*, 1991, **15**, 247). Because the unsaturated cyclopentadienyl ring is prone to form an aromatic 6 π-electron skeleton, the light absorption of the bridged dye (2.44) is strongly blue-shifted in comparison to the non-oxidized entitiy (2.43), in agreement with color rule 4. The longest wavelength absorption maximum of the respective dyes are given in Table 2.4. Color rule 2 comes additionally into play when the dimethylamino substituent is introduced in *meso*-position, which alternatingly perturbes the polymethine chain and, hence, causes blue-shift.

SCHEME 2.16

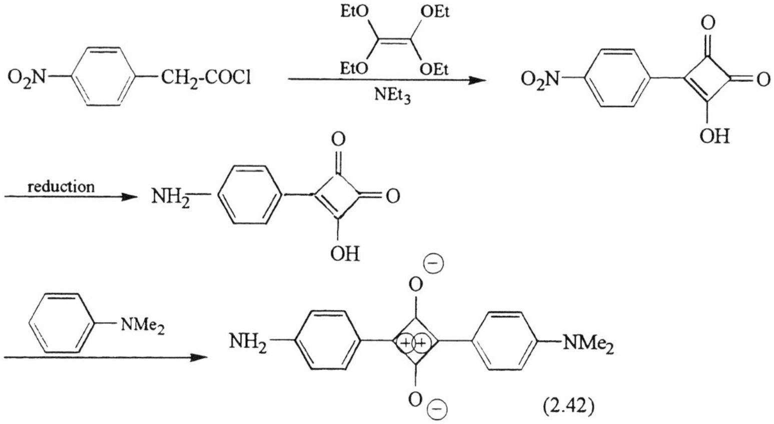

(2.42)

SCHEME 2.17

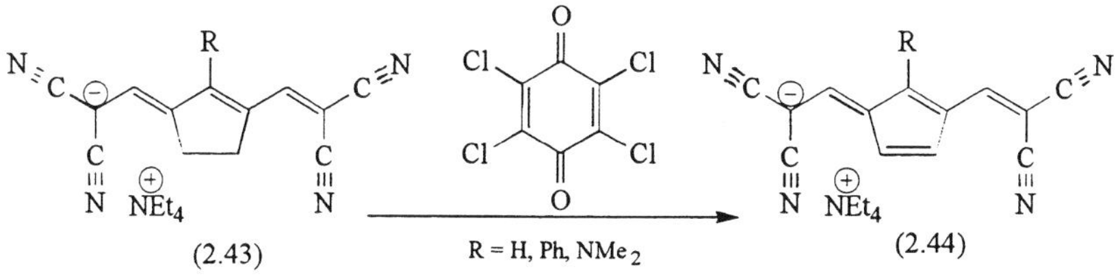

R = H, Ph, NMe$_2$

(2.43)   (2.44)

TABLE 2.4.
LONGEST WAVELENGTH ABSORPTION MAXIMUM, λ_{max}, OF DYES (2.43) AND (2.44)

R	(2.43) [nm]	(2.44) [nm]
H	584	486
Ph	594	491
NMe$_2$	540	456

Chain-bridging in combination with nonalternating substitution causes considerable red-shift of the absorption spectrum as was shown with NIR dyes in Sect. 1(*c*). But also strengthening the rigidity of the polymethine chain may cause appreciable red-shift as it was exemplified with dye (2.45), whose absorption maximum at 740 nm is shifted to 782 nm when the planarity of the chain is improved by introducing a central cyclohexene ring in (2.46) [J. Griffiths, *Chimia, (Aarau)*, 1991, **45**, 304].

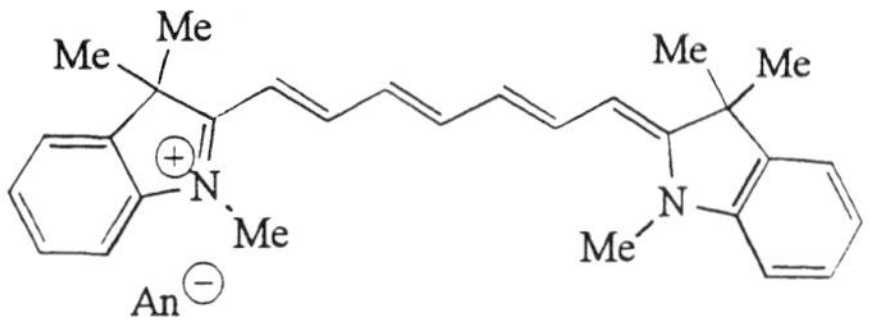

(2.45)

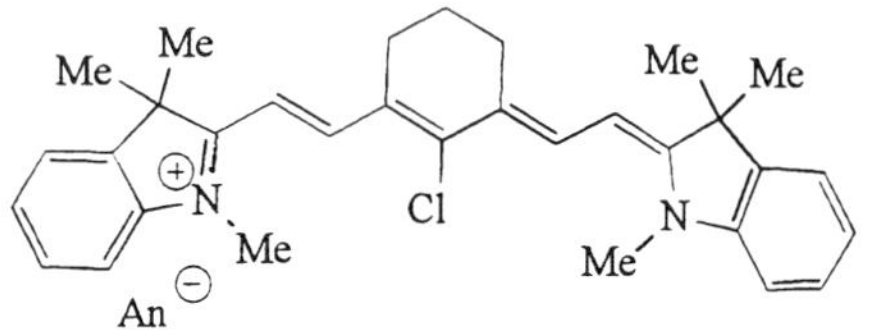

(2.46)

(d) Pyrylium, thiapyrylium, and selenapyrylium cyanines
A new, simple, and very variable synthesis of pyrylium cyanines (2.48) starts with the easily obtainable 2-(2-formylalkylidene)pyrans (2.47) according to Scheme 2.18

SCHEME 2.18

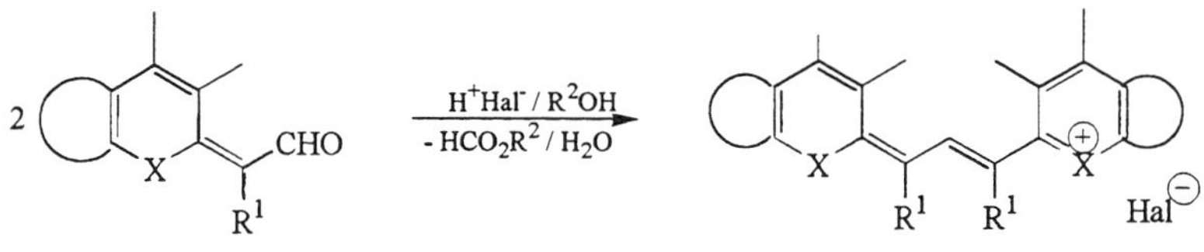

(2.47) X = O, S, Se; R^1 = Me, Ph; R^2 = Me, Et   (2.48) X = O, S, Se; R^1 = Me, Ph

Also merocyanines can be obtained by this procedure. As already mentioned in Sect. 1(*c*), such dyes are characterized by rather long-wavelength absorption bands in NIR region at 650-900 nm. (M. Weißenfels et al., *J. Prakt. Chem.*, 1989, **331**, 763).

(e) Reactions of cyanine dyes

Electrophilic protonation of cyanine dyes usually takes place at the negatively charged methine atoms, whereas nucleophilic attack is realized through addition of nucleophiles at the positively charged methine atoms, giving in case of OH⁻ ions the well-known carbinol bases. The equilibrium constants of protonation, pK_p, and hydrolysis, pK_h, were determined for many symmetrical cyanine dyes such as (2.49) and asymmetrical styryl dyes like (2.50) (I.S. Balog et al., *Zh. Analyt. Khim.*, 1990, **45**, 481; Ya.R. Bazel et al., *ibid.*, 1993, **48**, 631).

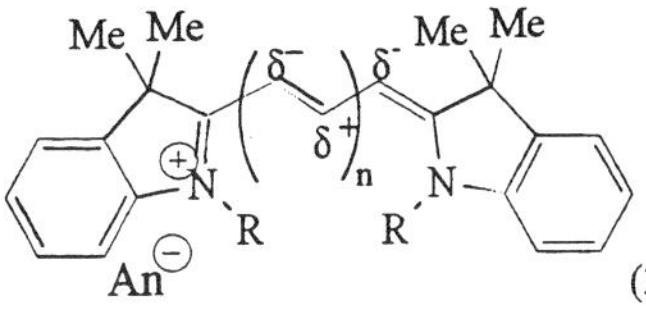

(2.49)

n = 1, 2; R = alkyl, substituted alkyl

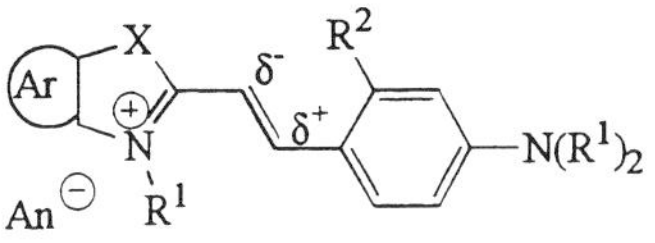

(2.50)

X = CMe$_2$, CH=CH, S;

Ar = benzo, substituted benzo, dibenzo;

R^1 = alkyl, substituted alkyl; R^2 = H, Me

Extremely fast deuteration of the negatively charged methine atoms is the consequence of their high nucleophilicity. The rate constant of proton-deuteron exchange is the greater the higher the localization energy at the methine atoms (R. Radeglia et al., *J. Prakt. Chem.* 1978, **320**, 539).

Electrophilic halogenation takes place at the negatively charged methine atoms likewise. For instance, monomethine cyanines (2.51; X = S, O, CMe$_2$, CH=CH) give with molecular halogens the meso-substituted derivatives (2.52) in di-*cis* configuration. The reaction is reversed by reducing agents or at pH values 3<pH>10 according to Scheme 2.19 (R. Steiger, *Phot. Sci. Eng.* 1981, **25**, 10):

SCHEME 2.19

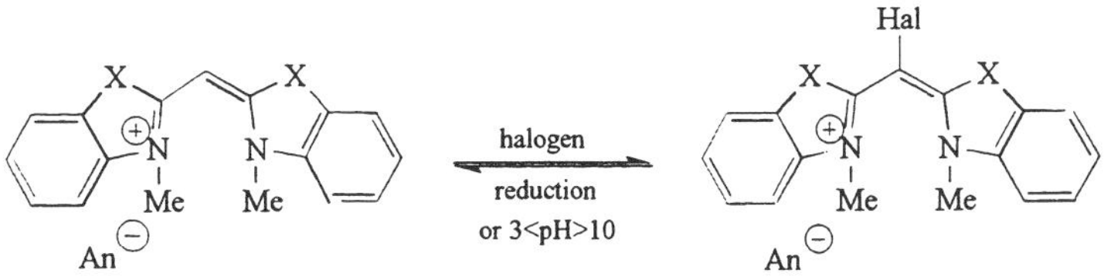

halogen
reduction
or 3<pH>10

(2.51)

(2.52)

416

The position of the long-wavelength absorption band in methanol and the anodic and cathodic half-wave potentials of some halogenated derivatives are given in Table 2.5.

TABLE 2.5.

ABSORPTION WAVELENGTH, λ_{max}, AND ELECTRO-CHEMICAL HALF-WAVE POTENTIALS[+], $E_{1/2}$, OF HALOGENATED MONO-METHINECYANINES (2.52) ACCORDING TO SCHEME 2.19

X	halogen	λ_{max} [nm]	$E_{1/2}$ [V] red	$E_{1/2}$ [V] ox
S	Cl	487	-0.3	>+1.2
S	Br	495	-0.3	>+1.2
CH=CH	Cl	595	-0.4	>+1.2
O	Br	415	-0.3	>+1.1
CMe_2,5-Cl	Br	510	not measurable	

[+] measured against Ag/AgCl reference electrode

Hydrogenation with sodium boron hydride of the vinylogous indocarbocyanine dyes (2.53; n = 1,2,3; X = CMe_2), possessing at least one 3,3-dimethylindoline cyclus gives, the derivatives (2.54) according to Scheme 2.20. The hydrogenated compounds were identified by ^1H nmr spectroscopy (Yu.L. Briks et al., *Zh. Org. Khim.*, 1990, **26**, 2591).

SCHEME 2.20

On the other hand, treatment of the indamonomethine cyanine (2.55) with sodium hydride causes splitting off NaAn yielding 1,3-diaminoallenes (2.56). The same product plus the anhydrocyanine dye (2.57) is generated through thermolysis (Scheme 2.21, W. Grahn, *Liebigs Ann. Chem.*, **1981**, 107). Both monomethinecyanine dyes (2.55) and allenes (2.56) easily react with N-halodiorganosulfonimides (Hal-N(SO$_2$R)$_2$, R = Me, Ph; Hal = F, Cl, Br, I) giving the halogenated derivatives (2.52; X = CMe$_2$; Hal = F, Cl, Br, I). Using an excess of halogen the terminal groups are halogenated in addition (W. Grahn et al., *Liebigs Ann. Chem.*, **1995**, 1003).

SCHEME 2.21

One-electron oxidation and reduction of symmetrical cyanine dyes (2.58) to dicationic and neutral dye radicals, respectively, can be deduced by cyclic voltammetry. As the radicals produced have no longer (N+1) or (N-1) π-electrons per N atoms, typical of polymethines, their longest wavelength absorption band is blue-shifted with respect to the original dye, in agreement with color rule 1 of triad theory [cf. Sect. 1(*b*), R.L Parton, J.R. Lenhard, *J. Org. Chem.*, 1990, **55**, 49; J.R. Lenhard, A.D. Cameron, *J. Phys. Chem.*, 1993, **97**, 4916]. Electrochemical one-electron oxidation and reduction has also been performed with pyrylium dyes [cf. Sect. 2(*d*)]. The dication radicals and uncharged radicals of which were characterized by their ESR spectra (M.G. Totikov et al., *Dokl. Akad. Nauk, SSSR*, 1989, **305**, 874).

After one-electron oxidation of carbo- and dicarbocyanine dyes (2.58; X = CMe$_2$, CH=CH, O, S, Se; n = 1, 2) that lack alkyl substituents, dimerization of the radical dications (2.59) occurs at the even-numbered

418

methine atoms having large π-electron density, see Scheme 2.22 (R.L. Parton, J.R. Lenhard, *loc. cit.*).

SCHEME 2.22

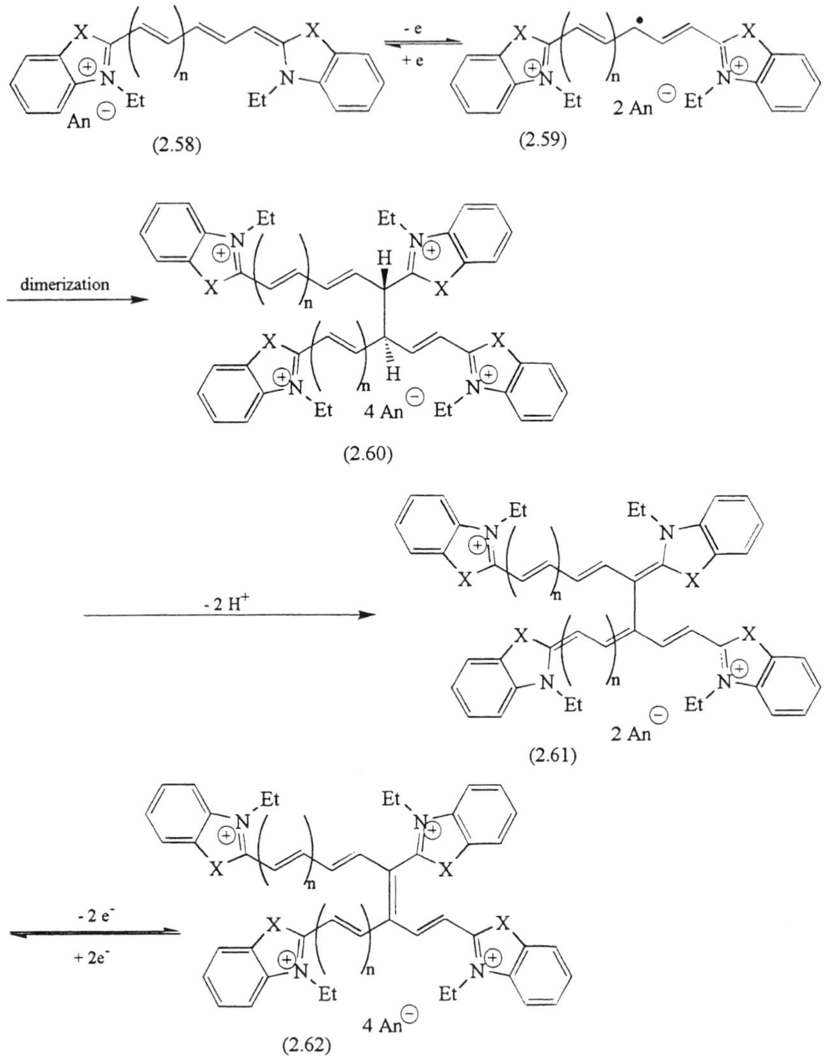

Deprotonation of the resultant uv-absorbing dimer (2.60) gives the dicationic bis-dye (2.61), the uv/vis spectrum of which is similar to that of the parent cyanine dye (2.58). The authors of the work proved by ^{1}H nmr spectroscopy that the bis-dye was further oxidized via a reversible two-electron mechanism yielding the cross-conjugated tetracationic species (2.62), which absorbs again at shorter wavelengths. Table 2.6 gives the absorption maxima of some cyanine dyes (2.58) and their reaction products.

TABLE 2.6.

LONGEST WAVELENGTH ABSORPTION MAXIMA, λ_{max}, OF CYANINE DYES (2.58) AND THEIR OXIDATION PRODUCTS (2.59) TO (2.62), AS SHOWN IN SCHEME 2.22

| cyanine dye | | (2.58) | (2.59) | (2.60) | (2.61) | (2.62) |
X	n	[nm]	[nm]	[nm]	[nm]	[nm]
CMe$_2$	2	639	-	-	639, 589	486, 406
CH=CH	2	706	-	344	692, 658	441, 377
O	2	580	-	-	574	450
S	2	650	-	345	638	472, 367
Se	2	660	-	-	650	496
CMe$_2$	1	544	480	-	-	-
S	1	554	-	-	545, 520	434
S-benzo	1	593	-	-	582, 552	504

3. Properties of cyanine dyes

(a) Physico-chemical properties in the ground state

(i) Electronic structure

The properties of cyanines and related compounds are mainly governed by the unique features of the ideal polymethine state, i.e. full bond lengths equalization and zero bond order alternation (BOA), respectively, combined with strong alternation of the π-electron densities through the polymethine chain [cf. Sect. 1(*a*)]. These properties were confirmed further by experimental data such as X-ray structural analyses of simple streptopolymethines (L. Daehne, G. Reck, *Z. Kristallogr.*, 1995, **210**, 40), as well as by nmr spectroscopy using more sophisticated techniques like nuclear Overhauser effect (NOE), and chemical shift-correlated spectroscopy (COSY) (N. Katayama et al., *J. Mol. Structure*, 1992, **274**, 171).

Also ir and Raman vibrational spectroscopy confirmed the characteristic features of ideal polymethines (L. Mitzinger, S. Daehne, *J. Prakt. Chem.*, 1982, **324**, 458; H. Sato et al., *J. Raman Spectrosc.*, 1988, **19**, 129; M. Pfeiffer et al., *Springer Series in Solid State Sci.*, 1992, **107**, 150; K. Iwata et al., *J. Phys. Chem.*, 1992, **96**, 10219). Surface-enhanced Raman scattering (SERS) and surface enhanced hyper Raman scattering (SEHRS) measurements proved to be useful to gain further information about vibrations and π-conjugation in polymethine chains (N.-T. Yu et al., *J. Raman Spectrosc.* 1990, **21**, 797). For instance, using the C=C stretching frequency of merocyanine dye (3.1) as indicator of the π-electron delocalization, it was shown by surface-enhanced resonance Raman scattering (SERRS) measurements of the dye adsorbed to silver electrodes, that the polar resonance structure (3.1b) is the better stabilized the more positive the electrode potential (Y. Mineo, K. Itoh, *J. Phys. Chem.*, 1991, **95**, 2451).

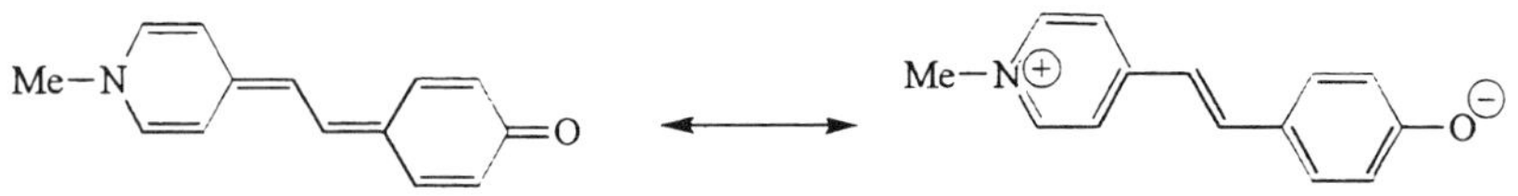

$$\text{Me}-\text{N} \diagup\hspace{-0.3em}\diagdown = \diagdown\hspace{-0.3em}\diagup = \diagdown\hspace{-0.3em}\diagup \text{O} \qquad \longleftrightarrow \qquad \text{Me}-\overset{\oplus}{\text{N}} \diagup\hspace{-0.3em}\diagdown = \diagdown\hspace{-0.3em}\diagup \text{O}^{\ominus}$$

(3.1a)   (3.1b)

Another method to judge the degree of conjugation and π-electron delocalization, respectively, in merocyanines is X-ray photoelectron spectroscopy (XPS) which allows determination of the partial charges localized at the heteroatoms of the dyes' terminal groups (M.M. Chehimi, M. Delamar, *J. Electron. Spectrosc. Relat. Phenom.* 1989, **49**, 213).

Ideal polymethines exist in *trans*-conformation which is the energetically most stable form. Sterical hindrance causes the formation of cis-isomers in the ground state. The factors which determine the conformational equilibrium were recently reviewed by A..M. Kolesnikov, F.A. Mikhailenko

(*Usp. Khim.*, 1987, **56**, 466) and H. Goerner, H.J. Kuhn (*Adv. Photochem.*, 1995, **19**, 1).

As the analytical characterization of cyanine dyes by nmr spectroscopy is limited by the low solubility of most dyes, some work was done to determine their structure with electron impact (EI) and fast atom bombardment (FAB) mass spectroscopy (R. Haessner et al., *Z. Chem.*, 1989, **29**, 65; *J. Prakt. Chem.* 1989, **331**, 859).

In case of triply branched cyanine dyes the influence of alternating branching of ideal polymethines on their electronic structure was investigated by X-ray structural analyses and ^{1}H and ^{13}C nmr spectroscopy (J. Dale et al., *Acta Chem. Scand.*, 1987, **B41**, 653; *ibid.* 1988, **B42**, 573).

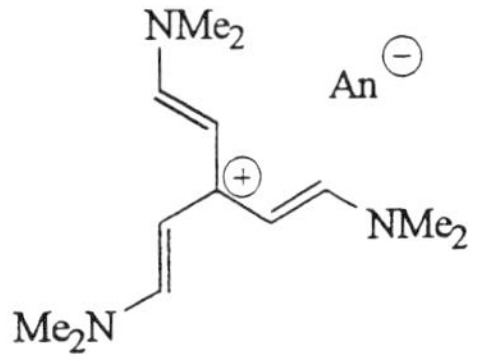

(3.2) 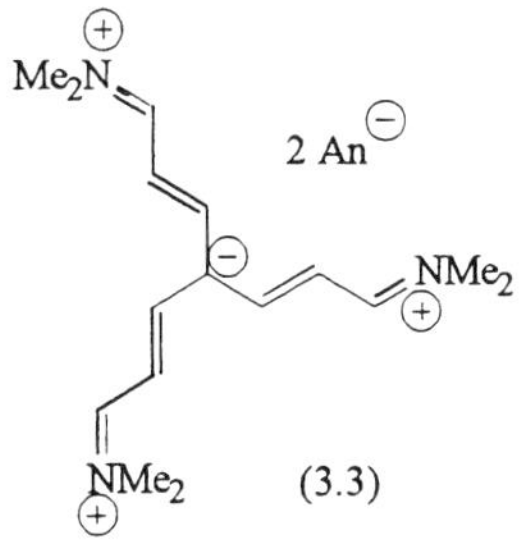   (3.3)

Whereas, for instance, monocation (3.2) possess a propellor-like structure, the homologous dication (3.3) is essentially planar. In agreement with the second basic structural principle [c.f. Sect. 1(b)], in both cations the symmetrical triple branching at the central methine atom weakens the conjugation within the ideal polymethine chain and, thus, induces slight bond length alternation in each branch. Similar results were obtained with triply branched vinylogous polymethine dyes having three indolenine terminal groups, which can be compared with the triphenyl methane dyes (C. Reichardt et al., *Chem. Ber.*, 1983, **116**, 1982).

A more detailed insight into the interaction of polymethine-like structures with aromatic and polyenic structural units provide compounds (3.4) to (3.6).

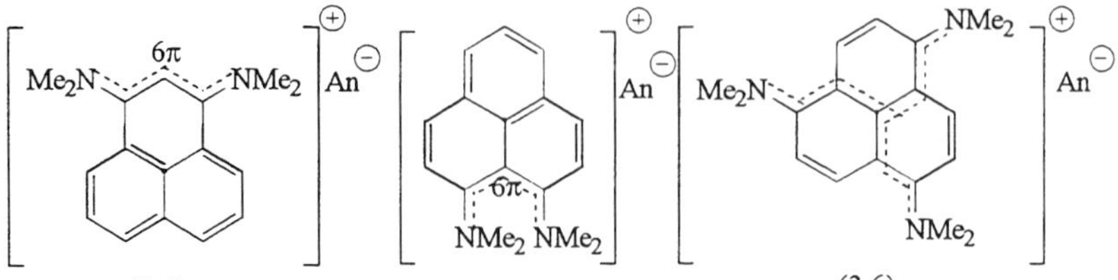

(3.4)   (3.5)   (3.6)

X-ray structural analyses gave direct evidence of the indicated structures (H.A. Staab et al., *Angew. Chem.*, 1991, **103**, *1003; Angew. Chem. Int. Ed. Engl.*, 1991, **30**, 1030). In (3.4), the trimethine chain is alternatingly linked to an aromatic 10 π-electron system. In (3.5), the aromatic phenyl ring alternatingly branches the trimethine cyanine group via one single bond and two vinylene groups. The phenalenium ion (3.6) contains a triply branched pentamethine cyanine system which exhibits conjugation over the central methine atom similarly to that of (3.2). For that reason the peripheral bonds form a vinylene structure.

Aromatization of the pentadienyl ring of the tetracyanoheptamethine dye (3.8), which is to be expected according to the second fundamental principle, was proved by nmr spectroscopy using lanthanide shift reagents. Whereas the negative charge of dye (3.7) which has the partly saturated cyclopentene nucleus is preferentially located at the cyano groups, in dye (3.8)it is concentrated in the aromatic pentadienyl ring as shown by (3.8b)(I.V. Komarov et al., *Teor. Eksp. Khim.*, **1991**, 197).

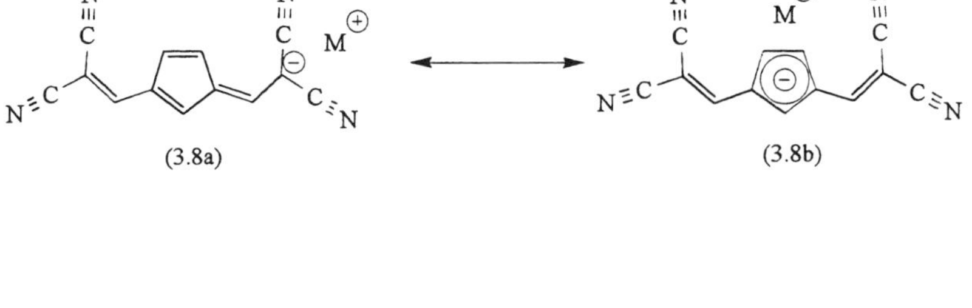

oxidation, $-H_2$

(3.7)

(3.8a)   (3.8b)

Nonalternating modification of polymethine structures causes strong red-shift of light absorption according to the third color rule [cf. Sect. 1(*b*)]. This was proved with dyes having two coupled polymethine chromophores which are linked under different angles (G.G. Dyadyusha et al., *Dyes Pigm.*, 1989, **10**, 111). Maximum nonalternating coupling happens in the homologous series (3.9) to (3.11) in which two mono-, tri-, and pentamethine chains, respectively, are parallel aligned (H. Bock et al., *Angew. Chem.*, 1989, **101**, 1715; *Angew. Chem. Int. Ed. Engl.*, 1989, **28**, 1684; K. Elbl-Weiser et al., *Angew. Chem.*, 1990, **102**, 183; *Angew. Chem. Int. Ed. Engl.*, 1990, **29**, 211) and related compounds (H. Bock et al., *Angew. Chem.*, *1989*, **101**, 1717; *ibid.* 1992, **104**, 564; *Angew. Chem. Int. Ed. Engl.*, 1989, **28**, 1685; *ibid.* 1992, **31**, 550).

$$\left[Me_2N \overset{4\,\pi}{\cdots} NMe_2 \;\; Me_2N \underset{4\,\pi}{\cdots} NMe_2 \right]^{2\oplus} \qquad \left[Me_2N \overset{6\,\pi}{\cdots} NMe_2 \;\; Me_2N \underset{6\,\pi}{\cdots} NMe_2 \right]^{2\oplus} \qquad \left[Me_2N \overset{8\,\pi}{\cdots} NMe_2 \;\; Me_2N \underset{8\,\pi}{\cdots} NMe_2 \right]^{2\oplus}$$

(3.9)   (3.10)   (3.11)

In accordance with the third fundamental principle of triad theory, the polymethine units are linked by single bonds up to 156 pm in length (indicated by thick bars), which are longer than normal sp^2-sp^2 single bond length of 148 pm, and owing to color rule 3, the light absorption is strongly red-shifted as compared to the constituent cyanine chromophores.

(ii) Energy levels

It is well established that there are linear relationships between the energy of the highest occupied molecular orbital (HOMO) of cyanine dyes, accessible by quantum chemical calculations, and their anodic halfwave potential, as well as between the dyes' energy of their lowest unoccupied molecular orbital (LUMO) and their cathodic halfwave potential (e.g. S. Daehne, F. Moldenhauer, *Progr. Phys. Org. Chem.*, 1985, **15**, 1).

The energy levels are also related to the cyanine dyes' susceptibility for spectral sensitization and desensitization in photoinduced electron transfer reactions [cf. Sect. 4(*e*)]. However, it turned out that all hitherto determined electrochemical halfwave potentials of cyanine dyes were distorted due to uncontrolled, irreversible electrode reactions. Precise

424

values were only obtained by phase-selected second-harmonic AC voltametry (J. Lenhard, *J. Imag. Sci.*, 1986, **30**, 27; T. Tani et al., *J. Electrochem. Soc.*, 1991, **138**, 1411; J.R. Lenhard et al., *J. Phys. Chem.*, 1993, **97**, 8269). The corrections to be made, however, were of minor importance, thus, the mentioned relationships are still valid. Additional support of their validity came from measurements of the dyes' energy levels by ultraviolet photoelectron spectroscopy (UPS) (T.B. Tang et al., *J. Phys. Chem.* 1989, **93**, 3970).

(iii) Solvatochromism and hydrophobicity
The present state of the art of solvatochromism has been comprehensively reviewed by C. Reichardt (*Solvents and Solvent Effects in Organic Chemistry*, 2nd Ed., VCH, Weinheim, 1988; *Chem. Rev.*, 1994, **94**, 2319) and E. Bunzel, S. Rajagopal (*Acc. Chem. Res.*, 1990, **23**, 226). It was experimentally proved (S. Daehne, K. Hoffmann, *Progr. Phys. Org. Chem.* 1990, **18**, 1) and described theoretically by a microstructural model (K.-D. Nolte, S. Daehne, *Adv. Mol. Relax. Interact. Processes*, 1977, **10**, 299; *Acad. Chim. Acad. Sci. Hung.*, 1978, **97**, 147) that the electronic structure of cyanine dyes is strongly altered in the ground state depending on the polarity of the solvent used. In case of merocyanines (3.12), the dipolar structure (c) is the more stabilized the greater the solvent polarity is. Thus, the dipole moment of merocyanine-like molecules reflects the larger contribution of the more dipolar resonance form (3.12c) over its nonpolar form (3.12a) when the solvent polarity is raised. On the other hand, due to basic color rule 5 [c.f. Sect. 1(*b*)], substances having a symmetrical π-electron structure (3.12b) possess the longest wavelength absorption.

(3.12a)   (3.12b)   (3.12c)

Therefore, positive solvatochromism, i.e. red-shift of the absorption maximum, occurs when merocyanine dyes having a more or less nonpolar structure (3.12a) are used. On the other hand, merocyanines which adopt already in nonpolar solvents a more dipolar structure than the symmetrical one (3.12b) exhibit negative solvatochromism, i.e. blue-shift of the light absorption with increasing solvent polarity. Consequently, for dyes in nonpolar solvents, having a structure only slightly less dipolar than that indicated by (3.12b), the sign of the solvatochromic behavior is reversed with rising solvent polarity. The vinylogous γ-pyridones (3.13; n = 0, 1, 2) are typical examples. The solvatochromic behaviour of dye (3.13; n = 1; R = *t*-Bu) is shown in Fig. 3.1 (J. Catalan et al., *J. Phys. Chem.*, 1992, **96**, 3615; L. da Silva et al., *J. Chem. Soc., Perkin Trans.* 2, **1995**, 483).

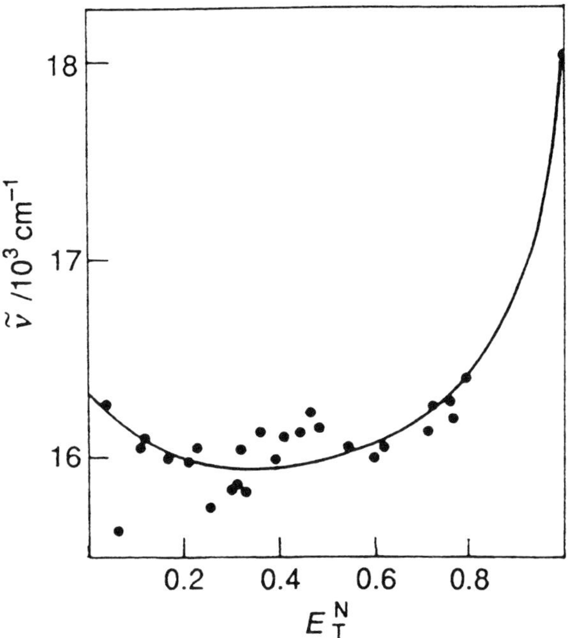

FIG. 3.1. DEPENDENCE OF THE POSITION OF THE LONGEST WAVELENGTH ABSORPTION MAXIMUM, λ_{MAX} , OF DYE (3.13 ; n = 1; R = *t*-Bu) ON THE POLARITY OF THE SOLVENT USED
($\tilde{v}$ = wavenumbers; E_T^N = Reichardt's solvent polarity parameter; reprinted from L. da Silva et al., *J. Chem. Soc., Perkin Trans.*, 2, **1995**, 483, with permission of the Royal Society of Chemistry).

$$\text{(3.13)} \qquad \text{(3.14)}$$

The extent of solvatochromism of chromophore (3.13) strongly depends on the substituents introduced at the positions R (M. Niedbalska, I. Gruda, *Can. J. Chem.*, 1990, **68**, 691). According to the basic color rules 2 and 3 in least polar solvents the light absorption of the unsubstituted chromophore at 618 nm is blue-shifted to 508 nm on alternating substitution R with nitro groups, and it is red-shifted to 656 nm on nonalternating substitution R with methoxy groups.

The polarity of chromophore (3.13) increases when the pyridine ring is substituted for a benzo[a]quinolinium system as in (3.14). Consequently, such dyes exhibit exclusively negative solvatochromism (S. Arai et al., *J. Chem. Soc., Perkin Trans.* 1, **1990**, 1915).

The hydrophobic character of cyanine dyes is decisive when they are used as probes of organized media in molecular biology [cf. Sect. 4(*b,ii*)]. The hydrophobic character can be controlled, e.g. by the type of the N,N' substituents at the cyanine chromophore. By means of the dyes' distribution coefficient (K_h) between the two immiscible phases water and n-octanol, the hydrophobicity of some 40 differently substituted thiacarbocyanine dyes (3.15) was determined (R.D. Raikhina et al., *Ukrain. Khim. Zh.*, 1991, **57**, 534).

$$\text{(3.15)}$$

As shown in Table 3.1, the hydrophobicity K_h is the higher the longer the N,N'-alkyl chains are. N,N'-Phenyl substituents are similar in hydrophobicity to saturated *n*-hexyl groups. When only one N-propylsulfonate substituent is introduced, the hydrophobicity resembles that of *n*-dibutyl substituents because in the betainic dye the negative charge of the sulfonate group is compensated by the dye's positive charge.

TABLE 3.1.
EFFECT OF N,N'-SUBSTITUENTS OF THIACARBOCYANINE DYES
(3.15) ON THEIR HYDROPHOBIC CHARACTER, K_h.

R^1	R^2	K_h
Me	Me	3.5
Et	Et	7.7
n-Pr	n-Pr	15.2
n-Bu	n-Bu	30.1
n-C$_5$H$_{11}$	n-C$_5$H$_{11}$	51.9
n-C$_6$H$_{13}$	n-C$_6$H$_{13}$	105.0
Ph	Ph	102.6
Et	(CH$_2$)$_3$-SO$_3^-$	33.6
(CH$_2$)$_3$SO$_3$H	(CH$_2$)$_3$SO$_3^-$	0.03

Only when two N,N'-dialkylsulfonate groups are introduced, the
hydrophobic character is strongly reduced due to the surplus negative
charge of one sulfonate group.

In order to record the voltage of biological membranes [cf. Sect. 4(b,iv)],
the lipophilicity of cyanine dyes, i.e. the balance between their hydrophobic
and hydrophilic properties, is of interest. As expected, this depends on the
chain length of the N,N'-alkyl substituents and was quantified in the case of
the hemicyanine dye (3.16; n = 1 to 6) through determination of the dyes'
binding constant to lipids (P. Fromherz, C. Roecker, *Ber,. Bunsenges. Phys.
Chem.*, 1994, **98**, 128).

(3.16)

(iv) Properties of photo-excited states
Increasing interest is found in the photophysical properties of cyanine dyes
after light excitation. Relationships between structure and fluorescence

428

behaviour of cyanine dyes have been comprehensively reviewed by A.A. Ishchenko (*Usp. Khim.*, 1991, **60**, 1708) who also reported on the influence of solvent polarity on the fluorescence behaviour, i.e. the solvatofluorochromism, of vinylogously lengthened indacarbocyanine dyes (3.17; n = 1, 2, 3; R = H) (A.A. Ishchenko et al., *Dyes Pigm.*, 1989, **10**, 85).

$$(3.17)$$

Whereas the absorption band of cyanine dyes is broadened on vinylogous elongation, the fluorescence band is narrowed as consequence of increased internal conversion due to enhanced vibronic interactions (A.A. Ishchenko et al., *Opt. Spektrosk.*, 1992, **72**, 110).

The singlet-triplet intersystem crossing (ISC) quantum yield of cyanine dyes, known to be rather small, was determined for a series of indacyanine dyes (3.17; n = 2, 3; R = H, Me, Cl; An = I, BF_4) at room temperature in solution. The range was 0.004 - 0.025 (C. Carré et al., *J. Chim. Phys.* 1987, **84**, 577; D.-J. Lougnot, *Proc. SPIE-Int. Soc. Opt. Eng.*, 1994, **2042**, 218).

Apart from J-aggregates which are described in Sect. 3(*b,ii*), the dephasing processes induced by coherent excitation of monomeric cyanine dyes are too fast to be directly measured. They were estimated indirectly from the width of holes burnt by laser experiments at low temperatures using the thiacarbocyanine (3.18) (M.V. Alfimov et al., *Dokl. Akad. Nauk*, 1992, **324**, 91).

$$(3.18)$$

An important excited state relaxation channel in cyanine dyes is *trans-cis* isomerization, which may undergo thermally activated isomerization back to the *trans* conformation in the ground state (A.M. Kolesnikov and F.A. Mikhailenko (*Usp. Khim.*, 1987, **56**, 466; H. Goerner, H.J. Kuhn, *Adv. Photochem.*, 1995, **19**, 1).

Whereas in the ground state the 2,2'-monomethine cyanine dyes usually exist in the *trans* (EE) conformation, 4,4',6,6'-tetra-*tert*-butyl-substituted pyrylo-2,2'-monomethine cyanines have a EZ conformation (3.19; X = Y = O) or ZZ conformation (3.19; X = Y = S, and X = O; Y = S) as was shown by X-rax structural analyses (A.I. Tolmachev et al., *Dyes Pigm.*, 1991, **17**, 71).

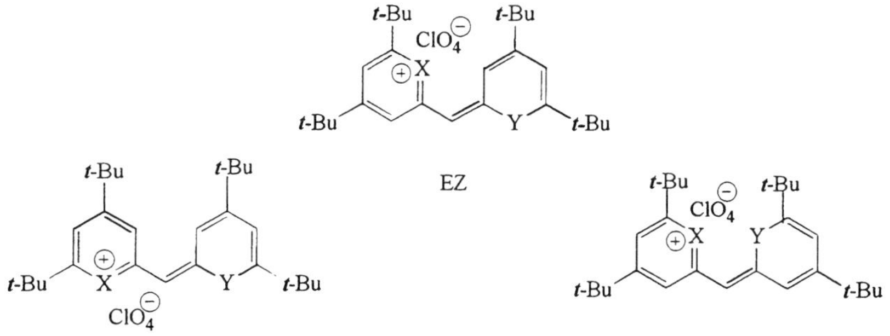

Recently, the existence of photoisomers in the excited state was directly proved by resonance Raman spectroscopy and resonance coherent antistokes Raman scattering (resonance CARS) (M. Katsumata et a., *Chem. Phys. Lett.*, 1988, **143**, 240). The dynamics of *cis-trans* photoisomerization in competition with internal conversion (IC) processes was quantitatively evaluated from the dependence of the fluorescence lifetime and fluorescence quantum yield on both the molecular weight and the viscosity of the solvent used. In these experiments, the viscosity was varied by high pressure measurements (S. Murphy et al., *J. Phys. Chem.*, 1994, **98**, 13476). In addition, the dynamics of *cis-trans* isomerization was studied on the 3,3'-bis(sulfopropyl)-5,5'-dichloro-9-ethylthiacarbocyanine (D.Noukakis et al., *J. Phys. Chem.*, 1995, **99**, 11860) and by introducing different substituents at the 5,5'-positions of indacarbocyanine dyes (3.17; n = 1; R = H, Me,

430

OMe, CF_3, SO_2CF_3, NO_2; An = PF_6)(S. Murphy, G.B. Schuster, *J. Phys. Chem.*, 1995, **99**, 8516). Also photoisomerization of some asymmetric carbo-, dicarbo, and tricarbocyanine dyes with indolino and benzimidazolo terminal groups was analyzed and compared with the parent symmetric cyanine dyes (A.S. Tatikolov et al., *Ber. Bunsenges. Phys. Chem.*, 1995, **99**, 763).

(b) Intermolecular interactions of cyanine dyes
As compared to aromatic and polyene-like compounds, some of the most superior properties of cyanine dyes are their high polarity and their easy electronic polarizability, which gives rise to strong intermolecular interactions, mainly caused by dispersion forces presumably combined with hydrophobic and Coulomb interactions. Therefore, cyanine dyes easily form ion pairs with counter ions as well as homo- and hetero-molecular aggregates in the ground state. In particular, it should be stressed that excited-state aggregation (i.e. excimer and exciplex formation) is preferred by aromatic and polyene-like molecules as their polarizability strongly increases on light excitation [cf. Sect. 1(*b*)].

(i) Ion pair formation
The formation of ion pairs, the mechanism of this process, and the structure of such entities is now well understood (e.g. B. Armitage et al., *J. Am. Chem. Soc.* 1993, **115**, 10786).

Using lanthanide shift reagents and quadrupolar magnetic nuclei in the anion, the spatial structure of the ion pairs (3.17; n = 1, 2, 3; R = H; An = NO_3) was determined by nmr spectroscopy. The nitrate anion is positioned nearby the nitrogen atom in the carbocyanine dye (3.17; n = 1), but it moves to the positively charged γ-position of the polymethine chain in the di- and tricarbocyanine dyes (3.17; n = 2, 3) (I.V. Komarov et al., *Dokl. Akad. Nauk. SSSR*, 1989, **306**, 1134; *Zh. Obshch. Khim.*, 1989, **59**, 2356).

The 1,7-bisdimethylaminoheptamethinium cation (3.20; n = 3) forms stable ion pairs with anions of low ionic strength like tetraphenylborate in nonpolar solvents. It was shown by uv/vis, ir, and ^{1}H nmr spectroscopy as well as by X-ray structural analyses, that an asymmetrical polarization of the polymethine chain occurs due to the asymmetrically positioned anion. Through this polarization the structure is brought near to that of (3.20) (L.

Daehne, *Angew. Chem.*, 1995, **107**, 735; *Angew. Chem. Int. Ed. Engl.*, 1995, **34**, 690).

$$(3.20)$$

Another consequence of ion pair formation in nonpolar solvents are changes of the excited state lifetime of 4,5-benzindapentacarbocyanine (3.21; $X,Y = CMe_2$; $n = 5$; $R^1 = Me$; $R^2 = 4,5$-benzo) and thiapentacarbocyanine (3.21, $X,Y = S$, $n = 5$; $R^1 = Et$; $R^2 = H$) depending on the anion used. For instance in case of the 4,5-benzindapentacarbocyanine dye, the lifetime measured by pump-probe laser spectroscopy increases from 50 to 100 ps in the sequence: $CF_3COO^- < BF_4^- < ClO_4^- < CF_3SO_3^- < I^- < p\text{-}CH_3C_6H_4SO_3^-$ (M.I. Demchuk et al., *Chem. Phys. Lett.*, 1988, **144**, 99).

$$(3.21)$$

In case of dyes which consist of both one cationic and one anionic cyanine chromophore, ion pair formation takes place depending on the polarity of the solvent used. This can be seen in a strong decrease of the fluorescence lifetime in nonpolar solvents (M.I. Demchuk et al., *Chem. Phys. Lett.*, 1990, **167**, 170). Also energy transfer between the anions and cations of such bichromophoric dye salts is observed, causing fluorescence emission of that partner which absorbs at the longest wavelengths when the counterion absorbing at shorter wavelengths is excited (A.S. Tatikolov, *Izv. Akad. Nauk, Ser. Khim.*, **1992**, 2524, 2531). Typical anionic chromophores are the dyes (3.22), (3.23; $X = CH_2$, O), and (3.24) which have been combined, e.g., with the simple streptocyanine dye (3.20; $n = 1$), thiacyanine dyes (3.21; $X,Y = S$; $n = 1, 2, 5$; $R^1 = Et$; $R^2 = H$) and indacyanine dyes (3.21; $X,Y = CMe_2$; $n = 1, 5$; $R^1 = Me$; $R^2 = H$, 4,5-benzo).

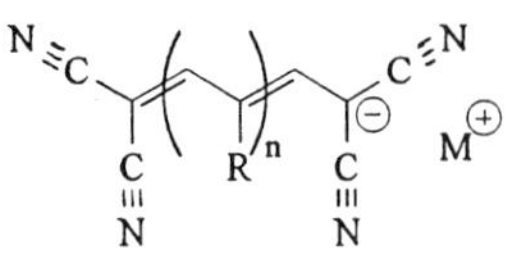

(3.22)

n = 1, 2; R = H, Me

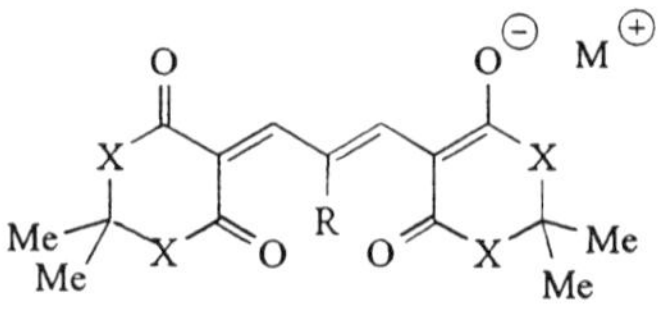

(3.23)

X = O, CH$_2$; R = H, Me

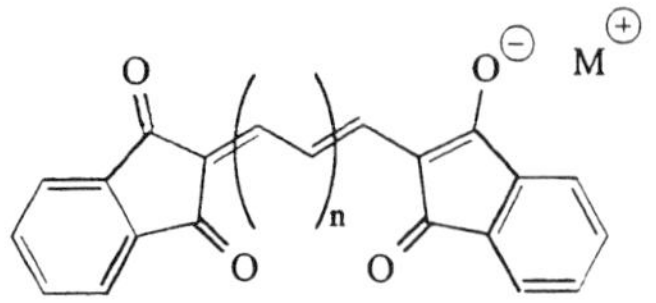

(3.24)

n = 0, 1, 2

(ii) Homomolecular aggregation of cyanine dyes
The effect of aggregation of polymethine-like molecules is well known. It brings about new absorption and emission bands which are red- and blue-shifted, respectively, in comparison with the long wavelength uv/vis absorption band of the monomeric molecules. According to exciton theory developed by M. Kasha and E.G. Mc Rae (*Physical and Chemical Mechanisms in Molecular Radiation Biology*, Eds. W.A. Glass, M.N. Varma, Plenum Press, New York, London, 1991, pp. 231-251), parallel aggregated dye molecules yield only one single absorption band which is hypsochromically (or blue-)shifted when the shift angle between two overlapping molecules is greater than 54^0, and which is red-shifted in case of smaller shift angles. Aggregates having blue-shifted bands are designated H-aggregates, whereas those with red-shifted bands are termed J- or Scheibe-aggregates after their discoverers E.E. Jelly (*Nature*, 1936, **138**, 1009) and G. Scheibe (*Angew. Chem.*, 1937, **50**, 51). If the molecules within the aggregate are not parallel aligned, an additional, so-called Davydov splitting happens with the aggregate absorption band giving rise to two or even more J- and/or H-bands (A.S. Davydov, *The Theory of Molecular Excitons*, Plenum Press, New York,1971).

New data about the thermodynamical behaviour of the aggregation process were summarized by T. Tanaka and T. Matsubara (*J. Soc. Phot. Sci. Technol. Jpn.*, 1991, **54**, 493; *J. Imag. Sci.*, 1993, **37**, 585). Association constants of oxacarbocyanine and oxadicarbocyanine dyes were conductometrically detected (D.H. Wiegand, P. Vanýsek, *J. Colloid Interface Sci.*, 1990, **135**, 272). The molecular orientation of individual J-aggregates of many cyanine dyes adsorbed to silver halide crystals was determined by polarized luminescence microscopy at low temperature (J.E. Maskasky, *Langmuir* 1991, 7, 407).

Owing to the small width of the J-absorption band of less than 200 cm^{-1}, the effect of resonance fluorescence of J-bands with high fluorescence quantum yield and without any Stokes-shift, the quenching of the fluorescence by extremely few molecules, and the effect of streaming anisotropy of the absorption and fluorescence band, G. Scheibe predicted as early as 1938 that there should be a strong coupling of the molecules and hence energy migration in the aggregates. This should cause shortening of the radiative lifetime of the aggregates down to less than picoseconds (G. Scheibe, L. Kandler, *Naturwissenschaften*, 1938, **24/25**, 412).

Since then, many authors have tried to prove this prediction using the J-aggregates of pseudoisocyanine (3.25) as model compound.

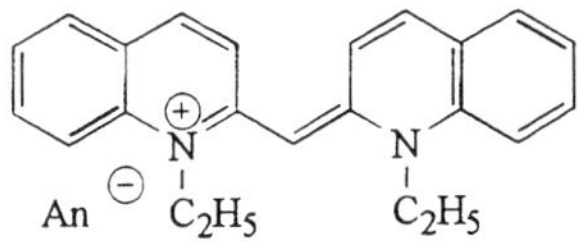

(3.25)

Indeed, in the 1970s fluorescence lifetimes of only a few ps had been measured when laser spectroscopic methods became available. However, in the 1980s it turned out that those results had been distorted by exciton-exciton annihilation processes due to the high laser excitation intensity used. This shortens the fluorescence lifetime with concurrent reduction of the fluorescence quantum yield (H. Stiel et al., *J. Lumin.*, 1988, **39**, 351). Only in 1989 D.A. Wiersma et al. (*Chem. Phys. Lett.* 187, **137**, 99), definitely proved by accumulated photon echo experiments that the radiative lifetime of pseudoisocyanine J-aggregates is shortened down to a few ps at low temperatures. Therefore, coherent excitonic energy migration actually takes place in J-aggregates. Since then, the energy migration phenomenon has been extensively investigated by many scientists. Recently, the challenging development has been reviewed in *J-Aggregates*, Ed. T. Kobayashi, World Scientific, Singapore, in press, and by D. Moebius (*Adv. Mater.*, 1995, 7,

437), H. Fidder, D.A. Wiersma [*phys. stat. sol.*, 1995, (b)188, 285], and S. Daehne (*J. Soc. Phot. Sci. Technol., Jpn.*, 1996, **59**, 250).

On the other hand, the photophysical properties of blue-shifted H-aggregates are mainly governed by radiationless decay processes. These have been carefully investigated taking indolocarbocyanine dyes (3.26) and 2,2'-quinadicarbocyanine iodide (3.21; n = 2, X,Y = CH=CH; R^1 = Et; R^2 = H; An = I) as examples (M. Van der Auweraer et al., *Chem. Phys.*, 1988, **119**, 355; S.-Y. Chen et al., *J. Phys. Chem.*, 1989, **93**, 3683).

(3.26)

Formation of dye aggregates also takes place when cyanine dyes are embedded in Langmuir Blodgett (LB) layers of lipophilic molecules where two-dimensional aggregate crystals are formed (R. Steiger, F. Zbinden, *J. Imag. Sci.* 1988, **32**, 64; F.-J. Schmitt et al., *Progr. Colloid Polymer. Sci.*, 1990, **83**, 136; K. Seito et al., *Jpn. J. Appl. Phys.*, 1991, **30**, 1836; K. Shinbo et al., *Thin Solid Films*, 1994, **243**, 630; D. Moebius, *loc. cit.*; A. Nabetani et al., *J. Chem. Phys.*, 1995, **102**, 5109). The molecular ordering of J-aggregates of carbocyanine dyes having negatively charged alkylsulfonate substituents, incorporated in positively charged lipid monolayers is herringbone-like, as was revealed by electron diffraction, fluorescence microscopy, and atomic force microscopy (AFM) (S. Kirstein, H. Moehwald, *Chem. Phys. Lett.*, 1992, **189**, 408; *Adv. Mater.*, 1995, 7, 460; V.N. Bliznyuk et al., *J. Phys. Chem.*, 1993, **97**, 569; V.V. Tsukruk et al., *Thin Solid Films*, 1994, **244**, 763). On the other hand, when the positively charged N,N'-distearylpseudoisocyanine (3.25; $C_{18}H_{37}$ instead of Et) is embedded in stearic acid monolayers, a brickstone arrangement of the two-dimensional crystals is formed (C. Duschl et al., *J. Phys. Chem.*, 1989, **93**, 4587).

The electrochromic behaviour of oxacyanine dye (3.21; X,Y = O; n = 0; R^1 = $C_{18}H_{37}$; R^2 = H) (N. Ohta et al., *Chem. Phys. Lett.*, 1994, **229**, 394) and the thiacarbocyanine (3.21; X,Y = S, n = 1; R^1 = $C_{18}H_{37}$; R^2 = H) (G.

Grewer, M. Loesche, *Macromol. Chem., Macromol. Symp.*, 1991, **46**, 79), incorporated in LB multilayers, gave independent proof that in J-aggregates the π-electrons are delocalized over more than one molecule because the quadratic Stark effect within the J-absorption band is dominated by large polarizability changes.

J-Aggregation was realized through interaction of betainic surfactants with the betainic tetrachloroimidacarbocyanine dye (3.27; $R^1 = (CH_2)_3SO_3Na$; $R^2 = Et$) (U. De Rossi et al., *Langmuir*, 1996, **12**, 1159). Also with betainic phospholipids, the J-aggregation tendency of cationic dyes, e.g. pseudoisocyanine (3.25) and 5,5'-dichloro-9-phenylthiacarbocyanine (3.28), is strongly enhanced (T. Sato et al., *J. Phys. Chem.*, 1989, **93**, 14).

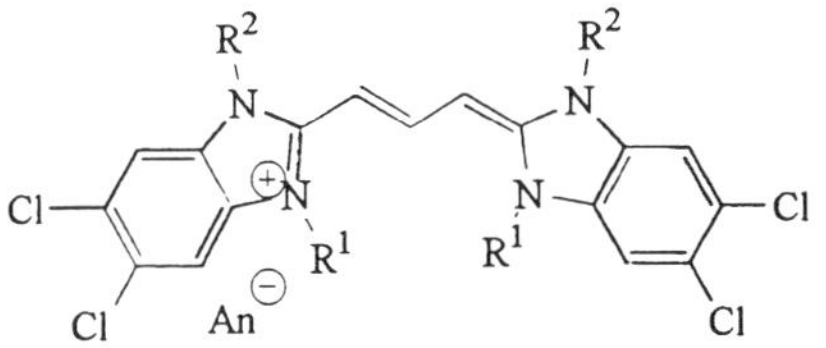

(3.27) 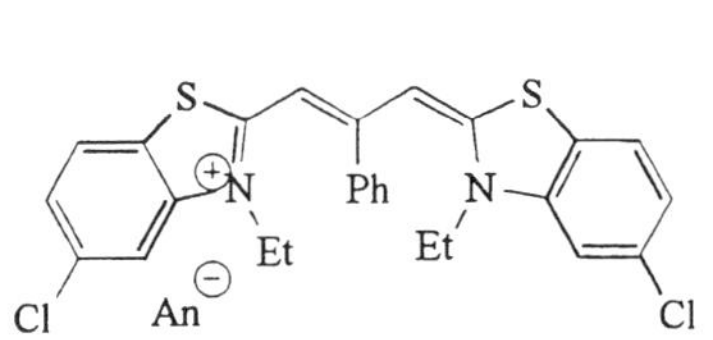   (3.28)

When mixtures of different dye aggregates embedded in liposomal surfaces were used, excitation energy transfer between the aggregates takes place on the condition that the fluorescence spectrum of the donor aggregates overlaps with the absorption spectrum of the acceptor aggregates. Optimum energy transfer was obtained using, for instance, the donor J-aggregates of 1,1'-diethyl-5,6:5',6'-dibenzo-2,2'-quinocyanine chloride (3.25; 5,6:5',6'-dibenzo-substituted; An = Cl) in liposomes having an absorption maximum at 600 nm in combination with the J-aggregates of dye (3.28), which acts as acceptor with an absorption maximum at 662 nm. On excitation of the donor, only the emission of the acceptor at 673 nm was observed (T. Sato et al., *Langmuir*, 1993, **9**, 3395). Energy transfer studies between J-aggregates of 3,3'-diethylthiadicarbocyanine (3.21; n = 2; X,Y = S; $R^1 = Et$; $R^2 = H$) and protein-containing macromolecules play an important part in the development of molecular devices in high technology fields (I. Hamachi et al., *Chem. Lett.*, **1993**, 1551).

J-aggregates are also obtained with dyes dispersed in polymer film matrices like pseudoisocyanine (3.25) (D.A. Higgins, P.F. Barbara, *J. Phys. Chem.*,

1995, **99**, 3) and oxadicarbocyanine iodide (3.21; n = 2; X,Y = O; R^1 = Et; R^2 = H) (T. Hiraga et al., *Jpn. J. Appl. Phys.*, 1994, **33**, 5051). With pseudoisocyanine, the physical extent of the excition migration domains has been determined to be about 100 nm by near field scanning optical microscopy (NSOM) combined with time-correlated single-photon-counting (TCSPC) detection (P.J. Reid et al., *J. Phys. Chem.*, 1996, **100**, 3892).

Another way to stabilize aggregated dye dimers is their inclusion into cyclodextrin cavities (M. Ohashi et al., *J. Am. Chem. Soc.*, 1990, **112**, 5824). The inclusion may even cause the formation of chiral aggregates from achiral monomeric molecules (V. Buss, C. Reichardt, *J. Chem. Soc., Chem. Commun.*, **1992**, 1636).

Strong intermolecular coupling occurs in the crystalline state of simple cyanine dyes like (3.20; n = 2, 3, 4), causing similar spectral features than in solution. The light absorption depends strongly on both the intermolecular structure of the aggregates and on the arrangement of the molecules with respect to the crystal faces. The result is that different crystal faces of one and the same dye appear in different colors (L. Daehne et al., *Chem. Phys.* 1995, **196**, 304). This phenomenon is even more pronounced in dye films of some nanometers thickness, which show chameleon-like coloration caused by both the formation of aggregates of different structure and their varying lay-out in respect to the film surface (L. Daehne, *J. Am. Chem. Soc.*, 1995, **117**, 12855).

By changing the N,N'-alkyl substituents at the 5,5',6,6'-tetrachloro-imidacarbocyanine chromophore (3.27a, b, c, d, e), the electronic structure and hence the absorption spectrum of the monomeric molecules (M) is scarcely changed, whereas the J-aggregates' structure is strongly altered as exemplified by their absorption spectrum (J) in Fig. 3.2 (S. Daehne et al., *J. Soc. Phot. Sci. Technol. Jpn.*, 1996, **59**, 250). The J-band may be red-shifted or blue-shifted, broadened or narrowed, and even split into several sub-bands, when both the 1,1'-disulfobutyl groups of dye (3.27a) are replaced by 1,1'-dicarboxypropyl groups and when its 3,3'-diethyl substituents are homologated. As indicated in Figure 3.2, the J-aggregates of 3,3'-didodecyl dye (3.27e) exist even in two different forms which can be reversibly converted into each other between 0 °C and -30 °C.

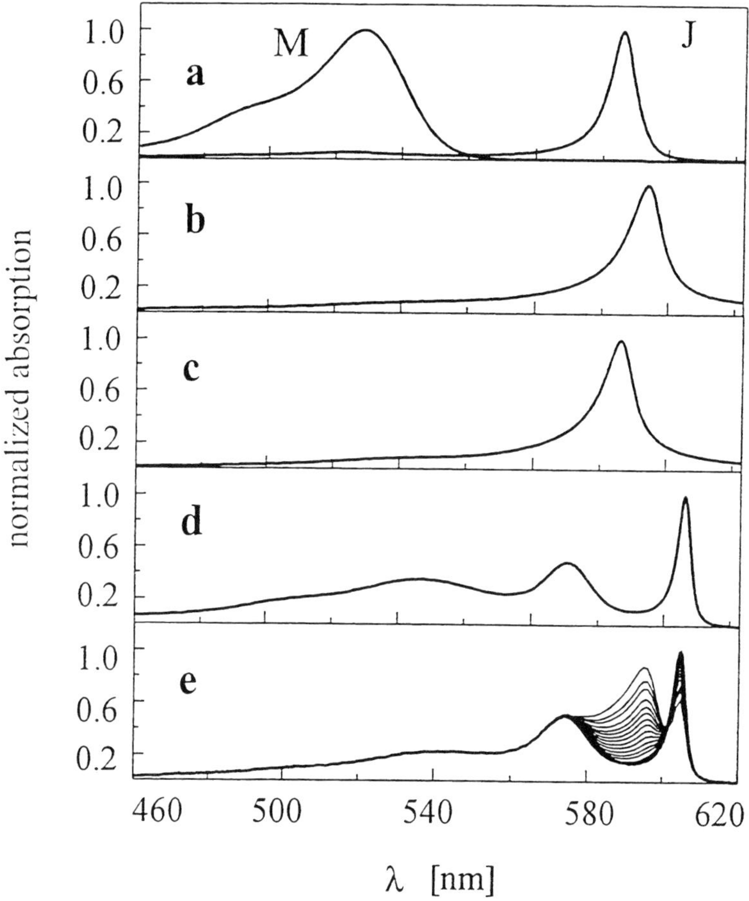

FIG. 3.2. ABSORPTION SPECTRA OF THE IMIDACARBOCYANINE DERIVATIVES (3.27) IN 0.01 N NaOH AT ROOM TEMPERATURE.
(3.27a): $R^1 = (CH_2)_4SO_3Na$; $R^2 = Et$
(3.27b): $R^1 = (CH_2)_3COONa$; $R^2 = Et$
(3.27c): $R^1 = (CH_2)_3COONa$; $R^2 = n\text{-}Bu$
(3.27d): $R^1 = (CH_2)_3COONa$; $R^2 = n\text{-}C_8H_{17}$
(3.27e): $R^1 = (CH_2)_3COONa$; $R^2 = n\text{-}C_{12}H_{25}$
(Taken from S. Daehne et al., *J. Soc. Phot. Sci. Technol. 1996*, **59**, 250, by courtesy of the Society of Photographic Science and Technology of Japan).

In the dioctyl- (3.27d) and didodecyl- (3.27e) derivatives, the halfwidth of the longest wavelength band is with 120 cm^{-1} the smallest ever observed in J-aggregates at room temperature. Using the exciton model from the ratio between the oscillator strength of the two split sub-bands, the angle between the transition dipole moments of two overlapping molecules has been estimated to be about 50^0 (U. De Rossi et al., *Angew. Chem.*, 1996, **108**, 827; *Angew. Chem Int. Ed. Engl.*, 1996, **35**, 760). The possible structure of such aggregates is depicted in Figure 3.3.

Surprisingly, the Davydov-split J-aggregates of dyes (3.27d and e) exhibit strong circular dichroism which by chance may be positive or negative. As the CD spectra gives a couplet having a positive and a negative branch, the two sub-bands in the absorption spectra must belong to the same J-aggregate which must be chiral. This can only be realized if the molecules in the aggregates are additionally tilted against each other by the dihedral angle δ forming a helix-like structure as it is indicated in Fig. 3.3 likewise. Obviously, combination of the dyes' aggregation tendency with the tendency of long alkyl groups to form vesicle-like structures provides the possibility for spontaneous generation of chiral structures. This finding may be helpful in the understanding of the wide distribution of chirality in life (U. De Rossi et al., *loc. cit.*).

Besides the J-aggregates of symmetrical cyanine dyes, the spectral behavior of J-aggregating merocyanine dyes (3.29; X = S; R^1 = $(CH_2)_4SO_3^-M^+$, or C_mH_{2m+1} with m = 2, 10, 12, 14, 16, 18, 20; R^2 = H, Me, Et, $C_{18}H_{37}$, Ph, CH_2COOH, $(CH_2)_oSO_3^-M^+$ with o = 3, 4) was also investigated (H. Nakahara, D. Moebius, *J. Colloid Interface Sci.*, 1986, **114**, 363; M. Kussler, H. Balli, *Helv. Chim. Acta*, 1989, **72**, 17, 295).

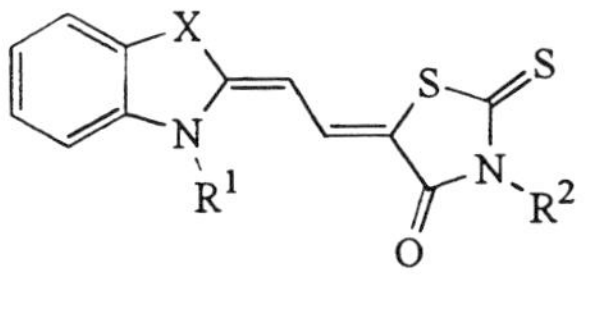

$$(3.29)$$

Structural variations of chromophore (3.29) influence its J-aggregation tendency which is surprisingly strong. Whereas dyes with nearly any donor group or atom, e.g. X = CMe_2, O, S, and Se, give J-aggregates, the aggregation tendency is rather sensitive to variations of the rhodaninyl acceptor unit.

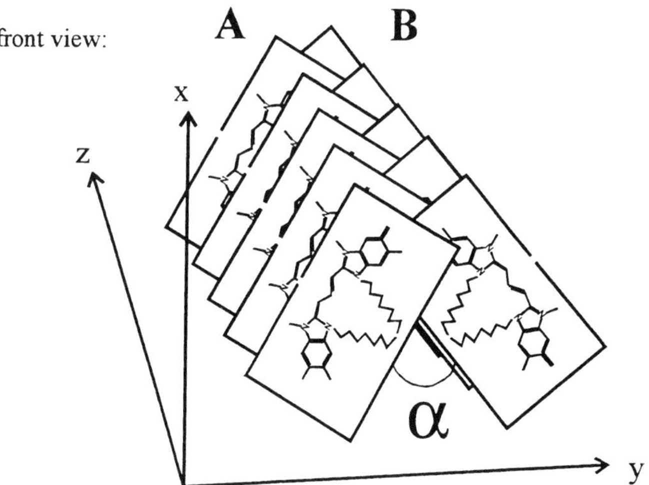

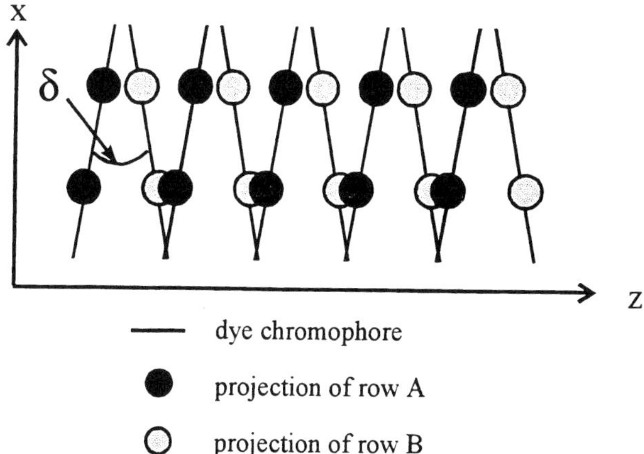

FIG. 3.3. POSSIBLE STRUCTURE OF SPONTANEOUSLY GENERATED CHIRAL J-AGGREGATES OF DYE (2.27d).
(Taken from U. De Rossi et al., *Angew. Chem.*, 1996, **108**, 827; *Angew. Chem. Int. Ed. Engl.*, 1996, **35**, 760, by courtesy of VCH Verlagsgesellschaft mbH, Weinheim).

For instance, the thiobarbituric acid group [3.30; $X = S, O$; $Y = S$; $R^1 = C_{18}H_{37}$; $R^2 = (CH_2)_2SO_3^-Na^+$] easily aggregates, but the barbituric acid group itself [3.30; $X = S, O$; $Y = O$; $R^1 = C_{18}H_{37}$; $R^2 = (CH_2)_2SO_3^-Na^+$] gives H-aggregates (E. Langhals, H. Balli, *Helv. Chim. Acta.*, 1985, **68**, 1782). On the other hand, even merocyanine dyes with open-chain acceptors such as (3.31; $X = S, Se$; $R^1 = C_{18}H_{37}$; $R^2 = (CH_2)_{3,4}SO_3^-K^+$) or (3.32; $X = S, Se$; $R = C_{18}H_{37}$) produce J-aggregates (E. Langhals, H. Balli, *loc. cit.*; M. Kussler, H. Balli, *Helv. Chim. Acta*, 1987, **70**, 1583; L. Wolthaus et al., *Chem. Phys. Lett.*, 1994, **225**, 322).

(3.30)

(3.31)

(3.32)

With dye (3.29; $X = S$; $R^1 = C_{18}H_{37}$; $R^2 = CH_2COO^- M^+$), depending upon the nature of the added cation M^+, up to five different J-aggregate conformers were obtained (A. Miyata et al., *Bull. Chem. Soc. Jpn.*, 1993, **66**, 999). This is probably due to binding the cations M^+ by formation of a chelate ring between the carboxyl group of R^2 and the carbonyl groups of the rhodaninyl nucleus (T. Kawaguchi, K. Iwata, *Thin Solid Films*, 1990, **191**, 173). From the decrease of the C=C stretching vibration frequency in the resonance Raman spectrum of this dye it was concluded that the π-electron delocalization in its J-aggregates is enhanced as compared to the monomeric molecules (Y. Ozaki et al., *Appl. Surf. Sci.*, 1988, **33/34**, 1317).

Monolayers of J-aggregates of dye (3.32; $X = CMe_2$; $R^1 = C_{18}H_{37}$) were investigated by scanning force microscopy (SFM). The dye molecules were found to be arranged in brickstone-like aggregates forming rectangular unit cells (L. Wolthaus et al., *Chem. Phys. Lett.*, 1994, **225**, 322).

Another way to design aggregates of definite structure is the covalent linkage of two chromophores via polymethylene groups as realized in (3.33) and (3.34) (A.A. Ishchenko et al., *Zh. Prikl. Spektrosk.* 1989, **50**, 237, 772).

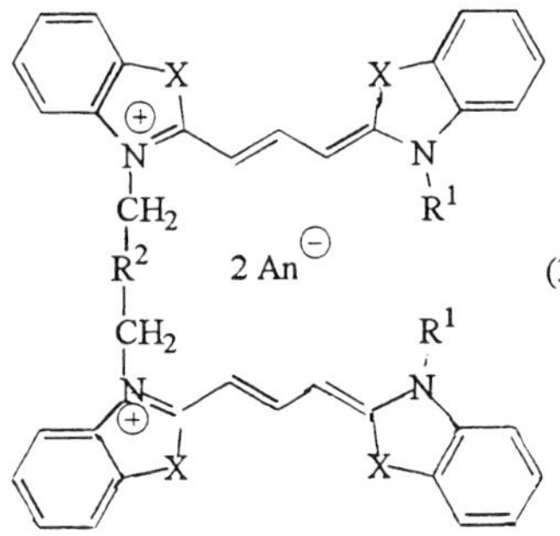

$$X = CMe_2, CH=CH; R^1 = Me, Et; R^2 = (CH_2)_n \text{ with } n = 0, 1, 2, 4; CH_2\text{-}O\text{-}CH_2$$

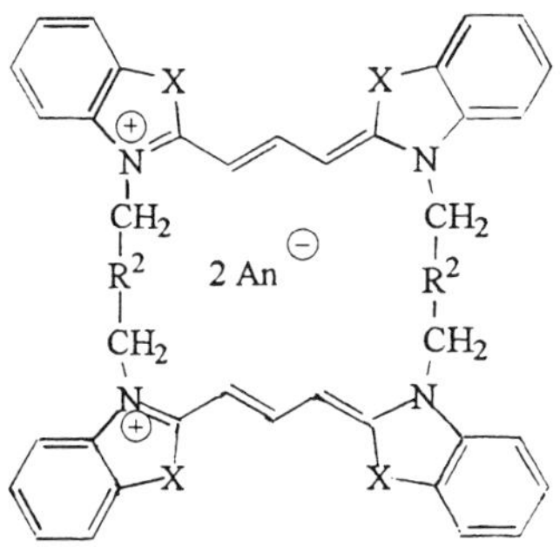

The compounds exhibit complicated static and dynamic spectral behaviour due to high intersystem crosssing rates which strongly depend on the length of the interconnecting polymethylene groups (A.K. Chibisov et al., *J. Phys. Chem.*, 1995, **99**, 886). However, the expected formation of absorption bands typical for aggregates was not observed.

(iii) Heteromolecular interaction of cyanine dyes
The formation of heteromolecular J-aggregates between partners which are both cationic cyanine dyes is well known and applied in order to supersensitize spectral sensitization of photographic silver halide emulsions by J-aggregates (e.g., B.I. Shapiro, *J. Soc. Phot. Sci. Technol. Jpn.*, 1992, **55**, 55; *Usp. Khim.*, 1994, **63**, 243)

Another type of intermolecular interaction takes place in mixed dye salts which contain both cationic and anionic cyanine chromophores and which usually form ion pairs as described in Sect. 3(*b,i*). Also stronger coupling of the partners in the sense of J-aggregation is noted with such dyes (Yu.L Slominski et al., *Dokl. Akad. Nauk, Ukr. SSR, Ser. B, Geol. Khim. Biol. Nauki*, **1987**, 50; A.S. Tatikolov et al., *Izv. Akad. Nauk, Ser. Khim.*, **1992**, 2524, 2532).

4. Applications of cyanine dyes

Cyanine dyes are increasingly applied in modern high technology fields, giving rise to the terminology "functional dyes" (see, *Chemistry of Functional Dyes*, Eds. Z. Yoshida, T. Kitao, Mita Press, Tokyo, 1989). As quite different physico-chemical properties of cyanine dyes are now used in these current applications, there is a trend to systematize cyanine dyes according to their physico-chemical features such as light absorption, light emission, light induced polarization, (photo)electric activity, and photochemical activity as shown in Fig. 4.1 (J. Griffiths [*Chimia (Aarau)*, 1991, **45**, 304].

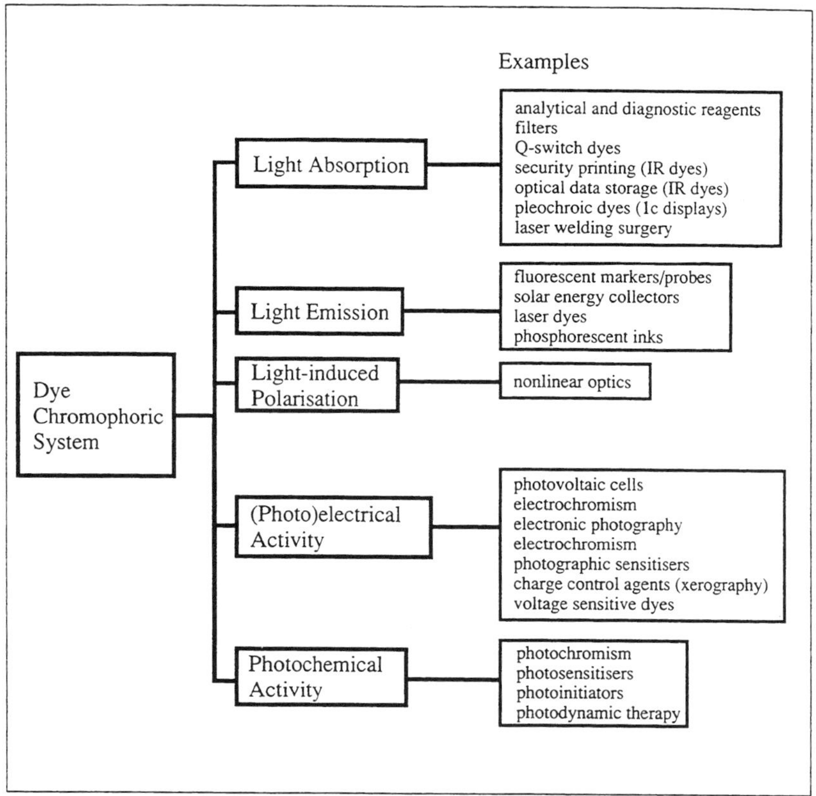

FIG. 4.1. MECHANISMS OF FUNCTIONAL DYES.
[Taken from J. Griffiths, *Chimia (Aarau)*, 1991, **45**, 304, by courtesy of the editorial staff of Chimia, the Schweizerischer Chemiker-Verband].

However, due to the vast amount of publications and patents which recently appeared in this field, the following Section places emphasis only on some selected topics.

(a) Analytical markers, probes, and sensors in chemistry and molecular biology
As outlined in Section 1(*c*), there are currently intense activities in the chemistry and application of cyanine dyes that absorb in the near infrared (NIR) region, because on one side cheap and tiny laser diodes emitting exclusively in the NIR region were introduced as light sources for analytical detection, and on the other in this region most of the probes to be analyzed are free of any absorption or emission band which may disturb analytical detection.

Presently the main fields of application are fluorescence detection for ion recognition (B. Valeur, *Probe design and chemical sensing*, in *Top. Fluor. Spectrosc.*, Ed. J.R. Lakowicz, Plenum Press, New York, London, 1994, pp. 21-48) and fluorescence labeling of biomolecules. The latter method is widely applied in analytical detection of amino acids, proteins, nucleic acids, living cells, and in immunoassay techniques (G. Patonay, M.D. Antoine, *Anal. Chem.*, 1991, **63**, 321A).

(i) *Probing solvent polarity and solvent hydrophobicity*
In case of merocyanines having a large dipole moment their solvatochromic behaviour [cf. Sect. 3(*a,iii*)] has been used for the development of several solvent polarity scales. Progress in this field has been reviewed, e.g. by C. Reichardt (*Solvents and Solvent Effects in Organic Chemistry*, 2nd Ed., VCH, Weinheim, 1988; *Chem. Rev.*, 1994, **94**, 2319) and E. Buncel, S. Rajagopal (*Acc. Chem. Res.*, 1990, **23**, 226).

When the merocyanine chromophore (4.1) is covalently attached via the N-alkyl group to polyelectrolytes it acts not only as spectroscopic probe for the environment's micropolarity but also as for probing the electrostatic surface potential (Y. Morishima et al., *J. Polym. Sci., Part A: Polym. Chem.*, 1991, **29**, 677; *ibid.* 1993, **31**, 373).

$$-O-(CH_2)_2-N^{\oplus}\text{-(pyridinium)-CH=CH-(phenyl)-}O^{\ominus} \qquad (4.1)$$

surface of polyelectrolytes

An interesting method to determine solvent hydrophobicity was developed by taking advantage of the cyanine dyes' aggregation tendency which strongly increases when the solvent hydrophobicity decreases. For example, the dimerization equilibrium of the thiatricarbocyanine dye (4.2) is a good measure of solvent hydrophobicity, which was determined both by changes in the absorption spectrum and by fluorescence quenching due to the dimerization effect (G. Patonay et al., *Appl. Spectrosc.*, 1991, **45**, 457).

$$(4.2)$$

(ii) Ion recognition

For cation detection most common complexation reagents are crown ethers linked with dye labels the absorption and emission wavelengths of which as well as fluorescence quantum yield and fluorescence lifetime change, when the cations are bound to the ethers (B. Valeur, *Probe design and chemical sensing*, in *Top. Fluor. Spectrosc.*, Ed. J.R. Lakowicz, Plenum Press, New York, London, 1994, pp. 21-48).

A typical example of such chromoionophores and fluoroionophores is the merocyanine dye DCM, linked with the macrocyclic monoaza-15-crown-5 (4.3), which gives complexes with alkaline and alkaline-earth-metal cations.

(4.3)

The absorption spectrum of (4.3) is drastically changed on complexation, whereas the fluorescence spectrum is only slightly blue-shifted and the fluorescence lifetime is almost unchanged (J. Bourson, B. Valeur, *J. Phys. Chem.*, 1989, **93**, 3871; L. Cazaux et al., *J. Photochem. Photobiol.*, 1994, **A77**, 217). Especially with Li^+ and Ca^{++}, complexes of (4.3) undergo photo-induced cation ejection under laser excitation giving insight into photo-release processes of biological interest like the photo-induced jump of cellular calcium ion concentration (M.M. Martin et. al., *Chem. Phys. Lett.*, 1993, **202**, 425).

In case of crown ether containing styryl dye chromophores such as (4.4; X = CMe_2, S; R = Me, Et, $(CH_2)_3SO_3^-M^+$), complexation with cations promotes photoisomerization of the dye, thus giving rise to photochromic behaviour (M.V. Alfimov et al., *Chem. Phys. Lett.*, 1991, **185**, 455).

(4.4)

Other examples are some pyrylo- and imidazolocyanine dyes (4.5; X = CH, O, N; R = Ph, p-NO_2Ph; An = Cl, I, CH_3COO) which are highly sensitive reagents for extraction and photometrical detection of thallium(III) ions (M.S. Chernov'yants et al., *Zh. Anal. Khim.*, 1991, **46**, 2214).

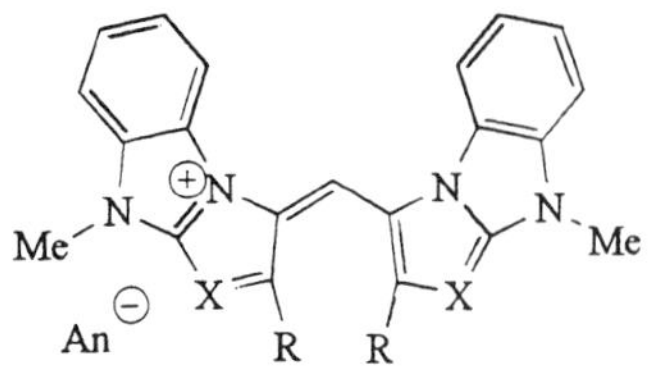

$$(4.5)$$

Also ion-pair formation of tellurium(IV) ions with some cyanine dyes was used for extraction and photometric determination (P.P. Kish et al., *Zh. Anal. Khim.*, 1990, **45**, 915).

(iii) Fluorescence labeling in molecular biology
Fluorescence labeling in molecular biology is realized both through noncovalent interaction of dye chromophores with biomolecules and by covalent linkages where cyanine dyes absorbing and fluorescing in the NIR region are preferred (G. Patonay, M.D. Antoine, *Anal. Chem.*, 1991, **93**, 321A; R.J. Williams et al., *Anal. Chem.*, 1993, **65**, 601).

Noncovalent labeling with carbocyanines was realized for low density lipoproteins, (M.F. Kleinherenbrink-Stins, *Lab. Invest.*, 1990, **63**, 73), membranes of living neurons (M.G. Honig, R.I. Hume, *Trends Neurosci.*, 1989, **12**, 333), lymphocytes (W. Bartlett et al., *J. Immunol.*, 1989, **143**, 1745), viruses (K. Kappe et al., *Biochem.* 1986, **25**, 8252), and *in vitro* as well as *in vivo* studies of liposomes (D.E. Wolf, *Biochem.*, 1985, **24**, 582; E. Claassen, *J. Immunol. Meth.*, 1992, **147**, 231).

For the noncovalent labeling of double-stranded DNA and RNA the lipophilic bisintercalating cyanine dyes YOYO [bis-oxazole yellow (4.6; X = O)] and TOTO [bis-thiazole yellow (4.6; X = S)] were developed.

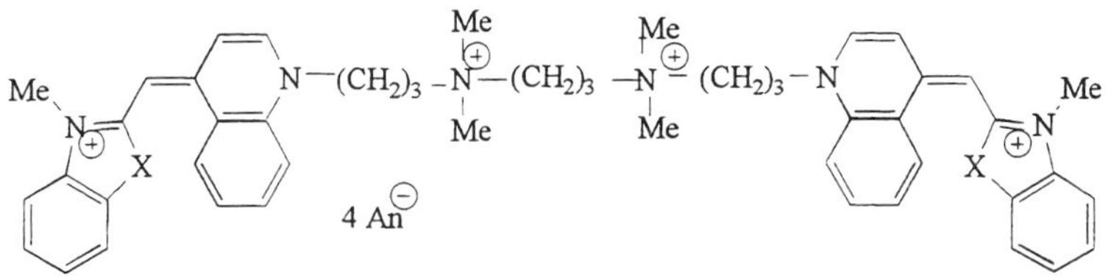

$$(4.6)$$

Through intercalation of the dyes between the nucleic acid double-strands their fluorescence intensity is enhanced by up to three orders of magnitude (H.S. Rye et al., *Nucleic Acids Res.*, 1992, **20**, 2803). The complexes are stable enough to allow the detection of pre-stained DNA restriction fragments after electrophoretic separation. As detection limits of some picogramms were reached the dyes are superior to all others (G.T. Hirons et al., *Cytometry*, 1994, **15**, 129).

Noncovalent fluorescence labeling of liposomes in living material, e.g. macrophages, lymphocytes, neurons, viruses, and bacteria was realized with "classic" cyanine dyes such as the N,N'-distearyl-substituted oxacarbocyanines (4.7; n = 1; X = O; R = $C_{18}H_{37}$) and 3,3,3',3'-tetramethylindacarbocyanines (4.7; n = 1; X = CMe_2; R = $C_{18}H_{37}$).

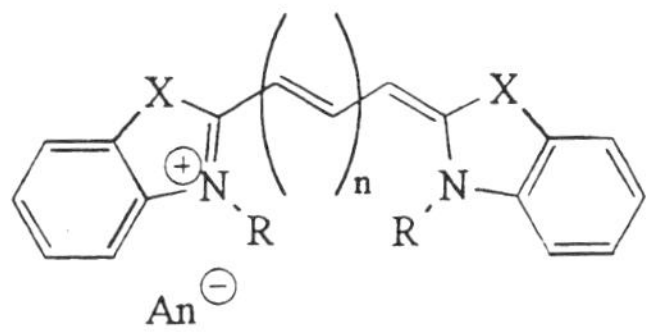

$$(4.7)$$

The liposomes were detected both spectrophotometrically, after their isolation from the living species, and directly by in vivo fluorescence microscopy. The latter method showed that the liposomes were localized in the marginal zone macrophages (E. Claassen, *J. Immunol. Methods*, 1992, **147**, 231).

An important application of noncovalent binding is the analytical determination of calcium binding sites in proteins with the 4,5:4',5'-dibenzothiacarbocyanine dye "STAINS-ALL" (4.8).

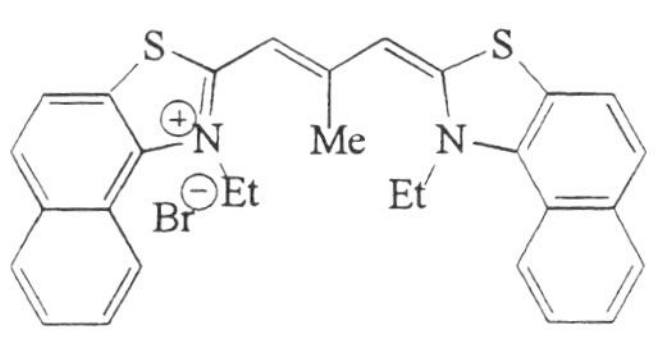

$$(4.8)$$

It gives characteristic absorption bands and induced circular dichroism through interaction with sites which can be reversibly displaced by calcium

448

ions (C.G. Caday et al., *Biopolymers*, 1986, **25**, 1579; Y. Sharma et al., *FEBS Lett.*, 1993, **326**, 59).

In order to effect *covalent binding* of dye labels to biomolecules carbo-, dicarbo-, and tricarbocyanine dyes such as (4.2) and (4.7; n = 1, 2, 3; X = CMe_2) are functionalized according to Scheme 4.1. Functionalized positions are either the nitrogen atoms (R^1), or the aromatic nuclei (R^2), or substitutions for the central chlorine atom. Reactive groups are, e.g. $NHCOCH_2I$, and $CH_2NHCOCH_2I$ (A.J.G. Mank et al. *Anal. Chem.* 1993, **65**, 2197), esters of N-hydroxysuccinimide (P.L. Southwick et al., *Cytometry* 1990, **11**, 418; R.B. Mujumdar et al., *Bioconjugate Chem.*, 1993, **4**, 105), as well as isothiocyanato groups (L. Strekowski et al., *J. Org. Chem.*, 1992, **57**, 4578; *Synth. Commun.*, 1993, **23**, 3087), the latter of which can be introduced using a new condensation method without any basic agent (N. Narayanan, G. Patonay, *J. Org. Chem.*, 1995, **60**, 2391).

SCHEME 4.1

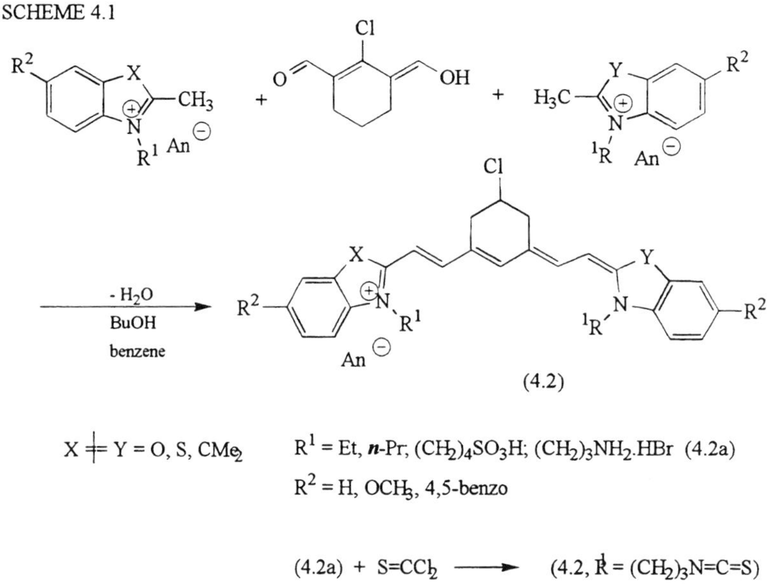

$$X \neq Y = O, S, CMe_2 \qquad R^1 = Et, \textit{n}\text{-Pr}, (CH_2)_4SO_3H; (CH_2)_3NH_2 \cdot HBr \ (4.2a)$$

$$R^2 = H, OCH_3, 4,5\text{-benzo}$$

$$(4.2a) + S{=}CCl_2 \longrightarrow (4.2, R = (CH_2)_3N{=}C{=}S)$$

$$(4.2) + HO\text{-Ph-N}{=}C{=}S \longrightarrow (4.2, Cl = O\text{-Ph-N}{=}C{=}S)$$

Among others, recent applications are high performance liquid chromatographic (HPLC) detection of human serum albumin (R.J. Williams et al., *Anal. Chem.*, 1993, **65**, 601), ultrasensitive NIR fluorescence detection in combination with electrophoresis of aminoacids (J.H. Flanagan, *Anal. Chem.*, 1995, **67**, 341), and DNA sequencing (L.R. Middendorf et al., *Electrophoresis*, 1992, **13**, 487), and the proof of gap junctional communication between living cells by the fluoro-immunoassay technique (M.Z. Hossain et al., *Neurosci. Lett.*, 1995, **184**, 71).

(b) Cyanine dyes for probing physico-chemical properties of organized media

Cyanine dyes are not only used as markers for biological purposes in general [cf. Sect. 4(*a,ii*)] but also as probes to investigate and characterize physico-chemical properties of organized media such as interfaces, Langmuir-Blodgett (LB) layers, vesicles, micelles, membranes, and living cells, which play an important part in molecular biology. Recent applications in the high technology fields are of great interest and are described in Sections 4(*c, d, e, f*).

(i) Structure of interfaces and Langmuir Blodgett (LB) layers

The structure of monomolecular LB films formed at air/water interfaces with the amphiphilic N-(n-alkyl)-4'-dimethylaminostilbazolium dyes (4.9; n = 1; R^1 = $(CH_2)_{m-1}$Me with m = 3 to 8; R^2 = Me) mentioned already in Section 3(*a,ii*) is strongly dependent on the length of the N-alkyl substituent

$$R^1\text{-}N^{\oplus} \quad \text{An}^{\ominus} \quad \cdots \quad \text{---}N(R^2)_2 \tag{4.9}$$

in such a way that dyes having odd-numbered alkyl groups (m = odd) exhibit a significantly higher standard free energy of adsorption and a higher minimum surface area demand than the even-numbered derivatives (K. Lunkenheimer, A. Laschewsky, *Progr. Colloid. Polym. Sci.*, 1992, **89**, 239).

The orientation of dyes (4.9; n = 1; R^1 = $C_{22}H_{45}$; R^2 = Me) in LB layers with and without dye-free interleaving layers was determined by polarized ir spectroscopy in order to improve the nonlinear optical properties [cf. Sect. 4(*d*)] of such layers (P. Stroeve et al., *Thin Solid Films*, 1989, **179**, 529).

An interesting insight into the structure of LB layers was obtained by means of uv/vis spectroscopy and X-ray diffraction when dye (4.9; R^1 = $C_{16}H_{33}$, $C_{22}H_{45}$; R^2 = Me, $(CH_2)_2OH$) was covalently bound to polymers through esterification with the N-hydroxyethyl substituents introduced at the dye's amino group (M.C.J. Young et al., *Thin Solid Films*, 1989 **182**, 319).

Ir and Raman spectroscopy was also useful in the determination of the orientation of N,N'-distearyl-substituted thiacarbocyanine and thiatricarbocyanine dyes (4.7; n = 1, 3; X = S; R = $C_{18}H_{37}$) in LB layers. The orientation was strongly dependent on the substrate used (N. Katayama et al., *Langmuir*, 1992, **8**, 2758).

When water-soluble cyanine dyes such as "STAINS ALL" (4.8) or pseudo-isocyanine (4.10) were adsorbed to arachidic acid LB layers two-dimensional single crystals of J-aggregates [cf. Sect. 3(*b,ii*)] were formed.

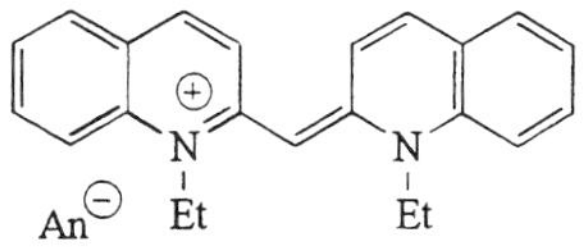

(4.10)

The structure and orientation of which were obtained by fluorescence and electron microscopy, electron diffraction, and X-ray reflection methods (C. Duschl et al., *Thin Solid Films*, 1988, **159**, 379). The crystallization of the J-aggregates causes "conditioning" of the LB-layers, i.e. an ordering of the fatty acid monolayer. The orientation of the LB layer was maintained when the dye aggregates were removed by dissolution. The memory effect was then read out by adsorption of a second J-aggregating cyanine dye the aggregates of which were oriented according to the pre-formed pattern of the LB-layer (R. Steiger, F. Zbinden, *J. Imag. Sci.*, 1988, **32**, 64; F.-J. Schmitt, W. Knoll, *Chem. Phys. Lett.*, 1990, **165**, 54).

(ii) Probing dielectric constants, lipophilicity, and microviscosity
The absorption maximum of the solvatochromic merocyanine dye (4.11; n = 2; m = 16), the vinylogous entity of dye (4.1), depends in linear relation to the dielectric constant of various organic solvent/water mixtures. In this way the effective dielectric constant of micelle-water interfaces between 15°C and 85°C was determined (G.G. Warr, D.F. Evans, *Langmuir*, 1988, **4**, 217).

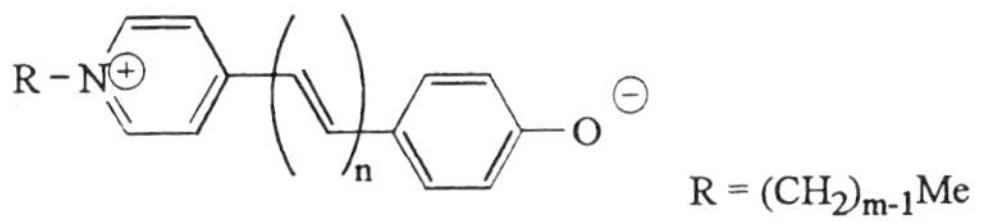

$$R-N^{(+)}\!\!=\!\!\left(\!\!\bigcirc\!\!\right)_{n}\!\!-\!\!\bigcirc\!\!-O^{(-)} \qquad (4.11)$$

$$R = (CH_2)_{m-1}Me$$

In order to probe micellar solubilization sites dyes (4.11; n = 1; m = 1 to 6) were used. On light excitation they are photochemically converted from their *trans* conformation to the *cis* isomers, which are additionally protonated at the oxygen atom in the first excited singlet state. After deprotonation in the ground state, the thermal back-reaction to the original *trans* conformation is strongly catalyzed by cationic and anionic micelles and increases with increasing hydrophobicity of the dyes, i.e. with lengthening of the dyes' alkyl group. Thus, providing measures of both binding constants and micellar rate constants (K.J. Dennis et al., *J. Phys. Chem.*, 1993, **97**, 8328). Structural variations of dye (4.11) have strong influence on its chemical and photophysical behaviour in polymeric films (D. Frackowiak et al., *J. Photochem. Photobiol.*, 1990, **A54**, 37).

Information about lipid bilayers was gained through variation of the N,N'-dialkyl group length of thiacarbocyanine dyes (4.7; n = 1; X = S; R = CH$_2$)$_{m-1}$Me with m = 2, 6, 8, 12, 18; An = Cl, I). It turned out that fluorescence quantum yield, triplet yield, and singlet oxygen quantum yield were the greater the longer the alkyl groups, i.e. the better they were penetrated into the lipid bilayer, whereas photoisomerization became more difficult in the liposomes (M. Krieg et al., *Biochim. Biophys. Acta*, 1993, **1151**, 168).

Photoinduced isomerization and solvent relaxation processes in the first excited singlet state were applied in the determination of the microviscosity of membranes and micelles which strongly influence the fluorescence features of asymmetrical tricarbocyanine dyes such as (4.12; n = 4, 5; X ≠ Y = CMe$_2$, NPh, S) and related derivatives.

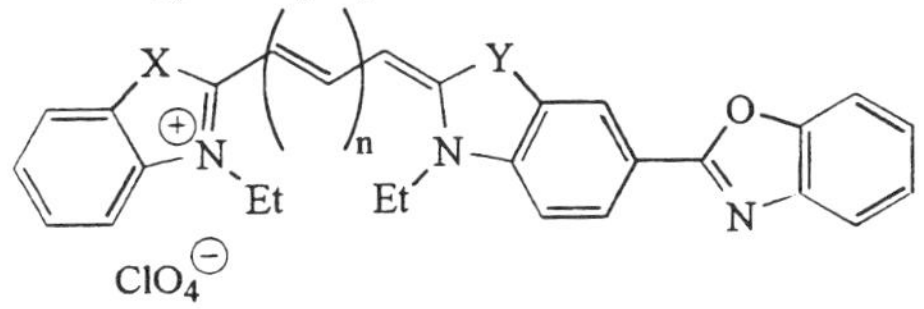

$$\qquad (4.12)$$

In non-viscous media only one fluorescence band is noted which, in contrast to the absorption spectrum, is not dependent on the polarity of the solvent used. Due to restrictions of the excited state relaxation processes in highly viscous media like biomembranes, whole living cells, as well as in solvents at low temperature, a second fluorescence band appears at shorter wavelengths. This spectral behaviour allows conclusions to be made concerning the microscopic viscosity of the molecular environment (M. Nakagaki et. al., *Chem. Pharm. Bull.*, 1986, **34**, 4486; Z.N. Volovik et al., *Ukr. Biokhim. Zh.*, 1988, **60**, 64; *Proc. SPIE-Int. Soc. Opt. Eng.*, 1994, **2137**, 600).

(iii) Determination of critical micelle concentration (cmc)
In order to determine the critical micelle concentration (cmc) of surfactants, J-aggregating cationic cyanine dyes [cf. Sect. 3(*b,ii*)] were used, the J-aggregates of which are destroyed through micelle formation (A.H. Herz, *Adv. Colloid Interface Sci.*, 1977, **8**, 237).

However, in case of the 5,5',6,6'-tetrachloroimidacarbocyanine dye (4.13) which was described already in Sect. 3(*b,ii*), presumably the electrostatic interaction of the betainic dye molecules with betainic surfactant molecules like (4.14; n = 10, 12, 14, 16) promotes J-aggregation (U. De Rossi, et al., *Langmuir* 1996, **12**, 1159).

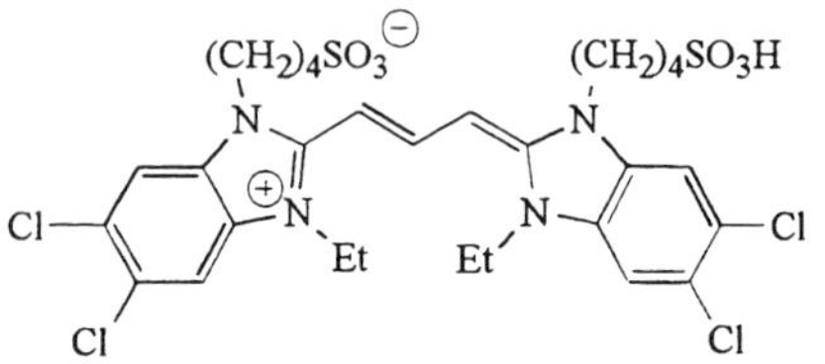

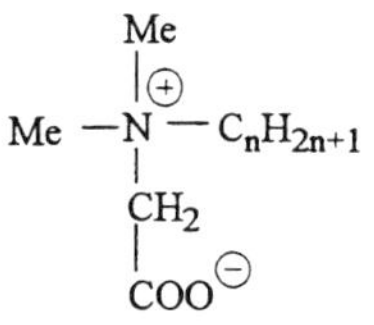

Therefore, caution is advised using J-aggregating dyes in cmc determinations.

Another molecular probe for studying cmc and other membrane properties is merocyanine M 540 (4.15) which shows both blue-shifted absorption bands of dimer and polymer H-aggregates and solvatochromism depending on the polarity of the solvent used. As the H-aggregates indicate polarity changes of the dye's microenvironment they were used too in the

$$(4.15)$$

determination of cmc. On the other side, the aggregation of the dye is strongly enhanced with increasing packing density of the lipid-hydrocarbon region near the interface and is thus very sensitive to changes in the vesicle curvature as well as to the presence of lipids which possess groups of different hydrophobicity (P. Kaschny, F.M. Goni, *Eur. J. Biochem.*, 1992, **207**, 1085).

Even better suited for cmc determination are dyes which exhibit different absorption maxima which depend on the hydrophobicity of the dye's microenvironment. The pyrenyl merocyanine-like dye (4.16), having an absorption maximum at 619 nm in water and nearby 570 nm in the hydrophobic environment of micelles, proved to be useful as it does not form aggregates in the concentration range used (A.E. Boyer et al., *Analyt. Lett.*, 1991, **24**, 701).

$$(4.16)$$

(iv) Determination of membrane potentials
Cyanine dyes were increasingly applied in the determination of the electrical potential of biological membranes in animal and bacterial cells, organelles, plant vesicles, plant protoplasts, etc.. Changes in the potential, which is dependent on the concentration ratio between the intra- and extracellular alkaline and alkaline-earth cations, induce either changes of the dyes' absorption spectrum or preferentially quench the dyes' fluorescence intensity.

454

Using cationic cyanine dyes which are permeant to membranes, the accumulation of the dyes takes place through cells which have large negative internal membrane potentials (A.S. Waggoner, *Soc. Gen. Physiol. Ser.* 1988, **43**, 209). Standard dyes are the lipophilic 3,3'-di-*n*-propyloxadicarbocyanine (4.7; n = 2; X = O; R = *n*-Pr; An = I) and 3,3'-di-*n*-propylthiadicarbo-cyanine (4.7; n = 2; X = S; R = *n*-Pr; An = I) (A.A. Eddy, *Methods in Enzymology*, 1989, **172**, Part S, 95) which were carefully tested by T. Kakutani et al. (*Bioelectrochem. Bioenerg.*, 1992, **28**, 221). Also several other cyanine and merocyanine dyes, such as the thiacarbocyanine (4.7; n = 1; X = S; R = *n*-Pr; An = I), indacarbocyanine (4.7; n = 1; X = CMe$_2$; R = *n*-Pr; An = I), and MC 540 (4.15) have been used (J.C. Freedman, T.S. Novak, *Methods in Enzymology*, 1989, **172**, Part S, 102).

When permeant anionic dyes, such as (4.17) are employed the dye molecules are ejected from cellular compartments and hence, the fluorescence signal is the stronger the more negative the membrane potential (A.S. Waggoner, *loc. cit.*; J.C. Freedman, T.S. Novak, *loc. cit*).

(4.17)

Most important are the impermeant membrane potential probes which are localized within the membranes. These undergo a potential-dependent movement between sites on the membrane or are even driven off the membrane when the potential inside of the cell becomes very negative (E.B. George et al., *J. Membrane Biol.*, 1988, **103**, 245, *ibid.*, 1988, **105**, 55). Dianionic oxonol dyes like (4.18), terminated by pyrazolone and barbituric acid mojeties, are suited for this purpose. Precondition to inhibit the dyes' crossing through membranes is a second negative group in the molecule which is provided by sulfonic acid substituents R^1 in the merocyanine dyes (4.18) (P.R. Pratap et al., *Biophys. J.*, 1990, **57**, 835).

$$(4.18)$$

X = O, S; R^1 = sulfoalkyl, sulfophenyl, etc.

R^2, R^3, R^4 = different alkyl groups varying in length

Also hemicyanines (4.9; n = 1, 2, 3; R^1 = $(CH_2)_4SO_3^-$; R^2 = C_mH_{2m+1} with m = 1 to 6) having a sulfonic acid substituent proved to be impermeant probes of membrane potential. Dyes of slightly different structure, however, gave quite different results when they were imbedded into membranes (P. Fromherz, C.O. Mueller, *Biochim. Biophys. Acta*, 1993, **1150**, 111).

Apart from the change of hydrophobicity in dependence on the alkyl group length (P. Fromherz, C. Roecker, *Ber. Bunsenges. Phys. Chem.*, 1994, **98**, 128), the main reason is that photoinduced *trans-cis* isomerization at the dyes' central double bond competes with the voltage-sensitive features. Therefore, some hemicyanine dyes without free double bonds, such as (4.9; n = 0; R^1 = $(CH)_2)_4SO_3^-$; R^2 = *n*-C_4H_9) and (4.19) were synthesized (E. Ephardt, P. Fromherz, *J. Phys. Chem.*, 1993, **97**, 4540).

$$(4.19)$$

Recently J-aggregates of the tetrachlorobenzimidacarbocyanine chromophore (4.13) proved to be sensitive against membrane potential changes in such a way that due to the formation of J-aggregates in the matrix of energized mitochondria the dye fluorescence alters reversibly from green to red with increasing membrane potentials (M. Reers et al., *Biochemistry*, 1991, **30**, 4480).

456

(v) Excitation energy transfer
Another common method to investigate both the "architecture" and the dynamics of supramolecular organized media is excitation energy transfer between different dyes, whereby the fluorescence spectrum of the donor dye has to overlap with the absorption spectrum of the acceptor dye. Most of the experiments were done with J-aggregating dyes which were described in Sect. 3(*b,ii*). Although energy transfer studies between monomeric dye molecules would be more precise, only few dyes are known which do not aggregate. These include the donor N,N'-distearyloxamonomethincyanine (4.7; n = 0; X = O; R = $C_{18}H_{37}$) combined with the acceptor N,N'-distearylthiamonomethincyanine (4.7; n = 0; X = S; R = $C_{18}H_{37}$). This combination was used to re-examine Foerster's theory of resonance energy transfer (P. Fromherz, G. Reinbold, *Thin Solid Films*, 1988, **160**, 347). Using indacarbocyanine (4.7; n = 1; X = CMe$_2$; R = $C_{18}H_{37}$) as donor combined with its vinylogous dicarbocyanine derivative (4.7; n = 2; X = CMe$_2$; R = $C_{18}H_{37}$) as acceptor an exceptionally long-range energy transfer up to distances of 100 nm was observed. The effect points to special mutual orientations between the donor and acceptor dye molecules (S. Draxler et al., *Chem. Phys. Lett.*, 1989, **159**, 231). Therefore, with respect to energy transfer experiments the determination of the orientation of cyanine dyes in LB layers using polarized absorption spectroscopy has attracted attention (N. Ohta et al., *Langmuir*, 1994, 10, 3909).

An interesting application of excitation energy transfer is the design of molecular switching devices based on sequential energy transport through three LB layers, which contain, e.g., the oxacarbocyanine (4.20) as absorber, the merocyanine (4.21) as transfer layer, and the 6,7:6',7'-dibenzindadicarbocyanine dye (4.22) as emitter. Switching off is then realized by photochemical conversion of the middle-positioned merocyanine dye to the colorless spiropyran (4.23) and *vice versa* according to Scheme 4.2 (T. Minami et al., *J. Phys. Chem.*, 1991, **95**, 3988; I. Yamazaki, N. Ohta, *Pure Appl. Chem.*, 1995, **67**, 209).

SCHEME 4.2

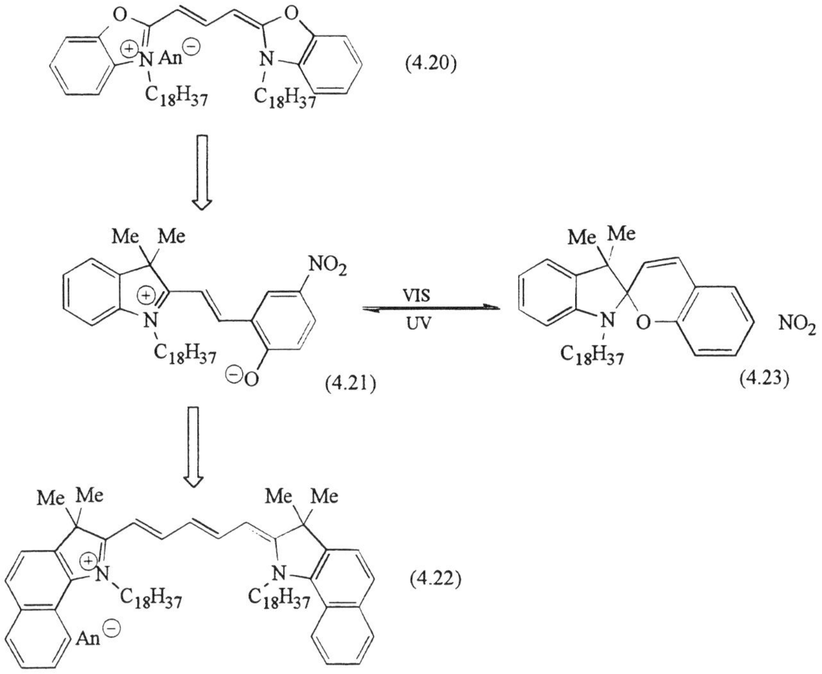

(4.20)

(4.21) (4.23)

(4.22)

(c) Cyanine dyes in laser technology

Cyanine dyes are most important for Q-switching, i.e. giant laser pulse generation, as well as for tunable dye laser generation. Both processes strongly depend on the absorption spectrum, excited state lifetime, and excited state relaxation processes like fluorescence, internal conversion, and intersystem crossing of the dyes used. Commonly applied dyes are composed by K.H. Drexhage, [*Top. Appl. Phys.*, 2nd. Ed. 1990, **1** (*Dye Lasers*), pp. 144-274] and R. Raue (*Ullmann's Encyclopedia of Industrial Chemistry*, 5th. Ed., 1990, VCH Weinheim, Vol. A15, pp. 151-164).

Current research work aims to increase the thermal and photochemical stability of laser dyes, especially in the NIR region, to evaluate the influences of photoisomerization processes as well as of vibronic and intermolecular interactions, and to achieve efficient lasing which necessiates large tuning ranges and large Stokes shifts between the absorption and fluorescence bands. (E.A. Luk'yanetz, *Mol. Cryst. Liq. Cryst., Sci. Technol., Sect. C*, 1992, **1**, 15; A.A. Ishchenko, *Ukr. Khim. Zh.*, 1991, **57**, 1166; *Kvant. Elektronika*, 1994, **21**, 513; P. Czerney et al., *J. Photochem. Photobiol.*, 1995, **A89**, 31). With the thiapyrylocyanine dye (4.24), containing a trisdecamethine chain and pumped by neodymium laser at 1056 nm, lasing wavelengths up to 1800 nm were realized (E.A. Luk'yanetz, *loc. cit.*).

(4.24)

Large Stokes shifts were preferentially obtained with dyes which undergo photoinduced proton transfer reactions as in the pentamethinoxonol unit contained in 3-hydroxyanthrone (4.25; Scheme 4.3) (M. Itoh et al., *J. Am. Chem. Soc.*, 1985, **107**, 4819; M. Kasha, *J. Chem. Soc., Faraday Trans. 2*, 1986, **82**, 2379).

SCHEME 4.3

(4.25)

In order to obtain light absorption over the large wavelengths ranges needed for flashlamp-pumped dye lasers, advantage is taken of energy transfer between mixtures of different cyanine dyes as explained in the preceding Section (B.M. Uzhinov et al., *Kvant. Elektronika*, 1991, **19**, 7).

Improvement of the stability of laser dyes was achieved with fluorine-containing cyanine dyes (B.M. Krasovitskii et al., *Dyes Pigm.*, 1988, **9**, 21). For the development of high performance dye lasers replacements of the usual solution techniques by aqueous-micellar dye solutions (M.G. Littman, *Proc. SPIE-Int. Soc. Opt. Eng.*, 1988, **912**, 56), as well as by polymeric media (V.I. Bezrodnyi, *Izv. Akad. Nauk SSSR, Ser. Fiz.*, 1990, **54**, 1476; A.A. Ishchenko, *Kvant. Elektronika*, 1994, **21**, 513), or by microporous glasses [G.B. Al'tshuler et al., *Kvant. Elektronika*, 1985, **12**, 1094; R. Reisfeld, *J. Phys.* IV, 1994, **4** (*C4 Laser M2P*), 281; *Chem. Abstr.* 1994, **121**, 310500z] proved to be useful.

(d) Application of light induced polarization. Nonlinear optical properties
Nonlinear optical (NLO) effects result from the polarization induced in matter by the intense electric field of laser radiation. Important applications are frequency conversion, integrated optical devices, and frequency-optical modulation of laser radiation by photorefractive materials for usages, e.g., in optical communication, optical image memory processing, and phase array control systems. For instance, single-mode optical wave-guides based on polymers with incorporated electrooptical dyes are easy to handle in order to produce multichip modules for high speed interconnects. Most recent reviews on the design of new NLO dyes were presented by W.E Moerner, S.M. Silence (*Chem. Rev.*, 1994, **94**, 127), D.R. Kanis et al. (*ibid.*, 1994, **94**, 195), and J.-L. Brédas et al. (*Chem Rev.*, 1994, **94**, 243; *Adv. Mater.*, 1995, 7, 263). The NLO behaviour of cyanine dyes embedded in organic polymers as well as inorganic sol-gel glasses was reviewed by D.M. Burland et al. (*Chem. Rev.*, 1994, **94**, 31).

In order to realize second harmonic generation (SHG) of laser radiation NLO materials should have large second order nonlinear optical susceptibilities $\chi^{(2)}$, i.e. high first hyperpolarizability values β. Correspondingly large third order optical susceptibilities $\chi^{(3)}$ and high second hyperpolarizability values γ, respectively, are necessary for third harmonic generation (THG). For a long time it has been established that in comparison to aromatic and polyenic compounds having the same number of π-electrons cyanine dyes possess the highest linear polarizability α (S. Daehne, K.-D. Nolte, *J. Chem. Soc., Chem. Commun.*, **1972**, 1056; *J. prakt. Chem.*, 1976, **318**, 643). Therefore, they are also prone to have high nonlinear optical susceptibilities and thus to be best suited for NLO materials [e.g. S.R. Marder, *Science (Washington)*, 1994, **265**, 632].

The first and second hyperpolarizability of symmetrical cyanine dyes is the greater the longer the polymethine chain is (B.M. Pierce, *Proc. SPIE-Int. Soc. Opt. Eng.*, 1991, **1560**, 148; I.D.L. Albert et al., *J. Opt. Soc. Am. B*, 1993, **B10**, 1365; K.D. Singer, J.H. Andrews, *Condens. Matter News*,

460

1994, **3**, 7; T. Johr et al., *Chem. Phys. Lett.*, 1995, **246**, 521). Additionally, in order to attain high first hyperpolarizability β, the structure of the molecules and the crystals, respectively, must not be centro-symmetrical. This is realized in merocyanine-like compounds which are also considered as vinylogous, donor-acceptor substituted polyenes (P.A. Cahill et al., *Proc. SPIE-Int. Soc. Opt. Eng.*, 1991, **1560**, 130; P.W. Perry et al., *ibid.*, 1991, **1560**, 302; R.D. Miller et al., *ibid.*, 1994, **2042**, 354; G.A. Lindsay et al., *ibid.*, 1994, **2143**, 88; M.B. Meinhardt et al., *ibid.*, 1994, **2143**, 110).

As with symmetrical cyanine dyes, the hyperpolarizability of merocyanine dyes increases with increasing polymethine chain length (S.R. Marder et al., *Proc. SPIE-Int. Soc. Opt. Eng.*, 1991, **1560**, 86; C.W. Spangler et al., *ibid.*, 1991, **1560**, 139; M. Blanchard-Desce et al., *ibid.*, 1994, **2143**, 20). On comparison with merocyanine dyes which have polymethine chains of equal length, the first and second hyperpolarizability is the greater the stronger the donor and acceptor strength of the terminal substituents, i.e. the higher the π-electron delocalization within the polymethine chain is (D.M. Burland et al., *Proc. SPIE-Int. Soc. Opt. Eng.*, 1991, **1560**, 111; X.-M. Duan et al., *ibid.*, 1994, **2143**, 41).

This is exemplified with merocyanines (4.26) to (4.31), where each is constituted of one heptamethine chain as shown in the formulae presented in Figure 4.2. The dyes have significant differences in their first and second hyperpolarizability values, which additionally are strongly dependent on the polarity of the solvent used. This behaviour is now well understood on the basis of triad theory [C.B. Gorman, S. R. Marder, *Proc. Natl. Acad. Sci. USA*, 1993, **90**, 11297; S.R. Marder et al., *Science (Washington)*, 1994, **265**, 632].

With the heptamethinemerocyanine (4.26) the effect of structural changes on the dyes' dipole moment and polarizability in the ground state was modeled by quantum chemical calculations using the INDO (intermediate neclect of differential overlap) method (S.R. Marder, *loc. cit.*). The structural changes were created by superposition the dye with an external electrical field and expressed in terms of polyenic bond order alternation BOA, which is about 0.6 in ideal polyenes and zero in ideal polymethines (S. Daehne, F. Moldenhauer, *Progr. Phys. Org. Chem.*, 1985, **15**, 1).

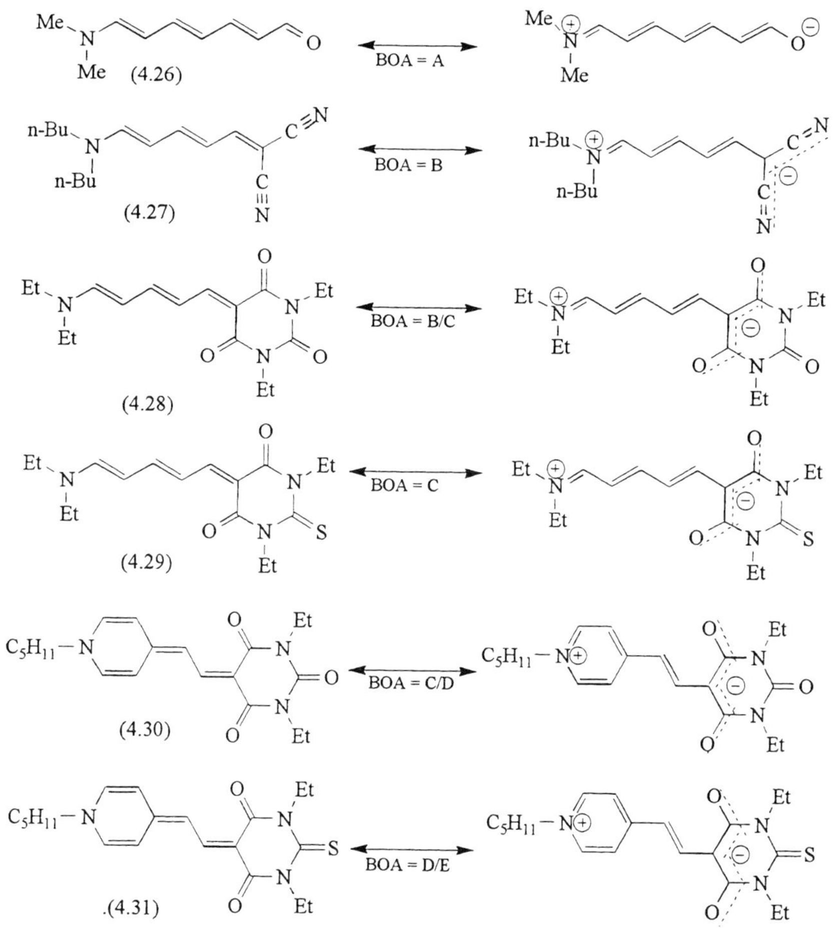

FIG. 4.2. LIMITING STRUCTURES OF MEROCYANINE DYES (4.26) to (4.31).

The region of bond order alternation (BOA) according to Figure 4.3 is indicated at each compound. [After S.R. Marder et al., *Science (Washington)*, 1994, **265**, 632].

The procedure corresponds to that applied in the description of the solvatochromic behaviour of merocyanine dyes [cf. Sect. 3(*a,iii*)].

462

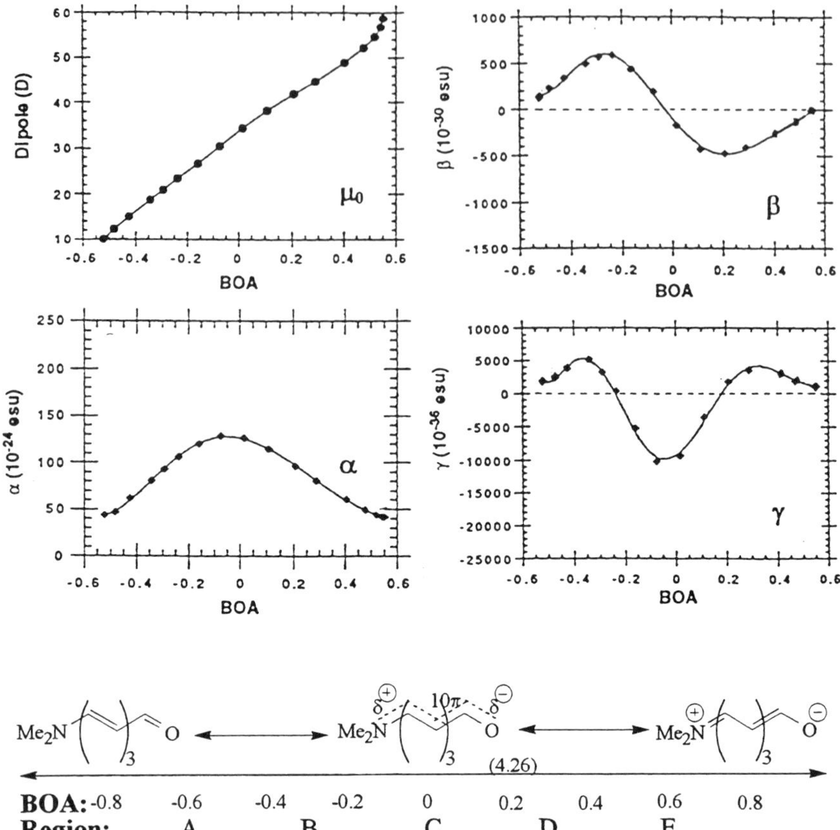

FIG. 4.3. MOLECULAR PARAMETERS OF MEROCYANINE DYE (4.26) IN DEPENDENCE ON ITS BOND ORDER ALTERNATION

μ_O: dipole moment in the ground state; α: electronic polarizability; β: first hyperpolarizability, and γ: second hyperpolarizability. The lower part indicates the assignment between the bond order alternation BOA and the region of resonance indicated in Fig. 4.2. (Excerpted with permission from S.R. Marder et al., *Science (Washington)*, 1994, **265**, 632. Copyright 1996 American Association for the Advancement of Science).

Fig. 4.3, part (μ_0), shows that with increasing field strength the dipole moment in the ground state μ_0 continously increases from a nonpolar polyene structure via a symmetrical polymethine structure up to a highly dipolar betaine which again has a polyene-like structure. This model exactly reflects the experimental features of dyes (4.26) to (4.31) where dipole moments increase in the same order. Each dye shown in Fig. 4.2 is designated by the BOA region A-E indicated in Fig. 4.3 to identify ist type.

Dyes of type A/B have a more polyenic structure of low polarity and high BOA, those of type C have a cyanine-like structure of intermediate polarity combined with symmetrical and, hence, highly delocalized π-electron system with zero BOA, whereas substances of type D/E possess again a polyenic structure with high dipole moment and as well as high BOA. The dipole moment of each dye can be fine-tuned by using solvents of different polarity same as with the dyes' solvatochromic behaviour [cf. Sect. 3(*a,iii*)].

Considering the linear polarizability α and the second hyperpolarizability γ in Fig. 4.3, maximum values are realized with dyes having a symmetrical π-electron structure in the ideal polymethine state (type C). On the other hand, the first hyperpolarizability β shown in Fig. 4.3 exhibits two peak positions (positive and negative, respectively) with dyes of type B and D and equals zero in the symmetrical, ideal polymethine state.

This behaviour was indeed observed with dyes (4.26) to (4.31) and could be further proved by variation of the solvent polarity. The experimentally determined β-values peak when the γ values are at minimum and *vice versa* (R.S. Marder, *loc. cit.*).

Additional changes of the hyperpolarizability are brought about on alternating and nonalternating modification of polymethine structures [S. Daehne, *Chimia (Aarau)*, 1991, **45**, 288; *J. Soc. Phot. Sci. Technol. Jpn.*, 1996, **59**, 250]. As exemplified in Figure 4.4, alternating substitutions of nitrogen atoms for methine groups in *para*-nitraniline derivatives slightly decreases not only the absorption wavelength λ_{max} but also the first hyperpolarizability β, whereas nonalternating replacement of methine groups with nitrogen atoms strongly increases both the λ_{max} and β values.

Nitranilines	λ_{max} [nm] (Calc.)	β [10^{-30} esu] (Calc.)	Substitution
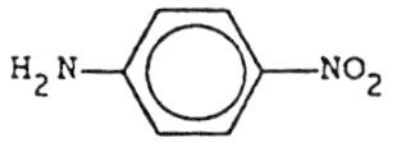 H_2N—⟨ring⟩—NO_2	325	35.3	none
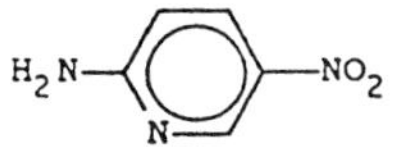 H_2N—⟨ring, N⟩—NO_2	310	33.9	alternating mono-aza
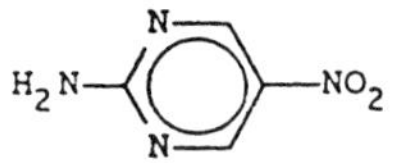 H_2N—⟨ring, N, N⟩—NO_2	283	25.6	alternating di-aza
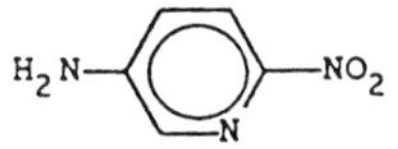 H_2N—⟨ring, N⟩—NO_2	355	46.7	non-alternating mono-aza
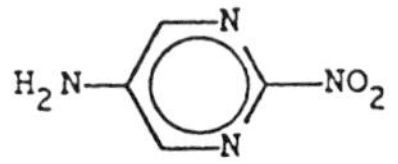 H_2N—⟨ring, N, N⟩—NO_2	354	47.6	non-alternating di-aza
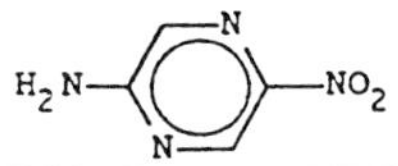 H_2N—⟨ring, N, N⟩—NO_2	349	34.8	alternating plus non-alternating di-aza

FIG. 4.4. EFFECT OF ALTERNATING AND NONALTERNATING AZA-SUBSTITUTION OF *para*-NITRANILINE DERIVATIVES ON THE ABSORPTION MAXIMUM, λ_{max}, AND THE FIRST HYPER-POLARIZABILITY, β.

PPP-SCF-CI calculation after J.F. Nicoud, R.J. Twieg, in *Nonlinear Optical Properties of Organic Molecules and Crystals*, Eds. D.S. Chemla, J. Zyss; Academic Press, Orlando; 1987, p. 254.

Further confirmation of strong enhancement of the first molecular hyperpolarizability through nonalternating substitutions comes from studies of the merocyanine dyes (4.32) to (4.35) shown in Fig. 4.5. Nonalternating substitution of one cyano group for hydrogen at the β-position in (4.33) and (4.35) as well as of two sulfur atoms for two vinylene structural units in the benzene rings of (4.34) and (4.35) shifts the light absorption to the red and increases the $\beta\mu_0$ values as compared to (4.32), (4.33), and (4.34, respectively. (For simplification the product $\beta\mu_0$ instead of β is taken: A.K.-Y. Jen et al., *Proc. SPIE-Int. Soc. Opt. Eng.*, 1994, **2143**, 30).

	λ_{max} [nm]	$\beta\mu_0$ at 1.907 μm [10^{-48} esu]
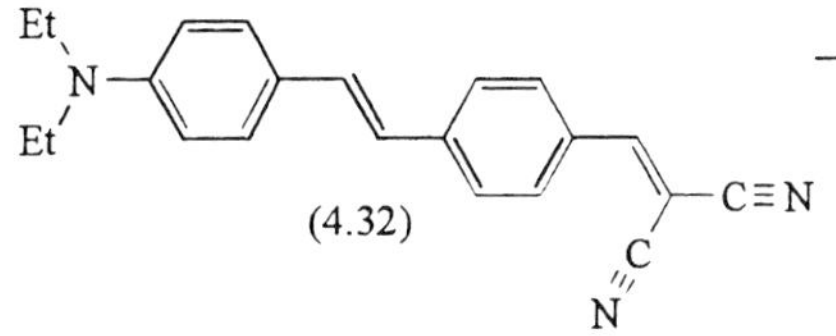 (4.32)	468	1100
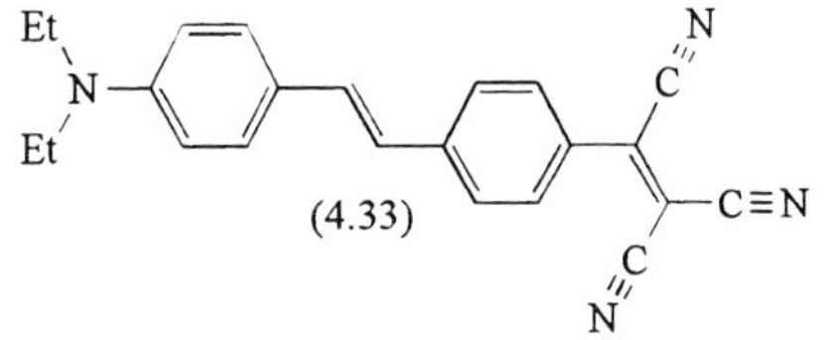 (4.33)	594	2700
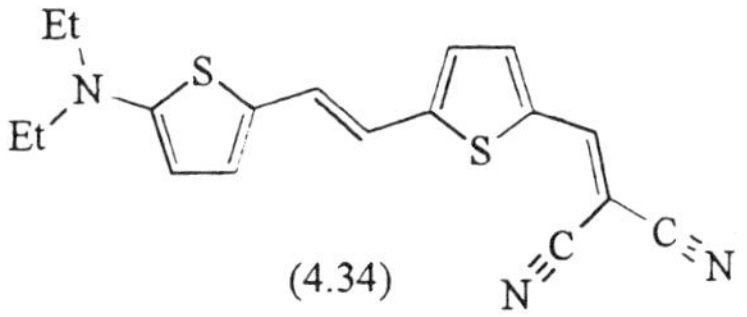 (4.34)	584	2600
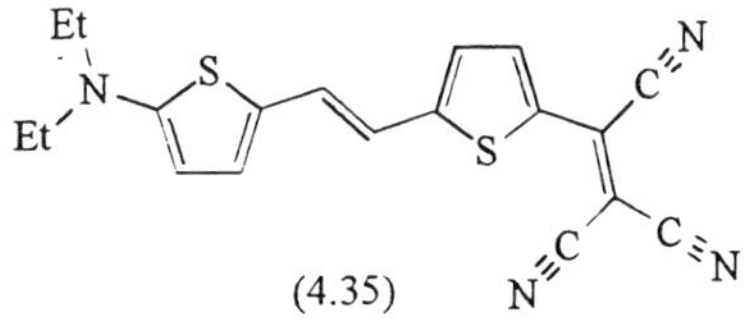 (4.35)	718	6900

FIG. 4.5. EFFECT OF NONALTERNATING SUBSTITUTION OF MEROCYANINE DYE (4.32) ON THE ABSORPTION MAXIMUM, λ_{max}, AND FIRST HYPERPOLARIZABILITY, $\beta\mu_0$ (Experimental values after A.K.-Y. Jen et al., *Proc. SPIE-Int. Soc. Opt. Eng.*, 1994, **2143**, 30).

Many cyanine-like substances, prone to high polarizability, have been synthesized and physico-chemically characterized, e.g. in the Series *Organic Materials for Nonlinear Optics* (Vols. I - IV, Royal Soc. Chem., London, 1989, 1991, 1993, 1995) and *Nonlinear Optical Properties of Organic Materials* [Vols. I - VIII, *Proc. SPIE-Int. Soc. Opt. Eng.*, 1988, **970** (I);

1989, **1127** (II); 1990, **1337** (III); 1991, **1560** (IV); 1992, **1775** (V); 1993, **2025** (VI); 1994, **2285** (VII), 1995, **2527** (VIII)]. The required high thermal stability is the greater the more positive the oxidation potential of the components is (J.R. Twieg et al., *Proc. SPIE-Int. Soc. Opt. Eng.*, 1994, **2143**, 2).

As many applications in nonlinear optics necessiates polymer film materials, the electrooptic merocyanine dyes were often incorporated in host-guest polymers (R.D. Miller et al., *Proc. SPIE-Int. Soc. Opt. Eng.*, 1994, **2042**, 354; R.J. Twieg et al., *ibid.*, 1994, **2143**, 2; G.A. Lindsay et al., *ibid.*, 1994, **2143**, 88; M.B. Meinhardt et al., *ibid.*, 1994, **2143**, 110; S. Gilmour et. al., *ibid.*, 1994, **2143**, 117; G.R. Moehlmann, *ibid.*, 1994, **2285**, 366; D.M. Burland et al., *Chem. Rev.*, 1994, **94**, 31; W.E. Moerner, S.M. Silence, *ibid.*, 1994, **94**, 127; K.D. Singer, J.H. Andrews, *Mol. Nonlinear Opt.*, **1994**, 245; J.-L. Brédas, *Adv. Mater.*, 1995, 7, 263).

The efficiency of SHG and the quality of the layer material is improved through covalent linkage of the cyanine and merocyanine dyes to polymers (K.J. Moon et al., *Mol. Cryst. Liq. Cryst.* 1994, **247**, 91; C.R. Moylan et al., *Proc. SPIE-Int. Soc. Opt. Eng.*, 1994, **2285**, 17; T. Hanemann et al., *Adv. Mater.*, 1995, **7**, 465) or to glass surfaces (D.S. Allan et al., *Proc. SPIE-Int. Soc. Opt. Eng.*, 1991, **1560**, 362; S. Yitzchaik et al., *ibid.*, 1994, **2285**, 282).

Also, imbedding of NLO substances into Langmuir Blodgett (LB) layers proved to be useful (T.L. Penner, *Proc. SPIE-Int. Soc. Opt. Eng.*, 1991, **1560**, 377; W.M.K.P. Wijekoon et al., *ibid.*, 1994, **2285**, 254). Further progress was achieved by inclusion of NLO active materials into molecular sieves, especially into zeolithe single crystals (F. Marlow et al., *J. Phys. Chem.*, 1993, **97**, 11286). In such components the state of order and the thermal stability of the dyes' is further improved.

Strong enhancement of the nonlinear optical susceptibilities of cyanine dyes is realized when the molecules are coupled in J-aggregates [cf. Sect. 3(b,ii)]. For instance, the J-aggregates of the merocyanine (4.36; n = 1; X,Y = S; R^1 = $C_{18}H_{37}$; R^2 = CH_2COOH; R^3 = H) embedded in LB layers give much stronger SHG signals as compared to the monomeric molecules (K.

Kajikawa et al., *Jpn. J. Appl. Phys.*, 1991, **30**, L1525; *Thin Solid Films*, 1994, **243**, 587).

$$\text{(4.36)}$$

Also, in case of symmetrical cyanine dyes like pseudoisocyanine (4.10), the second hyperpolarizability γ of their J-aggregates is much higher than the sum of the γ-values of the individual molecules, at least under resonance conditions within the dyes' absorption band (Y. Wang, *J. Soc. Opt. Am. B*, 1991, **B8**, 981; V.L. Bogdanov et al., *Pis'ma Zh. Eksp. Teor. Fiz. 1991*, **53**, 100; S. Kobayashi, F. Sasaki, *Nonlinear Optics*, 1993, **4**, 305; *Mol. Cryst. Liq. Cryst. Sci. Technol., Sect. B*, **1993**, 305; K. Misawa et al., *Proc. SPIE-Int. Soc. Opt. Eng.*, 1994, **2144**, 128). On the other hand, formation of blue-shifted H-aggregates of stilbazolium hemicyanine dye (4.9; n = 1, 2; R^1 = $(CH_2)_4SO_3^-$; $R^2 = C_{22}H_{45}$) diminuates the second hyperpolarizability (M.A. Carpenter et al., *J. Phys. Chem.*, 1992, **96**, 2801; R.A. Hall et al., *ibid.*, 1993, **97**, 11974; J.Y. Fang et al., *Thin Solid Films*, 1994, **243**, 450).

(e) Applications of (photo)electrical and photochemical activity
(i) Spectral sensitization
Spectral sensitization of photographic silver halide emulsions by cyanine dyes has been improved in many directions (cf. reviews from. S. Daehne, *Phot. Sci. Eng.*, 1979, **23**, 219; *J. Phot. Sci.*, 1990, **38**, 66; G. Ficken, *Chem. Ind.*, **1989**, 672; T. Tani, *J. Imag. Sci.*, 1990, **34**, 143; *J. Soc. Phot. Sci. Technol. Jpn.*, 1991, **54**, 489). Processes which diminuate spectral sensitivity by desensitization effects were recently reviewed by B.I. Shapiro (*Usp. Khim.*, 1994, **63**, 243).

From electrochemical halfwave potentials re-measured by phase selective second harmonic AC voltammetry [cf. Sect. 3(a,ii)], the linear relationship between E_{ox} and the ionization potential of cyanine dyes, determined by uv photoelectron spectroscopy (UPS), was confirmed (T. Tani et al., *J. Electrochem. Soc.*, 1991, **138**, 1411). Good correlations between E_{red} and the efficiency of the light-induced electron transfer from cyanine dyes to silver halide microcrystals as well as between E_{ox} and the concentration of positive holes trapped by cyanine dyes were obtained. These proved that photo-induced electron and hole transfer reactions are the main processes in spectral sensitization (T. Tani, *loc. cit.*; M.T. Spitler, *J. Imag. Sci.*, 1991, **35**, 351; J.R. Lenhard et al., *J. Phys. Chem.*, 1993, **97**, 8269)

Strong support of the existing concepts came from determinations of the interfacial electronic structure of some 10 merocyanine dyes, like (4.36; n = 0, 1, 2, 3; X = S, NEt; Y = S; R^1 = Et; R^2 = Et, C_6H_{13}; R^3 = H, Cl) and derivatives adsorbed to silver halide using uv photoemission spectroscopy (K. Seki et al., *J. Imag. Sci.*, 1993, **37**, 589; *Phys. Rev. B*, 1994, **49B**, 2760). X-Ray absorption near-edge structure (XANES) spectroscopy even revealed the orientation of the dye molecules on the silver halide surface (T. Araki et al., *Jpn. J. Appl. Phys.*, 1993, **32**, Suppl. 32-2, 434; K. Seki et al., *loc. cit.*).

Extension of the correlations between the dyes' energy levels and the band structure of other semiconductors, e.g. zinc oxide (ZnO), tin dioxide (SnO_2), and titanium dioxide (TiO_2) confirmed the existing theory of spectral sensitization (Y. Yonezawa et al., *J. Imag. Sci.*, 1990, **34**, 249).

Although spectral sensitization is satisfactorily described by the relationships between the energy levels of sensitizers and the relative positions of the valence and conduction band of the semiconductors from a physical point of view the effectiveness of spectral sensitization must be kinetically controlled (S. Daehne, *Phot. Sci. Eng.*, 1979, **23**, 219). First measurements of the rate constant of spectral sensitization of silver halide through J-aggregated cyanine dyes were undertaken by T. Tani et al., (*J. Phys. Chem.*, 1992, **96**, 2778) and A.Muenter et al., (*J. Phys. Chem.*, 1992, **96**, 2783). It turned out that the rate is the greater the smaller the size of the J-aggregates is. More precise data which distinguish between spectral sensitization through monomers and J-aggregates of the *meso*-ethylthiacarbocyanine dye (4.37) were recently published (B. Troesken et al., *Adv. Mater.*, 1995, **7**, 448).

$$R^1 = (CH_2)_3SO_3^{\ominus}; \qquad R^2 = (CH_2)_3SO_3K \qquad (4.37)$$

Spectral sensitization of photopolymerization of organic polymers for recording holograms was realized with some substituted indadicarbocyanine and indatricarbocyanine dyes (4.7; n = 2, 3; X = CMe$_2$; R = Me). Probably the long-living triplet state of the dyes is responsible for initiation the polymerization process. The holograms were recorded with high-intensity helium-neon or krypton gas lasers with emission wavelengths of 633-799 nm, and read out by NIR laser diodes emitting at 780-854 nm (D.-J. Lougnot, *Proc. SPIE-Int. Soc. Opt. Eng.*, 1994, **2042**, 218).

(ii) Photoinduced electron transfer (PET)

As with spectral sensitization photoinduced electron transfer (PET), processes are involved in all photoelectrochemical reactions which are used as light harvesting and charge separation systems (P. Fromherz, W. Arden, *J. Am. Chem. Soc.*, 1980, **102**, 6211). In order to determine the dependence of the PET rate on the distance between electron donors and electron acceptors, the molecules were incorporated in LB layers and separated by a spacer layer one molecule in thickness in which the length and the electronic nature of the surfactant molecules were varied. When conjugated organic molecules like *trans*-stilbene-derivatized fatty acids were used as spacers, instead of saturated fatty acids providing the same distance, an enlargement of the PET rate was observed (Y. Hsu et al., *J. Phys. Chem.*, 1992, **96**, 2790).

The efficiency of photoinduced electron transfer is enhanced when J-aggregates, e.g. from pseudoisocyanine (4.10), are used (T.L. Penner, *J. Chim. Phys.*, 1988, **85**, 1081; C. Koenigstein, R. Bauer, *Int. J. Hydrogen. Energy*, 1993, **18**, 735). Here the PET efficiency was drastically improved when the dye molecules were covalently linked by a pentamethylene spacer to viologen, giving compond (4.38) which likewise forms J-aggregates (C. Koenigstein, R. Bauer, *Sol. Energ. Mater. Sol. Cells*, 1994, **31**, 535).

470

$$(4.38)$$

Intramolecular charge transfer is probably the origin of electron spin resonance (ESR) signals in J-aggregates of the surface-active merocyanine dye (4.36; n = 1; X,Y = S; R^1 = $C_{18}H_{37}$; R^2 = CH_2COOH; R^3 = H). The signals are enhanced on light excitation giving support to the electron transfer mechanism. Using ^{15}N-enriched dye two different kinds of radicals, presumably the cation and anion radicals, were found. As the orientation of the J-aggregates, determined by uv/vis spectroscopical methods, coincides with the ESR spectroscopically examined orientation of the radicals the charge generation mechanism in the J-aggregates seems to be established (S. Kuroda et al., *Mol. Cryst. Liq. Cryst.*, 1990, **190**, 111; S. Kuroda, *Colloids Surf.*, 1993, **A72**, 127; S. Kuroda et al., *Thin Solid Films*, 1994, **242**, 96).

(iii) *Photovoltaic and solar cells, solar collectors*

The photosensitivity of semiconductor electrodes used in photovoltaic cells is extended to longer wavelengths when typical cyanine dyes such as (4.7; X = O, S, NR; R = optional) are used spectrally to sensitize the photoinduced separation of electrons and holes. Such sensitized electrodes are made among others from tungsten chalcogenides (M. Spitler, B.A. Parkinson, *Langmuir*, 1986, **2**, 549; D.V. Sviridov, A.I. Kulak, *New J. Chem.* 1991, **15**, 539), tin and titanium dioxides (Y. Yonezawa et al., *J. Imaging Sci.* 1990, **34**, 249; A. Haraguchi et al., *Photochem. Photobiol.*, 1990, **52**, 307; G. Biesmans et al., *Chem. Phys.* 1992, **160**, 97), and tin sulfide (B.A. Parkinson, *Langmuir* 1988, **4**, 967). Especially J-aggregates of cyanine dyes proved to be very effective spectral sensitizers (A. Haraguchi, *loc. cit.*; D.V. Sviridov et al., *J. Photochem. Photobiol.*, 1992, **A67**, 377).

Photovoltaic cells, which exclusively consist of dye films deposited by evaporation in high vacuum in between two metal electrodes, have some advantages (H. Boettcher, *J. Prakt. Chem.*, 1992, **334**, 14). In such devices relationships between the dyes' capability to generate photocurrent and their molecular structure were established from studies of more than 100 merocyanine dyes by D.L. Morel et al. (*J. Phys. Chem.*, 1984, **88**, 923), A.P. Piechowski et al., *J. Phys. Chem.*, 1984, **88**, 934), and H.O. Yadav et al., (*Sol. Energy Mater. Sol. Cells*, 1994, **35**, 347).

Voltage generation in photovoltaic cells is improved when stilbazolium merocyanine dye molecules (4.39; n = 6, 11) are ordered in a nematic liquid crystal (LC) which is sandwiched between two semiconducting (In_2O_3) electrodes (J. Goc, D. Frackowiak, *J. Photochem. Photobiol.*, 1991, **A59**, 233).

$$HO-(CH_2)_n-N^{\oplus} \text{—CH=CH—} \quad \textit{t}\text{-Bu}, \ O^{\ominus}, \ \textit{t}\text{-Bu} \tag{4.39}$$

Schottky-type photoelectric cells with good efficiency were also obtained with merocyanine dye (4.36; n = 1; X,Y = S; $R^1 = C_{18}H_{37}$; $R^2 = CH_2COOH$) embedded in LB layers (K. Saito, H. Yokoyama, *Thin. Solid Films*, 1994, **243**, 526).

An additional improvement of solar cells is realized through combination of the cells with *solar collectors*. These are polymeric plates which contain any highly fluorescent cyanine dye. The sunlight absorbed by the dyes is converted to fluorescent light, which is nearly completely emitted at the edges of the plates due to strong internal reflexion at the surface of the wave-guiding polymeric plates. The collected light is then converted to electric power by photovoltaic cells positioned at the edges of the plate. (R. Raue et al., *Heterocycles*, 1984, **21**, 167)

(iv) Optical recording
Modern information recording and storage are based on electronic processess, especially on laser optical memory systems such as compact disc (CD) records and videos, because they enable digital sound recording

472

of high quality (*Chemistry of Functional Dyes*, Eds. Z. Yoshida, T. Kitao, Mita Press, Tokyo, 1989, Chapt. 8, pp. 341-398; P. Gregory, *Chem. Brit.*, **1989**, 47; F. Matsui, in *Infrared Absorbing Dyes, Top. Appl. Chem.*, Ed. M. Matsuoka, Plenum Press, New York, London, 1990, pp. 117-140; M. Matsuoka, *Mol. Cryst. Liq. Cryst.*, 1993, **224**, 85; G.H.W. Buning, in *Organic Materials for Photonics, Science and Technology*, Ed. G. Zerbi, Elsevier Sci. Publ., Amsterdam, 1993, pp. 367-397).

In the "direct read after write" (DRAW) optical discs the light of semiconductor lasers, having emission wavelengths longer than 630 nm, is either intensely or weakly reflected by the support and than properly modulated. This is realized by pattern-structured polymeric supports which are covered with thin dye films of high reflectivity. The NIR cyanine dyes described in Section 1(*c*) are best suited for this purpose as was first shown by H. Oba et al. (*Appl. Opt.*, 1986, **25**, 4023). As NIR dyes usually lack lightfastness, due to autoxidation processes, combination of the dye cations with anions which act as singlet oxygen quencher proved to be useful. The indatricarbocyanine cations combined with anions of the dithiolene nickel complex as quencher shown in formula (4.40) are presently contained in commercially available CD's (H. Nakazumi et al., *J. Soc. Dyers Colour.*, 1989, **105**, 26)

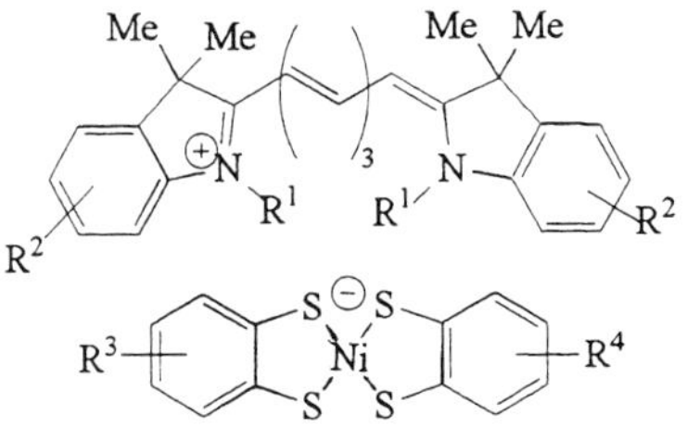

$$R^1 = \text{alkyl, aryl, aralkyl}$$
$$R^2, R^3, R^4 = \text{H, or benzo-anellation} \tag{4.40}$$

Merocyanine dyes are also used in double-layered magneto-optical memory systems which consist of a magnetic layer and a merocyanine dye layer. The latter strongly enhances the Kerr rotation angle of the magnetic layer in case of plasma resonance near the dye's absorption edge (T. Kitaguchi et al., *Jpn. J. Appl. Phys.*, 1991, **30**, 3377).

Presently erasable discs are under development as reversible optical storage systems. These allow overwriting of information many times (M. Matsuoka,

Mol. Cryst. Liq. Cryst., 1993, **224**, 85; B.L. Feringa et al., *Tetrahedron*, 1993, **49**, 8267). Suitable phenomena to induce photochromism are *cis-trans* isomerization, photocyclization reactions, keto-enol tautomerism, chiroptical molecular switching, and photoinduced electron transfer (PET) (B.L. Feringa, *loc. cit.*). Among the candidates for photocyclization the spiropyrans (4.41) have a good chance of commercialization. Upon uv radiation they give colored merocyanine derivatives (4.42) (Scheme 4.4).

SCHEME 4.4

(4.41) — uv light / red light, ΔH — (4.42)

$X = O, S;\ R^1 = alkyl;\ R^2 = H, NO_2;$

$R^3 = H,$ substituted alkyl; $R^4 = H, Cl,$ or NH-polysiloxane

Here the thoroughly examined reaction is known to be reversed by visible light or heat (E. Pottier et al., *Helv. Chim. Acta*, 1990, **73**, 303; S.-R. Keum et al., *Can. J. Chem.*, 1991, **69**, 1940).

The dynamics and mechanism of the photochromic reaction are altered when the spiropyran dyes are incorporated into liquid crystal (LC) layers (T. Seki, K. Ichimura, *Macromolecules*, 1990, **23**, 31), or when they are covalently linked to polyacrylic polymers (S. Yitzchaik et al., *Macromolecules*, 1990, **23**, 707).

After photoinduced ring cleavage of nitrospiropyrane derivatives (4.41; X = O; $R^1 = C_{18}H_{37};$ $R^2 = NO_2;$ $R^3 = CH_2OCOC_{21}H_{43};$ $R^4 = H$), embedded in LB films or in liquid crystalline layers, red-shifted J-aggregates (E. Ando et al., *Thin Solid Films*, 1988, **160**, 279) or blue-shifted H-aggregates (I. Cabrera, V. Krongauz, *Nature (London)*, 1987, **326**, 582) are formed. The latter are preferred when the dye is covalently bound to polysiloxanes (I. Cabrera et al., *Angew. Chem.*, 1987, **99**, 1204; *Angew. Chem. Int. Ed. Engl.*, 1987, **26**, 1178). In this way three primary colors of one and the same photochromic compound are realized (M. Matsuoka, *Mol. Cryst. Liq. Cryst.*, 1993, **224**, 85).

During photoinduced ring cleavage the stereogenic center at the spiro position causes a reversible change of optical rotation in chiral cholesteric liqid crystalline polymers. The effect is used likewise for optical data storage processes (Y. Suzuki et al., *Polym. Bull.*, 1987, **17**, 285).

In order to multiply the storage density of memory systems future aims are multiple wavelengths optical recording. This is realized, e.g., by multi-layered LB films containing J-aggregates of different dyes. Owing to their narrow absorption band an overlap of the absorption spectra can be avoided as was shown with the J-aggregates of the dyes (4.43) in combination with that of (4.44) (C. Ishimoto et al., *Appl. Phys. Lett.*, 1986, **49**, 1677).

$R^1 = (CH_2)_3SO_3^{\ominus}; R^2 = (CH_2)_3SO_3H$

(4.43)   (4.44)

Recording was then achieved with proper laser light which bleaches the J-absorption band by disordering the aggregate structure. The absorption is restored by placing the medium in an environment of high humidity.

In the full color hard copy system (cycolor system), cyanine dyes having triphenylalkylborate anions are used as spectral sensitizers which are included in microcapsules together with polymerizable acrylic acid monomers and different leuco dyes. On light excitation, the cyanine dye cations spectrally sensitize photoinduced electron transfer [PET, cf. Sect. 4(*e,ii*)] to borate anions through which alkyl and phenyl radicals are produced. These polymerize the acrylic acid monomers and hence inactivate the irradiated microcapsules. When the non-polymerized microcapsules are then broken and come into contact with acid on the recording paper the complementary colors (yellow, magenta, or cyan) are generated (R.F. Wright, in *Chemistry of Functional Dyes*, Eds. Z. Yoshida, T. Kitao, Mita

Press, Tokyo, 1989, p. 473; M. Matsuoka, *Mol. Cryst. Liq. Cryst.*, 1993, **224**, 85).

Photoinduced electron transfer (PET) is also used to decolorize toner systems. For instance, according to Scheme 4.5, irradiation of dye (4.45) generates both neutral phenyl and neutral dye radicals which recombine to give the colorless leuco dye (4.46) (M. Matsuoka, *loc. cit.*).

SCHEME 4.5

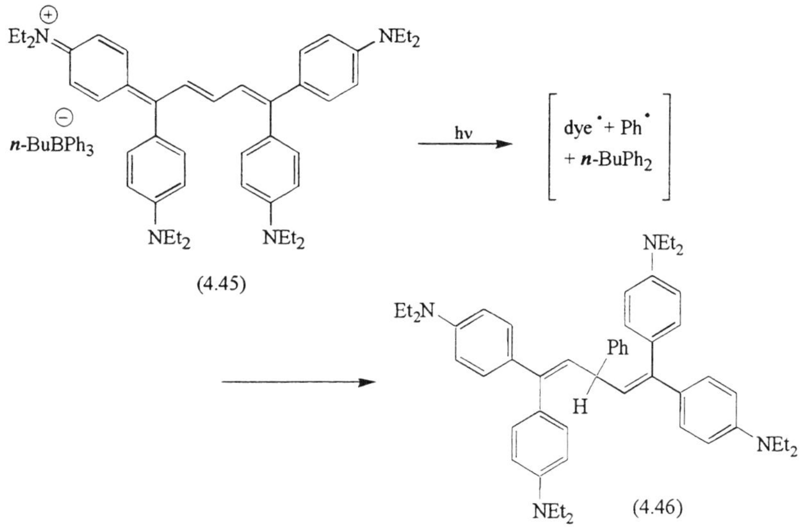

(v) Charge control agents for electrophotographic processes
In electrophotographic (xerographic) copy systems and in laser printers organic photoconductors are the key materials, which consist of a charge generation layer and a charge transport layer (A. Kakuta, in *Infrared Absorbing Dyes, Top. Appl. Chem.*, Ed. M. Matsuoka, Plenum Press, New York, London, 1990, pp. 155-172). Whereas the charge transport materials are mostly electron donors like hydrazine and triphenyl amine derivatives prone to transport positive holes, for the photoinduced charge generation materials cyanine dyes are used. Typical examples are the squarylium dyes described in Section 1(c) and hemicyanine dyes, e.g. (4.47), with an azulenium terminal group (K. Katagiri et al., *Nippon Kagaku Kaishi*, **1986**,

387; *Chem. Abstr.*, 1986, **105**, 88567k). The latter gives quite broad photosensitivity up to 850 nm in the crystalline state which is probably due to the formation of J-aggregates. J-Aggregation also improves photoinduced charge generation of the thiapyrylium dye (4.48) (P.M. Borsenberger, D.C. Hoesterey, *J. Appl. Phys.*, 1980, **51**, 4248; K. Kakuta, *loc cit.*).

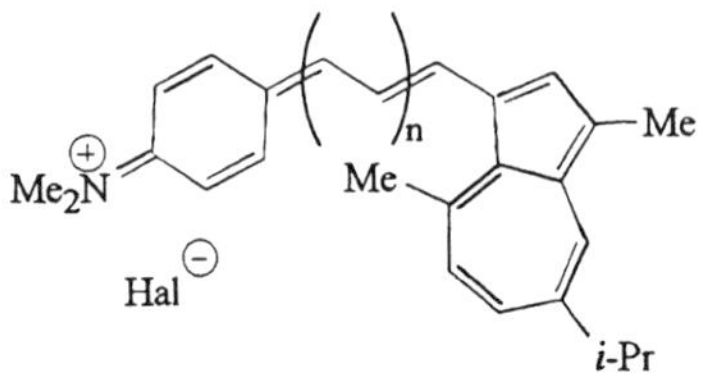

$$(4.47) \qquad (4.48)$$

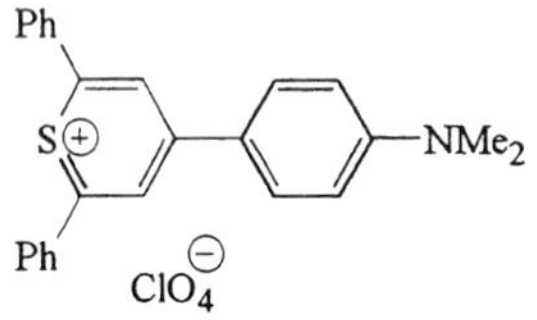

(vi) Electroluminescence

In light-emitting diodes (LED's) electrons and holes are generated in the conduction and valence band, respectively, of semiconductors through electrical pulses. The electron-hole recombination results in the emission of visible light [N.C. Greenham et al., *Nature (London)*, 1993, **365**, 628; S. Kirstein et al., *Synthetic Metals*, 1995, **69**, 415]. As the emission wavelengths of inorganic semiconductor and organic polymer materials are usually limited to the red or even NIR region of the spectrum, multilayered LB-films, incorporated with J-aggregates of the oxacarbocyanine dyes (4.49), were used to generate electroluminescence at shorter wavelengths near 550 nm (M. Era et al., *J. Chem. Soc., Chem. Commun.*, **1985**, 557).

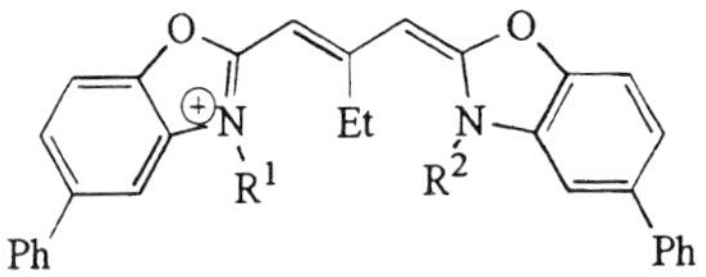

$$R^1 = (CH_2)_3SO_3^{\ominus}$$
$$R^2 = (CH_2)_3SO_3Na$$

$$(4.49)$$

(vii) Electronic photography

In order to obtain color copies or prints from magnetic tapes or discs of video cameras, the information stored must be converted into yellow, magenta, and cyan color dots by printers (P. Gregory, *Chem. Brit.*, **1989**, 47; F. Jones, *Rev. Progr. Coloration*, 1989, **19**, 20; Y. Nagae, in *Infrared Absorbing Dyes, Top. Appl. Chem.*, Ed. M. Matsuoka, Plenum Press, New

York, London, 1990, pp. 141-154). One possibility is ink jet printing in which the electronic signals are controlling the jets containing colored inks by piezo-electric crystals. Many cyanine dyes, e.g. (4.50; n = 3; X = CMe$_2$, O, S, Se, Te; R = Me, Et, Pr, Bu, sulfoalkyl) were claimed to be suitable for this purpose in recent patent literature (e.g., K. Aoki et al., *Jpn. Kokai Tokkyo Koko*, JP, 05,171,079, 9.7.1993; *Chem. Abstr.*, 1994, **120**, 79669c; C. Yamamoto et al., *Jpn. Kokai Tokkyo Koko*, JP, 05,140,493, 8.6.1993; *Chem. Abstr.*, 1994, **120**, 247526z).

(4.50)

Most promising for printing are thermal writing displays using the dye diffusion thermal transfer process. In the three-color dye diffusion thermal process the information is written onto a dye sheet, which contains the colors yellow, magenta, and cyan, and transferred to the receiver layer by heat-induced diffusion whereby the heat is provided by controlled heat pulse current or laser beams (P. Gregory, *Chem. Brit.*, **1989**, 47). Only dyes which are extremely stable against heat and autoxidation fulfill the requirements for this process. Merocyanine-like dyes are suitable candidates for yellow (4.51), and magenta (4.52) color. The red-shift of (4.52) as compared to (4.51) is realized by nonalternating substitution the α-position of the *para*-aminostyrene skeleton with an additional cyano group.

(4.51)

(4.52)

More promising are thermal writing processes based on liquid-crystal (LC) displays. Also here the energy of controlled laser beams is converted to heat bringing about changes of LC phases, which are additionally controlled by the electro-optical signals of the video tape. Most common absorbing media for cyan color are cyanine dyes which are known as laser dyes for NIR lasing, for instance, squarylium dye (4.53) (J. Murata et al., *Jpn. Kokai*

478

Tokkyo Koko, JP, 05,162,460, 29.6.1993; *Chem. Abstr.,* 1994, **120**, 10534n; Y. Nagae, in *Infrared Absorbing Dyes, Top. Appl. Chem.,* Ed. M. Matsuoka, Plenum Press, New, York, London, 1990, 141-154).

$$(4.53)$$

(f) Application of cyanine dyes in molecular biology and medicine
Some applications of cyanine dyes in molecular biology, e.g., noncovalent and covalent fluorescence labeling, was already described in Sect. 4(*a,iii*). The determination of membrane potentials in living cells with cyanine dyes is discussed in Section 4(*b,iv*). Most recent research aims for controlling structure and function of biomolecules through exploiting certain photochemical activities of cyanine dyes, like *cis-trans* isomerization and photocyclization reactions. It is even to be expected that photo-switchable biopolymeres will be in future applied in information processing, information storage, and information transmission described in Section 4(*e*) (I. Willner, S. Rubin, *Angew. Chem.,* 1996, **108**, 419; *Angew. Chem. Int. Ed. Engl.,* 1996, **35**, 367).

(i) Probing transport phenomena
Typical applications of cyanine dyes in medicine are the usage of carbo- and dicarbocyanine dyes such as (4.7; n = 1, 2; X = CMe_2, O, S; R = Et, C_5H_{11}, C_6H_{13}) in order to investigate neural transport mechanisms by neural labeling and pathway tracing (M.G. Honig, R.I. Hume, *Trends Neurosci.,* 1989, **12**, 333), amino acid transport in renal tissue [K.E. Joergensen et al., *J. Physiol. (London),* 1990, **422**, 41], and ion transport in enterocytes (Y. Kinoshita, A. Irimajiri, *Jpn. J. Physiol.,* 1988, **38**, 659). The same dyes are used to visualize vasculature (M.J. Trotter et al., *Brit. J. Cancer,* 1989, **59**, 706), and to localize endoplasmic reticulum in living cells by fluorescence probing (M. Terasaki et al., *Cell (Cambridge, Mass.),* 1984, **38**, 101).

More recently, some of the oxacyanine dyes (4.7; n = 1, 2; X = O; R = Et, C_5H_{11}, C_6H_{13}) were shown to be cytotoxic and inhibit mitochondrial

NADH-ubiquinone reductase activity (W.M. Anderson et al., *Biochem. Pharmacol.*, 1993, **45**, 2115). Therefore, care should be taken in using the dyes to examine biological processes.

The inhibition of transport processes by cyanine dyes was exploited for therapeutical purposes in order to suppress extraneural noradrenaline transport in the sympathetic nervous system. Especially pseudoisocyanine (4.10) and the isocyanines (4.54) having different alkyl substituents, like $R^1 \neq R^2$ = Me, Et, *i*-Pr, proved to be candidates as antidepressants (H. Russ et al., *J. Med. Chem.*, 1993, **36**, 4208).

(4.54)

(ii) Identification of active oxygen species

Active oxygen species such as hydroxyl radicals (OH$^\bullet$), superoxide ($O_2^{-\bullet}$), and hydroxy peroxide (H_2O_2) are involved in both normal cellular processes and in a wide spectrum of pathologies. The bleaching reaction of the thiazolodicarbocyanine dyes (4.55; R = Me, Et, *n*-Bu, *n*-C$_7$H$_{15}$) through OH$^\bullet$ and OH$^{-\bullet}$ are used to probe this species in phenomena like lung injury, carcinogenesis, and aging (H. Hori et al., *Adv. Exp. Med. Biol.*, 1992, **317**, 255).

(4.55)

480

(iii) Photodynamic therapy by singlet oxygen generation through cyanine dyes

The main application of cyanine dyes in medicine comes from their photochemical activity to generate singlet oxygen, which is cytotoxic and thus used in photodynamic therapy, e.g. in dermatology and cancer therapeutics, to destroy viruses, microbes, and neoplastic cells. The merocyanine dye MC 540 (4.15) and its derivatives (4.56; X = O, S; Y = S; R^1 = H, *n*-Bu; R^2,R^3 = 4,5-, or 5,6-, or 6,7-benzoanellation; M = Na, $HNEt_3$) exhibit high-selective affinity for leukemia and lymphoma cells (W.H.H. Guenther et al., *Semin. Hematol.*, 1992, **29**, 88). MC 540 is also used for blood sterilization through light excitation (F. Sieber et al., *Semin. Hematol.* 1992, **29**, 79).

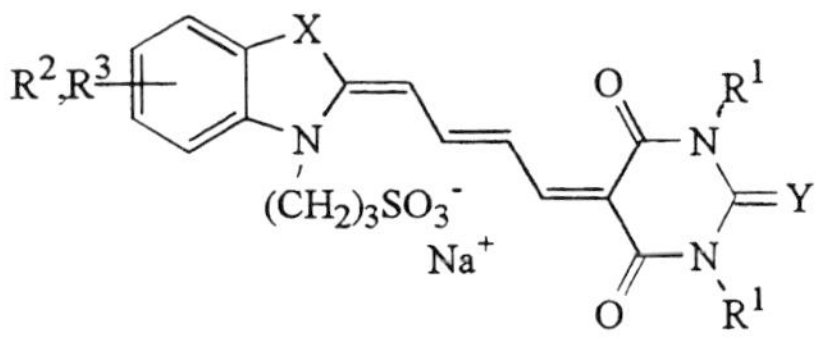

R^2,R^3 X   N $(CH_2)_3SO_3^-$ Na^+ O R^1 N   Y   N   O R^1

(4.56)

Systematical examination of structural modification of dye (4.56) showed that increased lipophilicity leads to enhanced localization of the dye molecules within the cells and hence to more efficient cell killing. This was realized by both increasing the chain length of the R^1 alkyl substituents, omitting the hydrophilic sulfonic acid group, and extending the aromaticity of the benzoxazole system by further benzoanellation (A.C. Benniston et al., *J. Chem. Soc. Faraday Trans.*, 1994, **90**, 953).

As the singlet oxygen yield is the greater the higher the intersystem crossing (ISC) rate to the triplet state is, efforts were undertaken to increase the ISC rate by heavy atom effects. Among 14 differently substituted MC 540 derivatives those having a selenone group (4.56; Y = Se) in the barbituric acid system were found to be the most effective photosensitizers with more than 100-fold higher photogeneration efficiency of singlet oxygen (W.H.H. Guenther et al., *Phosphorus, Sulfur, Silicon*, 1992, **67**, 417).

Other photosensitizers in photodynamical therapy are the thiacarbocyanine dyes (4.7; n = 1; X = S; R = Et, C_6H_{13}, C_8H_{17}, $C_{12}H_{25}$, $C_{18}H_{37}$; An = Cl, I).

Their singlet oxygen yield increases when the dyes are incorporated into liposomes (M. Krieg et al., *Biochim. Biophys. Acta*, 1993, **1151**, 168).

Favorable in photodynamic therapy are the chalcogenopyrylium dyes (4.57; X = Y, or X ≠ Y = O, S, Se, Te; R^1,R^2 = H, Me, or R^1 ≠ R^2 = H, Me; An = ClO_4, PF_6).

(4.57)   X,Y = O, S, Se, Te R^1, R^2 = H, Me

(4.58)

Their absorption maximum lies in the NIR region at wavelengths longer than 700 nm, which allows a very good tissue penetration by light because there are no competing tissue absorptions. Due to the heavy atom effect the dyes with the heaviest chalcogen atom tellurium display the highest ISC rate and thus the greatest singlet oxygen yield (M.R. Detty, P.B. Merkel, *J. Am. Chem. Soc.*, 1990, **112**, 3845). Special advantage of the tellurapyrylium dyes is their self-sensitized bleaching through subsequent oxidation by the hydrogen peroxide formed during photosensitization, which gives the colorless dihydroxytelluranes (4.58) (M.R. Detty, *Organometallics*, 1992, **11**, 2310). In this way, the dye will no longer act as absorbing chromophor in the treated tissue and thus no longer screen the light from reacting deeper positioned tumor cells (M.R. Detty, P.B. Merkel, *loc. cit.*).

Guide to the Index

This index is constructed in a similar manner to the volume indexes of the first edition of the Chemistry of Carbon Compounds. However, to make the index easier to use, more descriptive entries have been made for the commonly occurring individual, and groups of chemicals.

The indexes cover primarily the chemical compounds mentioned in the text, and also include reactions and techniques, where named, and some sources of chemical compounds such as plant and animal species, oils, etc.

Chemical compounds have been indexed alphabetically under the names used by authors, editing being restricted to ensuring uniformity of entries under the same heading. In view of the alternative nomenclature that can often be used, a limited amount of cross-referencing has been done where it is considered to be helpful, but attention is particularly drawn to Convention 2 below.

For this and the succeeding volumes, the indexing conventions listed below have been adopted.

1. Alphabetisation

(a) A letter by letter alphabetical sequence is followed for entries, firstly for the main entry, followed by the descriptive entry.

(b) The following prefixes have not been counted for alphabetising:

n-	*o-*	*as-*	*meso-*	*C-*	*E-*
	m-	*sym-*	*cis-*	*O-*	*Z-*
	p-	*gem-*	*trans-*	*N-*	
	vic-			*S-*	
		lin-		*Bz-*	
				Py-	

Some prefixes and numbering have been omitted in the index, where they do not usefully contribute to the reference.

(c) The following prefixes have been alphabetised:

Allo	Epi	Neo
Anti	Hetero	Nor
Bis	Homo	Pseudo
Cyclo	Iso	

2. Cross references

In view of the many alternative trivial and systematic names for chemical compounds, the indexes should be searched under any alternative names which may be indicated in the main body of the text. Only a limited amount of cross-referencing has been carried out, where it is considered that it would be helpful to the user.

3. Derivatives

Simple derivatives are not normally indexed if they follow in the same short section of the text.

4. Collective and plural entries

In place of "– derivatives" the plural entry has normally been used. Plural entries have occasionally been used where compounds of the same name but differing numbering appear in the same section of the text.

5. Main entries

The main entry of the more common individual compounds is indicated by heavy type. Multiple entries, such as headings and sub-headings over several pages are shown by "–", e.g., 67–74, 137–139, etc.

Index

508